THE BIG PICTURE

GROSS ANATOMY

THE BIG PICTURE

GROSS ANATOMY

David A. Morton, PhD
Associate Professor
Anatomy Director
Department of Neurobiology and Anatomy
University of Utah School of Medicine
Salt Lake City, Utah

K. Bo Foreman, PhD, PT
Assistant Professor
Anatomy Director
University of Utah College of Health
Salt Lake City, Utah

Kurt H. Albertine, PhD
Professor of Pediatrics, Medicine, and Neurobiology and Anatomy
Associate Dean of Faculty Administration
Editor-In-Chief, *The Anatomical Record*
University of Utah School of Medicine
Salt Lake City, Utah

New York   Chicago   San Francisco   Lisbon   London   Madrid   Mexico City
Milan   New Delhi   San Juan   Seoul   Singapore   Sydney   Toronto

The McGraw·Hill Companies

The Big Picture: Gross Anatomy

1 2 3 4 5 6 7 8 9 0   CTP/CTP   15 14 13 12 11

ISBN 978-0-07-147672-0
MHID 0-07-147672-5

This book was set in Minion by Aptara, Inc.
The editors were Michael Weitz, Susan Kelly, and Brian Kearns.
The production supervisor was Catherine Saggese.
The illustration manager was Armen Ovsepyan.
Project management was provided by Shikha Sharma, Aptara, Inc.
The designer was Alan Barnett; the cover designers were Alan Barnett and Elizabeth Pisacreta.
China Translation & Printing, Ltd., was printer and binder.

Library of Congress Cataloging-in-Publication Data

Morton, David A.
  Gross anatomy : the big picture / David A. Morton, K. Bo Foreman, Kurt
H. Albertine.
     p. ; cm.
  ISBN-13: 978-0-07-147672-0 (pbk. : alk. paper)
  ISBN-10: 0-07-147672-5 (pbk. : alk. paper)
  1. Human anatomy–Outlines, syllabi, etc. 2. Human
anatomy–Examinations, questions, etc. I. Foreman, K. Bo (Kenneth Bo)
II. Albertine, Kurt H. III. Title.
  [DNLM: 1. Anatomy–Examination Questions. 2. Anatomy–Outlines. QS
18.2 M889g 2011]
  QM31.M67 2011
  612–dc22

2010025929

DEDICATION

To Jared, Ireland, Gabriel, Max, and Jack; and their cousins Lia, Sophia, Joshua, Cayden, Ethan, Nathan, Kelsey, Robert, Stefani, Ella, Reid, Roman, Marcus, Jared, Hannah, Tanner, Liam, Maia, Riley, Sydney, Luke, Cole, Desiree, Celeste, Connlan, Isabelle, Nathan, Simon, Thomas, Alexandre, Lyla, Logan and Andilynn. I could not ask for a better family.

—*David A. Morton*

To my devoted family: my wife, Cindy, and our two daughters Hannah and Kaia.

—*K. Bo Foreman*

To Erik, Kristin, and Laura. Thank you for your patience with and understanding of my efforts to contribute to biomedical education and research.

—*Kurt H. Albertine*

CONTENTS

PREFACE

If you were asked to give a friend directions from your office to a restaurant down the street, your instructions may sound something like this—turn right at the office door, walk to the exit at the end of the hall, walk to the bottom of the stairs, take a left, exit out of the front of the building, walk across the bridge, continue straight for two blocks passing the post office and library, and you will see the restaurant on your right. If you pass the gas station, you have gone too far. The task is to get to the restaurant. The landmarks guide your friend along the way to complete the task.

Now, imagine if an anatomist were to give directions from the office to the restaurant in the same way most anatomy textbooks are written. Details would be relayed on the dimensions of the office, paint color, carpet thread count, position and dimensions of the desk in relation to the book shelf along the wall, including the number, types, and sizes of books lining the shelves, and door dimensions and office door material in relation to the other doors in the same building. This would occur over the course of 10 pages—and the friend still would not have left the office. The difference between you giving a friend directions to a restaurant and the anatomist giving directions to the same restaurant can be compared with the difference between many anatomy textbooks and this Big Picture textbook—either getting to the restaurant with succinct relevant directions or taking a long time to get to the restaurant or possibly not finding it.

The purpose of this textbook is to provide students with the necessary landmarks to accomplish their task—to understand the big picture of human anatomy in the context of health care, bypassing the minutia. The landmarks used to accomplish this task are text and illustrations. They are complete, yet concise and both figuratively and literally provide the "Big Picture" of human anatomy.

The format of the book is simple. Each page-spread consists of text on the left-hand page and associated illustrations on the right-hand page. In this way, students are able to grasp the big picture of individual anatomy principles in bite-sized pieces, a concept at a time.

- Key structures are highlighted in bold when first mentioned.
- Bullets and numbers are used to break down important concepts.
- Approximately 450 full-color figures illustrate the essential anatomy.
- High-yield clinically relevant concepts throughout the text are indicated by an icon.
- Study questions and answers follow each section.
- A final examination is provided at the end of the text.

We hope you enjoy this text as much as we enjoyed writing it.

—*David A. Morton*

—*K. Bo Foreman*

—*Kurt H. Albertine*

ACKNOWLEDGMENTS

Early in his life my father, Gordon Morton, went to art school. He purchased a copy of Gray's Anatomy to help him draw the human form. That book sat on our families book shelf all throughout my life and I would continually look through its pages in wonder of the complexity and miracle of the human body. After I completed high school my father gave me that book which I have kept in my office ever since. I would like to acknowledge and thank my father and my mother (Gabriella Morton) for their influence in my life to bring this book to publication. Thank you to my coauthors, Dr. Foreman and Dr. Albertine—they are a joy to work with and I look forward to many years of collaborating with them.

I express a warm thank you to Michael Weitz. His dedication, help, encouragement, vision, leadership, and friendship were key to the successful completion of this title. I also express great thanks to Susan Kelly. She was a joy to work with over the past few years through rain, shine, snow, tennis competitions, Olympics, and life in general—I thank her for her eagle eye and encouraging telephone conversations and e-mails. Thank you to Karen Davis, Armen Ovsepyan, Brian Kearns, John Williams, and to the folks at Dragonfly Media Group for the care and attention they provided in creating the images for this title. Finally, a warm thank-you to my wife and best friend Celine. Her unyielding support and encouragement through long nights of writing were always there to cheer me on. I adore her.

—David A. Morton

I thank my parents, Ken Foreman and Lynn Christensen, as well as my mentor and friend, Dr. Albertine. A special thank you to Cyndi Schluender and my students for their contributions to my educational endeavors. I also express a great thanks to Dr. Morton for his continued encouragement and support in writing this text book.

—K. Bo Foreman

Many medical educators and biomedical scientists contributed to my training that helped lead to writing medical education textbooks such as this one. Notable mentors are CCC O'Morchoe, S. Zitzlsperger, and N.C. Staub. For this textbook, however, I offer my thanks to my coauthors, Dr. Morton and Dr. Foreman. Coauthoring this textbook with them has been a thrill because now my doctoral degree students are my colleagues in original educational scholarship. What better emblem of success could a mentor ask for? So, to David and Bo, thank you! I look forward to watching your careers flourish as medical educators.

—Kurt H. Albertine

Aerial view of University of Utah campus, Salt Lake City, Utah. Photo taken by Kurt Albertine, educator and author.

ABOUT THE AUTHORS

David A. Morton completed his undergraduate degree at Brigham Young University, Provo, Utah, and his graduate degrees at the University of Utah School of Medicine, Salt Lake City. He currently serves as Anatomy Director and is a member of the Curriculum Committee at the University of Utah School of Medicine. He was awarded the Early Career Teaching Award and the Preclinical Teaching Award. Dr. Morton is an adjunct professor in the Physical Therapy Department and the Department of Family and Preventive Medicine. He also serves as a visiting professor at Kwame Nukwame University of Science and Technology, Kumasi, Ghana, West Africa.

K. Bo Foreman completed his undergraduate degree in physical therapy at the University of Utah and his graduate degree at the University of Utah School of Medicine. Currently, he is an Assistant Professor in the Department of Physical Therapy and serves as the Anatomy Director and Director of the Motion Analysis Core Facility at the University of Utah School of Medicine. In addition, he is an adjunct assistant professor in the Division of Plastic Surgery, the Department of Mechanical Engineering, and the Department of Neurobiology and Anatomy. Dr. Foreman has been awarded the Early Career Teaching Award from the University of Utah and the Basmajian Award from the American Association of Anatomists.

Kurt H. Albertine completed his undergraduate studies in biology at Lawrence University, Appleton, Wisconsin, and his graduate studies in human anatomy at Loyola University of Chicago, Stritch School of Medicine. He completed postdoctoral training at the University of California, San Francisco, Cardiovascular Research Institute. He has taught human gross anatomy for 34 years. Dr. Albertine established the Human Anatomy Teacher-Scholar Training Program in the Department of Neurobiology & Anatomy at the University of Utah School of Medicine. The goal of this training program is to develop teacher-scholars of human anatomy to become leaders of anatomy teachers on a national level, contribute teaching innovations, and design and perform teaching outcomes research for upcoming generations of medical students. Graduates of this training program include Dr. Morton and Dr. Foreman.

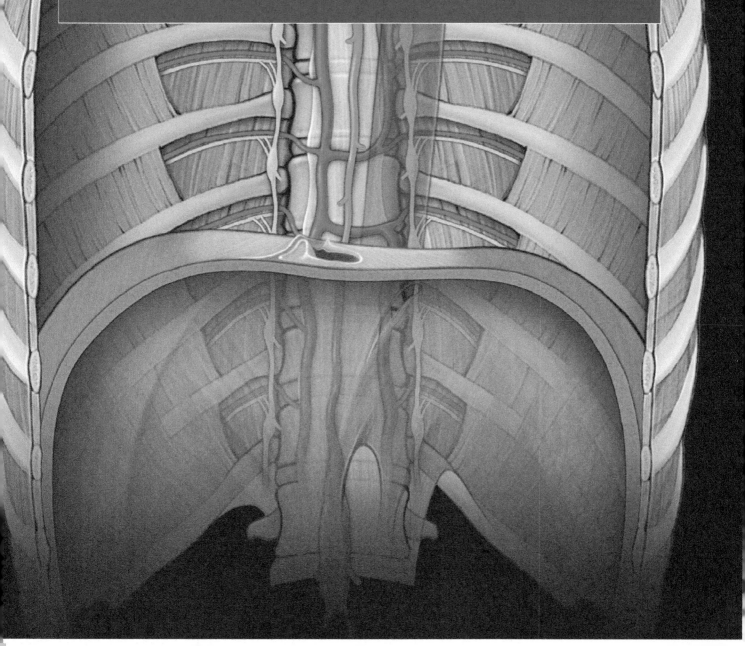

SECTION 1

BACK

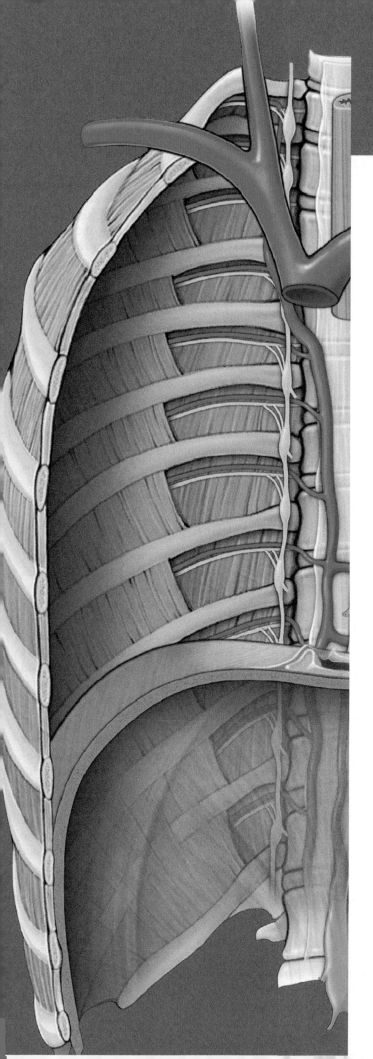

CHAPTER 1

BACK

SKIN OF THE BACK

THE BIG PICTURE

The skin of the back is thick, with increasing thickness toward the nape of the neck. The cutaneous innervation of the back is segmentally innervated through the dorsal rami of spinal nerves. The spinous processess of the vertebrae and other osteologic landmarks are palpable, which enable localization of spinal levels through surface landmarks.

CUTANEOUS INNERVATION

The skin of the back is segmentally innervated by cutaneous nerves that originate from the dorsal rami (Figure 1-1A). Dorsal rami contain both motor and sensory neurons as they branch from each level of the spinal cord and course posteriorly in the trunk. The motor neurons terminate in the deep back muscles (e.g., erector spinae muscles), where they cause muscle contraction. The sensory neurons, however, continue on and terminate in the skin where they provide cutaneous sensations such as pain, touch, and temperature at each dermatomal level of the back (Figure 1-1B). There is some segmental overlap of the peripheral sensory fields from adjacent dermatomes.

It should be noted that

- The dorsal ramus of C1 carries only motor neurons to the suboccipital muscles.
- The dorsal ramus of C2 carries only sensory neurons to the back of the scalp.
- The lateral aspects of the back are innervated by lateral cutaneous nerves, which are derived from the ventral rami segments of spinal nerves at each level (Figure 1-1A).
- S5 and Co 1 only carry sensory neurons.

OSTEOLOGY AND SURFACE ANATOMY

The following structures are easily palpated through the skin (Figure 1-1A):

- **Cervical vertebrae.** The first prominent spinous process that is palpable is **C7 (vertebra prominens)**. Cervical spines 1 to 6 are covered by the **ligamentum nuchae**, a large ligament that courses down the back of the neck and connects the skull to the spinous processes of the cervical vertebrae.
- **Thoracic vertebrae.** The most prominent spine is the T1 vertebra; other vertebrae can be easily recognized when the trunk is flexed anteriorly. Thoracic vertebrae have long spines that point downward so that each spinous process is level with the body of the inferior vertebra. The spines can be counted downward from C7 and T1, or from a line joining the iliac crests at the L4 vertebral level and then counting upward from that site.

- **Lumbar vertebrae.** The body of L3 is approximately at the center of the body from superior to inferior and medial to lateral. When a person is in an upright posture, most vertebral spines are not obvious because they are covered by the erector spinae muscles in the cervical and lumbar regions.
- **Sacrum.** The spines of the sacrum are fused together in the midline to form the median sacral crest. The crest can be felt beneath the skin in the uppermost part of the gluteal cleft between the buttocks. The sacral hiatus is situated on the posteroinferior aspect of the sacrum and is the location where the extradural space terminates. The hiatus lies approximately 5 cm above the tip of the coccyx and beneath the skin of the natal cleft.
- **Coccyx.** The tip of the coccyx is located in the upper part of the natal cleft and can be palpated approximately 2.5 cm posterior to the anus. The anterior surface of the coccyx can be palpated via the anal canal.
- **Ilium.** The crest of the ilium is located at the L4 vertebral level and is easily palpated. Regardless of the size or weight of a person, the posterior iliac spines can be palpated because they lie beneath a skin dimple at the S2 vertebral level and the middle of the sacroiliac joint.
- **Scapulae.** The scapulae overlie the posterior portion of the thoracic wall, covering the upper seven ribs. The superior angle can be palpated at the T1 vertebral level, the spine of the scapula at T3, and the inferior angel at T7. The acromion forms the lateral end of the scapular spine and is easily palpated.

VERTEBRAL CURVATURES

In utero, the entire vertebral column has a primary or kyphotic curvature (concave anteriorly) (Figure 1-1C). After birth, with the demands of walking, weightbearing, and gravity, the cervical and lumbar regions form secondary or lordotic curvatures (concave posteriorly) (Figure 1-1D). **Primary (kyphotic) curvatures** occur in the thoracic and sacral regions, where the vertebrae curve posteriorly. This allows increased space for the heart and lungs in the thorax and birth canal in the sacral region. **Lordotic or secondary curvatures** occur in the cervical and lumbar regions, where the vertebrae curve anteriorly.

Abnormal primary curvatures are referred to as **kyphosis (excessive kyphosis)**, whereas abnormal secondary curvatures are referred to as **lordosis (excessive lordosis)**. Patients may present with abnormal lateral curvatures (**scoliosis**), which may be due to muscular dominance of one side over the other or to poor posture or congenital problems. To diagnose scoliosis, the physician may ask the patient to bend forward to determine if one side of the thorax is higher than the other due to asymmetry of the spine. ▼

Figure 1-1: A. Surface anatomy of the back showing bony landmarks on the left and cutaneous nerves on the right. **B**. Axial section of the back showing the dorsal rami transmitting sensory neurons from the skin of the back to the spinal cord. Normal curvatures of the vertebral column in a newborn (**C**) and in an adult (**D**).

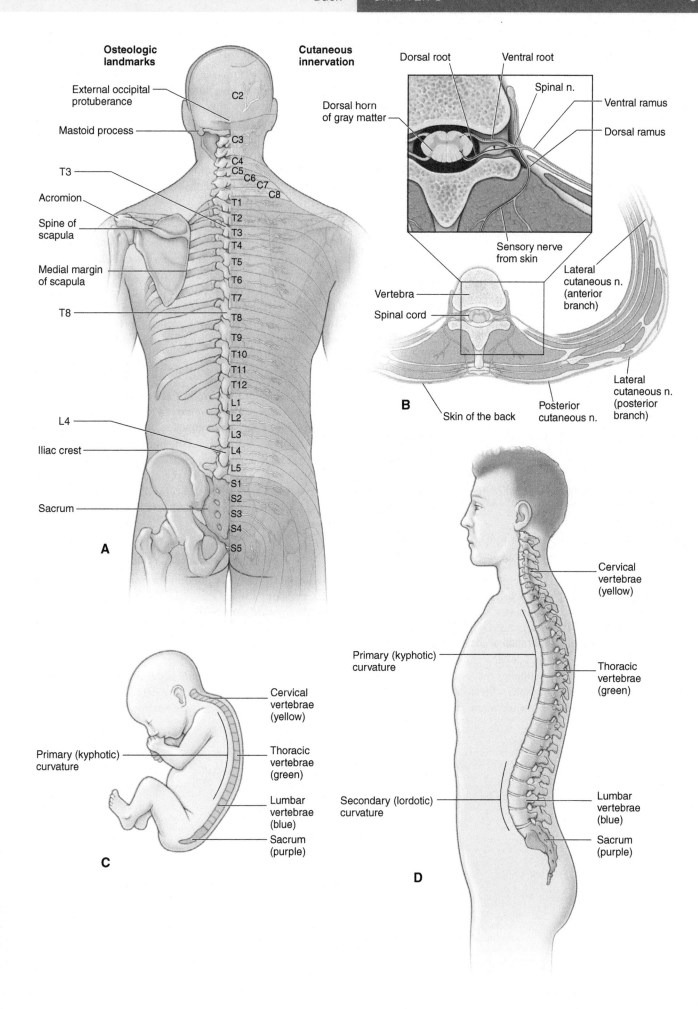

Osteologic landmarks

External occipital protuberance

Mastoid process

T3

Acromion

Spine of scapula

Medial margin of scapula

T8

L4

Iliac crest

Sacrum

C2
C3
C4
C5
C6
C7
C8
T1
T2
T3
T4
T5
T6
T7
T8
T9
T10
T11
T12
L1
L2
L3
L4
L5
S1
S2
S3
S4
S5

A

Cutaneous innervation

Dorsal root

Ventral root

Dorsal horn of gray matter

Spinal n.

Ventral ramus

Dorsal ramus

Sensory nerve from skin

Vertebra

Spinal cord

Lateral cutaneous n. (anterior branch)

Lateral cutaneous n. (posterior branch)

Skin of the back

Posterior cutaneous n.

B

Cervical vertebrae (yellow)

Primary (kyphotic) curvature

Thoracic vertebrae (green)

Lumbar vertebrae (blue)

Sacrum (purple)

C

Cervical vertebrae (yellow)

Primary (kyphotic) curvature

Thoracic vertebrae (green)

Secondary (lordotic) curvature

Lumbar vertebrae (blue)

Sacrum (purple)

D

SUPERFICIAL BACK MUSCLES

THE BIG PICTURE

The superficial back muscles consist of the trapezius, levator scapulae, rhomboid major, rhomboid minor, and latissimus dorsi muscles (Figure 1-2A; Table 1-1). Although these muscles are located in the back, they are considered to be muscles of the upper limbs because they connect the upper limbs to the trunk and assist in upper limb movements directly or indirectly via the scapula or humerus. The superficial back muscles are located in the back region and receive most of their nerve supply from the ventral rami of spinal nerves (primarily the brachial plexus) and act on the upper limbs. These muscles are discussed in much greater detail in Section VI, Upper Limb, but are included here because a brief discussion must precede the primary discussion of the deeper muscles and structures of the back.

TRAPEZIUS MUSCLE

The trapezius muscle is the most superficial back muscle. It attaches to the occipital bone, nuchal ligament, spinous processes of C7–T12, scapular spine, acromion, and clavicle. The trapezius muscle has a triangular shape and has the following muscle fiber orientations:

- **Superior fibers.** Course obliquely from the occipital bone and upper nuchal ligament to the scapula, causing scapular elevation and upward rotation.

- **Middle fibers.** Course horizontally from the lower nuchal ligament and thoracic vertebrae to the scapula, causing scapular retraction.

- **Inferior fibers.** Course superiorly from the lower thoracic vertebrae to the scapula, causing scapular depression and upward rotation.

In addition, the multiple fiber orientation of the trapezius muscle fixes the scapula to the posterior wall of the thorax during upper limb movement. It is innervated by the **spinal accessory nerve (CN XI)**. The **superficial branch of the transverse cervical artery** supplies the trapezius muscle, whereas the deep branch of the transverse cervical artery supplies the levator scapulae and rhomboid muscles. In some instances, the dorsal scapular artery will replace the deep branch of the transverse cervical artery.

LEVATOR SCAPULAE MUSCLE

The levator scapula muscle is located deep to the trapezius muscle and superior to the rhomboids. The levator scapula muscle attaches to the cervical vertebrae and the superior angle of the scapula, causing elevation and downward rotation of the scapula. The nerve supply is from branches of ventral rami from spinal nerves C3–C4 and occasionally from C5, via the **dorsal scapular nerve**, and vascular supply is from the deep branch of the transverse cervical artery.

RHOMBOID MAJOR AND MINOR MUSCLES

The rhomboid minor is superior to the rhomboid major, with both positioned deep to the trapezius muscle. The rhomboid muscles attach to the spinous processes of C7–T5 and the medial border of the scapula, resulting in scapular retraction. They are innervated by the dorsal scapular nerve (i.e., the ventral ramus of C5) and the vascular supply from the deep branch of the transverse cervical artery.

LATISSIMUS DORSI MUSCLE

The latissimus dorsi is a broad, flat muscle of the lower region of the back. It attaches to the spinous processes of T7, inferior to the sacrum via the **thoracolumbar fascia**, and inserts laterally into the intertubercular groove of the humerus. The latissimus dorsi acts on the humerus (arm) causing powerful adduction, extension, and medial rotation of the arm. It is innervated by the **thoracodorsal nerve** (ventral rami of C6–C8) and receives its blood supply from the **thoracodorsal artery** (branch off the axillary artery).

SCAPULAR MOVEMENTS

The scapula glides over the thoracic wall, but there are no distinct anatomic joints (Figure 1-2B). Instead, muscles pull the scapula forward and backward (protraction and retraction, respectively) and up and down (elevation and depression, respectively) and rotate it so that the glenoid fossa moves superiorly (upward rotation) or inferiorly (downward rotation). Rotation of the scapula is defined by the direction that the glenoid fossa faces. For more details describing the upper limb and scapular muscles, see Section VI.

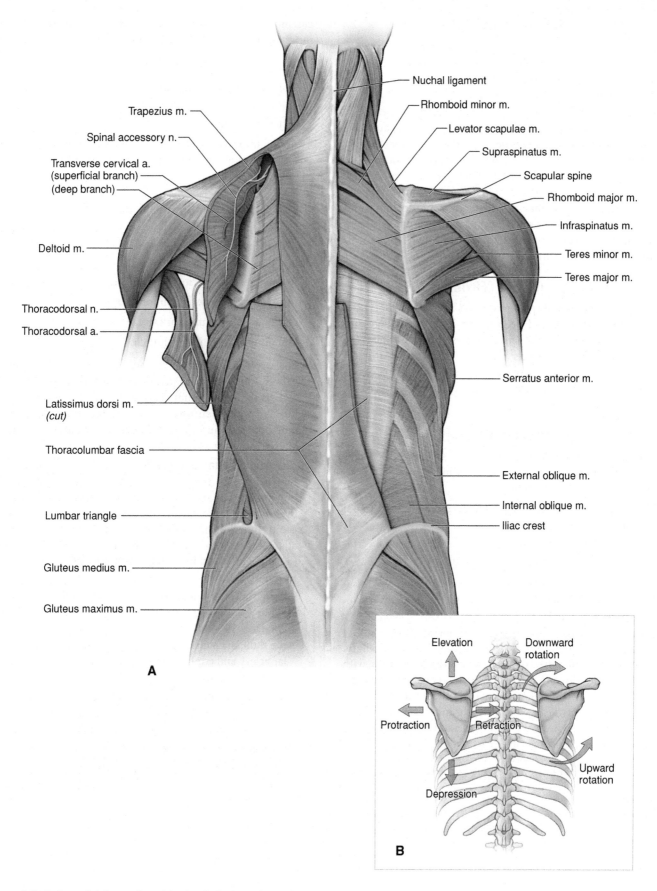

Figure 1-2: A. Superficial muscles of the back. **B**. Scapular actions.

DEEP BACK MUSCLES

THE BIG PICTURE

The deep back muscles consist of the splenius, erector spinae, transversospinalis, and suboccipital muscles (Table 1-2). These deep back muscles are segmentally innervated by the dorsal rami of spinal nerves at each vertebral level where they attach. It is not important to know every detailed attachment for the deep back muscles; however, you should realize that these muscles are responsible for maintaining posture and are in constant use during locomotion. The erector spinae muscles course medially to superolaterally, and they extend the vertebral column and rotate the trunk ipsilaterally. The transversospinalis group course laterally to superomedially and extend the vertebral column and rotate the trunk contralaterally.

SPLENIUS CAPITIS AND CERVICIS MUSCLES

The splenius capitis and splenius cervicis muscles lie superficial to the erector spinae and deep to the levator scapulae and rhomboid muscles. The splenius capitis extends and rotates the head, whereas the splenius cervicis rotates and extends the cervical spine. Both muscles, acting unilaterally (only one side contracts while the other side relaxes), produce lateral flexion of the head and neck.

ERECTOR SPINAE MUSCLES

From lateral to medial, the erector spinae muscles include the iliocostalis, the longissimus, and the spinalis muscles (Figure 1-3A and B).

- Acting bilaterally, the erector spinae extend the vertebral column.
- Acting unilaterally, the erector spinae laterally flex the vertebral column and rotate to the ipsilateral side.

The deep muscles of the back are involved in the control of posture; they contract intermittently during swaying movements, such as walking.

The erector spinae muscles ascend throughout the length of the back as columns composed of fascicles of shorter length. Each column is composed of a rope-like series of fascicles, with various bundles arising as others are inserting; each fascicle spans from 6 to 10 segments between bony attachments. The erector spinae muscles are segmentally innervated by dorsal rami. The motor neurons in the dorsal rami terminate in and innervate the erector spinae muscles (Figure 1-3B).

TRANSVERSOSPINALIS MUSCLES

The transversospinalis muscles are deep to the erector spinae muscles. From superficial to deep, the transversospinalis mus-cles include the semispinalis, multifidus, and rotatores. A defining characteristic of this muscle group is that their muscle fibers ascend superomedially, crossing from one to six vertebral levels, and attach inferiorly from a transverse process and superiorly to the spinous process of neighboring vertebrae.

- Acting bilaterally, the transversospinalis group extends the vertebral column.
- Acting unilaterally, they rotate the vertebral column to the opposite side on which the transversospinalis muscle is located.

SUBOCCIPITAL MUSCLES

The suboccipital muscles are inferior to the occipital bone and deep to the semispinalis capitis muscle (Figure 1-3C). The suboccipital muscles are as follows:

- Rectus capitis posterior major muscle
- Rectus capitis posterior minor muscle
- Obliquus capitis superior muscle
- Obliquus capitis inferior muscle

These muscles attach to the occipital bone and C1 and C2 vertebrae, and are innervated by the dorsal ramus of C1 (also known as the suboccipital nerve). The suboccipital nerve carries only motor neurons. These **suboccipital muscles** are mainly postural muscles, but they may also extend and rotate the head.

- The **suboccipital triangle** is the area in the suboccipital region between the rectus capitis posterior major and the obliquus capitis superior and inferior muscles. It is covered by a layer of dense fibro-fatty tissue, deep to the semispinalis capitis muscle. The floor is formed by the posterior occipito-atlantal membrane and the posterior arch of the C1 vertebrae (atlas). **Contents of the suboccipital triangle** include the following structures:

  - **Vertebral artery.** Exits the transverse foramen and courses along the top surface of the posterior arch of C1, prior to entering the skull via the foramen magnum. Once inside the skull, the paired vertebral arteries join to form the basilar artery and contribute to the collateral arterial supply of the brain.

  - **Suboccipital nerve.** Emerges between the occipital bone and the atlas, by the vertebral artery, and branches to provide motor innervation to the suboccipital muscles.

  - **Greater occipital nerve** (dorsal ramus of C2 nerve). Carries only sensory neurons as it emerges below the obliquus capitis inferior muscle and courses upward, piercing the semispinalis capitis and trapezius muscles to supply sensory innervation to the back of the scalp.

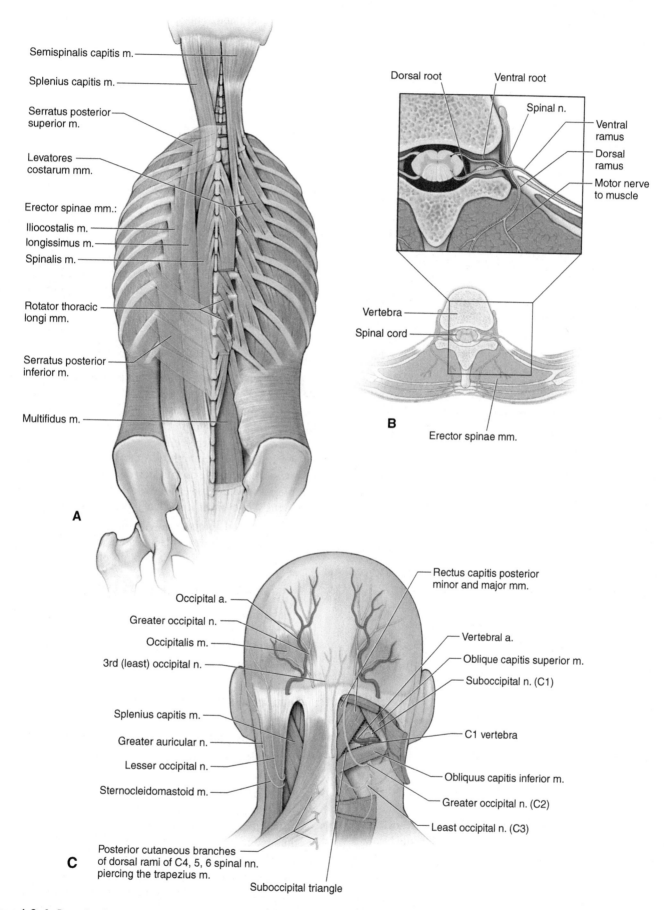

Figure 1-3: A. Deep back muscles with erector spinae muscles on the left and deeper transversospinalis muscles on the right. **B**. Axial section of the back showing the dorsal rami transmitting motor neurons to the erector spinae muscles. **C**. Suboccipital region; step dissection on the right side revealing the underlying suboccipital triangle.

VERTEBRAL COLUMN

THE BIG PICTURE

The vertebral column consists of 33 vertebrae (7 cervical, 12 thoracic, 5 lumbar, 5 sacral, and 3–4 coccygeal). These vertebrae, along with their ligaments and intervertebral discs, form the flexible, protective, and supportive vertebral column that contains the spinal cord.

VERTEBRAL COLUMN

The vertebral column consists of 33 vertebrae, which are divided into cervical (C), thoracic (T), lumbar (L), sacral (S), and coccygeal (Co) regions (Figure 1-4A; Table 1-3). To simplify their descriptions, each vertebra is referred to by the first letter of its region. For example, the fourth cervical vertebra is referred to as the C4 vertebra.

- **Seven cervical vertebrae.** The C1 vertebra, also called the "**atlas**," articulates with the skull. The C2 vertebra is also called the "**axis**" because, through its synovial joint articulation with C1, it provides a side-to-side rotation, as in the head movement indicating "no." A unique characteristic of cervical vertebrae is that each transverse process has its own transverse foramen for the transmission of the **vertebral arteries** from the subclavian artery to the brain.
- **Twelve thoracic vertebrae.** On each side of a thoracic vertebra, there is a rib that articulates with the vertebral body and transverse process. Thus, there are 12 pairs of ribs articulating with 12 thoracic vertebrae.
- **Five lumbar vertebrae.** These massive vertebrae form the support to the lower part of the back.
- **Five sacral vertebrae.** These vertebrae are fused into one bone called the **sacrum**.
- **Coccygeal vertebrae.** There are three to four fused coccygeal vertebrae that form the **coccyx**, or "tailbone."

A TYPICAL VERTEBRA

A typical vertebra consists of the following (Figure 1-4B):

- **Body.** The anterior vertebral region that is the primary weight-bearing component of the vertebrae.
- **Vertebral arch.** Arches posteriorly from the vertebral body and forms the vertebral foreman, which contains the spinal cord.
- **Pedicle.** Part of the vertebral arch that joins the vertebral body to the transverse process.
- **Transverse processes.** Extend laterally from the junction of the pedicle and lamina.
- **Superior and inferior articular facets.** Form synovial zygapophyseal (facet) joints with the vertebrae above and below.
- **Laminae.** The paired posterior segments of the vertebral arch that connect the transverse processes to the spinous process.
- **Spinous processes.** The posteriorly projecting tip of the vertebral arch. These processes are easily palpated beneath the skin.

- **Vertebral foramen.** The space in which the spinal cord, its coverings, and the rootlets of the spinal nerves lie. The series of vertebral foramina form the **vertebral canal**.
- **Intervertebral foramina.** Bilateral foramina that form the space between pedicles of adjacent vertebrae for passage of spinal nerves.

VERTEBRAL ARTICULATIONS

There are two types of articulations that enable the 33 vertebrae to not only provide support, protection, and absorb shock, but also to provide flexibility. These two articulations are the intervertebral discs and the zygapophyseal joints.

- **Intervertebral discs.** Located between vertebrae, from the second cervical disc down to the first sacral vertebra (Figure 1-1D). The discs permit a limited amount of movement between adjacent vertebrae. Intervertebral discs are composed of an annulus fibrosus and a nucleus pulposus. The **anulus fibrosus** is the tough hyaline cartilaginous rim, whereas the **nucleus pulposus** is the softer fibrocartilaginous core. The disc can bear weight because spread of the nucleus pulposus is constrained by the anulus fibrosus.

▽ **Damage to the anulus fibrosus** may allow the softer nucleus pulposus to bulge or herniate, which may compress the segmental nerve roots. This is often referred to as a "slipped disc." The symptoms of this condition will depend upon the level at which the rupture occurs and the structures that are affected. For example, compression of the L4 nerve may cause weakness in ankle dorsiflexion due to the L4 innervation of the anterior tibialis muscle in the leg, or lack of cutaneous sensation in the skin over the knee due to the L4 dermatome. ▽

- **Zygapophyseal joints.** Synovial joints located between the superior and inferior articular facets of adjacent vertebrae. These joints provide varying amounts of flexibility.

VERTEBRAL LIGAMENTS AND JOINTS

The vertebral column is stabilized by ligaments that run between vertebrae (Figure 1-4C). Some of the **ligaments that limit vertebral flexion** (bending forward) include the following:

- **Ligamentum flavum.** Connects paired laminae of adjacent vertebrae.
- **Supraspinous ligament.** Connects the apices of the spinous processes from C7 to the sacrum.
- **Interspinous ligament.** Connects adjoining spinous processes.
- **Nuchal ligament.** Extends from the external occipital protuberance along the spinous processes of C1–C7.
- **Posterior longitudinal ligament.** Courses longitudinally, down the posterior surface of the vertebral bodies within the vertebral canal. It supports the intervertebral disc posteriorly, thus reducing the incidence of herniations that may compress the spinal cord and cauda equina.
- **Anterior longitudinal ligament.** Courses longitudinally along the anterior surface of the vertebral bodies limiting vertebral extension.

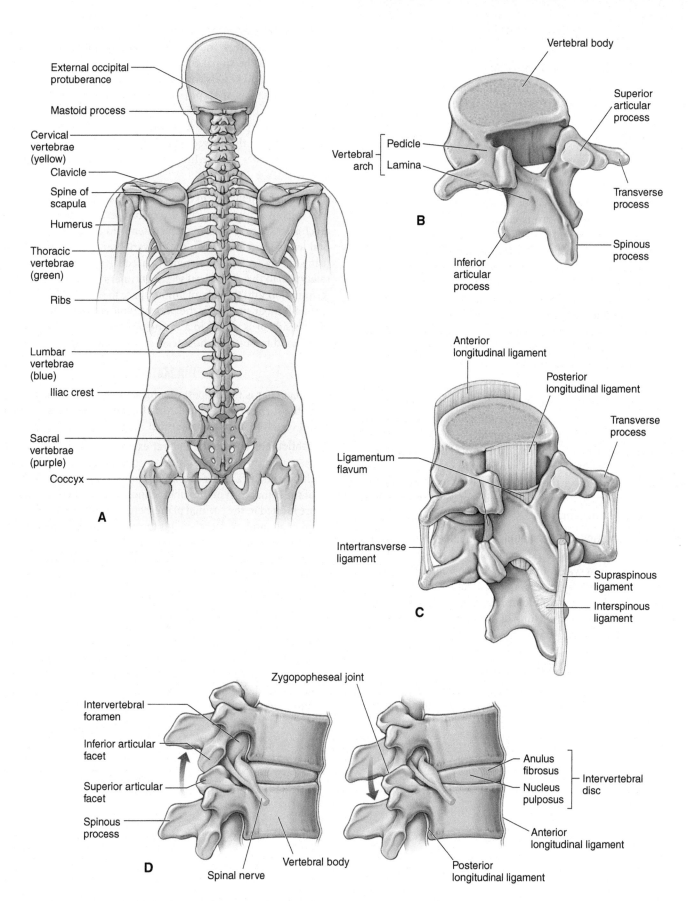

Figure 1-4: A. Posterior view of the vertebral column. **B**. A typical thoracic vertebra. **C**. Two articulated vertebrae showing the ligaments. **D**. Lateral view of two vertebrae demonstrating intervertebral discs as shock absorbers. Observe how the facet joints facilitate flexion and extension of the vertebral column.

SPINAL MENINGES

BIG PICTURE

The brain and spinal cord are surrounded and protected by three layers of connective tissue meninges called the dura mater, the arachnoid mater, and the pia mater (Figure 1-5A and B). The meninges are supplied by the general sensory branches of the cranial and spinal sensory nerves.

DURA MATER

The dura mater is the most superficial layer of the meninges and is composed of dense fibrous connective tissue. It forms a sheath around the spinal cord that extends from the internal surface of the skull to the S2 vertebral level where the dura mater forms the coccygeal ligament, which covers the pial filum terminale. The dura mater evaginates into each intervertebral foramen and becomes continuous with the connective tissue covering (**epineurium**) around each spinal nerve. The nerve roots in the subarachnoid space lack this connective tissue covering and, therefore, are more fragile than spinal nerves. The dura mater defines the epidural space and the subdural space.

▽ The **epidural space** is located between the dura mater and the vertebral canal. An anesthetic agent can be injected into the epidural space to anesthetize the spinal nerve roots, and is particularly useful for procedures involving the pelvis and perineum (e.g., during childbirth). To administer an anesthetic agent into the epidural space, a needle passes through the skin of the back overlaying the L4–L5 vertebrae and through the supportive ligaments between the adjacent vertebrae, such as the ligamentum flavum. ▽

▽ **Inflammation of the meninges (meningitis)** is painful because, unlike visceral organs such as the stomach, the meninges are innervated by general sensory neurons, which include pain receptors. For example, if a patient with meningitis tries to touch his/her chin to his/her chest, he/she may experience pain due to the stretching of the meninges surrounding the cervical spinal cord. ▽

ARACHNOID MATER

The arachnoid mater is the intermediate meningeal layer. It is attached to the underlying pia by numerous **arachnoid trabec-** ulae. The **subarachnoid space** is the cavity between the arachnoid and pial layers and contains **cerebrospinal fluid (CSF)**. CSF suspends the spinal cord, brain, and nerve roots. In addition, large blood vessels course within the subarachnoid space. The dural sac is that portion of the subarachnoid space between the conus medullaris (approximately the L1 vertebral level) and the point at which the coccygeal ligament begins (approximately the S2 vertebral level). The dural sac contains only spinal roots and the filum terminale, which are suspended in CSF (Figure 1-5C).

▽ **Lumbar puncture to obtain CSF.** The spinal cord terminates in an adult at the L1–L2 vertebral level, whereas the subarachnoid space containing CSF extends to about the S2 vertebral level. Therefore, when the L4 vertebral spine is identified, using the iliac crest as a reference, a needle is passed with relative safety two to three vertebral segments inferior to the spinal cord termination into the subarachnoid space to sample CSF. To perform a lumbar puncture, the patient typically is asked to lie on his/her side or is placed in a sitting position so that the spine is fully flexed to open up the intervertebral space. ▽

PIA MATER

The pia mater is the deepest meningeal layer and is inseparable from the surface of the spinal cord. It contains a plexus of small blood vessels that supply the spinal cord.

▪ **Denticulate ligaments.** Lateral extensions of the pia mater that support the entire spinal cord by attaching to the arachoid and dura mater and maintaining a centralized location of the spinal cord in the subarachnoid space. These ligaments are located in the coronal plane, between the ventral and dorsal roots, and project through the arachnoid mater to attach to the dura mater in a series of sawtooth projections. In this manner, the spinal cord is tethered to the dura mater while floating within the CSF.

▪ **Filum terminale.** An extension of the pia mater beyond the tip of the spinal cord (conus medullaris) that attaches to the coccyx in the vertebral canal. It is as if the prolongation of the pia mater were occupied by the spinal cord *in utero* but, as the spinal cord withdraws upward during development and maturation, the sleeve of the pia mater that surrounded the distal end of the spinal cord collapsed upon itself becoming the filum terminale.

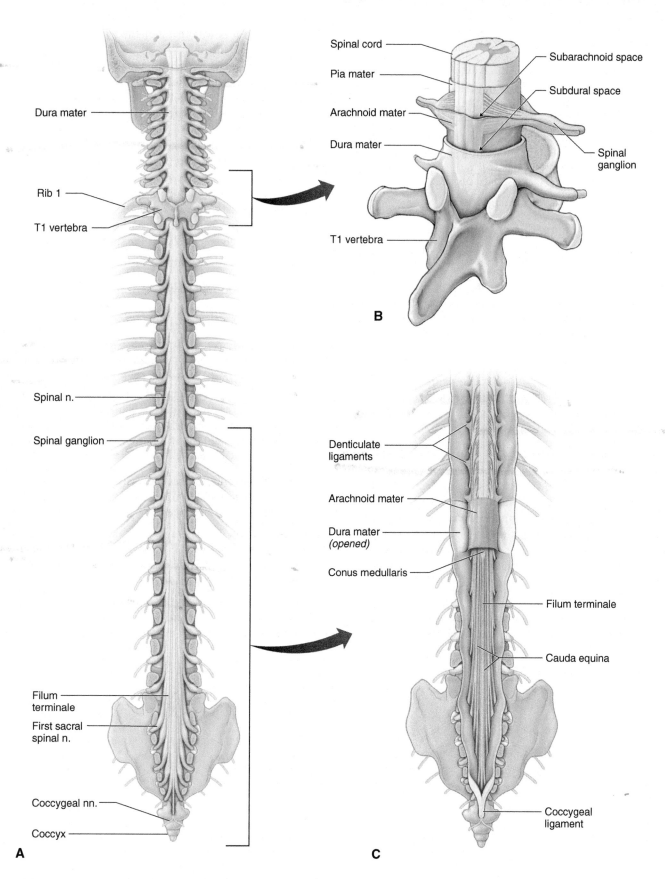

Figure 1-5: A. Coronal section of the vertebral column through the pedicles from a posterior view revealing the dura mater surrounding the spinal cord. **B**. T2 segment of the spinal cord showing spinal meningeal layers. **C**. Caudal spinal cord with meninges.

SPINAL CORD

THE BIG PICTURE

The spinal cord is a part of the central nervous system (CNS) and receives sensory input from the body tissues via spinal nerves. These messages are processed within the CNS and appropriate motor responses are sent out through spinal nerves. The spinal cord consists of both white and gray matter. The spinal cord is located within the vertebral (spinal) canal, where it is protected by bones, ligaments, and the three layers of connective tissue known as the meninges.

TOPOGRAPHY AND OVERVIEW

The spinal cord is located within the vertebral (spinal) canal and extends from the medulla oblongata at the C1 vertebral level and terminates at the conus medullaris at the L1 and L2 vertebral level (Figure 1-6A). In a newborn, the spinal cord terminates at the L3 and L4 vertebral level; in a fetus, it continues on to the sacrum. The vertebral canal serves as a physical protector to the delicate spinal cord.

In cross-section, the spinal cord consists of white matter surrounding gray matter (Figure 1-6B). The white matter consists of neuronal axons, and the myelin appears white. The gray matter consists of aggregates of neuronal cell bodies, which do not contain myelin and thus appears gray.

▽ **Contrasting vertebral and spinal cord levels.** The vertebral canal is longer than the spinal cord in adults as a result of unequal growth during development. Therefore, a patient with a **C3 vertebral fracture** potentially could have a bone fragment that would impinge upon the C3 spinal cord segment. However, a patient with a **T10 vertebral fracture** potentially could have a bone fragment that would impinge upon the L1 segment of the spinal cord. ▼

WHITE MATTER OF THE SPINAL CORD

White matter is composed of vertical columns of axons arranged so that those that perform similar functions are grouped together to form a **tract**. These tracts are not sharply demarcated from each other and, therefore, there may be some overlap between them.

Bundles of axons in the white matter carry impulses up to the brain from sensory tracts and, conversely, bundles of axons carry impulses in the white matter down from the brain to neurons in the gray matter of the spinal cord from motor tracts.

The amount of white matter increases at each successive higher spinal segment. Cervical spinal cord levels, therefore, contain more white matter because all neurons descending from the brain inferiorly or from body tissues to the brain pass through the cervical spinal cord. As a result, the sacral spinal cord has the least amount of white matter because most ascending or descending fibers have entered or exited the spinal cord superior to the caudal spinal cord region (compare Figures 1-6B–E).

GRAY MATTER OF THE SPINAL CORD

Spinal cord gray matter forms a letter "H" in cross-section and consists of the following regions:

Ventral horn. Contains cell bodies of motor neurons innervating skeletal muscle. The ventral horns are largest in those parts of the spinal cord that serve regions of the body with many muscles. For example, the spinal cord segments C5–T1, which serve the upper limbs (brachial plexus), have a cervical spinal cord swelling, whereas L4–S3 spinal cord segments serving the lower limbs (lumbosacral plexus) form a lumbar spinal cord swelling.

▽ **Poliomyelitis** causes paralysis of voluntary muscle by attacking the neurons in the ventral horn of the gray matter in the spinal cord. ▼

Lateral horn. The lateral horn is only present in the spinal cord between the T1 and L2 vertebral levels and contains cell bodies of preganglionic sympathetic motor neurons of the autonomic nervous system.

Dorsal horn. Receives sensory impulses entering via the dorsal root. The thoracic and upper lumbar levels have relatively small amounts of gray matter because these vertebral levels only innervate the thoracic and abdominal regions.

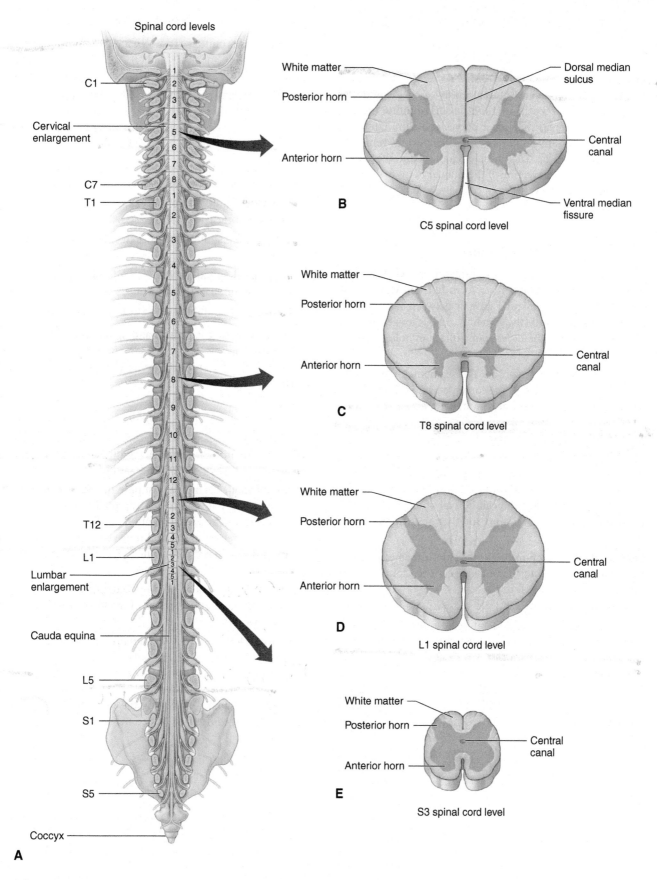

Figure 1-6: A. Posterior view of the coronal section of the vertebral canal. Compare and contrast the spinal cord, spinal nerve, and vertebral levels. **B–E**. C5, T8, L1, and S3 cross-sections of the spinal cord, respectively. Compare and contrast gray and white matter at the various levels.

SPINAL ROOTS, SPINAL NERVES, AND RAMI

THE BIG PICTURE

At each spinal cord segment, ventral and dorsal spinal roots join to form spinal nerves that bifurcate into ventral and dorsal rami. Spinal roots carry sensory (dorsal root) or motor (ventral root) neurons, whereas the spinal nerves and rami contain a mixture of sensory and motor neurons. The dorsal rami segmentally innervate deep back muscles (motor) and the skin of the back (sensory). This section of the chapter helps to differentiate among vertebral levels, spinal cord levels, and spinal nerve levels.

SPINAL ROOTS

At each spinal cord segment, paired **dorsal** and **ventral roots** exit the lateral sides of the cord to form **left** and **right spinal nerves**. As a result of unequal growth between the vertebral canal and the spinal cord (the vertebral canal is longer than the spinal cord in adults), the nerve roots follow an oblique course from superior to inferior (Figure 1-7A).

- Only in the cervical region are the segments of the spinal cord at the same level with the corresponding cervical vertebrae. Inferior to the cervical region, each spinal nerve from the thoracic, lumbar, and sacral spinal cord segments exits inferior to its similarly numbered vertebra.
- **Dorsal roots** convey sensory (afferent) information from body tissues to the spinal cord (i.e., skin to spinal cord) (Figure 1-7B). The dorsal root ganglion is a swelling in the dorsal root and houses the cell bodies of all sensory neurons entering the spinal cord for that specific body segment.
- **Ventral roots** convey motor (efferent) information away from the spinal cord to the body tissues (i.e., spinal cord to the biceps brachii muscle).
- The spinal cord terminates at the L1 vertebral level in adults. Therefore, the lumbar and sacral nerve roots descending in the vertebral canal below the L1 vertebral level form a mass of nerve roots that resembles the tail of a horse; hence, the name **cauda equina** (Figure 1-7C). Because the cauda equina floats in the CSF, a needle introduced into the subarachnoid space will displace the roots with little possibility of puncture damage.

SPINAL NERVES

The spinal roots unite in or near the intervertebral foramen to form a spinal nerve (Figure 1-7B and C). The 31 pairs of spinal nerves formed by the dorsal and ventral roots are organized as follows:

- **Eight cervical spinal nerves.** The first seven cervical spinal nerves, C1–C7, exit the vertebral canal superior to each respective cervical vertebra. The last cervical nerve, C8, exits inferior to the seventh cervical vertebra.

All of the remaining spinal nerves segmentally exit the spinal cord inferior to their respective vertebra, as follows:

- **Twelve thoracic spinal nerves.** Exit inferior to the 12 thoracic vertebrae.
- **Five lumbar spinal nerves.** Exit inferior to the five lumbar vertebrae.
- **Five sacral spinal nerves.** Exit inferiorly through the dorsal sacral foramina of the sacrum.
- **One coccyx spinal nerve.** Exits by the coccyx bone.

- ▽ **Contrasting vertebral and spinal nerve levels.** The C4 spinal nerve exits between the C3 and C4 vertebrae. However, the L4 spinal neve exits between the L4 and L5 vertebrae. ▽

RAMI

The spinal nerve exits the vertebral canal through the intervertebral foramen. Each spinal nerve bifurcates into a dorsal primary ramus and a ventral primary ramus (Figure 1-7B).

- **Dorsal rami.** Segmentally supply the skin of the back (in a dermatomal pattern) as well as provide motor innervation to the deep vertebral muscles of the median portion of the back (e.g., the erector spinae and transversospinalis muscles). The dorsal rami do not contribute to the innervation of the limbs or face.
- **Ventral primary rami.** Supply the dermatomes and myotomes of the anterolateral portions of the torso as well as the upper and lower limbs.

  - *Note: Dorsal and ventral roots are not the same as dorsal and ventral rami. Dorsal roots convey sensory impulses, whereas ventral roots convey motor impulses. Once these roots unite to form the spinal nerve, the majority of subsequent branches, including the rami, convey both sensory and motor impulses (mixed nerves).*

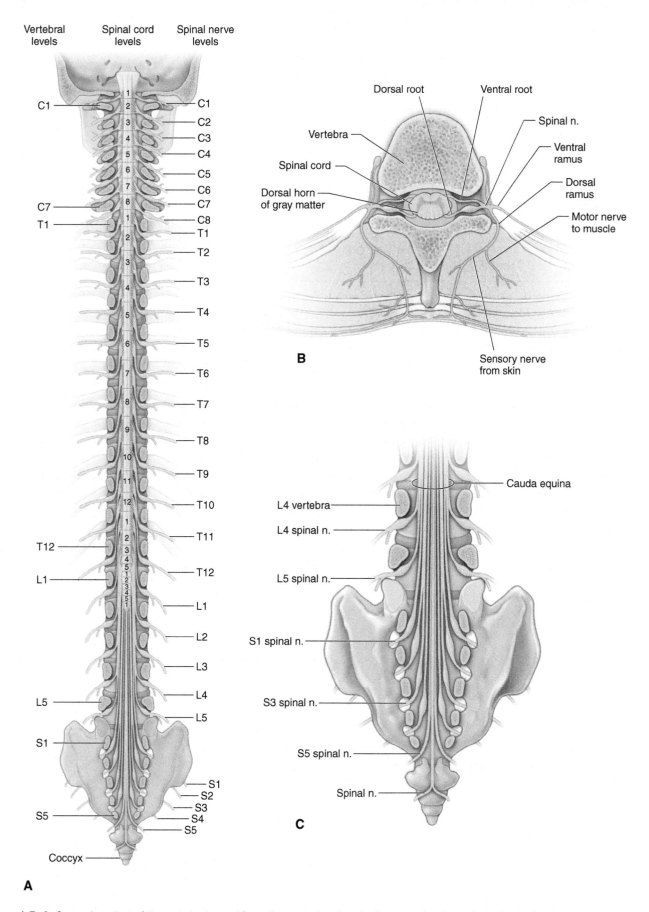

Figure 1-7: A. Coronal section of the vertebral canal from the posterior view. **B**. Cross-section through the back showing spinal roots, nerves, and rami. Spinal nerves branch into a posterior ramus (mixed), which transports sensory neurons from the skin of the back to the spinal cord and motor neurons from the spinal cord to the erector spinae muscles. **C**. Caudal end of the vertebral canal revealing spinal roots forming the cauda equina before exiting the vertebral canal through the intervertebral and sacral foramina.

TABLE 1-1. Superficial Muscles of the Back

Muscle	Proximal Attachment	Distal Attachment	Action	Innervation
Trapezius	Occipital bone, nuchal ligament, C7–T12 vertebrae	Clavicle, acromion, and spine of scapula	Elevates, retracts, depresses, and rotates scapula	Spinal root of accessory n. (CN XI) and cervical nn. (C3–C4)
Levator scapulae	Transverse processes of C1–C4 vertebrae	Superior angle of scapula	Elevates and rotates scapula; inclines neck to same side of contraction	Cervical nn. (C3–C4) and dorsal scapular n. (C5)
Rhomboid major	Spinous processes of T2–T5 vertebrae	Medial margin of scapula	Retract and rotate scapula	Dorsal scapular n. (C4–C5)
Rhomboid minor	Spinous processes of C7–T1 vertebrae	Medial margin of scapula	Retract and rotate scapula	Dorsal scapular n. (C4–C5)
Latissimus dorsi	T7 sacrum, thoracolumbar fascia, iliac crest, and inferior 3 ribs	Intertubercular groove of humerus	Extends, adducts, and medially rotates humerus	Thoracodorsal n. (C6–C8)

TABLE 1-2. Deep Muscles of the Back

Muscle	Proximal Attachment	Distal Attachment	Action	Innervation
Splenius capitis	Nuchal ligament, spinous processes of C7–T4 vertebrae	Mastoid process of temporal bone and occipital bone	Acting alone, laterally bends and rotates head Acting together, extend head and neck	Dorsal rami of middle and lower cervical spinal nn. at each vertebral level where they attach
Splenius cervicis	Spinous processes of T3–T6	Transverse processes of C2–C3	Acting alone, laterally bends and rotates head Acting together, extend head and neck	

Erector spinae group (a series of muscles that extends from the sacrum to the skull)

• Iliocostalis	Iliac crest, sacrum, ribs	Thoracolumbar fascia, ribs, cervical vertebrae	Bilaterally, extend vertebral column Unilaterally, lateral flexion of vertebral column	Segmentally innervated by dorsal primary rami of spinal nn. at each vertebral level where they attach
• Longissimus	Thoracodorsal fascia, transverse and cervical vertebrae	Vertebrae and mastoid process of temporal bone		
• Spinalis	Spinous processes of vertebrae	Spinous processes of vertebrae		

Transversospinalis group

• Semispinalis	Transverse processes of thoracic vertebrae	Spinous processes of thoracic and cervical vertebrae and occipital bone	Bilaterally, extends vertebral column and unilaterally rotates vertebral column contralaterally	Segmentally innervated by dorsal primary rami of spinal nn. at each vertebral level where they attach
• Multifidus	Sacrum and transverse processes of lumbar, thoracic, and cervical vertebrae	Spinous processes of lumbar, thoracic, and lower cervical vertebrae	Bilaterally, extends vertebral column and unilaterally rotates vertebral column contralaterally	
• Rotatores	Transverse processes of C2 vertebra to the sacrum	Lamina immediately above the vertebra of origin		

Note: Each of the deep muscles of the back receives its blood supply segmentally from branches of the posterior cutaneous vessels.

TABLE 1-3. Regions of the Spinal Cord and Vertebral Column

Region	Number of Spinal Nerves	Number of Vertebrae
Cervical	8	7
Thoracic	12	12
Lumbar	5	5
Sacral	5	5 (fused)
Coccygeal	1	3–4

STUDY QUESTIONS

Directions: Each of the numbered items or incomplete statements is followed by lettered options. Select the **one** lettered option that is **best** in each case.

1. A 48-year-old man goes to his physician because of pain and paresthesia along the lateral aspect of the leg and the dorsum of the foot. The patient's symptoms suggest impingement of the L5 spinal nerve resulting from a herniated intervertebral disc. The L5 spinal nerve most likely exits between which of the following vertebrae?

   A. L3–L4 vertebrae

   B. L4–L5 vertebrae

   C. L5–S1 vertebrae

   D. S1–S2 vertebrae

2. The muscles of the posterior aspect of the thigh, or hamstring musculature, are responsible for flexing the knee joint. Beginning with the motor neuron cell bodies in the gray matter of the spinal cord, identify the most likely pathway that axons would travel from the spinal cord to the hamstring muscles?

   A. Ventral horn, ventral root, ventral ramus

   B. Ventral horn, ventral root, dorsal ramus

   C. Lateral horn, ventral root, ventral ramus

   D. Lateral horn, dorsal root, dorsal ramus

   E. Dorsal horn, dorsal root, ventral ramus

   F. Dorsal horn, dorsal root, dorsal ramus

3. A 27-year-old man is brought to the emergency department after being involved in an automobile accident. Radiographic imaging studies indicate that he has sustained a fracture of the L1 vertebral arch and has a partially dislocated bone fragment impinging upon the underlying spinal cord. Which spinal cord level is most likely compressed by this bone fragment?

   A. C1

   B. L2

   C. S3

   D. T4

4. A 50-year-old man is diagnosed with flaccid paralysis limited to the right arm, without pain or paresthesias. No sensory deficits are noted. Laboratory studies reveal that the patient is infected with West Nile virus. The target that the virus has infected resulting in this patient's symptoms is most likely the

   A. Ventral horn of spinal cord gray matter

   B. Ventral rami of spinal nerves

   C. Dorsal horn of spinal cord gray matter

   D. Dorsal rami of spinal nerves

5. A 6-year-old boy is stung by a wasp between his shoulder blades. Identify the pain sensation pathway the axons would travel to course from the skin of his back to the spinal cord.

   A. Ventral horn, dorsal root, dorsal ramus

   B. Ventral horn, dorsal root, ventral ramus

   C. Ventral horn, ventral root, dorsal ramus

   D. Ventral horn, ventral root, ventral ramus

   E. Dorsal ramus, dorsal root, dorsal horn

   F. Dorsal ramus, dorsal root, ventral horn

   G. Dorsal ramus, ventral root, ventral horn

   H. Dorsal ramus, ventral root, dorsal horn

6. Which of the following paired muscles of the back is primarily responsible for extension of the vertebral column?

   A. Iliocostalis

   B. Latissimus dorsi

   C. Levator costarum

   D. Rhomboid major and minor

   E. Trapezius

7. A 44-year-old woman is suspected of having meningitis. To confirm the diagnosis, a lumbar puncture is ordered to collect a sample of the cerebrospinal fluid (CSF). Identify the last layer of tissue the needle will traverse in this procedure before reaching CSF.

   A. Arachnoid mater

   B. Dura mater

   C. Ligamentum flavum

   D. Pia mater

   E. Skin

ANSWERS

1—C: Spinal nerves in the thoracic and lumbar vertebral region exit the vertebral canal below their associated vertebra. Therefore, the L5 spinal nerve exits below L5, between L5 and S1.

2—A: The ventral horn of the spinal cord gray matter houses motor neuron cell bodies and conveys motor neurons out through the ventral root into the spinal nerve. All muscles of the limbs and body wall are innervated by ventral rami. Although the muscles are present along the posterior aspect of the thigh, muscles are still innervated by the ventral rami. Dorsal rami innervate the skin of the back and the deep back muscles, such as the erector spinae.

3—C: In an adult, the caudal end of the spinal cord is at the L1–L2 vertebral level. Therefore, a bone fragment from the L1 vertebra would have the potential of touching the caudal end of the spinal cord, not the L1 spinal cord level. C1, L2, and T4 are spinal cord levels superior to the fracture.

4—A: The patient has no sensory deficits and presents with only motor deficits. Therefore, the virus affects the ventral horn of the gray matter because that is the location of the motor neuron cell bodies.

5—E: All skin of the back is segmentally innervated by the dorsal rami branches of spinal nerves. Sensory information is then conducted through the dorsal root into the dorsal horn of the gray mater of the spinal cord.

6—A: The paired iliocostalis muscles, part of the erector spinae musculature, are postural muscles that help to extend the vertebral column and thus keep the spine erect. The latissimus dorsi, rhomboids, and trapezius muscles act primarily on the upper limb. The levator costarum muscles help to elevate the ribs during inspiration but will not extend the vertebral column.

7—A: A lumbar puncture collects cerebrospinal fluid and, therefore, the needle has to enter the subarachnoid space, which is located between the arachnoid and pia mater. Therefore, the last layer of tissue the needle would traverse to enter the subarachnoid space is the arachnoid mater.

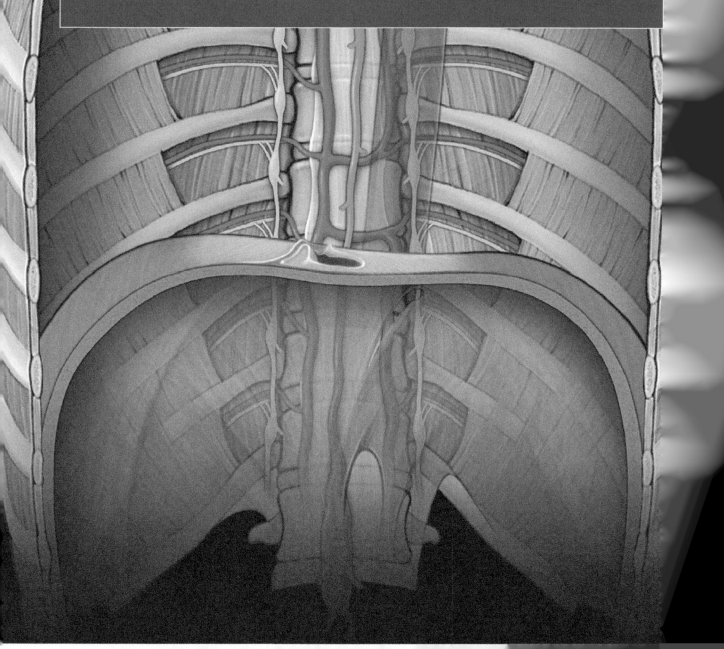

SECTION 2

THORAX

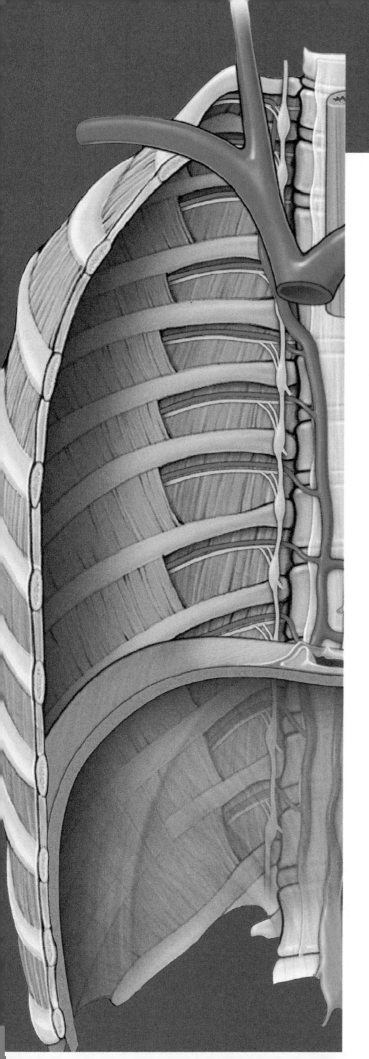

ANTERIOR THORACIC WALL

SURFACE ANATOMY

BIG PICTURE

The clavicle and parts of the thoracic cage provide prominent surface landmarks. Directional terms are used to help orient the reader to the thorax. The cutaneous innervation of the anterior portion of the thoracic wall is from lateral and anterior cutaneous nerves via ventral rami, forming a segmental dermatomal pattern.

BONY LANDMARKS

Palpable bony landmarks and reference lines are important to use for anatomic orientation as well as a guide to locate deep structures. You should become familiar with the following structures of the anterior portion of the thoracic wall (Figure 2-1A; Table 2-1):

- **Clavicles.** Course transversely along the superior portion of the chest wall from the manubrium to the acromion of the scapula and, therefore, are easily palpable. However, the clavicle, overly rib 1, which makes rib 1 impalpable.
- **Jugular notch of the sternum.** Located between the medial ends of the clavicles.
- **Sternal angle (of Louis).** Articulation between the manubrium and sternal body. The sternal angle serves as the location for the articulation of rib 2 with the sternum at the T4–T5 vertebral level. The sternal angle is often visible and palpable and serves as an important surface landmark for several underlying structures.
- **Xiphoid process.** An inferior pointed projection of the sternal body. The xiphoid process lies at the level of the T9 vertebral body.
- **Costal margins.** Formed by costal cartilages 7–10.

DIRECTIONAL TERMS

The following imaginary vertical lines provide anatomic and clinical descriptions for orientation (Figure 2-1A):

- **Midsternal line.** Dropped through the center of the sternum. It is a component of the median plane.
- **Midclavicular line.** Dropped through the middle of the clavicle, just medial to the nipple.
- **Anterior axillary line.** Dropped through the anterior axillary fold formed by the pectoralis major muscle.
- **Midaxillary line.** Dropped through the middle of the axilla, along the lateral border of the thoracic wall. The midaxillary line is bounded by the anterior axillary fold (pectoralis major muscle) and the posterior axillary fold (latissimus dorsi muscle).
- **Posterior axillary line.** Dropped through the posterior axillary fold formed by the latissimus dorsi muscle.

CUTANEOUS INNERVATION

Lateral and anterior cutaneous branches of the intercostal nerves innervate the skin of the thorax in a segmental pattern. The following notable landmarks are helpful to identify the following dermatomes (Figure 2-1B; Table 2-2):

- **Nipple**–T4 dermatome
- **Xiphoid process**–T6 dermatome
- **Umbillicus**–T10 dermatome

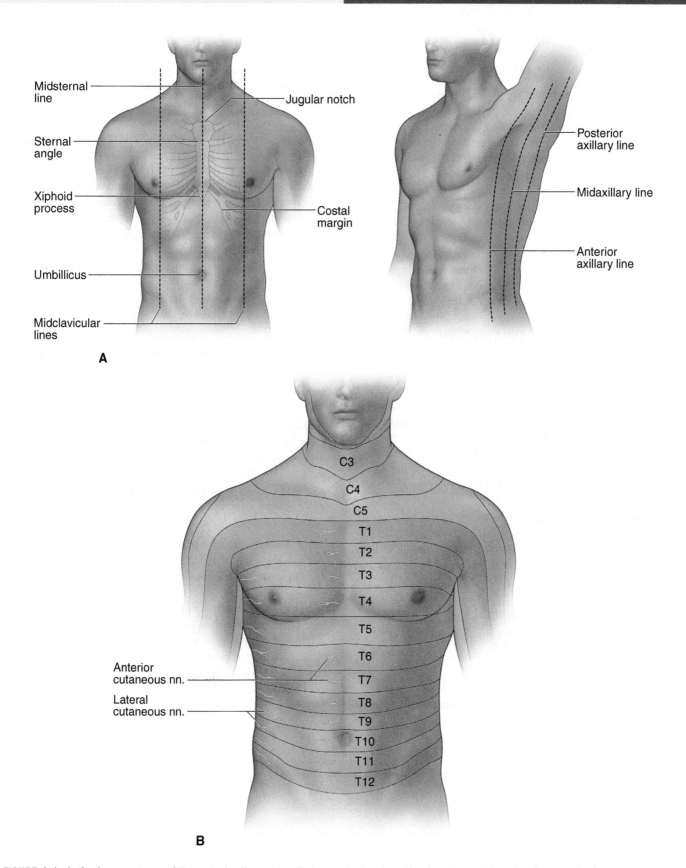

FIGURE 2-1: A. Surface anatomy of the anterior thoracic wall demonstrating bony landmarks and directional terms. **B**. Cutaneous nerves revealing dermatomal pattern of the thorax.

THE BREAST

BIG PICTURE

The breast contains the **mammary gland**, a subcutaneous gland that is specialized in women for the production and secretion of milk and, as a result, enlarges during the menstrual cycle and during pregnancy. The mammary gland overlies ribs 2 to 6 between the sternum and midaxillary line. The mammary gland is located within the superficial fascia, surrounded by a variable amount of adipose tissue that is responsible for the shape and size of the breast. The breast lies on the deep fascia related to the **pectoralis major** and **serratus anterior muscles**. The **retromammary space** is a layer of loose connective tissue that separates the breast from the deep fascia and provides some degree of breast mobility over underlying structures. A prolongation of each breast, called the **axillary tail**, extends superolaterally along the inferior border of the pectoralis major into the axilla.

MAMMARY GLAND STRUCTURE

Each gland consists of 15 to 20 radially aligned lobes of glandular tissue. A **lactiferous duct** drains each lobe (Figure 2-2A and B). The lactiferous ducts converge and open onto the **nipple**. The nipple is positioned on the anterior surface of the breast and is surrounded by a somewhat circular hyperpigmented region, the **areola**. **Sebaceous glands** within the areola enlarge to form swollen tubercles during pregnancy as the areola darkens. Small collections of smooth muscle located at the base of the nipple may cause erection of the nipple when breast feeding or sexually aroused.

Fibrous and adipose tissues occupy the spaces between the lobes of glandular tissue. To help support the weight of the breast, strong fibrous bards called **suspensory (Cooper's) ligaments** separate the lobes and attach between the dermis and deep layer of the superficial fascia.

ARTERIES, VEINS, AND LYMPHATICS OF THE BREAST

To simplify an understanding of vascular supply and lymphatic drainage, the breast's anatomy can be best understood by division into medial and lateral regions (Figure 2-2C and D).

■ **Medial region of the breast.** Derives its blood supply from perforating branches of the **internal thoracic artery** (medial mammary branches from the second through fourth intercostal spaces) and is drained by **internal thoracic veins**. Lymphatic vessels exiting the medial side of the breast drain into the **parasternal lymph nodes**, which lie beside the internal thoracic veins.

■ **Lateral region of the breast.** Derives its blood supply from branches of the **lateral thoracic artery** (axillary artery origin) and mammary branches from the 2nd through the 5th **posterior intercostal arteries** (thoracic aorta origin) and is drained by the **lateral thoracic** and **intercostal veins**, respectively. The lymphatic vessels draining the lateral side of the breast lead primarily into the **pectoral group of axillary lymph nodes** and account for most of the lymph drained from the breast. Figure 2-2D shows lymph from the right breast draining into the **right lymphatic duct** at the junction of the right brachiocephalic vein. In contrast, the left breast (not shown) drains into the **thoracic lymphatic duct** at the left brachiocephalic vein junction.

Some breast lymphatics drain into the infraclavicular group of axillary nodes, as well as the supraclavicular group of deep cervical nodes.

▽ **Breast augmentation** is a procedure that is performed to enhance the shape and size of the breast. The procedure is frequently performed for cosmetic reasons or for breast reconstruction. The implant (filled with saline or silicone) may be inserted from various incision sites. Regardless of the incision site, the implant is placed in one of two areas, either subglandular (between the pectoralis major muscle and the mammary gland) or submuscular (deep to the pectoralis major muscle). ▽

▽ **Most breast adenocarcinomas** are lactiferous duct carcinomas that begin as painless masses, most commonly in the upper lateral quadrant. ▽

INNERVATION OF THE BREAST

The sensory innervation of the skin overlying the breast has a segmental arrangement and is supplied by branches of the 2nd through 7th intercostal nerves. However, physiologic changes in the breast are not mediated by nerves but by circulating hormones. For example, high prolactin levels result in milk production, and oxytocin causes the milk "letdown" reflex.

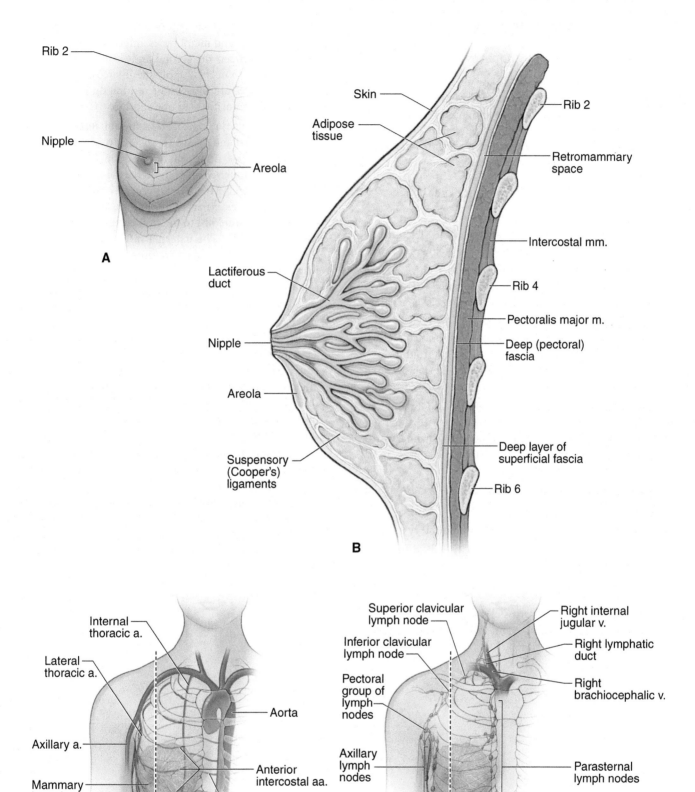

FIGURE 2-2: A. Anterior view of the breast. **B**. Sagittal section of the breast. **C**. Arterial supply of the breast. **D**. Lymphatic drainage of the breast.

THORACIC MUSCLES

BIG PICTURE

For simplicity, we will divide muscles of the anterior thoracic wall into two groups. The first group, superficial thoracic muscles, originate on the thoracic skeleton and inserts on the upper limb therefore producing motion of the upper limb. The second group, the deep thoracic muscles located within the intercostal space, act primarily with the thoracic cage (Table 2-3).

SUPERFICIAL THORACIC MUSCLES

The superficial muscles attached to the anterior portion of the thorax primarily act on the scapula and humerus. These muscles are shown in Figure 2-3A and include the following:

- **Pectoralis major muscle.** Attaches to the sternum, clavicle, and costal margins and laterally to the intertubercular groove of the humerus. The pectoralis major muscle is a prime flexor, adductor, and medial rotator of the humerus.

- **Pectoralis minor muscle.** Attaches anteriorly on the thoracic skeleton to ribs 3 to 5 and superiorly to the coracoid process of the scapula. The pectoralis minor muscle stabilizes the scapula against the thoracic wall.

- **Serratus anterior muscle.** Attaches to ribs 1 to 8 along the midaxillary line and courses posteriorly to the medial margin of the scapula. The serratus anterior muscle is a primary protractor of the scapula as well as a stabilizer of the scapula against the thoracic wall.

- **Subclavius muscle.** Attaches to the clavicle and rib 1 and moves the clavicle inferiorly.

These muscles also function as accessory respiratory muscles by helping to expand the thoracic cavity when inspiration is deep and forceful.

INTERCOSTAL MUSCLES

The area between ribs is the intercostal space (Figure 2-3B). The eleven intercostal spaces each contain three overlapping layers of tissue consisting of the external, internal, and innermost intercostal muscles and membranes. Intercostal nerves and vessels, coursing between the second and third layers of muscles (from superficial to deep), supply the intercostal muscles. The intercostal muscles are muscles of respiration and are named according to their positions to one another.

- **Superficial muscle layer.** The **external intercostal muscles** are the first and most superficial layer in the intercostal space and course in an oblique fashion from superior to inferior. The muscle fills the intercostal spaces posteriorly from the tubercles of the ribs to the costochondral joint anteriorly. The external intercostal muscle continues from the costochondral joint to the sternum as the external intercostal membrane. The external intercostal muscles move the ribs superiorly and, therefore, are most active during inspiration.

- **Middle muscle layer.** The **internal intercostal muscles** are the second and, therefore, the middle layer in the intercostal space and course at a right angle to the external intercostal muscles. The muscle fills each intercostal space anteriorly from the sternum to the angle of the ribs posteriorly. The internal intercostal muscle continues from the angle of the ribs to the vertebral column as the internal intercostal membrane. The internal intercostal muscles move the ribs inferiorly and, therefore, are most active during expiration.

- **Deep muscle layer.** Small, thin muscles form the third and the deepest layer of the intercostal spaces. The occurrence and location of these muscles is variable. In general, the muscles occur in three groups: an anterior group called the **transverses thoracis muscles**; a lateral group called the **innermost intercostal muscles**; and a posterior group called the **subcostal muscles**. This group of muscles most likely aids in depressing the ribs, facilitating expiration.

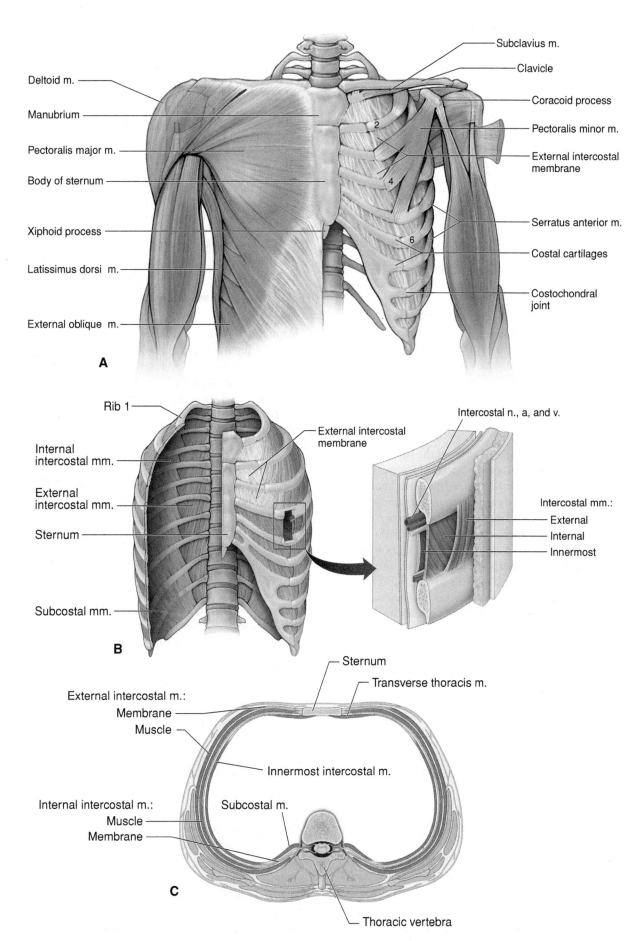

FIGURE 2-3: A. Muscles of the anterior thoracic wall. **B**. Intercostal muscles. Note the step dissection showing the intercostal muscles. **C**. Axial illustration of the intercostal muscle group.

THORACIC SKELETON

BIG PICTURE

The thoracic skeleton consists of the thoracic vertebrae posteriorly, the ribs laterally, and the sternum and costal cartilages anteriorly. The costal cartilages secure the ribs to the sternum. The thoracic cage forms a protective cage around vital organs such as the heart, lungs, and great blood vessels. The thoracic skeleton provides attachment points for the muscles of the back and chest that allow support of the shoulder girdle (scapula and clavicle) and movement of the upper limbs.

THORACIC VERTEBRAE

The 12 thoracic vertebrae articulate with the 12 pairs of ribs. Thoracic vertebrae typically bear two **costal facets** on each side, one at the superior edge and the other at the inferior edge of the vertebral body, where they receive the heads of the ribs (Figure 2-4A). The bodies of T10 to T12 vary from this pattern, however, by having only a single facet for their respective ribs. In addition, the T1 to T10 transverse processes have costal facets that articulate with the **tubercles of the ribs**.

RIBS

Twelve pairs of ribs form the flared sides of the thoracic cage and generally extend anteriorly from the thoracic vertebrae to the sternum (Figure 2-4B). Rib pairs 1 to 7 are known as the **true ribs** and attach directly to the sternum by individual **costal cartilages**. The remaining five pairs of ribs are called **false ribs** because they either attach indirectly to the sternum or lack a sternal attachment entirely. Rib pairs 8 to 10 attach to the sternum indirectly by joining each other via costal cartilages immediately above. Rib pairs 11 and 12 are called **floating ribs** because they have no anterior attachments; instead, they are embedded in the muscles of the lateral body wall.

▽ **Oblique course of ribs.** The ribs course in an oblique, inferior direction from their thoracic vertebral articulation to their anterior sternal articulation. For example, rib 2 articulates with the T2 vertebra posteriorly but with the sternal angle at the T4 vertebral level anteriorly. Therefore, an axial section of the thorax, such as the one seen in a CT scan, intersects several ribs. ▽

Ribs 2 to 10 are considered **typical ribs** and have the following bony landmarks (Figure 2-4C):

Head and neck. Form costovertebral joints by articulating with the **costal demifacets** of adjacent thoracic vertebral bodies and intervertebral discs.

Tubercle. Articulates with the **costal facets** of adjacent thoracic vertebral transverse processes (with the exception of ribs 11 and 12).

Shaft. The long portion of the rib consisting of a smooth superior border and a sharp, thin inferior border possessing a costal groove housing the intercostal veins, arteries, and

nerves. The distal end of the rib shaft articulates with the costal cartilage, forming **costochondral joints**.

▽ **Fracture of a rib** commonly occurs just anterior to the angle, the weakest point of the rib, and may puncture the parietal pleura, resulting in a **pneumothorax**. ▽

Ribs 1, 11, and 12 are **atypical ribs**. Rib 1 is not palpable because it lies deep to the clavicle. It has the scalene tubercle on the upper surface for the inferior attachment of the anterior scalene muscle. The groove for the subclavian vein is anterior to the scalene tubercle; the groove for the subclavian artery is posterior to the tubercle. Ribs 11 and 12 do not articulate with the sternum and receive the name "floating ribs."

▽ The term "costal" means rib or rib-like part. ▽

STERNUM

The sternum, or **breastbone**, is flat and consists of the following structures (Figure 2-4D):

Manubrium. Along its superior border, the manubrium contains the **jugular notch** at the T2 vertebral level. The notch is flanked on each side by a synovial **sternoclavicular joint**, which provides the only bony attachment between the upper limb and axial skeleton. Although the other joints between the **costal cartilages** and the sternum are synovial, the first **sternochondral joint** with rib 1 is cartilaginous. The upper half of rib 2 attaches to the body of the sternum at the manubriosternal joint, better known as the sternal angle.

Sternal angle. The sternal angle (of Louis) is the junction between the manubrium and the sternal body. The junction can be palpated as a prominent transverse ridge largely because the posterior facing angle between the manubrium and the body is less than 180 degrees. The sternal angle is an important clinical landmark for two reasons. First, rib 2 articulates with the sternum at the level of the sternal angle. Accordingly, palpation of the sternal angle permits bilateral identification of rib 2 and all of the lower ribs. Second, the **sternal angle marks** the following important clinical levels:

- The level of the T4 and T5 vertebrae in an axial plane.
- The start and end of the aortic arch.
- The tracheal bifurcation into the right and left primary bronchi.
- The azygos vein courses over the right primary bronchus to join the superior vena cava.
- The border between the superior and middle mediastinum is demarcated.
- The thoracic lymphatic duct transitions from the right to the left side of the thoracic cavity.

Sternal body. The sternal body articulates with the second through seventh costal cartilages, the manubrium, and the xiphoid process.

Xiphoid process. The shape of the xiphoid process varies. It is located at the T9 vertebral level and articulates with the sternal body. The linea alba, a thickened vertical cord of connective tissue in the midline of the abdominal wall, attaches to the xiphoid process.

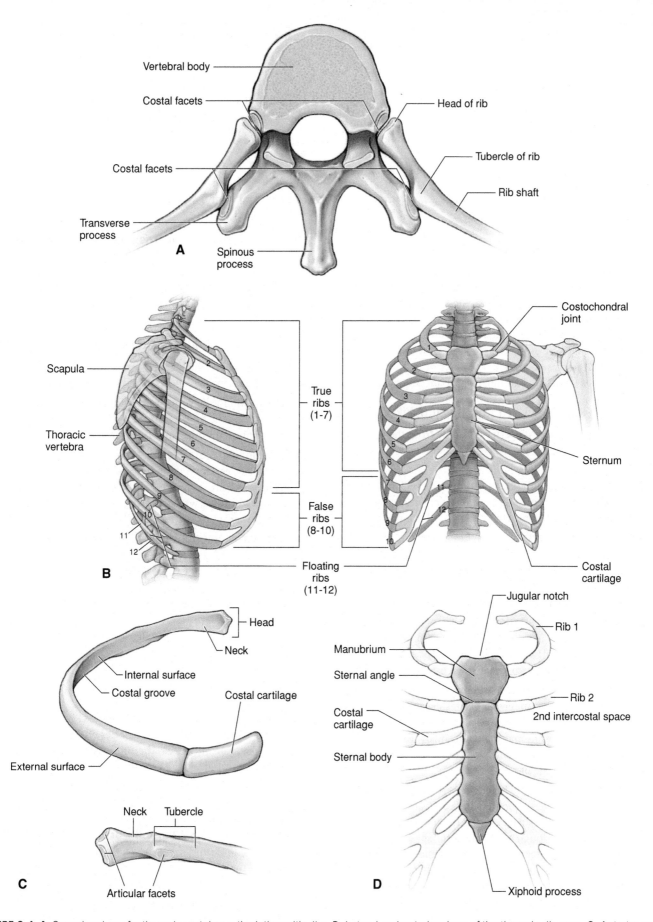

FIGURE 2-4: A. Superior view of a thoracic vertebra articulating with ribs. **B**. Lateral and anterior views of the thoracic rib cage. **C**. Anterior and posterior views of a typical rib. **D**. Anterior view of the sternum.

VESSELS AND LYMPHATICS OF THE THORACIC WALL

BIG PICTURE

The nerve and blood supply of the thoracic wall consists largely of the neurovascular elements that course through the intercostal spaces. The major elements in each space consist of an intercostal vein, artery, and nerve. The primary neurovascular bundle courses along the costal groove of the upper rib, between the internal and innermost intercostal muscles. The anterior and posterior intercostal arteries supply the intercostal spaces and form anastomoses with each other.

INTERCOSTAL NERVES

The intercostal nerves are derived from the ventral rami of the first 11 thoracic spinal nerves and are named accordingly (Figure 2-5A and B). For example, the right fifth intercostal nerve is derived from the ventral ramus of the fifth thoracic spinal nerve and courses through the right fifth intercostal space under the partial cover of the costal groove of the right fifth rib. The intercostal nerves give rise to the anterior and lateral cutaneous branches (sensory) and muscular branches (motor).

ARTERIAL SUPPLY TO THE THORACIC WALL

The thoracic wall receives its arterial supply from the following branches of the subclavian artery and thoracic aorta (Figure 2-5B and D):

- **Superior (supreme) intercostal arteries.** Branch off the costocervical trunk, a branch of the subclavian artery, giving rise to the first and second posterior intercostal arteries.
- **Internal thoracic artery.** Branches from the subclavian artery, where it travels along the internal surface of the ribcage just lateral to the sternum, giving rise to the anterior intercostal arteries. The internal thoracic artery branches into the musculophrenic and superior epigastric arteries.
- **Anterior intercostal arteries.** Arise from the internal thoracic (upper thoracic wall) and the musculophrenic arteries (lower thoracic wall) and travel posteriorly between the ribs, anastomosing with the posterior intercostal arteries.
- **Anterior perforating arteries.** Perforate the internal intercostal muscles in the upper intercostal spaces with the anterior cutaneous nerves. These arteries provide partial vascular supply to the pectoralis major muscle and overlying skin. The second through fourth branches provide the **medial mammary branches** to the breast.
- **Musculophrenic artery.** The lateral terminal branch of the internal thoracic artery. It follows the costal arch on the inner surface of the costal cartilages, giving rise to two anterior arteries in the seventh through ninth intercostal spaces. It perforates the diaphragm and terminates at about the 10th intercostal space, where it anastomoses with the **deep circumflex iliac artery.** The musculophrenic artery supplies the pericardium, the diaphragm, and the muscles of the abdominal wall.

- **Superior epigastric artery.** The medial terminal branch of the internal thoracic artery. It descends on the deep surface of the rectus abdominis muscle within the rectus sheath and anastomoses with the inferior epigastric artery.
- **Posterior intercostal arteries.** The first two posterior intercostal arteries arise from the superior (supreme) intercostal arteries, and the remaining arteries arises from the thoracic aorta. The posterior intercostal arteries course anteriorly between the internal and innermost intercostal muscles and anastomose with the anterior intercostal arteries.
- **Subcostal artery.** Arises from the aorta and travels anteriorly below the 12th rib.

▽ When the **aortic arch is constricted (coarctated)** just beyond the origin of the left subclavian artery, the anastomoses between the anterior and posterior intercostal arteries enable blood in the internal thoracic arteries to reach the descending aorta, bypassing the coarctation. ▼

VENOUS DRAINAGE OF THE THORACIC WALL

Venous drainage of the thoracic wall is from the following tributaries of the subclavian and azygos system of veins (Figure 2-5C):

- **Internal thoracic vein.** Accompanies the internal thoracic artery along the internal surface of the rib cage just lateral to the sternum, draining into the brachiocephalic vein.
- **Anterior intercostal veins.** Accompany the anterior intercostal arteries between the internal and innermost intercostal muscles and are tributaries to the musculophrenic and internal thoracic veins, which in turn are tributaries of the brachiocephalic veins.
- **Posterior intercostal veins.** Accompany the posterior intercostal arteries between the internal and innermost intercostal muscles. The upper two posterior intercostal veins drain into the brachiocephalic veins. The nine lowest spaces are tributaries of the azygos system, which ultimately drains into the superior vena cava.

LYMPHATIC DRAINAGE OF THE THORAX

Lymphatic drainage of the thoracic cavity is primarily through the following lymph nodes:

- **Parasternal (internal thoracic) nodes.** Located along the internal thoracic artery. These nodes receive lymph from the medial side of the breast, the intercostal spaces, diaphragm, and supraumbilical region of the abdominal wall. They drain into the junction of the internal jugular and subclavian veins.
- **Intercostal nodes.** Lie near the heads of the ribs and receive lymph from the intercostal spaces and parietal pleura. They drain into the **cisternal chili** or the **thoracic lymphatic duct.**
- **Phrenic nodes.** Lie on the surface of the diaphragm and receive lymph from the pericardium, diaphragm, and liver. They drain into the mediastinal nodes.

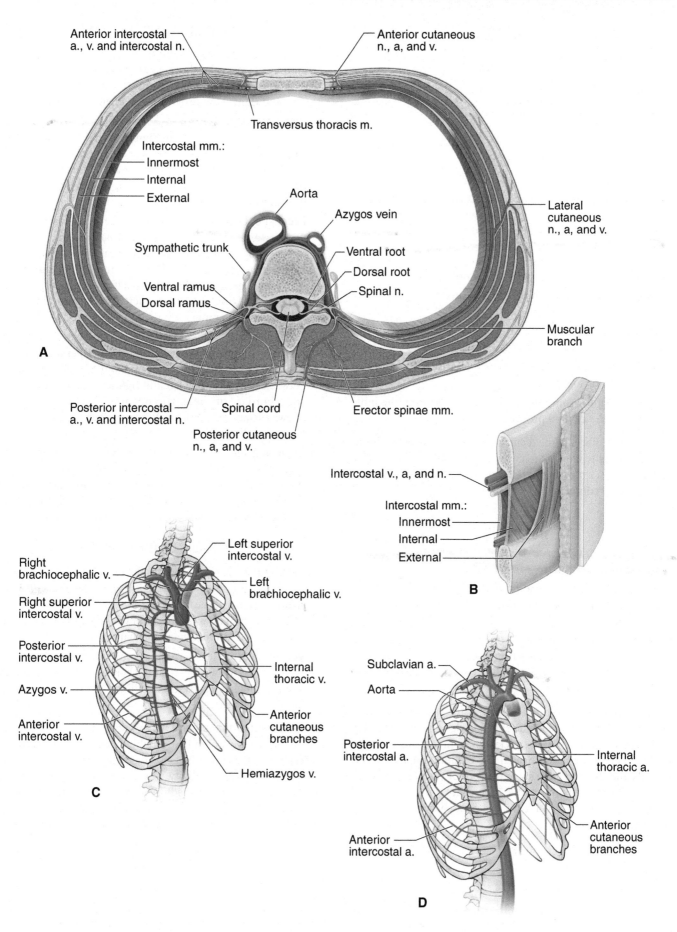

FIGURE 2-5: A. Axial section through the thorax showing the vascular supply. **B**. Intercostal structures. Schematic illustrating the thoracic veins (**C**) and the thoracic arteries (**D**).

DIAPHRAGM

BIG PICTURE

The diaphragm muscle divides the thoracic and abdominal cavities and serves as the principal muscle for inspiration and expiration.

ATTACHMENTS

The diaphragm is a musculotendinous structure that separates the thoracic cavity from the abdominal cavity. The diaphragm is composed of a peripheral muscular portion and a central tendon (Figure 2-6A). The diaphragm is dome shaped, and upon contraction of its muscular portion, it descends (flattens). The right dome, resting atop the liver, is generally higher than the left dome, which rests atop the fundus of the stomach.

The muscular portion has three regions of origin (Figure 2-6B):

- **Lumbar origin.** Two crura originate from the bodies of the upper two (left crus) or three (right crus) lumbar vertebrae.
- **Costal origin.** From muscle fibers that arise from the inner surfaces of the lower six ribs.
- **Sternal origin.** From muscle fibers that arise from the inner surface of the xiphoid process.

The muscle fibers of the diaphragm extend centrally to insert into its central tendon. The central tendon is the structure that the diaphragm's muscle fibers primarily pull upon when they concentrically contract.

INNERVATION

The right and left phrenic nerves provide sensory and motor innervation to the diaphragm as follows:

- **Sensory.** Sensory innervation of the pericardium, mediastinal and diaphragmatic pleurae, and the diaphragmatic peritoneum.
- **Motor.** Motor innervation of the diaphragm is through the phrenic nerve, which arises from the cervical plexus via ventral rami branches of C3, C4, and C5 ("*C3, C4, and C5 keep the diaphragm alive*").

The phrenic nerves course along the anterior surface of the anterior scalene muscle in the side and descend into the thoracic cavity between the subclavian vein and artery. The phrenic nerves are accompanied by the **pericardiacophrenic vessels** and descend anterior to the root of each lung between the mediastinal pleura and pericardium en route to the diaphragm. The vagus nerve, by contrast, courses posterior to the root of the lung.

APERTURES IN THE DIAPHRAGM

The diaphragm has several apertures that permit the passage of structures between the thorax and the abdomen (Figure 2-6A and B). The caval, esophageal, and aortic openings are large, and the left and right crura provide other small openings.

- **Caval opening.** Located at the T8 vertebral level within the central tendon of the diaphragm, just right of the midline. The caval opening allows passage for the inferior vena cava and branches of the right phrenic nerve. Branches of the left phrenic nerve pass through the diaphragm by piercing the central tendon on the left side.
- **Esophageal opening.** Located to the left of the midline at the T10 vertebral level. The opening usually splits the muscle fibers of the right crus. The esophageal opening allows passage for the esophagus and the left and right vagus nerves. The esophageal branches of the left gastric artery and the esophageal tributaries of the left gastric vein also passes through this opening.
- **Aortic opening.** Located at the T12 vertebral level, behind the two crura. Strictly speaking, the aortic opening is not an opening through the diaphragm but rather a large gap between the crura. The left and right crura form the left and right borders of the aortic opening. The aortic opening allows passage for the aorta, the azygos vein, and the thoracic lymphatic duct.
- **Right and left crura.** The right and left greater and lesser splanchnic nerves course from the thoracic cavity, deep to the right and left crura en route to the prevertebral ganglia of the abdomen. In addition, the left crus also allows passage for the hemiazygos vein.

FUNCTIONS OF THE DIAPHRAGM

- **Respiration.** The diaphragm is the principal muscle of inspiration. It flattens upon contraction, thus increasing the vertical dimensions of the thoracic cavity (Figure 2-6C). The roles of the diaphragm and other thoracic muscles of respiration are discussed in more detail in Chapter 3.
- **Venous return.** The alternating contraction and relaxation of the diaphragm causes pressure changes in the thoracic and abdominopelvic cavity that facilitate the return of venous blood to the heart.

▽ **Valsalva maneuver.** When taking and holding a deep breath, an individual forcibly contracts the diaphragm inferiorly on the abdominal viscera, thereby increasing the pressure in the abdominal cavity. This is done to help expel vomit, feces, and urine from the body by increasing the intra-abdominal pressure and by preventing gastric reflux by exerting pressure on the esophagus as it passes through the esophageal hiatus. ▽

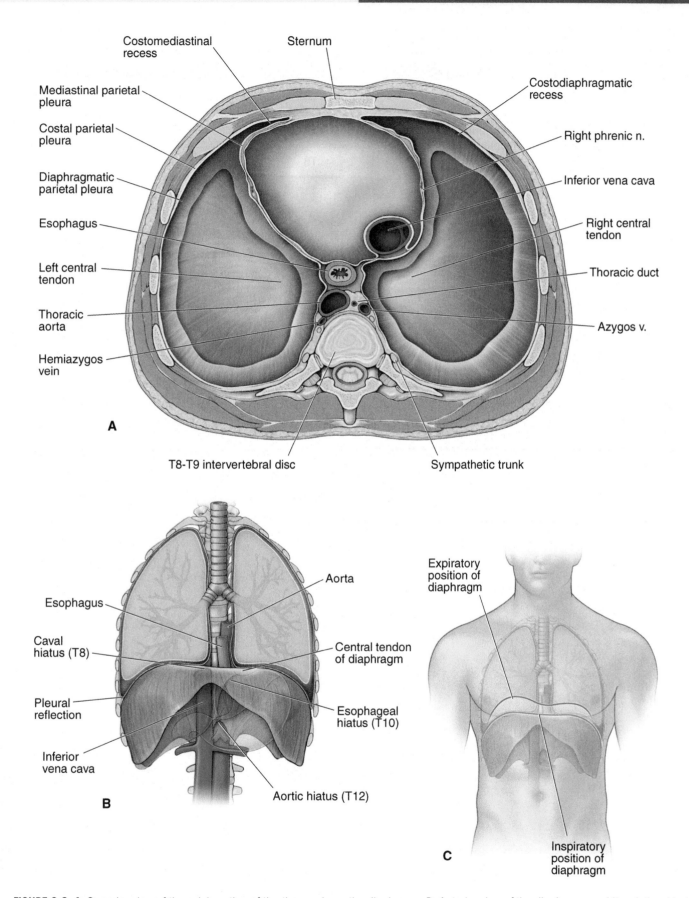

FIGURE 2-6: A. Superior view of the axial section of the thorax above the diaphragm. **B**. Anterior view of the diaphragm and its relationship to the lungs. **C**. Position of the diaphragm during inspiration and expiration.

TABLE 2-1. Important Vertebral Landmarks of the Thorax

Vertebral Level	Landmark
T2	Jugular notch
T3	Base of spine of scapula; junction of brachiocephalic veins to form superior vena cava
T4	Sternal angle (second costal cartilage, tracheal bifurcation, beginning and end of aortic arch, beginning of thoracic aorta, azygos vein arch over right primary bronchus, thoracic lymphatic duct crosses from right to left side of thoracic cavity)
T7	Inferior angle of scapula
T8	Caval hiatus of diaphragm
T9	Xiphoid process
T10	Esophageal hiatus of diaphragm
T12	Aortic hiatus of diaphragm

TABLE 2-2. Important Dermatomes of the Thorax

Dermatome	Landmark
C4	T-shirt neckline
T4	Nipple line
T6	Xiphoid process

TABLE 2-3. Muscles of the Thoracic Region

Muscle	Proximal Attachment	Distal Attachment	Action	Innervation
Pectoralis major	Clavicle, sternum, and ribs	Lateral intertubercular groove of humerus	Flexion, adduction, medial rotation of humerus	Medial (C8–T1) and lateral (C5–C7) pectoral nn.
Pectoralis minor	Ribs 3–5	Coracoid process of scapula	Protraction and stabilization of scapula	Medial pectoral n. (C8–T1)
Serratus anterior	Lateral border of ribs 1–8	Medial margin of scapula	Protraction and stabilization of scapula	Long thoracic n. (C5–C7)
Subclavius	Rib 1	Clavicle	Stabilize clavicle	Nerve to subclavius (C5–C6)
Intercostals				
• **External** • **Internal** • **Innermost**	Inferior border of ribs	Superior border of inferior rib	Elevate ribs Depress ribs	Segmental innervation by intercostal nn. (T1–T11) and subcostal n. (T12)
Transversus thoracis	Posterior inferior portion of sternum and xiphoid process	Deep surface of costal cartilages 2–6		Segmental innervation by intercostal nn. (T1–T11) and subcostal n. (T12)
Subcostalis	Deep surface of lower ribs	Superior borders of ribs 2 and 3	Elevates ribs	

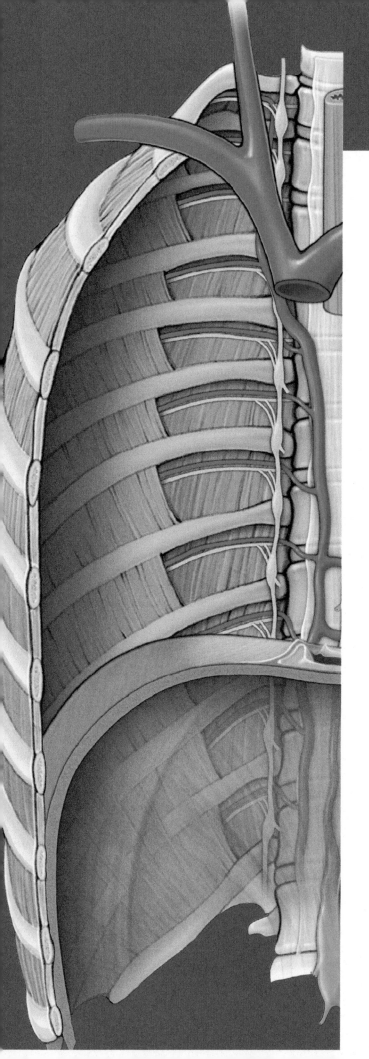

CHAPTER 3

LUNGS

PLEURA

BIG PICTURE

The lungs exchange oxygen and carbon dioxide and are the functional organs of the respiratory system. To serve that vital function, the lungs are located adjacent to the heart within the pleural sacs. The pleurae are serous membranes that line the internal surface of the thoracic cage and the outside surface of the lungs. The plurae secrete fluid that decreases resistance against lung movement during breathing.

DESCRIPTION OF PLEURAL SACS

Each lung (right and left) is contained within a serous membrane called the **pleural sac** (Figure 3-1A). The right and left pleural sacs occupy most of the thoracic cavity and flank both sides of the heart. Each pleural sac is composed of **two layers of serous (secretory) membrane**, the parietal pleura and the visceral pleura (Figure 3-1B).

- **Parietal pleura.** The external serous membrane lining the internal surface (wall) of the thoracic cavity.
- **Visceral pleura.** The internal serous membrane intimately attached to the surface of each lung.
- **Pleural fluid.** A layer of fluid located between the parietal and visceral pleurae in what is called the pleural space.

REGIONS OF THE PARIETAL PLEURA

As previously described, the parietal pleura lines the internal surface (wall) of the thoracic cavity, including the lateral sides of the mediastinum (Figure 3-1C and D). The parietal pleura is separated from the thoracic wall by the **endothoracic fascia**, a thin layer of connective tissue located between the parietal pleura and the innermost intercostal muscles and membrane. The parietal pleura is assigned specific names, depending on the structures that it lines.

- **Mediastinal parietal pleura.** Lines the lateral surface of the mediastinum (the space between the lungs where the heart is located).
- **Costal parietal pleura.** Lines the internal surface of the ribs.
- **Diaphragmatic parietal pleura.** Lines the superior surface of the diaphragm.
- **Cervical parietal pleura (cupula).** Extends above rib 1 to the root of the neck.

INNERVATION AND VASCULAR SUPPLY OF THE PARIETAL PLEURA

Intercostal nerves supply the costal parietal pleura and the peripheral portion of the diaphragmatic parietal pleura. Phrenic nerves supply the central portion of the diaphragmatic parietal pleura and the mediastinal parietal pleura. The parietal pleura is innervated by general sensory neurons, and therefore, it is **sensitive to pain.**

The parietal pleura receives its vascular supply via branches of the internal thoracic, superior phrenic, posterior intercostal, and superior intercostal arteries.

VISCERAL PLEURA

The visceral pleura is intimately attached to each lung and follows the contour of the lobes of the lung (Figure 3-1C and D). The visceral pleura is contiguous with the parietal pleura at the site where bronchi, vessels, nerves, and lymphatics pass from the mediastinum into the lung (the root, or hilum, of the lungs). In contrast to the parietal pleura, the visceral pleura is **insensitive to pain** because its visceral sensory neurons originate from the autonomic vagus nerve (CN X). The visceral pleura receives its blood supply via the bronchial arteries, whereas its venous drainage is via the pulmonary veins.

PLEURAL SPACE

Each pleural cavity has a pleural space. The pleural space is located between the parietal and visceral pleurae and is a closed space, which means that there is no communication between the right and left pleural spaces. The pleural space contains a thin film of pleural fluid that lubricates the surface of the pleurae and facilitates the movement of the lungs across the thoracic wall and diaphragm during inspiration and expiration.

During expiration, air flows out of the lungs, causing the internal pressure of the lungs to decrease and the lung tissue to potentially collapse. However, the thin layer of pleural fluid, working in conjunction with surfactant in the alveoli, keeps the lungs from collapsing by keeping the visceral pleura coupled to the parietal pleura and the alveoli open, respectively. As a result, the lungs remain inflated, even at the end of a deep expiration.

▽ If air is introduced into the pleural space as a result of chest trauma (e.g., a knife wound), the coupling between the parietal pleura and visceral pleura may be broken, causing the lung to collapse. This pathology is called a **pneumothorax**. When blood fills the pleural space, the pathology is called a **hemothorax.** ▽

SIGNIFICANCE OF THE PLEURAL REFLECTIONS AND RECESSES

In quite respiration, the lung, with its covering of visceral pleura, does not fill the entire pleural sac. The locations where the lung does not completely fill the pleural sac are called pleural recesses. There are **two clinically important recesses** for the right and left lungs that are termed the same as the reflection next to it (Figure 3-1C and D):

- **Costodiaphragmatic recess.** The recess where the costal parietal pleura meets the diaphragmatic parietal pleura. This recess is located at the inferior limit of the pleural sac.
- **Costomediastinal recess.** The recess where the costal parietal pleura meets the mediastinal parietal pleura anteriorly, near the midline.

The pleural recesses are sites where pleural fluid accumulates during quiet breathing. When a deep breath is taken, the greatly expanded lungs push into the recesses, allowing lung volume to increase and, consequently, the pleural fluid becomes displaced around each lung.

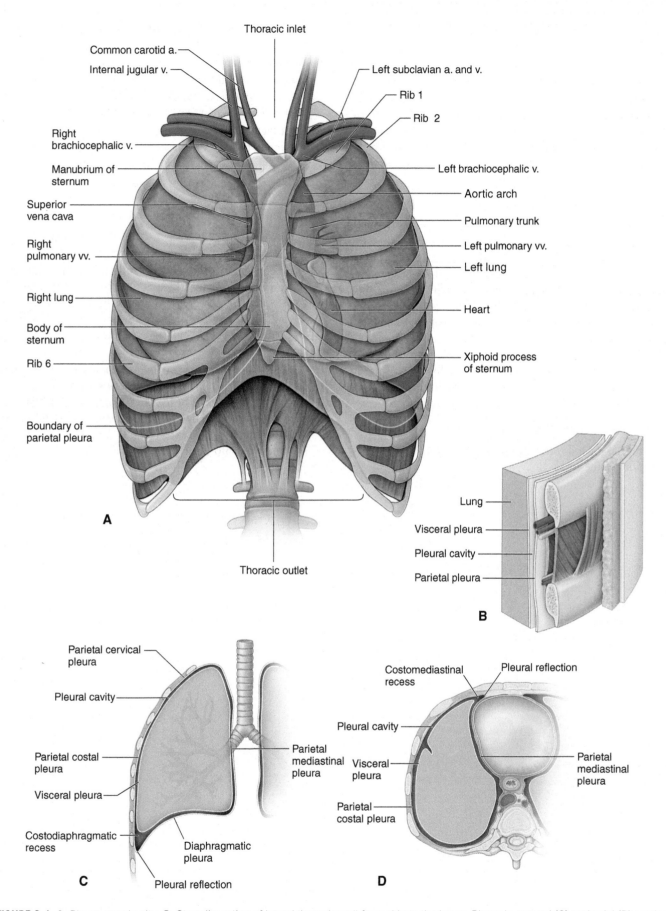

FIGURE 3-1: A. Pleura sacs in situ. **B.** Step dissection of lateral thoracic wall from skin to the lungs. Pleura in coronal (**C**) and axial (**D**) sections.

ANATOMY OF THE LUNG

BIG PICTURE

The lungs are the organs of gas exchange. To serve that function efficiently, the lungs are attached to the trachea and the heart. The pulmonary circulation supplies the respiratory tissues with deoxygenated blood, whereas the bronchial vessels nourish the nonrespiratory tissues with oxygenated blood. Parasympathetic innervation to the lung causes bronchoconstriction, whereas sympathetic innervation causes with oxygenated blood.

TRACHEA AND BRONCHI

The trachea, or windpipe, courses from the larynx into the thorax, where the trachea bifurcates into the right and left primary (principal or mainstem) bronchi at the T4–T5 vertebral level (Figure 3-2B). The **right primary bronchus** divides into **superior, middle, and inferior secondary (lobar) bronchi**, corresponding to superior, middle, and inferior lobes of the right lung, respectively. The **left primary bronchus** divides into **superior and inferior secondary bronchi**, corresponding to superior and inferior lobes of the left lung, respectively. Each secondary bronchus further divides into **tertiary (segmental) bronchi**, which further divide. The smallest bronchi give rise to bronchioles, which terminate in alveolar sacs where the exchange of gases occurs.

REGIONS OF THE LUNG

Each lung has the following regions corresponding to its respective regions of the chest:

- **Costal lung surface.** The large convex area related to the inner surface of the ribs.
- **Mediastinal lung surface.** The medial concave surface containing the **root (hilum)** of the lungs related to the heart. This surface contains a mainstem bronchus and a pulmonary artery and pulmonary veins, nerves, and lymphatics.
- **Diaphragmatic lung surface.** The base of the lungs, which are convex because they rest on the domed diaphragm.
- **Apex of the lung.** This area projects into the root of the neck and is crossed anteriorly by the subclavian artery and vein.

LOBES OF THE LUNG

Each lung is divided into lobes with its own covering of visceral pleura.

- The right lung has **three lobes (superior, middle, and inferior)**, which are divided by a **horizontal** and an **oblique fissure** (Figure 3-2A). The right lung is shorter and wider than the left because of the higher right dome of the diaphragm and because the heart bulges more into the left side of the thorax.
- The left lung has only **two lobes (superior and inferior)**, which are divided by an **oblique fissure** along the sixth rib (Figure 3-2C). Instead of having a middle lobe, the left lung has a space occupied by the heart. Therefore, the left lung has a **cardiac notch** as well as the **lingula**, an extension of the left superior lobe into the left costomediastinal recess.

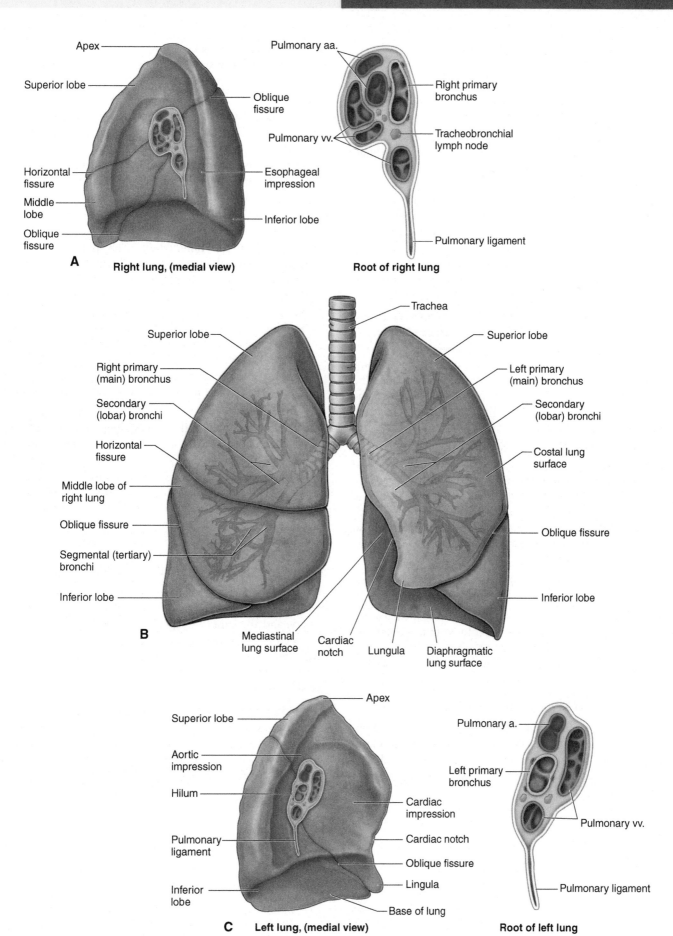

FIGURE 3-2: A. Right lung in medial view. **B**. Bronchial tree and lungs. **C**. Left lung in medial view.

HILUM OF THE LUNG

BIG PICTURE

For gas exchange to occur, the lung must be connected to the heart so that oxygenated blood and deoxygenated blood flow between both organs. The location where blood vessels and other structures enter and leave the lungs is called the hilum of the lung.

VASCULATURE OF THE LUNG

The pulmonary and bronchial arteries and veins provide dual vascular supply. These two supplies can be confusing. In a nutshell, here is the information you need to know about the vascular supply: *Pulmonary arteries and veins deal with gas exchange and the circulation of blood between the heart and lungs, whereas the bronchial arteries and veins are the vascular supply to the structural elements of the lungs, such as the bronchial tree.*

- **Pulmonary arteries.** Branch from the pulmonary trunk, which receives blood from the right ventricle of the heart (Figure 3-3A). Pulmonary arteries deliver deoxygenated blood from the systemic circulation to exchange carbon dioxide with oxygen in the lungs. At the bifurcation of the pulmonary trunk into the pulmonary arteries, there is a connection to the aortic arch via the **ligamentum arteriosum**, the fibrous remnant of the fetal ductus arteriosus.

- **Pulmonary veins.** Transport oxygenated blood from the pulmonary capillaries to the left atrium of the heart. They do not accompany the bronchi or the segmental arteries within the lung parenchyma. Two pulmonary veins exit the left lung and three pulmonary veins exit the right lung (one for each lobe), but the right upper and middle veins usually join so that usually only four pulmonary veins enter the left atrium.

- **Bronchial arteries.** Branch from the thoracic (descending) aorta and supply the bronchial tree. There is usually one bronchial artery for the right lung and two for the left lung.

- **Bronchial veins.** Drain the bronchi. The bronchial veins receive blood from the larger subdivisions of the bronchi and empty into the azygos vein on the right and the accessory hemiazygos vein on the left. Some bronchial veins become tributaries to the pulmonary veins.

LYMPHATICS OF THE LUNG

Lymph from lobes of the lungs drains into **pulmonary** and **bronchopulmonary (hilar) nodes** and then into the **tracheobronchial (carinal) nodes** and into the **paratracheal nodes** en route to drain into either the **right lymphatic duct** (for the right lung) or the **thoracic duct** (for the left lung). It should be noted that no lymphatics are present in the walls of the alveolar sacs.

INNERVATION OF THE LUNG

The **pulmonary plexus** follows the trachea and bronchial tree, providing parasympathetic and sympathetic innervation to the smooth muscle and glands of the lungs (Figure 3-3B). The pulmonary plexus is divided and named according to its position to the root of the lung, where the **anterior pulmonary plexus** lies anterior and the **posterior pulmonary plexus** lies posterior to the corresponding bronchus. Branches of the pulmonary plexus accompany the blood vessels and bronchi into the lung as well.

- **Parasympathetic innervation.** The pulmonary plexus receives **preganglionic parasympathetic** and **visceral sensory** innervation via the **vagus nerves**. The vagus nerves are the tenth pair of cranial nerves and have the widest field of distribution to the body. They innervate structures in the head, neck, thorax, and abdomen. In the thorax, the vagus nerves provide all the parasympathetic innervation of the viscera. Parasympathetic innervation causes bronchoconstriction of the smooth muscle of the bronchial tree, vasodilation of the pulmonary vessels, and secretion from bronchial glands.

- **Sympathetic innervation. Postganglionic sympathetic fibers** from the T1 to T4 levels of the **sympathetic trunk** and cervical sympathetic ganglia contribute to the pulmonary plexus. Sympathetic innervation causes bronchodilation, vasoconstriction of the pulmonary vessels, and inhibition of secretion from the bronchial glands. Visceral sensory fibers from the visceral pleura and bronchi may also accompany sympathetic fibers.

BRONCHOPULMONARY SEGMENTS

A bronchopulmonary segment consists of a lobar (segmental or tertiary) bronchus, a corresponding branch of the pulmonary artery, and the supplied segment of the lung tissue, all surrounded by a connective tissue septum. A bronchopulmonary segment refers to the portion of the lung supplied by each segmental bronchus and segmental artery. The pulmonary veins lie between bronchopulmonary segments.

- **Surgical removal of lung segments.** Bronchopulmonary segments are clinically important because they serve as the anatomic, functional, and "surgical" unit of the lungs. A surgeon may remove a bronchopulmonary segment of the lung without disrupting the surrounding lung parenchyma. ▼

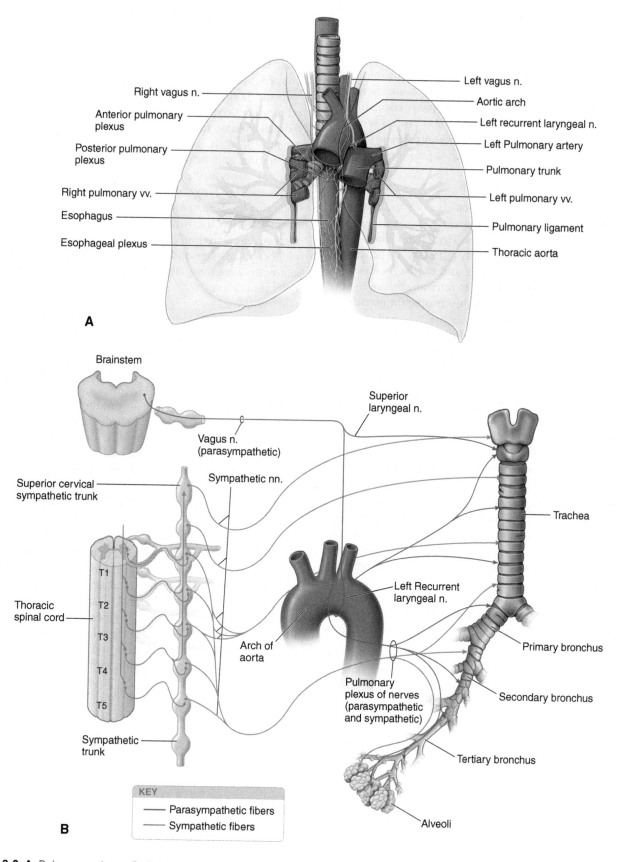

FIGURE 3-3: A. Pulmonary plexus. **B.** Autonomic innervation of the lung.

RESPIRATION

BIG PICTURE

Respiration is the vital exchange of oxygen and carbon dioxide as blood circulates through the lungs. The thoracic skeleton, thoracic wall muscles, bronchial tree, and pulmonary circulation all aid in this process. Breathing is a mechanical process resulting from volume changes in the thoracic cavity with inverse changes in pleural pressure. Pressure changes lead to gas flow.

INSPIRATION

The process of inspiration (**inhalation**) is easily understood if you visualize the thoracic cavity as a closed box with a single opening at the top called the trachea. The trachea allows air to move in and out of the thoracic cavity ("the box"). The volume of the thoracic cavity is changeable and can be increased by enlarging all its diameters (superior to inferior, anterior to posterior, and medial to lateral), thereby decreasing the pressure in the pleural space (pleural pressure) (Figure 3-4A and B). This, in turn, causes air to rush from the atmosphere (positive pressure relative to the lungs) into the lungs (negative pressure relative to the atmosphere) because gas flows down its pressure gradient. The muscles that primarily expand the thoracic cavity during inspiration are the diaphragm and the intercostal muscles.

- **Diaphragm.** In the relaxed state, the diaphragm is dome shaped. When the diaphragm contracts, it flattens, increasing the vertical dimensions and thus the volume of the thoracic cavity.

- **Intercostal muscles.** Contraction of the external intercostal muscles lifts the rib cage and pulls the sternum anteriorly. Because the ribs curve downward as well as forward around the chest wall, the broadest lateral and anteroposterior dimensions of the rib cage are normally directed downward. However, when the ribs are raised and drawn together, they also swing outward, expanding the diameter of the thorax both laterally and in the anteroposterior plane. This is similar to the action that occurs when a curved bucket handle is raised away from the bucket (Figure 3-4C).

Although these actions expand the thoracic dimensions by only a few millimeters along each plane, this expansion is sufficient to increase the volume of the thoracic cavity by approximately 0.5 L, the approximate volume of air that enters the lungs during normal inhalation. The diaphragm is by far the most important structure that brings about the pressure, gas flow, and volume changes that lead to normal inhalation.

As the thoracic dimensions increase during inspiration, pleural pressure becomes more negative and "pulls" on the lungs as thoracic volume increases. The consequence is that the lungs expand (fill with gas) and intrapulmonary volume increases. Inspiration ends when thoracic volume ceases to increase, resulting in no further reduction in pleural pressure. Gas flow ceases and thus lung volume does not change.

During the deep or forced inspirations that occur during vigorous exercise, the volume of the thoracic cavity is further increased by activation of the accessory muscles. Accessory respiratory muscles, including the scalenes, sternocleidomastoid, and pectoralis minor, elevate the ribs more than occurs during quiet inspiration.

EXPIRATION

Quiet expiration (**exhalation**) is largely a passive process that depends more on the natural elasticity of the thoracic wall and lungs than on muscle contraction. In contrast, forced expiration is an active process (Figure 3-4A–C).

- **Quiet expiration.** As the inspiratory muscles relax, the diaphragm ascends, the rib cage descends, and the stretched elastic tissue of the lungs recoils. Thus, both thoracic and lung volumes decrease. Decreased lung volume compresses the alveoli, resulting in increases above atmospheric pressure, thereby forcing gas flow out of the lungs. For example, when the diaphragm relaxes, it passively moves superiorly. Consequently, the vertical dimension of the thorax is decreased and thus the volume of the thoracic cavity decreases.

- **Forced expiration.** When the expiratory muscles (e.g., the external and internal oblique and transverse and rectus abdominis) contract, they increase intra-abdominal pressure. This forces the abdominal organs superiorly against the diaphragm, raising it. The same muscles depress the rib cage. Both actions forcibly reduce the volume in the thoracic cavity, increasing pleural volume and thence pressure in the lungs, forcing air to move from the lungs and out of the trachea.

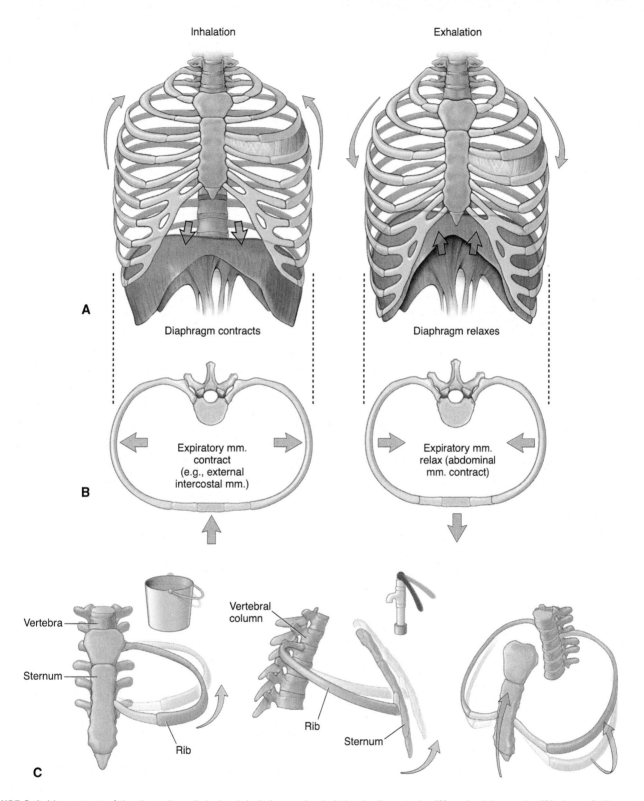

FIGURE 3-4: Movements of the thoracic wall during inhalation and exhalation in the anterior (**A**) and axial superior (**B**) views. **C**. Thoracic wall movements during respiration. The bucket and water-pump handle are analogies for the movement of the rib cage when acted upon by respiratory muscles.

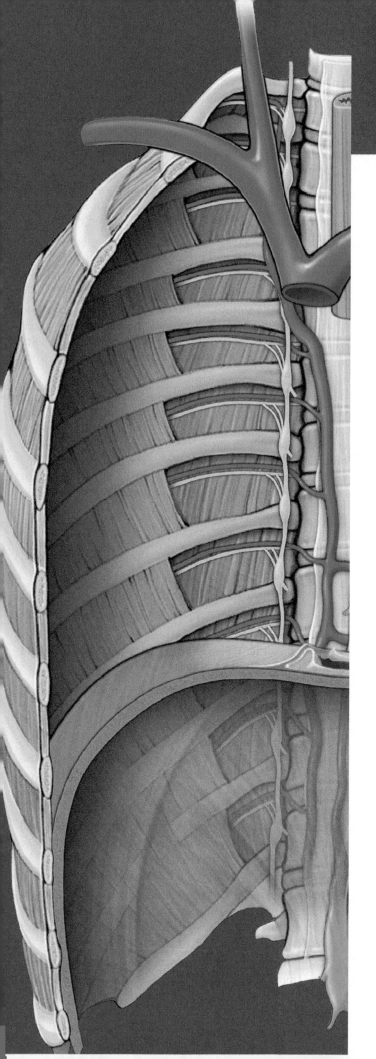

HEART

PERICARDIUM

BIG PICTURE

The pericardium is a sac that encloses the heart, akin to the pleura that encloses the lungs. The pericardium and the heart are located in the middle of the thorax, between T4 and T8 vertebrae. The pericardium has parietal and visceral layers.

PERICARDIAL SAC

The **parietal pericardium** is composed of an external fibrous layer (**fibrous pericardium**) and an internal serous layer (**serous pericardium**) (Figure 4-1A and B).

- **Fibrous pericardium.** A strong, dense collagenous tissue that blends with the tunica externa of the great vessels and the central tendon of the diaphragm.
- **Serous pericardium.** Lines the inner surface of the fibrous pericardium.
- **Pericardial space.** Lies between the serous layer of the parietal pericardium and the visceral pericardium.
- **Visceral pericardium (epicardium).** A serous layer that intimately follows the contours of the heart surface. At the root of the heart, the visceral pericardium is contiguous with the serous pericardium, analogous to the visceral and parietal pleura of the hilum of each lung.

The **pericardial sinuses** are subdivisions of the pericardial sac and consist of the following spaces:

- **Transverse sinus.** Lies posterior to the ascending aorta and the pulmonary trunk and anterosuperior to the left atrium and the pulmonary veins.
- **Oblique sinus.** Lies posterior to the heart and is surrounded by the reflection of the serous pericardium around the right and left pulmonary veins and the inferior vena cava.

The pericardium receives its blood supply from the pericardiacophrenic vessels and branches from the bronchial and esophageal vessels. It is innervated by visceral sensory fibers from the phrenic and vagus nerves and the sympathetic trunks.

OVERVIEW OF THE HEART

BIG PICTURE

The heart has the following **three layers** (Figure 4-1B):

- **Epicardium.** The outer layer, also known as the visceral pericardium.
- **Myocardium.** The middle layer, consisting of cardiac muscle responsible for contraction of the heart.
- **Endocardium.** The inner layer, consisting of endothelial cells that line the lumen of the four chambers.

ANATOMY OF THE HEART

The heart, including its left and right sides, surfaces, borders, and sulci (Figure 4-1C to F), art described in a variety of ways.

RIGHT AND LEFT SIDES OF THE HEART The expression "right side of the heart" refers to the right atrium and the right ventricle, which collect systemic deoxygenated blood and pump it into the lungs. By comparison, the "left side of the heart" refers to the left atrium and the left ventricle, which collect oxygenated blood from the pulmonary circulation and pump it into the body.

SURFACES OF THE HEART The shape of the heart is roughly that of a cone with a flat base and a pointed apex. The surfaces of the heart consist of the following:

- **Base.** Is flat, faces posteriorly, and is formed mainly by the pulmonary veins of the left atrium. The base faces posteriorly toward the T6–T9 vertebral bodies, intervened by the pericardium, the esophagus, and the aorta.
- **Sternocostal surface.** Is part of the heart's cone shape and is named for its location deep to the sternum and costal cartilages of the five uppermost ribs.
- **Diaphragmatic surface.** Is also part of the heart's cone shape and is named for the heart surface directly superior to the central tendon of the diaphragm. Note that the heart, as it lies within the thorax, rests upon the diaphragmatic surface but not on its base.

The apex is formed by the tip of the left ventricle and points anteriorly, inferiorly, and to the left. The apex lies deep to the left fifth intercostal space medial to the midclavicular line. In an adult, the apex of the heart is approximately one hand's breadth from the median plane, near the midclavicular (nipple) line. This location is useful clinically because auscultation of the **mitral valve** is best heard over the apex.

BORDERS OF THE HEART

- The **superior border** is formed by the right and left auricles. The ascending aorta and pulmonary trunk emerge from this border, and the superior vena cava enters the right atrium.
- The right ventricle forms the **inferior border**.
- The **left border** is formed by the **left ventricle and auricle of the left atrium**.
- The **right border** is formed by the **superior vena cava, right atrium**, and **inferior vena cava**.

SULCI OF THE HEART Sulci (**grooves**) mark the outer surface of the heart at the sites where one chamber meets another. The **cardiac sulci** are as follows:

- **Sulcus terminalis.** A groove on the external surface of the right atrium. This sulcus marks the junction of the primitive sinus venosus with the atrium in the embryo, and corresponds to a ridge on the internal surface of the right atrium, the crista terminalis.
- **Atrioventricular (AV) groove (coronary sulcus).** An external demarcation of the atria from the ventricles. The AV groove contains the right coronary, the left coronary, and the circumflex coronary arteries and the coronary sinus.
- **Anterior interventricular groove.** Is demarcates the left ventricle from the right ventricle on the anterior surface of the heart and contains the left anterior descending artery and the great cardiac vein.
- **Posterior interventricular groove.** Is demarcates the left ventricle from the right ventricle on the posterior surface of the heart and contains the posterior interventricular artery and the middle cardiac vein.

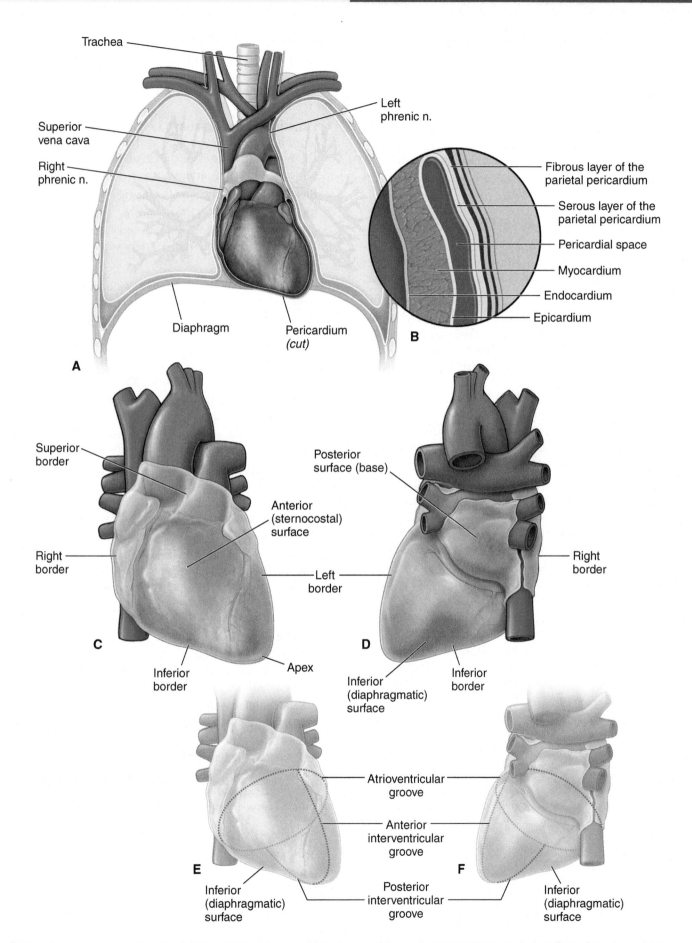

Figure 4-1: A. Coronary section through the thorax. **B**. Layers of the pericardial sac. **C**. Anterior (sternocostal) surface of the heart. **D**. Posterior (base) and inferior (diaphragmatic) surface of the heart. **E**. Coronary grooves (anterior view). **F**. Coronary grooves (posterior view).

CORONARY CIRCULATION

BIG PICTURE

Although blood fills the chambers of the heart, the myocardium is so thick that it requires its own artery–capillary–vein system, called the "coronary circulation," to deliver and remove blood to and from the myocardium. The vessels that supply oxygenated blood to the myocardium are known as **coronary arteries**. The vessels that remove the deoxygenated blood from the heart muscle are known as **cardiac veins**.

CORONARY ARTERIES AND ASSOCIATED BRANCHES

The coronary arteries and branches course along the epicardium in the cardiac sulci and interventricular grooves (Figure 4-2). Each coronary artery sends branches to the heart muscle.

▽   **Myocardial infarction.** Coronary vessels are classified as an "end circulation"; that is, they may not anastomose with each other. Therefore, blockage of any of these vessels is detrimental because once a coronary artery is blocked cardiac tissue supplied by that vessel is damaged. Blood flow in the coronary arteries is maximal during diastole (ventricular relaxation) and minimal during systole (ventricular contraction) because of the compression of the blood vessels in the myocardium during systole. All coronary arteries branch from either the left or the right coronary arteries. ▽

LEFT CORONARY ARTERY The left coronary artery arises from the aorta, superior to the **left cusp** of the aortic valve, and is shorter than the right coronary artery. However, the branches from the left coronary artery distribute blood to a larger area of myocardium. The left coronary artery supplies most of the left ventricle, left atrium, bundle of His, and the anterior aspect of the interventricular septum. The left coronary artery gives rise to the following **two sizable branches**:

- **Left anterior descending artery.** Also called the anterior interventricular artery and is often referred to by the acronym LAD. It supplies the anterior region of the left ventricle, including the anterolateral myocardium, apex, anterior interventricular septum, and the anterolateral papillary muscle.
- **Left circumflex artery.** Wraps around the left side to the posterior side of the heart. The circumflex artery supplies the posterolateral side of the left ventricle and gives off the **left marginal branches** which also supply the left ventricle.

RIGHT CORONARY ARTERY The right coronary artery arises from the aorta, superior to the **right cusp** of the aortic valve.

The right coronary artery travels along the right AV groove, between the root of the pulmonary trunk and the right auricle, and supplies the right atrium, right ventricle, the sinuatrial (SA) node, and the AV node. The right coronary artery gives rise to the following branches:

- **Posterior descending artery.** Supplies the inferior wall, posterior interventricular septum, and the posteromedial papillary muscle. In a few cases, the circumflex artery gives off the posterior descending artery.
- **Right marginal artery.** Supplies the right ventricular wall.
- **SA nodal artery.** Passes between the right atrium and the opening of the superior vena cava and supplies the SA node. In a few cases, the circumflex artery supplies the SA nodal artery.

CORONARY DOMINANCE Coronary dominance is determined by which of the left and right coronary arteries gives rise to the posterior interventricular artery, which supplies the posterior interventricular semptum and part of the left ventricle. The dominant artery is usually the right coronary artery.

CARDIAC VEINS

The cardiac veins and associated tributaries are the major veins of the coronary circulation and run parallel to the arteries (Figure 4-2). They drain blood from the heart wall. The **cardiac veins** are as follows:

- The **coronary sinus** is the largest vein draining the heart muscle and lies in the coronary sulcus. The coronary sinus collects most of the venous return from the great, middle, and small cardiac veins and returns the venous blood to the right atrium. The coronary sinus opening in the right atrium is superior to the septal leaflet of the tricuspid valve.
  - **Great cardiac vein.** Begins at the apex of the heart and ascends in the anterior interventricular groove, parallel to the left anterior descending artery, and drains into the coronary sinus.
  - **Middle cardiac vein.** Begins at the apex of the heart and ascends in the posterior interventricular sulcus, parallel to the posterior interventricular artery, and drains into the coronary sinus.
  - **Small cardiac vein.** Courses along the acute margin of the heart, along with the marginal artery, and then courses posteriorly into the coronary sinus.
- The **anterior cardiac veins** drain the anterior portion of the right ventricle, cross the coronary groove, and empty directly into the right atrium. Anterior cardiac veins do not drain into the coronary sinus.

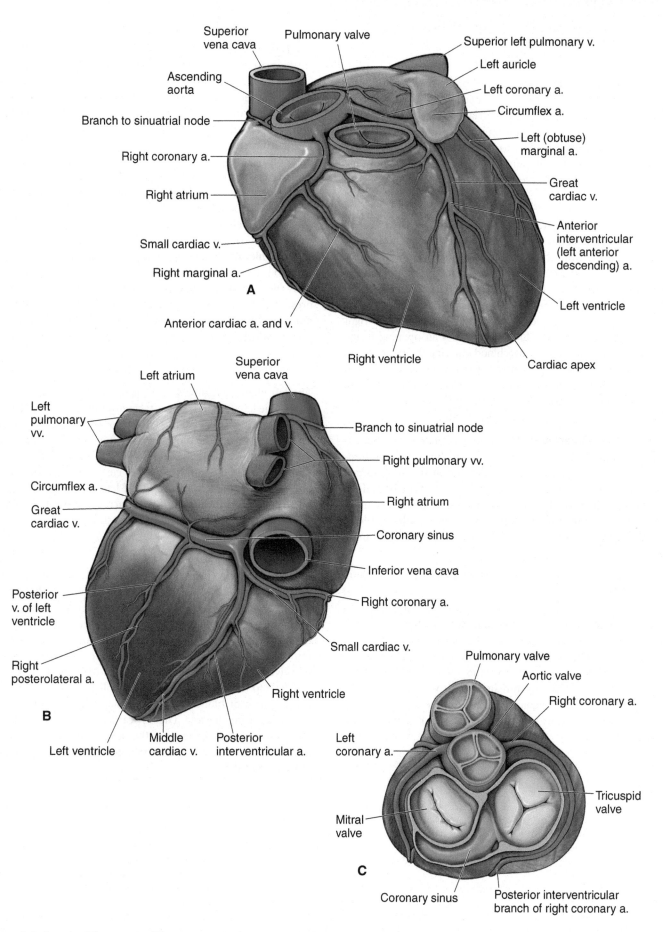

Figure 4-2: Anterior (**A**), posterior (**B**), and superior (**C**) views of the coronary arteries and cardiac veins.

CHAMBERS OF THE HEART

BIG PICTURE

The human heart is a four-chambered pump composed of cardiac muscle. Two of the heart chambers, called **atria**, are relatively thin walled. The function of atria is to receive blood from organs outside the heart and pump it under low pressure to the corresponding ventricle. The two ventricles have thicker muscular walls that function to pump blood to organs, through capillary networks, which add resistance to blood flow.

RIGHT ATRIUM

The right atrium (Figure 4-3A) receives deoxygenated venous blood from the systemic circulation via the superior and inferior venae cavae and from the coronary circulation via the coronary sinus. The right atrium pumps blood into the right ventricle through the **tricuspid (right AV) valve**. The tricuspid valve, located between the right atrium and the right ventricle, ensures unidirectional flow of blood from the right atrium to the right ventricle. In the fetus, the **foramen ovale** is an opening in the interatrial septum, which allows blood entering the right atrium from the venae cavae to pass directly to the left side of the heart, bypassing the lungs. Blood flow to the lungs is bypassed because the placenta is responsible for gas exchange in utero. In an adult, this foramen, termed the **fossa ovalis**, is shut.

▽ A **patent foramen ovale** is a congenital defect in the septal wall between the two atria. Before birth, a patent foramen ovale is normal; however, following birth, the foramen ovale seals shut because of increased pressure in the left side of the heart. Failure of the foramen ovale to seal shut occurs in approximately 20% of the population. This results in blood bypassing the lungs for filtering. Most cases of patent foramen ovale are asymptomatic. ▼

On the superior aspect of the right atrium is the **auricle**, which is an outpouching of tissue derived from the fetal atrium, with rough myocardium on its internal surface known as **pectinate muscles**. The **sinus venarum**, which has a smooth wall, is on the internal surface of the remainder of the right atrium. The **crista terminalis** is the internal vertical ridge that separates the rough portion from the smooth portion of the right atrium. The crista terminalis extends vertically from the superior to the inferior vena cava. The SA node of the conducting system of the heart is located in the superior part of the crista terminalis.

RIGHT VENTRICLE

The right ventricle (Figure 4-3B) forms the greater part of the anterior surface of the heart and lies anterior to much of the left ventricle. The right ventricle receives blood from the right atrium via the tricuspid valve. Outflow during systole is to the pulmonary trunk via the pulmonary semilunar valve. The function of the pulmonary valve is to ensure unidirectional flow of blood from the right ventricle to the pulmonary trunk, thus preventing retrograde return of blood from the pulmonary trunk back into the right ventricle. The internal surface of the right ventricle has projecting ridges called **trabeculae carneae**, with **papillary muscles** attached to the ventricular wall. Fibrous chords called chordae tendineae attach papillary muscles to the cusps of the tricuspid valve, preventing prolapse.

The **moderator band** (septomarginal trabecula) is a large ridge of heart muscle that is attached from the anterior and septal papillary muscles to the septal and anterior ventricular walls. When present (which occurs in approximately 60% of hearts), the moderator band conveys within it the right branch of the AV bundle, which is part of the conducting system of the heart.

LEFT ATRIUM

The left atrium (Figure 4-3C) forms most of the base of the heart. The left atrium is in contact with the esophagus. The left atrium receives oxygenated blood from two left and two right pulmonary veins. The blood in the left atrium is pumped into the left ventricle through the mitral (bicuspid, or left AV) valve. The function of the mitral valve is to ensure unidirectional flow of blood from the left atrium to the left ventricle during diastole, thus preventing retrograde return of blood from the left ventricle into the left atrium during systole.

LEFT VENTRICLE

The left ventricle (Figure 4-3C) is longer and more conically shaped than the right ventricle and lies **anterior to the left atrium**. Most of the left ventricle is located on the posterior side of the heart. The left ventricle receives oxygenated blood from the left atrium through the mitral valve, and it pumps blood into the aorta through the aortic valve. The function of the aortic valve is to ensure unidirectional flow of blood from the left ventricle to the aortic arch during systole and to prevent retrograde return of blood from the aortic arch back into the left ventricle during diastole. The myocardial wall of the left ventricle is much thicker than that of the right ventricle, which is necessary to generate higher pressure to pump blood to the organs, muscles, and skin.

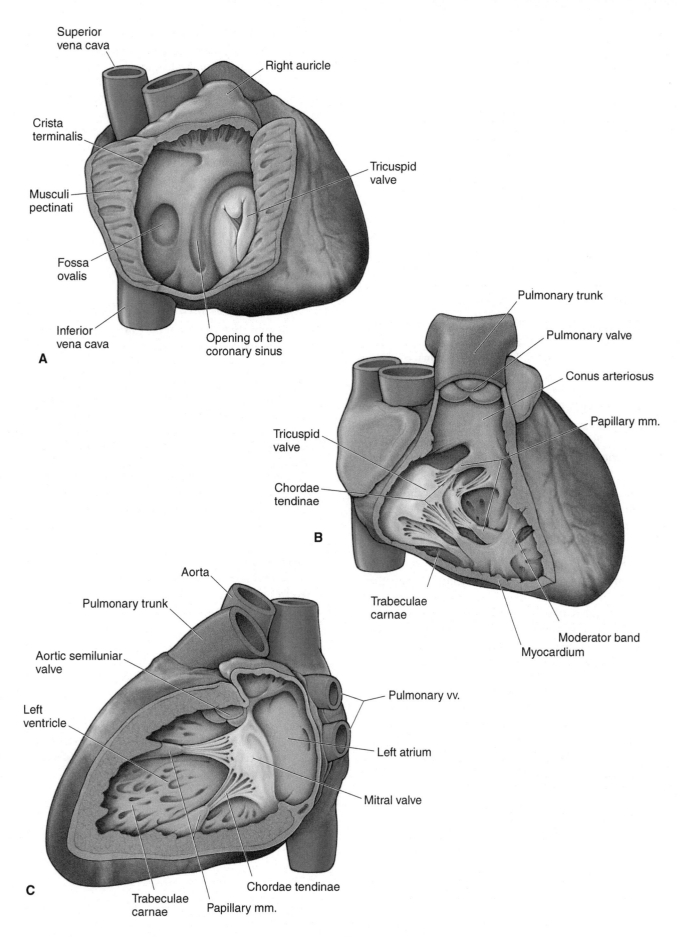

Figure 4-3: A. Right atrium. **B**. Right ventricle. **C**. Left atrium and ventricle.

INNERVATION OF THE HEART

BIG PICTURE

Control of heart contractions is mediated through a combination of autonomic innervation via the cardiac plexus as well as the conduction pathway through the heart muscle itself. To understand heart innervation, we must study the autonomic contribution to the cardiac plexus and how it influences the conduction system of the heart.

CARDIAC PLEXUS

The cardiac plexus is divisible into a superficial cardiac plexus, between the aortic arch and the pulmonary artery, and a deep cardiac plexus, between the aortic arch and the tracheal bifurcation. The plexuses receive a combination of both sympathetic and parasympathetic fibers in the manner discussed below (Figure 4-4).

SYMPATHETIC CONTRIBUTION Preganglionic sympathetic fibers originate bilaterally in the lateral horns of the gray matter of the spinal cord between the T1 and the T5 spinal cord levels and enter the sympathetic chain via the white rami communicantes. After entering the sympathetic chain, fibers travel to the cardiac plexus via **two possible routes:**

- Preganglionic sympathetic fibers synapse in the superior parts of the thoracic sympathetic chain and send **postganglionic sympathetic fibers** directly from the sympathetic ganglia to the cardiac plexuses via **thoracic cardiac nerves.**

- The preganglionic sympathetic fibers ascend through the sympathetic chain and synapse in either the **superior, middle, or inferior cervical ganglia** before sending off **postganglionic sympathetic fibers** via **cervical cardiac nerves** to the cardiac plexuses.

▽ **Referred pain.** Both sympathetic and parasympathetic fibers carry visceral sensory fibers from the heart to the spinal cord and brain, respectively. However, the visceral sensory fibers within the cardiac branches from the cervical and superior five thoracic sympathetic ganglia are sensitive to **ischemia** (tissue damage due to lack of oxygen). These sensory fibers mediate the visceral pain associated with angina pectoris and myocardial infarctions. Such myocardial ischemic pain is often referred to regions of the T1–T4 dermatomes simply because the visceral sensory fibers enter the spinal cord at the same levels of the segments for the superior four thoracic spinal nerves. The brain cannot differentiate between sensory input from the spinal nerves and that from the visceral nerves and thus refers ischemic pain to the same dermatome. ▼

PARASYMPATHETIC CONTRIBUTION Preganglionic parasympathetic (vagal) fibers in the left and right vagus nerves originate in the medulla oblongata and descend through the neck and into the thorax to the cardiac plexuses. The synapse of vagal preganglionic and **postganglionic parasympathetic fibers** occurs either in the cardiac plexus or in the walls of the heart near the SA node of the right atrium. Therefore, the cardiac plexus serves as a conduit not only for parasympathetic preganglionic and postganglionic and visceral sensory fibers but also for sympathetic postganglionic fibers.

In summary, mixed nerves from the cardiac plexus supply the heart with sympathetic fibers, which increase the heart rate and the force of contraction and cause dilation of the coronary arteries, and parasympathetic fibers, which decrease the heart rate, reduce the force of contraction, and constrict the coronary arteries.

CONDUCTING SYSTEM OF THE HEART

The autonomic branches from the cardiac plexus help to regulate the rate and force of heart contractions, through influencing the SA node and the AV node, as follows:

- **SA node.** The rhythm of the heart is normally controlled by the SA node, a group of automatically depolarizing specialized cardiac muscle cells located at the superior end of the crista terminalis, where the right atrium meets the superior vena cava. The SA node is considered the "**pacemaker of the heart**" and initiates the heart beat, which can be altered by autonomic nervous stimulation (sympathetic stimulation speeds it up, whereas vagal stimulation slows it down). The wave of depolarization sweeps down the walls of the atria, stimulating them to contract, and eventually reaches the AV node.

- **AV node.** The AV node is located in the interatrial septum just superior to the opening of the coronary sinus. This node receives impulses from the SA node and passes them to the AV bundle (of His).

- **AV bundle (of His).** The AV bundle begins at the AV node and descends through the fibrous skeleton of the heart before dividing into the left and right bundles (of His), corresponding to the left and right ventricles, respectively. This divergent pathway ensures that ventricular contraction begins in the region of the apex. Conduction ends near the aortic and pulmonic valves. Impulses also pass from the left and right bundle branches to the papillary muscles in the corresponding ventricles. In the right ventricle, the moderator band (septomarginal trabeculum) contains the right bundle branch.

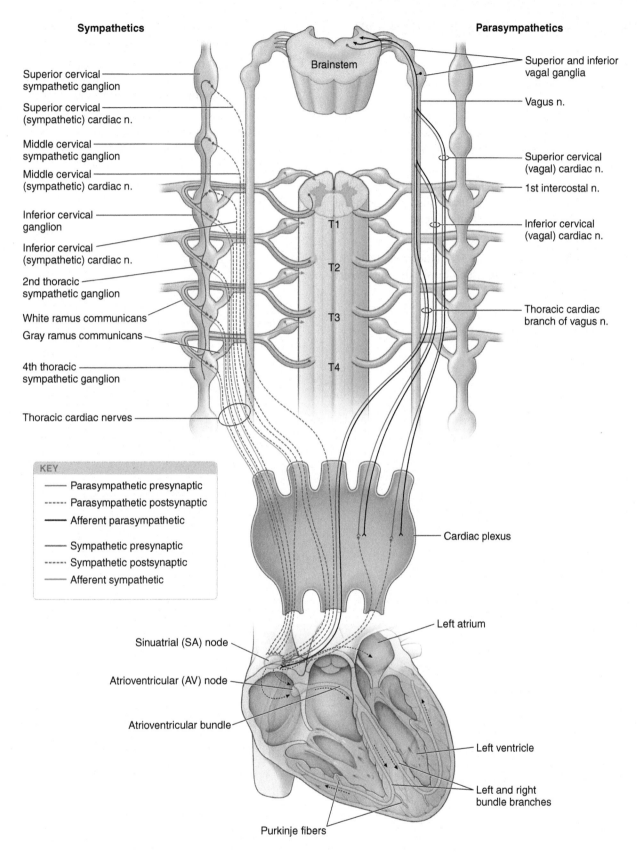

Sympathetics

Superior cervical sympathetic ganglion

Superior cervical (sympathetic) cardiac n.

Middle cervical sympathetic ganglion

Middle cervical (sympathetic) cardiac n.

Inferior cervical ganglion

Inferior cervical (sympathetic) cardiac n.

2nd thoracic sympathetic ganglion

White ramus communicans

Gray ramus communicans

4th thoracic sympathetic ganglion

Thoracic cardiac nerves

Parasympathetics

Superior and inferior vagal ganglia

Vagus n.

Superior cervical (vagal) cardiac n.

1st intercostal n.

Inferior cervical (vagal) cardiac n.

Thoracic cardiac branch of vagus n.

Brainstem

T1

T2

T3

T4

KEY

——— Parasympathetic presynaptic

- - - - Parasympathetic postsynaptic

━━━ Afferent parasympathetic

——— Sympathetic presynaptic

- - - - Sympathetic postsynaptic

——— Afferent sympathetic

Cardiac plexus

Sinuatrial (SA) node

Atrioventricular (AV) node

Atrioventricular bundle

Left atrium

Left ventricle

Left and right bundle branches

Purkinje fibers

Figure 4-4: Autonomic innervation and conducting system of the heart.

CIRCULATORY PATHWAY THROUGH THE HEART

CARDIAC CYCLE

The flow of blood through the heart is described in the following steps (Figure 4-5):

A. All blood returning to the heart from the tissues enters the atria. Blood from the systemic and coronary circulation fills the right atrium, whereas blood from the pulmonary circulation fills the left atrium.

B. As the left and right atria are filled, pressure rises against the AV valves. The atria contract simultaneously and both the right and left AV valves are forced open (diastole).

C. Blood fills the left and right ventricle.

D. Ventricles contract (systole), forcing the blood against the AV valve cusps and closing them, which causes the "lub" heart sound. The chordae tendineae and papillary muscles prevent prolapse of the valve leaflets into the atria.

E. Intraventricular pressure continues to rise and forces the pulmonary and aortic semilunar valves open. As a result blood flows into the pulmonary trunk and aorta, respectively.

F. As the ventricles relax (diastole) and intraventricular pressure decreases, blood flows back into the pulmonary trunk and aorta, filling the cusps of their semilunar valves and forcing the semilunar valves to close, which results in the "dub" heart sound.

5. The cycle begins again and repeats itself.

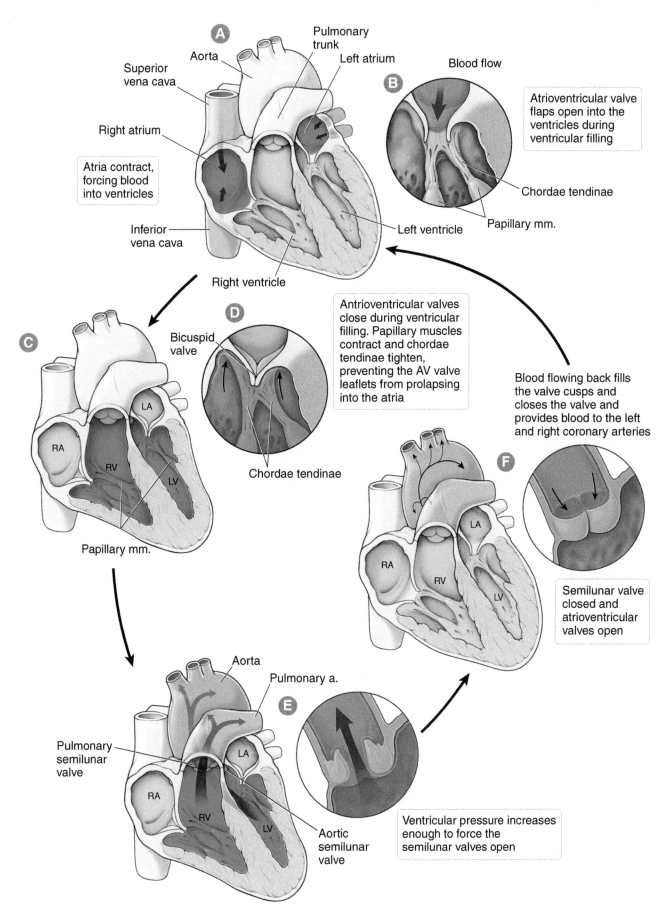

Figure 4-5: Blood flow through the heart.

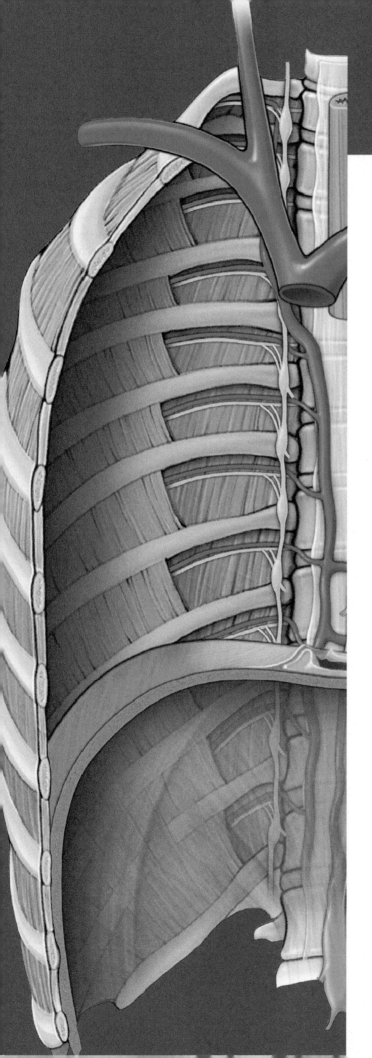

CHAPTER 5

SUPERIOR AND POSTERIOR MEDIASTINA

DIVISIONS OF THE MEDIASTINA

BIG PICTURE

The mediastinum is the anatomic region medial to the pleural sacs between the sternum, vertebral column, rib 1, and the diaphragm. The mediastinum is further divided into inferior and superior parts by a horizontal plane passing through the sternal angle to the T4–T5 intervertebral disc (Figure 5-1A). The inferior mediastinum is classically subdivided into anterior, middle, and posterior parts. Therefore, the four subregions of the mediastinum are as follows:

- **Anterior mediastinum.** The region between the sternal angle, the deep sternal surface, the pericardial sac, and the diaphragm. The anterior mediastinum contains fat and areolar tissue and the inferior part of the thymus or its remnant.

- **Middle mediastinum.** This region contains the pericardial sac and heart (see Chapter 4 for further details).

- **Posterior mediastinum.** The region containing anatomic structures deep to the pericardial sac, including the thoracic portion of the descending aorta, the azygos system of veins, the thoracic duct, the esophagus, and the vagus and sympathetic nerves (Figure 5-1B). This chapter will focus on the structures located in the posterior mediastinum and their projection into the superior mediastinum.

- **Superior mediastinum.** The region superior to the sternal angle containing the aortic arch and its three branches, the superior vena cava (SVC) and the brachiocephalic veins, the trachea, the esophagus, and the phrenic and vagus nerves. The superior mediastinum also contains the thymus; however, in an adult, the thymus is usually atrophied and presents as a fatty mass.

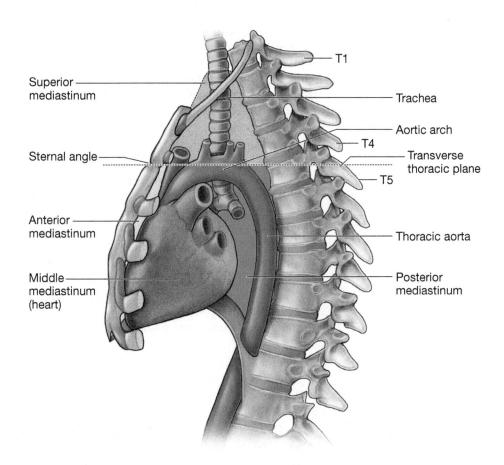

A

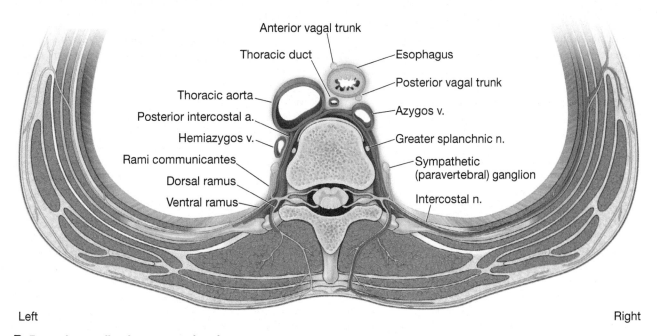

Left   Right

B Posterior mediastinum; superior view

Figure 5-1: A. The lateral view of the thorax illustrating the mediastinal subdivisions. **B**. The posterior mediastinum in axial section.

SYMPATHETIC TRUNK AND ASSOCIATED BRANCHES

BIG PICTURE

The autonomic nervous system of the thorax consists of both sympathetic and parasympathetic motor neurons through which cardiac muscle, smooth muscle, and the glands of the thorax and the abdomen are innervated. Autonomic innervation involves two types of neurons: preganglionic neurons and postganglionic neurons. The sympathetic nerves in the thorax and in other areas of the body include visceral sensory fibers that course along the general sensory neurons.

SYMPATHETIC NERVES OF THE THORAX

The **thoracic sympathetic chain or trunk** courses vertically across the heads of the ribs along the posterior thoracic wall, deep to the parietal pleura (Figure 5-2A). This location parallels the vertebral column; the sympathetic chain, therefore, is also referred to as the **paravertebral ganglia**. The sympathetic chain descends posterior to the diaphragm to continue its descent in the abdominal cavity.

The thoracic portion of the sympathetic trunk typically has 12 paravertebral ganglia connected to adjacent thoracic spinal nerves by **white and gray rami communicantes**. White rami have myelinated nerve fibers and thus appear white. Gray rami have unmyelinated nerve fibers and, therefore, appear gray. The sympathetic ganglia are numbered according to the thoracic spinal nerve with which they are associated. The chain is composed primarily of ascending and descending preganglionic sympathetic fibers and visceral afferent fibers. Each paravertebral ganglion houses the nerve cell bodies for postganglionic sympathetic nerve fibers. Internodal fibers connect sympathetic ganglia vertically, resulting in the chain or trunk.

The cell bodies for preganglionic sympathetic fibers originate in the lateral horn of the spinal cord gray matter. The axons exit the ventral root into the ventral ramus where white rami convey the **preganglionic sympathetic fibers** from the **ventral ramus** to a **paravertebral ganglion**. Once the preganglionic sympathetic fibers enter the paravertebral ganglion, the following possible pathways may occur (Figure 5-2B):

1. Preganglionic sympathetic fibers synapse with postganglionic sympathetic fibers within the ganglion, and the segmental gray rami communicantes carry the postganglionic sympathetic fibers back to the ventral ramus at the same vertebral level. The postganglionic sympathetic fibers innervate blood vessels, sweat glands, and arrector pili muscles of hair follicles within the associated dermatome.

2. Preganglionic sympathetic fibers enter **internodal fibers**, ascending to a higher ganglion (e.g., the superior cervical ganglion), where they synapse with a postganglionic fiber. Postganglionic sympathetic fibers exit the ganglion en route to the cardiac plexus (see Chapter 4, Innervation of the Heart).

3. Preganglionic sympathetic fibers en route to abdominal organs become thoracic splanchnic nerves. Thoracic splanchnic nerves are formed from the following levels (Figure 5-2A):

   - T5–T9 unite to form the **greater splanchnic nerve**, which perforates the crus of the diaphragm or occasionally passes through the aortic hiatus and ends in the celiac or superior mesenteric ganglion.

   - T10–T11 unite to form the **lesser splanchnic nerve**, which pierces the crus of the diaphragm and ends in the aorticorenal ganglion.

   - T12 becomes the **least splanchnic nerve**, which pierces the crus of the diaphragm and ends in the renal plexus.

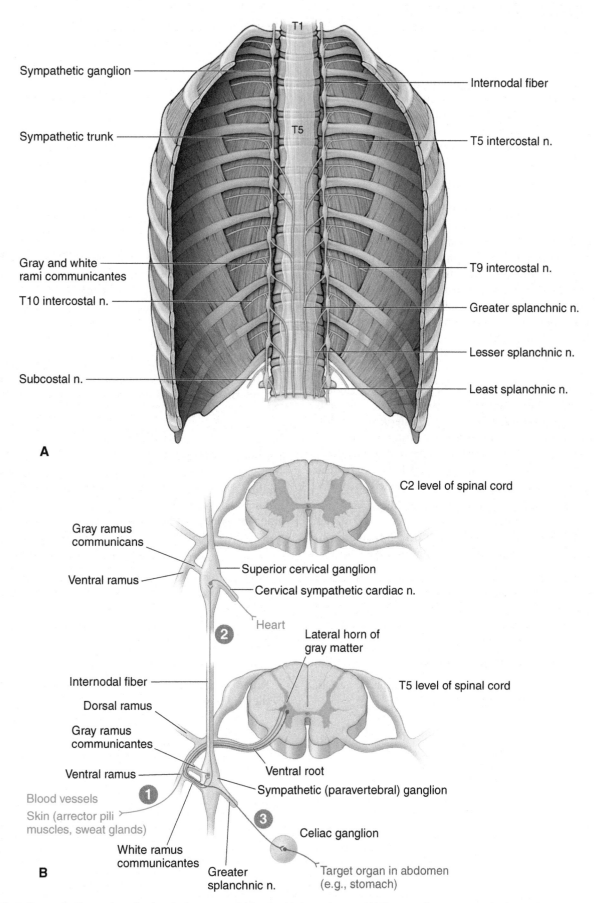

Figure 5-2: A. Sympathetic trunk and splanchnic nerves. **B**. Sympathetic pathways: (1) Synapse in a paravertebral ganglion at the same level; (2) synapse in a paravertebral ganglion at a different level; (3) synapse in a prevertebral ganglion (i.e., celiac ganglion) via a splanchnic nerve.

AZYGOS VEINS, THORACIC DUCT, AND THORACIC AORTA

BIG PICTURE

The structures superficial to the sympathetic chain are the azygos system of veins (which drain the posterior thoracic and abdominal walls), the thoracic lymphatic duct (which courses between the thoracic aorta and esophagus), and the thoracic aorta (located left of the midline, along the anterior surface of the thoracic vertebrae).

AZYGOS SYSTEM OF VEINS

The azygos system of veins is formed through the convergence of the following unpaired veins (Figure 5-3):

Azygos vein. The right ascending lumbar and right subcostal veins unite to form the azygos vein and ascend into the thorax through the aortic opening of the diaphragm. The azygos vein receives blood directly from the right posterior intercostal veins and indirectly via the left-sided connections from the hemiazygos and accessory hemiazygos. The azygos vein terminates by coursing anteriorly at the T4 vertebral, to arch over the right primary bronchus and converge with the superior vena cava.

Hemiazygos vein. The union of the left ascending lumbar and left subcostal veins forms the hemiazygos vein. It ascends on the left side of the vertebral bodies, posterior to the thoracic aorta, and receives the lower four posterior intercostal veins. The hemiazygos vein joins the azygos vein via tributaries that cross the vertebral column from left to right. The hemiazygos vein often communicates with the left renal vein.

Accessory hemiazygos vein. The accessory hemiazygos vein begins at the fourth intercostal space and receives posterior intercostal veins four through eight. It descends anterior to the posterior intercostal arteries and joins the azygos vein.

The first posterior intercostal vein on each side drains into the corresponding brachiocephalic vein. Posterior intercostal veins two through four join to form the superior intercostal vein, which drains into the azygos vein on the right side and the brachiocephalic vein on the left side.

LYMPHATIC DRAINAGE

Lymph in the thoracic region drains via two lymphatic vessels, the thoracic duct and the right lymphatic duct.

1. The **thoracic (lymphatic) duct**
   - Is the main lymphatic channel that receives lymph from the entire body, with the exception of the right upper limb, the right side of the head and neck, and the right upper thorax.
   - May have a beaded appearance because of its numerous valves.
   - Begins in the abdomen just inferior to the diaphragm at a dilated sac created by convergence of the intestinal and lumbar lymphatic trunks, called the **cisternal chili**.
   - Ascends deep to the diaphragm by coursing through the aortic hiatus.
   - Continues superiorly between the esophagus and thoracic aorta.
   - Shifts to the left of the esophagus at the level of the sternal angle (T4 vertebral level).
   - Curves laterally at the root of the neck and empties into the junction of the left internal jugular and subclavian veins.

2. The **right lymphatic duct**
   - Drains the right side of the thorax, the upper limb, and the head and neck and empties into the junction of the right internal jugular and subclavian veins.

THORACIC AORTA

The thoracic (**descending**) aorta begins at the T4 vertebral level and courses anterior to the vertebral column, just left of the midline. The thoracic aorta enters the abdominal cavity through the aortic hiatus at vertebral level T12. The following **branches arise from the thoracic aorta:**

Paired posterior intercostal arteries (3–11); the first two intercostal arteries arise from the costocervical trunk.

Bronchial arteries (usually one to the right bronchus and two to the left bronchus) for supply of the nonrespiratory tissues in the lungs.

Multiple esophageal branches to the middle third of the esophagus.

Paired subcostal arteries.

Paired superior phrenic arteries to the posterior regions of the diaphragm.

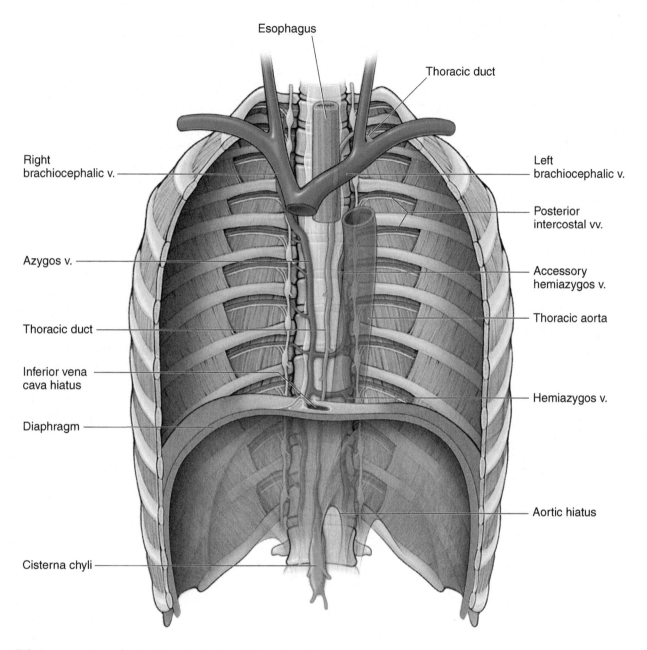

Figure 5-3: Azygos system of veins, thoracic duct, and thoracic aorta.

ESOPHAGUS

BIG PICTURE

The esophagus is a muscular tube that is continuous with the pharynx in the neck and enters the thorax posterior to the trachea. The esophagus is located against the upper thoracic vertebrae in the midline and descends to the left of the aorta (near vertebral level T4). Because of its location deep to the pericardium, the esophagus has its anterior surface pressed by the left atrium. At vertebral level T10, the esophagus exits the thorax through the **esophageal hiatus of the diaphragm**. Note that the esophageal hiatus is formed by the **right crus** of the diaphragm, splitting to wrap around the esophagus (the so-called esophageal sphincter).

VASCULATURE OF THE ESOPHAGUS

The esophagus receives its blood supply in the thorax from branches of the thoracic aorta (e.g., the esophageal and bronchial arteries). Additional contributing arteries include the left gastric and inferior phrenic arteries.

VAGUS NERVE

The right and left vagus (**parasympathetic**) nerves supply the esophagus via the esophageal plexus. Preganglionic parasympathetic fibers originate bilaterally from the medulla oblongata and travel through the vagus nerve through the thorax, posterior to the root of the lung (Figure 5-4).

- **Right vagus nerve.** Descends into the thoracic cavity anterior to the right subclavian artery and gives rise to the right recurrent laryngeal nerve, which hooks around the right subclavian artery and ascends back into the neck en route to intrinsic laryngeal muscles. The right vagus nerve continues inferiorly by coursing posterior to the superior vena cava and right primary bronchus and curves to the posterior surface of the esophagus, becoming the posterior vagal trunk. The right vagus nerve also contributes to the cardiac and pulmonary plexuses.

- **Left vagus nerve.** Enters the thorax between the left common carotid and subclavian arteries and gives rise to the left recurrent laryngeal nerve, which hooks around the aortic arch by the ligamentum arteriosum and ascends back into the neck en route to the intrinsic laryngeal muscles. The left vagus nerve continues on to the anterior surface of the esophagus, becoming the anterior vagal trunk. The left vagus nerve also contributes to the cardiac and pulmonary plexuses.

The **anterior and posterior vagal trunks** exchange fibers, creating an **esophageal plexus (web) of nerves**. The vagus nerve carries visceral sensory fibers whose sensory cell bodies are located in the inferior vagal ganglion. The visceral afferents from the vagus nerves transmit information to the brain about normal physiologic processes and visceral reflexes. They do not relay pain information.

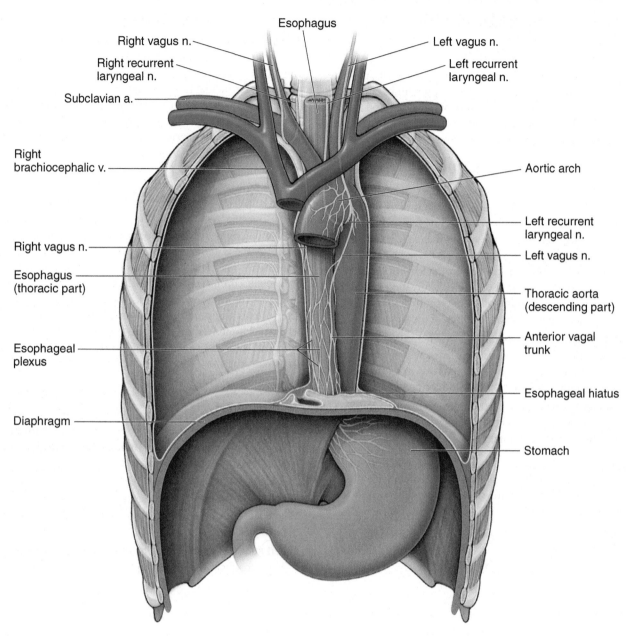

Figure 5-4: Esophagus.

SUPERIOR MEDIASTINUM

BIG PICTURE

The superior mediastinum is a thoroughfare for vessels, nerves, and lymphatics to pass between the neck, upper limbs, and thorax. The trachea, bronchial tree, pulmonary arteries, and aortic arch are important for oxygen transport and exchange from the external environment to internal structures.

TRACHEA AND BRONCHIAL TREE

The trachea begins at the cricoid cartilage, at the C6 vertebral level, and has 18 to 20 incomplete hyaline cartilaginous rings, which are open posteriorly (Figure 5-5). The cartilaginous rings prevent the trachea from collapsing while a person is exhaling. The trachea is anterior to the esophagus and bifurcates into the right and left primary (principal or mainstem) bronchi at the level of the sternal angle (T4–T5 vertebral level). The point of bifurcation, called the **carina**, is marked in the inside of the airway by a cartilaginous wedge that projects upward into the lumen of the airway.

- **Right primary bronchus.** Is shorter, wider, and more vertical than the left primary bronchus. The right primary bronchus courses inferior to the arch of the azygos vein and divides into **superior, middle, and inferior secondary (lobar) bronchi.**

- **Left primary bronchus.** Crosses anterior to the esophagus and divides into two **secondary (lobar) bronchi: the superior and inferior lobar bronchi**.

Each lobar bronchus further divides into segmental (tertiary) bronchi, which further divide. The smallest bronchi give rise to bronchioles, which terminate in alveolar sacs, where the exchange of gases takes place.

▽ Because of its wider, more vertical orientation the right bronchus usually has inhaled foreign objects fall into it from the trachea (e.g., peanut). ▼

PULMONARY ARTERIES

The pulmonary trunk and pulmonary arteries are located approximately at vertebral level T4, where the arteries are inferior to the aortic arch and azygos vein. The pulmonary arteries follow the bronchial tree throughout the lungs. The relation of the pulmonary arteries to bronchi at the root of the lungs can be remembered by the following mnemonic: _Right pulmonary artery is Anterior to the right primary bronchus, and the Left pulmonary artery is Superior to the left primary bronchus (RALS)._

AORTIC ARCH AND ASSOCIATED BRANCHES

The ascending aorta arises from the heart at the T4 vertebral level and ascends into the superior mediastinum over the pulmonary vessels and left primary bronchus to become the aortic arch. A fibrous connective tissue cord, called the **ligamentum arteriosus**, connects the deep surface of the aortic arch to bifurcation of the pulmonary trunk in the adult. The ligamentum arteriosus is a remnant of the ductus arteriosus, which, during fetal development, shunted blood from the pulmonary trunk to the aorta, similar to the patent foramen ovale. The aortic arch terminates at the T4 vertebral level to become the thoracic (descending) aorta.

The aortic arch has the following three branches (Figure 5.5):

- **Brachiocephalic artery.** The first branch of the aortic arch. The brachiocephalic artery courses superiorly to the right, where it bifurcates into the **right common carotid** and **right subclavian arteries**, supplying the right side of the head and neck and upper limb, respectively.

- **Left common carotid.** The second branch of the aortic arch. The left common carotid artery supplies the left side of the head and neck.

- **Left subclavian arteries.** The third branch of the aortic arch. The left subclavian artery supplies the left upper limb.

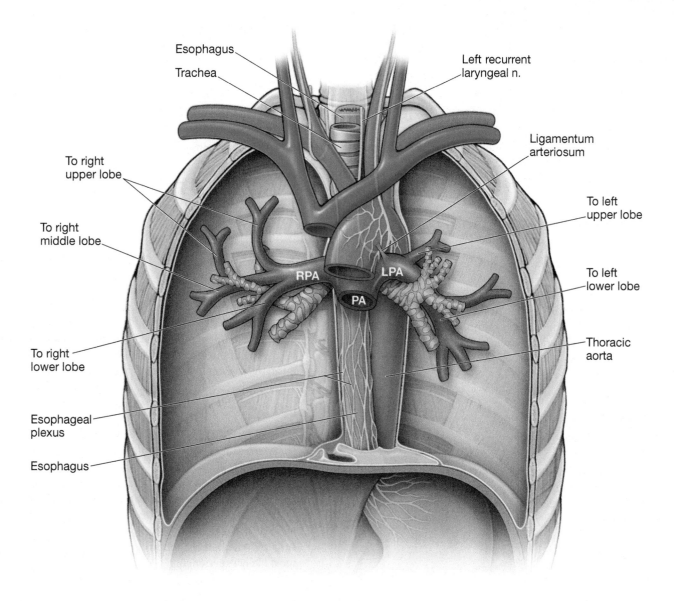

Figure 5-5: Structures of the superior mediastinum. RPA, right pulmonary artery; LPA, left pulmonary artery; PA, pulmonary artery.

STUDY QUESTIONS

Directions: Each of the numbered items or incomplete statements is followed by lettered options. Select the **one** lettered option that is **best** in each case.

1. During the autopsy of a trauma victim, the pathologist noted a tear at the junction of the superior vena cava and the right atrium. Which of the following structures would most likely have been damaged by the tear?

   A. Atrioventricular (AV) bundle

   B. AV node

   C. Left bundle branch

   D. Right bundle branch

   E. Sinuatrial (SA) node

2. A 62-year-old man is brought to the emergency department after experiencing a myocardial infarction. His heart rate is 40 beats/min. Further examination reveals an occlusion of the patient's right coronary artery. Which of the following structures is most likely affected by this blockage?

   A. AV node

   B. Bundle of His

   C. Mitral valve

   D. Tricuspid valve

3. A contrast study of the pulmonary vessels will most likely reveal several pulmonary veins entering the left atrium. How many pulmonary veins entering the left atrium will most likely be seen?

   A. Two

   B. Three

   C. Four

   D. Five

   E. Six

4. Which of the following structures typically arises from the musculophrenic arteries?

   A. Anterior intercostal arteries for the intercostal spaces 7 to 9

   B. Inferior phrenic artery

   C. Lumbar arteries

   D. Posterior intercostal arteries for intercostal spaces 3 to 11

   E. Subcostal artery

5. The opening of the coronary sinus is located in which of the following structures?

   A. Left atrium

   B. Left ventricle

   C. Right atrium

   D. Right ventricle

6. Which of the following vessels is responsible for transporting oxygenated blood from the lungs to the heart?

   A. Ascending aorta

   B. Cardiac veins

   C. Left coronary artery

   D. Pulmonary arteries

   E. Pulmonary veins

7. The azygos vein is located in which division of the mediastinum?

   A. Anterior mediastinum

   B. Middle mediastinum

   C. Posterior mediastinum

   D. Superior mediastinum

8. Ebstein's anomaly is a congenital heart defect where one or two of the tricuspid valve leaflets forms abnormally low because of misalignment. The heart becomes less efficient. What type of murmur would most likely be associated with this type of anomaly?

   A. Diastolic murmur with regurgitation

   B. Diastolic murmur with stenosis

   C. Systolic murmur with regurgitation

   D. Systolic murmur with stenosis

9. After surgery, a 62-year-old patient began experiencing complications. After examination, the physician determined that an important structure located immediately behind the ligamentum arteriosum was damaged during surgery. Which of the following symptoms was the patient most likely experiencing?

   A. Partially paralyzed diaphragm

   B. Heart arrhythmia

   C. Hoarseness of voice

   D. Jaundice

   E. Loss of cutaneous sensation along T4 dermatome

10. An intercostal artery is identified in a 44-year-old man who is undergoing thoracic surgery. This artery would most likely be located between which two structures?

    A. External and internal intercostal muscles

    B. Endothoracic fascia and parietal pleura

    C. Innermost intercostal muscles and endothoracic fascia

    D. Internal and innermost intercostal muscles

    E. Skin and external intercostal muscles

11. The ganglia associated with the sympathetic trunk typically contain which of the following cell bodies?

    A. Postganglionic parasympathetic cell bodies

    B. Postganglionic sympathetic cell bodies

    C. Preganglionic parasympathetic cell bodies

    D. Preganglionic sympathetic cell bodies

12. The greater, lesser, and least splanchnic nerves are examples of which of the following nerves?

A. Cervical splanchnic nerves

B. Lumbar splanchnic nerves

C. Pelvic splanchnic nerves

D. Sacral splanchnic nerves

E. Thoracic splanchnic nerves

13. Which of the following structures, along with the esophagus travels through the esophageal hiatus from the thoracic cavity into the abdominal cavity?

A. Abdominal aorta

B. Inferior vena cava

C. Lesser splanchnic nerves

D. Paravertebral ganglia

E. Prevertebral ganglia

F. Vagus nerves

14. In a healthy person, blood from the pulmonary trunk will flow next into which of the following structures?

A. Aortic arch

B. Left atrium

C. Left ventricle

D. Pulmonary arteries

E. Pulmonary veins

F. Right atrium

G. Right ventricle

15. A Doppler echocardiogram evaluates blood flow, speed, and direction of blood within the heart and also screens the four valves for any leakage. If a patient's heart function during diastole is being studied, which valves would the Doppler detect to be open?

A. Mitral and aortic valves

B. Mitral and pulmonary valves

C. Mitral and tricuspid valves

D. Pulmonary and aortic valves

E. Pulmonary and mitral valves

F. Pulmonary and tricuspid valves

16. During thoracocentesis, the needle is pushed in the intercostal space superior to the rib to prevent damage to the intercostal nerve, artery, and vein. Beginning with the external intercostal muscles and ending with the pleural space, which thoracic wall layers, from superficial to deep, does the needle penetrate?

A. Endothoracic fascia, internal intercostal muscles, costal parietal pleura, and pleural cavity

B. Internal intercostal muscles, innermost intercostal muscles, mediastinal parietal pleura, endothoracic fascia, and pleural cavity

C. Internal intercostal muscles, innermost intercostal muscles, costal parietal pleura, endothoracic fascia, and pleural cavity

D. Internal intercostal muscles, innermost intercostal muscles, endothoracic fascia, costal parietal pleura, and pleural cavity

E. Innermost intercostal muscles, internal intercostal muscles, endothoracic fascia, costal parietal pleura, and pleural cavity

17. A 19-year-old man is admitted to the emergency department after being stabbed in the chest with a pocketknife with a blade 5-cm long. The stab wound was in the left intercostal space just lateral to the sternal body. Which part of the heart is most likely injured?

A. Left atrium

B. Left ventricle

C. Right atrium

D. Right ventricle

ANSWERS

1—E: The SA node, or pacemaker, lies within the right atrial wall, where the right atrium is joined by the superior vena cava.

2—A: A myocardial infarction in the inferior wall involving the right coronary artery may affect the AV node, resulting in bradycardia.

3—C: At the wall of the left atrium four pulmonary veins deliver oxygenated blood into the left atrium.

4—A: The musculophrenic artery supplies the anterior intercostal arteries for intercostal spaces 7 to 9.

5—C: The coronary sinus collects venous blood from the coronary circulation and returns the blood to the right atrium. Therefore, the coronary sinus opens into the right atrium.

6—E: Oxygenated blood is transported from the lungs to the left atrium via the pulmonary veins.

7—C: The azygos system of veins, along with the thoracic duct, thoracic aorta, esophagus, vagus nerves, sympathetic trunk, and the greater and least splanchnic nerves are located within the posterior mediastinum.

8—C: The function of the tricuspid valve is to ensure unidirectional flow of blood from the right atrium to the right ventricle. In other words, when the right ventricle contracts (systole), blood flows into the pulmonary trunk and not back into the right atrium. Therefore, if the valve is malformed and does not function correctly (as in Ebstein's anomaly), blood will regurgitate back into the right atrium during systolic contraction of the right ventricle.

9—C: The left vagus nerve gives rise to the recurrent laryngeal nerve, located immediately behind the ligamentum arteriosum. The recurrent laryngeal nerve innervates laryngeal muscles that are associated with speaking. Therefore, if the recurrent laryngeal nerve is damaged, the patient will experience a raspy voice or hoarseness.

10—D: The intercostal arteries and veins course between the internal and innermost intercostal muscles.

11—B: Synapses occur with postganglionic sympathetic neurons within the paravertebral ganglia of the sympathetic trunk for sympathetics en route to blood vessels, sweat glands, and arrector pilae muscles in the associated dermatome. Preganglionic sympathetic cell bodies are located in the lateral horn gray matter of the T1–L2 spinal cord levels.

12—E: The greater, lesser, and least splanchnic nerves all arise from thoracic spinal nerves (T5–T9 form the greater splanchnic nerves; T10–T11 form the lesser splanchnic nerves; and T12 forms the least splanchnic nerves). The cervical sympathetic nerves course from the superior, middle, and inferior cervical ganglia and course to the pulmonary and aortic plexuses. The lumber and sacral splanchnics are located in the abdominal cavity and serve the abdominal viscera. The pelvic splanchnics originate from the S2–S4 ventral rami and transport preganglionic parasympathetic neurons.

13—F: In the thoracic cavity, the vagus nerves (CN X) form a plexus on the surface of the esophagus and then form the anterior and posterior vagal trunks. These vagal trunks course through the esophageal hiatus to enter the abdominal cavity.

14—D: Deoxygenated blood from the right ventricle is pumped into the pulmonary trunk, which bifurcates into the right and left pulmonary arteries before coursing to the lungs.

15—C: The segment of the cardiac cycle when the ventricles relax and the atria contract is known as diastole. When the atria contract, they pump blood through the AV valves (mitral and tricuspid) into the ventricles.

16—D: The layers of the lateral thoracic wall in the intercostal spaces that the needle would pass through during thoracocentesis are skin, superficial fascia, external intercostal muscle, internal intercostal muscle, innermost intercostal muscle, endothoracic fascia, parietal pleura, and pleural cavity.

17—D: The anterior surface of the heart is formed primarily by the right ventricle. Therefore, a stab wound such as the one that occurred in this patient, would injure the right ventricle of the heart.

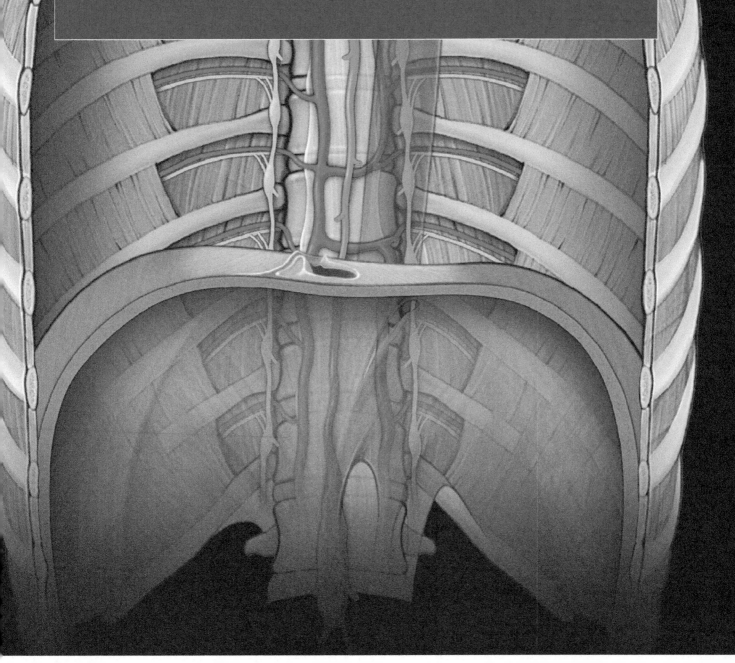

SECTION 3

ABDOMEN, PELVIS, AND PERINEUM

SECTION 3

ABDOMINAL PEDIS
AND PERINEUM

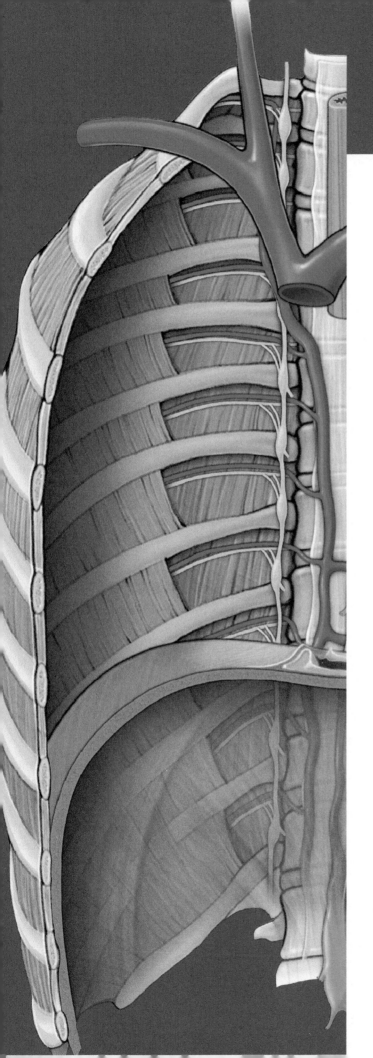

CHAPTER 6

OVERVIEW OF THE ABDOMEN, PELVIS, AND PERINEUM

OSTEOLOGIC OVERVIEW

BIG PICTURE

In the adult, the pelvis (os coxae) is formed by the fusion of three bones: ilium, ischium, and pubis (Figure 6-1A and B). The union of these three bones occurs at the acetabulum. The paired os coxae articulate posteriorly with the sacrum and anteriorly with the pubic symphysis.

PELVIC BONE

The following structures are formed within the fused os coxa (Figure 6-1A–C):

- **Acetabulum.** A cup-shaped socket into which the ball-shaped head of the femur fits firmly.
- **Obturator foramen.** Covered by a flat sheet of connective tissue called the **obturator membrane**. A small opening located at the top of the membrane provides a route through which the obturator nerve, artery, and vein course.
- **Greater sciatic notch.** Located between the posterior inferior iliac spine and the ischial spine. The sacrospinous ligament converts the notch into the **greater sciatic foramen**, where the piriformis muscle, sciatic nerve, and pudendal neurovascular structures course.
- **Lesser sciatic notch.** Located between the ischial spine and the ischial tuberosity. The sacrotuberous ligament converts the notch into the lesser sciatic foramen.
- **Pubic symphysis.** Fibrocartilage connecting the two pubic bones in the anterior midline of the pelvis.
- **Pelvic inlet.** The superior aperture of the pelvis. The pelvic inlet is oval shaped and bounded by the ala of the sacrum, arcuate line, pubic bone, and symphysis pubis. The pelvic inlet is traversed by structures in the abdominal and pelvic cavities.
- **Pelvic outlet.** The inferior aperture of the pelvis. The pelvic outlet is a diamond-shaped opening formed by the pubic symphysis and sacrotuberous ligaments. Terminal parts of the vagina and the urinary and gastrointestinal tracts traverse the pelvic outlet. The perineum is inferior to the pelvic outlet.

The pelvic bone is formed by the fusion of three bones: **ilium, ischium, and pubis**.

ILIUM

- **Iliac crest.** Thickened superior rim.
- **Iliac fossa.** Concave surface on the anteromedial surface.
- **Anterior superior iliac spine.** Anterior termination of the iliac crest. Serves as an attachment site for the sartorius and tensor fascia lata muscles.
- **Anterior inferior iliac spine.** Serves as an attachment site for the rectus femoris muscle.
- **Posterior superior iliac spine.** Posterior termination of the iliac crest.
- **Posterior inferior iliac spine.** Forms the posterior border of the ala of the sacrum.

ISCHIUM

- **Ischial tuberosity.** A large protuberance on the inferior aspect of the ischium for attachment of the hamstring muscles and for supporting the body when sitting.
- **Ischial spine.** A pointed projection that separates the greater and lesser sciatic notches.
- **Ischial ramus.** A bony projection that joins with the inferior pubic ramus to form the **ischiopubic ramus (conjoint ramus)**.

PUBIS

- **Pubic tubercle.** A rounded projection on the superior ramus of the pubis.
- **Superior pubic ramus.** A bony projection that forms a bridge from the acetabulum to the ischiopubic ramus, and thus the ischium. The crest on the superior aspect of the superior pubic ramus is the **pectineal line**, which serves as part of the border for the pelvic inlet and as an attachment site for muscles.
- **Inferior pubic ramus.** A bony projection that forms a bridge from the superior pubic ramus to the ischial ramus. The inferior pubic ramus serves as an attachment site for muscles of the lower limb.

SEX DIFFERENCES IN THE PELVIS

The female pelvis differs from the male pelvis because of its importance in childbirth.

- **Pelvic inlet.** A typical female pelvic inlet is usually more circular-shaped compared to the typically heart-shaped male pelvic inlet.
- **Pelvic outlet.** A typical female pelvic outlet is wider and has shorter and straighter ischial spines compared to the typical male pelvis. In addition, the ischial spines project more medially in males than in females.
- **Pubic arch.** The pubic arch is the angle between adjacent ischiopubic rami. A typical female pubic arch is usually larger (85 degrees) than the male pubic arch (60 degrees). The angle formed by the female pubic arch can be estimated by the angle between the thumb and the forefinger; in contrast, the male pubic arch is estimated by the angle between the index and the middle fingers (Figure 6-1C and D).

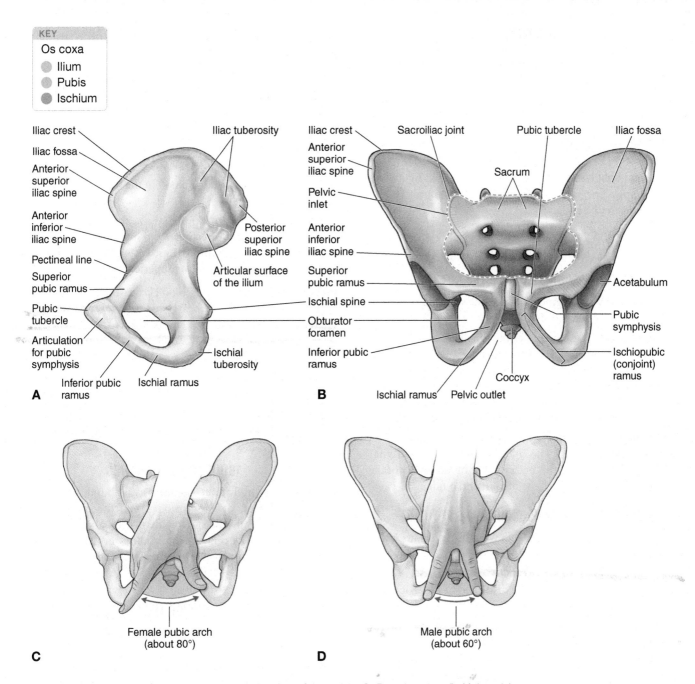

KEY
Os coxa
Ilium
Pubis
Ischium

A

Iliac crest
Iliac fossa
Anterior superior iliac spine
Anterior inferior iliac spine
Pectineal line
Superior pubic ramus
Pubic tubercle
Articulation for pubic symphysis
Inferior pubic ramus

Iliac tuberosity
Posterior superior iliac spine
Articular surface of the ilium
Ischial tuberosity
Ischial ramus

B

Iliac crest
Anterior superior iliac spine
Pelvic inlet
Anterior inferior iliac spine
Superior pubic ramus
Ischial spine
Obturator foramen
Inferior pubic ramus
Ischial ramus   Pelvic outlet

Sacroiliac joint   Pubic tubercle   Iliac fossa
Sacrum
Acetabulum
Pubic symphysis
Ischiopubic (conjoint) ramus
Coccyx

C

Female pubic arch (about 80°)

D

Male pubic arch (about 60°)

Figure 6-1: A. Medial view of the os coxa. **B.** Anterior view of the pelvis. **C.** Female pelvis. **D.** Male pelvis.

GUT TUBE

BIG PICTURE

The gut tube is subdivided into segments based on structural (lumen diameter) and functional (vascular supply) characteristics.

- **Structural subdivision.** The diameter of the lumen in the small intestines is smaller than that in the large intestines.
- **Functional subdivision.** The gut tube is classified by its vascular supply. The foregut, midgut, and hindgut all receive their own vascular supply.

Both methods of classifying regions of the gut tube are used in this text book, as well as in clinical medicine.

STRUCTURAL SUBDIVISION OF THE GUT TUBE

The gut tube is divided segmentally into the small intestine and the large intestine (Figure 6-2A).

- **Small intestine.** The small intestine functions mainly in the chemical breakdown of food and its subsequent absorption into the blood stream. The veins of the small intestine transport the absorbed nutrients to the liver for processing and ultimately to all other parts of the body. The small intestine consists of three parts: duodenum, jejunum, and ileum.
- **Large intestine (colon).** Receives its name because of its large luminal diameter. The colon absorbs water and vitamins and houses numerous bacteria. The blood supply of the colon overlaps with that of the midgut and hindgut.

FUNCTIONAL SUBDIVISION OF THE GUT TUBE

The gut tube is divided into the following three regions based on their primary arterial supply (Figure 6-2B):

- **Foregut.** Supplied primarily by the celiac trunk. This region of the gut tube extends from the distal end of the esophagus to the proximal half of the duodenum.
- **Midgut.** Supplied primarily by the superior mesenteric artery. This region of the gut tube extends from the distal half of the duodenum to the splenic flexure of the colon.
- **Hindgut.** Supplied primarily by the inferior mesenteric artery. This region of the gut tube extends from the splenic flexure of the colon to the rectum.

ABDOMINAL VENOUS DRAINAGE

Blood in the abdomen drains back to the heart via two routes: caval drainage and portal drainage (Figure 6-2C).

CAVAL DRAINAGE Venous blood that is returned to the heart from the anterior and posterior abdominal walls and the retroperitoneal organs via the superior or inferior vena cava.

- **Inferior epigastric veins.** Return blood to the heart via the inferior vena cava.
- **Intercostal veins.** Return blood to the heart via the superior vena cava.
- **Lumbar veins.** Return blood to the heart directly via the inferior vena cava or indirectly via the superior vena cava (lumbar veins may drain into the ascending lumbar veins to the azygos system of veins to the superior vena cava).

PORTAL DRAINAGE Venous blood from the gut tube and its derivatives returns to the heart via the **hepatic portal vein** to the liver. In other words, venous blood from the gut tube reaches the inferior vena cava after coursing through the liver.

- **Foregut.** Branches from the gastric and splenic veins to the portal vein.
- **Midgut.** Branches from the superior mesenteric vein to the portal vein.
- **Hindgut.** Branches from the inferior mesenteric vein to the portal vein.

ABDOMINAL LYMPHATICS

Lymphatics generally follow neurovascular bundles throughout the body. Clusters of lymph nodes, which are important in monitoring the immune system, are found along the course of the lymphatics. The central lymph nodes in the abdomen are named according to their associated artery. For example, the lymph nodes clustered at the origin of the celiac trunk are called celiac lymph nodes.

INNERVATION OF THE GUT TUBE

The regions of the gut tube receive the following autonomic innervation:

- **Foregut.** Sympathetics from the greater splanchnic nerves (T5–T9). Parasympathetics from the vagus nerves.
- **Midgut.** Sympathetics from the lesser splanchnic nerves (T10–T11). Parasympathetics from the vagus nerves.
- **Hindgut.** Sympathetics from the lumbar splanchnic nerves. Parasympathetics from the pelvic splanchnics (S2–S4 spinal cord levels).

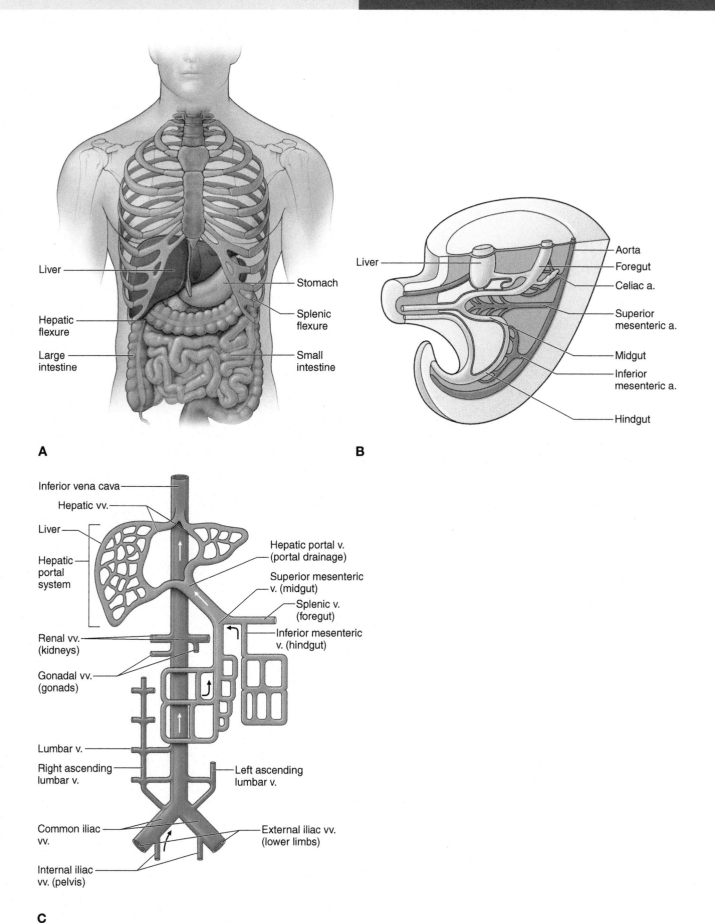

Figure 6-2: A. Gut tube in situ. **B.** Embryonic development of the gut tube, demonstrating the foregut, midgut, and hindgut. **C.** Caval (purple) and portal venous (turquoise) drainage of the abdomen, pelvis, and perineum.

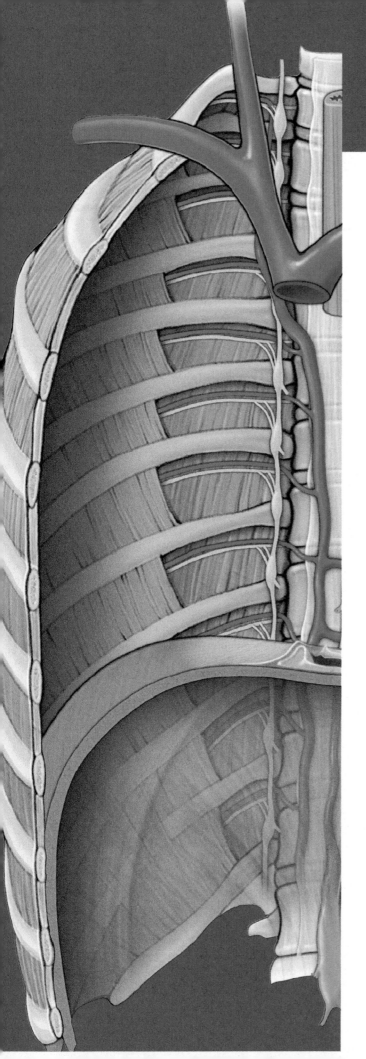

CHAPTER 7

ANTERIOR ABDOMINAL WALL

PARTITIONING OF THE ABDOMINAL REGION

BIG PICTURE

The abdomen typically is described topographically using two methods. The first method partitions the abdomen into four quadrants. The second method partitions the abdomen into nine regions.

4 QUADRANT PARTITIONS

The most direct method of partitioning the abdomen is through an imaginary **transverse (transumbilical) plane** that intersects with a **sagittal midline plane** through the **umbilicus** between the L3 and L4 vertebral levels (Figure 7-1A). The two intersecting planes divide the abdomen into four quadrants, described as **right and left upper and lower quadrants**. The four-quadrant system is straightforward when used to describe anatomic location. For example, the appendix is located in the lower right quadrant of the abdomen.

9 REGIONAL PARTITIONS

For a more precise description, the abdomen is partitioned into nine regions created by two imaginary vertical planes and two imaginary horizontal planes (Figure 7-1B).

- **Vertical planes.** Paired vertical planes correspond to the **midclavicular lines**, which descend to the midinguinal point.
- **Subcostal (upper horizontal) plane.** Transversely courses inferior to the **costal margin**, through the level of the **L3 vertebra**. The L3 vertebra serves as an important anatomic landmark in that it indicates the level of the inferior extent of the third part of the duodenum and the origin of the inferior mesenteric artery.
- **Transtubercular (lower horizontal) plane.** Transversely courses between the two **tubercles of the iliac crest**, through the level of the **L5 vertebra**.

▽ The **transpyloric plane** is an imaginary horizontal line through the L1 vertebra, a line that is important when performing radiographic imaging studies. The pylorus of the stomach, the first part of the duodenum, the fundus of the gallbladder, the neck of the pancreas, the origin of the superior mesenteric artery, the hepatic portal vein, and the splenic vein are all located along the level of the transpyloric plane. ▼

SURFACE LANDMARKS

The following structures are helpful anatomic surface landmarks on the anterior abdominal wall (Figure 7-1C):

- **Xiphoid process.** The xiphoid process is the inferior projection of the sternum. It marks the dermatome level of T7.
- **Umbilicus.** The umbilicus lies at the vertebral level between the L3 and L4 vertebrae. However, the skin around the umbilicus is supplied by the thoracic spinal nerve T10 (T10 dermatome) A helpful mnemonic is "T10 for belly but-ten".
- **Inguinal ligament.** The inferior border of the **external oblique muscle and aponeurosis** has an attachment between the anterior superior iliac spine and the pubic tubercle. This fascial attachment is known as the **inguinal ligament** and is evident superficially as a crease on the inferior extent of the anterior abdominal wall.

▽ **McBurney's point** is the name given to a point on the lower right quadrant of the abdomen, approximately one-third the distance along an imaginary line from the anterior superior iliac spine to the umbilicus. McBurney's point roughly corresponds to the skin overlying the most common attachment of the appendix to the cecum. ▼

SUPERFICIAL LAYERS OF THE ANTERIOR ABDOMINAL WALL

BIG PICTURE

Multiple layers of fascia and muscle form the anterior abdominal wall (Figure 7-1D). The layers, from superficial to deep, are skin, two layers of superficial fascia, three layers of muscles and their aponeuroses, transversalis fascia, extraperitoneal fat, and the parietal peritoneum.

SKIN

The skin receives its vascular supply via the intercostal and lumbar vessels and its segmental innervation via the ventral rami of the intercostal and lumbar spinal nerves.

SUPERFICIAL FASCIA

The superficial fascia of the anterior abdominal wall consists of two layers: an external layer of adipose tissue (**Camper's fascia**) and an internal layer of dense collagenous connective tissue (**Scarpa's fascia**). Camper's fascia is absent in the perineum. In contrast, Scarpa's fascia continues into the perineum, but the nomenclature is changed relative to the region in which it is located. For example, Scarpa's fascia becomes **Colles' fascia** when surrounding the roots of the penis and clitoris; it becomes **superficial penile (or clitoral) fascia** when it surrounds the shaft of the penis (or clitoris); and it becomes **dartos fascia** in the scrotum.

Embedded in the adipose tissue of Camper's fascia are the **superficial epigastric veins**, which drain the anterior abdominal wall. These cutaneous veins drain into the **femoral** and **paraumbilical veins**.

▽ A patient diagnosed with **cirrhosis (fibrotic scarring) of the liver** may present with **portal hypertension**. Blood pressure within the portal vein increases because of the inability of blood to filter through the diseased (cirrhotic) liver. In an attempt to return blood to the heart, small collateral (**paraumbilical veins**) veins expand at and around the **obliterated umbilical vein** to bypass the hepatic portal system. These paraumbilical veins form tributaries with the veins of the anterior abdominal wall, forming a **portal–caval anastomosis**, and drain into the **femoral or axillary veins**. In patients with chronic cirrhosis, the paraumbilical veins on the anterior abdominal wall may swell and distend as they radiate from the umbilicus and are termed **caput medusae** because the veins appear similar to the head of the Medusa from Greek mythology. ▼

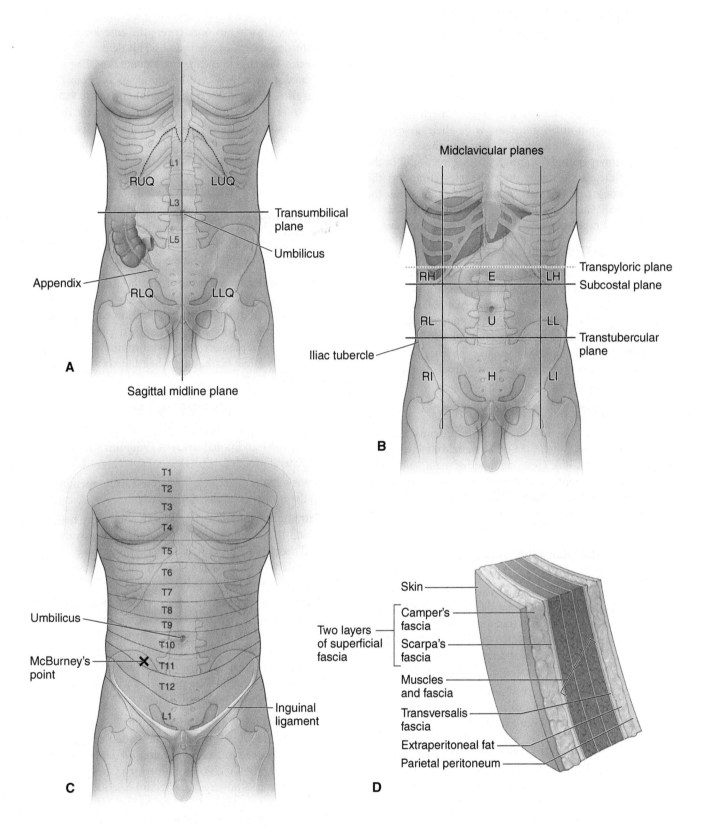

Figure 7-1: A. Quadrant partitioning: right upper quadrant (RUQ); left upper quadrant (LUQ); right lower quadrant (RLQ); and left lower quadrant (LLQ). **B**. Regional partitioning: right hypochondriac (RH); right lumbar (RL); right iliac (RI); epigastrium (E); umbilical (U); hypogastrium (H); left hypochondriac (LH); left lumbar (LL); and left iliac (LI). **C**. Surface anatomy and dermatome levels. **D**. Fascial layers of the anterior abdominal wall.

DEEP LAYERS OF THE ANTERIOR ABDOMINAL WALL

BIG PICTURE

Five paired anterior abdominal wall muscles are deep to the superficial fascia. The external oblique, internal oblique, and transverse abdominis muscles, with their associated aponeuroses, course anterolaterally, whereas the rectus abdominis and tiny pyramidalis muscles course vertically in the anterior midline. Collectively, these muscles compress the abdominal contents, protect vital organs, and flex and rotate the vertebral column. Each muscle receives segmental motor innervation from the lower intercostal and lumbar spinal nerves.

EXTERNAL OBLIQUE MUSCLE

The external oblique muscle is the most superficial of the anterolateral muscles and attaches to the outer surfaces of the lower ribs and iliac crest (Figure 7-2A). The external oblique muscle continues anteriorly as the **external oblique aponeurosis**, which courses anteriorly to the **rectus abdominis muscle** and inserts into the **linea alba**. The inferior border of the external oblique aponeurosis, between the anterior superior iliac spine and the pubic tubercle, is called the **inguinal ligament**.

INTERNAL OBLIQUE MUSCLE

The internal oblique muscle is the intermediate muscle of the anterolateral muscles (Figure 7-2A). The muscle attaches to the thoracolumbar fascia, iliac crest, inguinal ligament, and lower ribs. The internal oblique muscle continues anteriorly as the **internal oblique aponeurosis**, which splits around the rectus abdominis muscle to insert into the linea alba, with some inferior attachments to the pubic crest and pectineal line.

TRANSVERSE ABDOMINIS MUSCLE

The transverse abdominis muscle is the deepest of the anterolateral muscles (Figure 7-2A). This muscle attaches to the thoracolumbar fascia, iliac crest, inguinal ligament, and the costal cartilages of the lower ribs. The transverse abdominis muscle continues anteriorly as the **transverse abdominis aponeurosis**, which courses deep to the rectus abdominis muscle and inserts into the linea alba, the pubic crest, and the pectineal line.

- Intercostal and lumbar nerves, arteries, and veins course along the anterolateral abdominal wall between the internal oblique and transverse abdominis muscles.

RECTUS ABDOMINIS MUSCLE

The rectus abdominis muscle is the anteriorly positioned vertical strap muscle (Figure 7-2A). The rectus abdominis muscle attaches inferiorly to the pubic bone and symphysis and superiorly to the xiphoid process and lower costal cartilages. The external oblique, internal oblique, and transverse abdominis aponeuroses envelope the rectus abdominis muscle in a fascial sleeve known as the **rectus sheath**. The **linea alba** is a vertical mid line of fascia that separates the paired rectus abdominis muscles and is formed by the fusion of the three pairs of anterolateral aponeuroses (Figure 7-2B and C). The rectus sheath completely encloses the superior three-fourths of the rectus abdominis muscle but only covers the anterior surface of the inferior one-fourth of the muscle. This demarcation region in the rectus sheath is known as the **arcuate line**. The arcuate line is located midway between the umbilicus and pubic bone and serves also as the site where the **inferior epigastric vessels** enter to the rectus sheath. Inferior to the arcuate line, the rectus abdominis muscle is in direct contact with the transversalis fascia because the rectus sheath only covers the anterior surface of the rectus abdominis muscle (Figure 7-2B).

PYRAMIDALIS MUSCLE

The pyramidalis muscle is a small, pyramidally shaped muscle anterior to the rectus abdominis muscle (Figure 7-2A). The pyramidalis muscle attaches to the pubic bone and linea alba, and its only action is to tense the linea alba.

▽ The **Valsalva maneuver** is performed by forcibly exhaling against a closed airway (closed vocal folds). When the maneuver is completed, the contraction of abdominal wall muscles increases the intra-abdominal pressure. Increased intra-abdominal pressure assists with vomiting, urinating, defecating, and vaginal birth, and, when the vocal folds are open, with exhaling. ▼

DEEP FASCIAL LAYERS

In addition to the superficial layers of the abdominal fascia, there are three additional layers of abdominal fascia deep to the anterolateral muscle layers (Figure 7-2B,C).

- **Transversalis fascia.** A thin, aponeurotic membrane, deep to the transverse abdominis muscle.
- **Extraperitoneal fat.** A thin layer of connective tissue and fat lining the abdominal wall between the transversalis fascia and the parietal peritoneum. The extraperitoneal fat is more abundant in the posterior abdominal wall, especially around the kidneys and in the pelvic floor.
- **Parietal peritoneum.** Parietal peritoneum is a **serous membrane** lining the internal surface of the abdominal wall. The parietal peritoneum forms the mesentery that suspends the abdominal viscera and is continuous with the **visceral peritoneum**. The parietal peritoneum is innervated segmentally by the ventral rami of the spinal (somatic) nerves that course in the abdominal body wall.

▽ A **Caesarean section ("C-section")** is a surgical procedure for which incisions are made through a pregnant woman's abdomen to access the uterus for delivery of the infant. The most common incision location for a C-section is the lower uterine section (known as the "bikini-line incision"), where a transverse cut is made superior to the pubis and bladder, through all layers of the anterior abdominal wall. From superficial to deep, the layers cut through during a C-section are the skin, Camper's fascia, Scarpa's fascia, rectus sheath, pyramidalis muscle, rectus abdominis muscle, transversalis fascia, extraperitoneal fascia, and, finally, the parietal peritoneum. An alternative location is a midline incision through the linea alba, allowing a larger opening for delivery of the infant. ▼

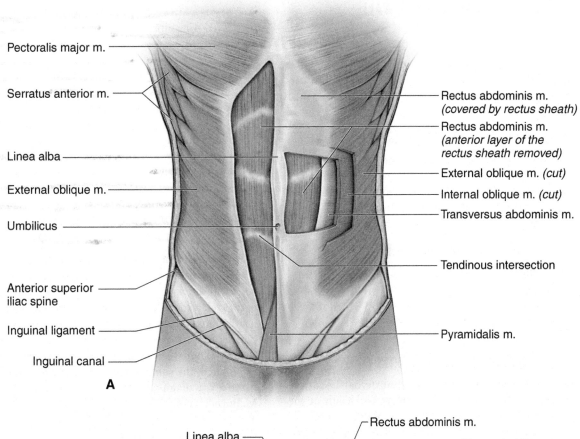

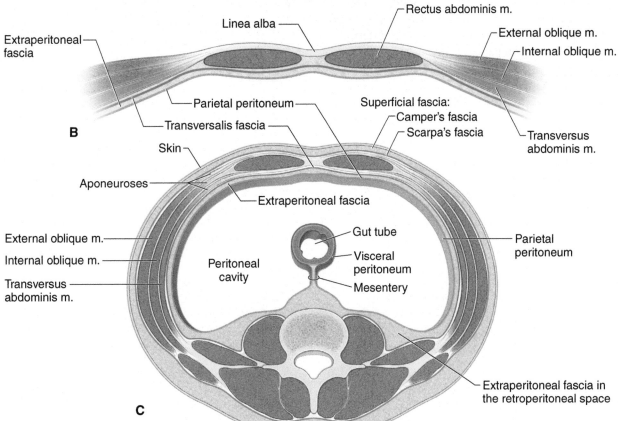

Figure 7-2: A. Step dissection of the anterior abdominal wall muscles. **B**. Horizontal section of the rectus sheath inferior to the arcuate line. **C**. Fascial and muscular layers of the abdomen in horizontal section superior to the arcuate line.

VASCULAR SUPPLY AND INNERVATION OF THE ANTERIOR ABDOMINAL WALL

BIG PICTURE

The neurovascular supply to the body wall courses between the second and third muscle layers. Intercostal, subcostal, lumbar, and epigastric arteries supply the skin of the anterolateral abdominal wall (Figure 7-3).

INTERCOSTAL AND LUMBAR ARTERIES

The inferior posterior intercostal arteries and the lumbar arteries from the descending aorta supply the lateral part of the anterior abdominal wall. These vessels course between the internal oblique and the transverse abdominis muscles and may anastomose with the inferior and superior epigastric arteries.

SUPERIOR EPIGASTRIC ARTERY

The superior epigastric artery arises from the internal thoracic artery, enters the rectus sheath, and descends on the deep surface of the rectus abdominis muscle. The superior epigastric artery anastomoses with the inferior epigastric artery within the rectus abdominis muscle.

INFERIOR EPIGASTRIC ARTERY

The inferior epigastric artery arises from the external iliac artery above the inguinal ligament, enters the rectus sheath at the arcuate line, and ascends between the rectus abdominis muscle and the posterior layer of the rectus sheath. The inferior epigastric artery anastomoses with the superior epigastric artery, providing collateral circulation between the subclavian and external iliac arteries.

INNERVATION OF THE ANTERIOR ABDOMINAL WALL

The nerves of the anterior abdominal wall are the ventral rami of the T6–L1 spinal nerves. These nerves course downward and anteriorly between the internal oblique and the transverse abdominis muscles. They segmentally supply cutaneous innervation to the skin and parietal peritoneum and are the motor supply to the anterolateral abdominal wall muscles. The lower intercostal nerves and the subcostal nerve pierce the deep layer of the rectus sheath and course through to the skin to become the anterior cutaneous nerves of the abdomen. The first lumbar nerve bifurcates into the **iliohypogastric** and **ilioinguinal nerves**, which do not enter the rectus sheath. Instead, the iliohypogastric nerve pierces the external oblique aponeurosis superior to the superficial inguinal ring, whereas the ilioinguinal nerve passes through the inguinal canal to emerge through the superficial inguinal ring.

▽　When the **appendix becomes inflamed**, the visceral sensory fibers are stimulated. These fibers enter the spinal cord with the sympathetic fibers at spinal cord level T10. The pain is referred to the dermatome of T10, which is in the umbilical region. The visceral pain is diffuse, not focal; each time a peristaltic wave passes through the ileocecal region, the pain recurs. When the parietal peritoneum eventually becomes inflamed, the somatic pain is sharp and focal. This is recognized by a diagnostic test for acute appendicitis: push on McBurney's point. The knee and hip on the inflamed side flex abruptly. Although appendicitis is common, other disorders of the bowel and pelvis may present with similar symptoms. ▽

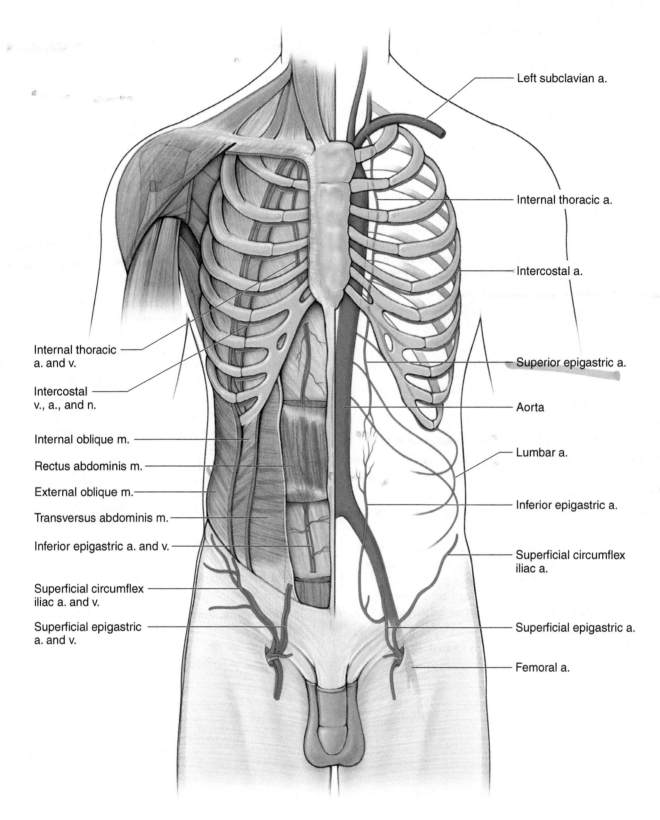

Figure 7-3: Neurovascular structures of the anterior abdominal wall. The left side of the figure shows a step dissection detailing the location of the neurovascular structures. The right side of the figure shows a schematic of arterial supply.

INGUINAL CANAL

BIG PICTURE

The inguinal canal is an oblique passage through the inferior region of the anterior abdominal wall. The inguinal canal is clinically more important in males because it is the passageway for structures to course to and from the testis to the abdomen. The presence of the spermatic cord in the male predisposes males to indirect inguinal harnias. In females, the inguinal canal is less important in that the only structure that traverses through the inguinal canal is the round ligament of the uterus. The thinness of the round ligament of the uterus, compared to the spermatic cord in males, is associated with lower frequency of indirect inguinal hernia in females. The inguinal canal is approximately 5-cm long and extends from the deep inguinal ring downward and medially to the superficial inguinal ring. The inguinal canal lies parallel to and immediately superior to the inguinal ligament. In males, the spermatic cord and scrotum consist of the same layers of muscle and fascia as does the anterior abdominal wall.

INGUINAL CANAL STRUCTURE

The inguinal canal is much like a rectangular tube in that it consists of four walls with openings at both ends, described as follows (Figure 7-4):

- **Anterior wall.** Formed by the external oblique aponeurosis.
- **Posterior wall.** Formed by the conjoint tendon of the internal oblique and the transverse abdominis muscles and the transversalis fascia.
- **Roof.** Formed by arching fibers of the internal oblique and the transverse abdominis muscles.
- **Floor.** The medial half of the inguinal ligament forms the inferior wall of the inguinal canal. This rolled-under, free margin of the external oblique aponeurosis forms a gutter or trough on which the contents of the inguinal canal are positioned. The **lacunar ligament** reinforces most of the medial part of the floor.
- **Deep inguinal ring.** Formed by an opening in the transversalis fascia. The deep inguinal ring is located superior to the inguinal ligament, lateral to the inferior epigastric vessels, and halfway between the pubic bone and the anterior superior iliac spine.
- **Superficial inguinal ring.** Formed by an opening in the external oblique aponeurosis superior and medial to the pubic tubercle.

The contents of the inguinal canal include the **genital branch of the genitofemoral nerve (L1–L2)**, the **spermatic cord** in males, and the **round ligament of the uterus** in females. Additionally, in both males and females, the **ilioinguinal nerve (L1)** passes through part of the canal. The ilioinguinal nerve courses between the internal oblique and the transverse abdominis muscles and enters in the middle of the inguinal canal in both males and females. The ilioinguinal nerve exits the inguinal canal through the superficial inguinal ring with other contents that course through the inguinal canal.

▽ When the anterior abdominal wall muscles contract, intra-abdominal pressure increases (e.g., forceful exhalation; coughing). This increase in pressure pushes the diaphragm up, forcing air out of the lungs. The inguinal canal, with its openings in the anterior abdominal wall, serves as a potential weakness when intra-abdominal pressure increases. When the posterior wall the inguinal canal weakens (e.g., in the elderly), an increase in intra-abdominal pressure may force the small intestine into the inguinal canal, resulting in a **hernia**. To check for the presence of a hernia in males, the healthcare provider will insert a finger up into the scrotum to the superficial inguinal ring. The patient is instructed to increase intra-abdominal pressure by coughing. If the physician feels contact on the fingertip, a hernia is most likely present.

Hernias are classified as **direct** or **indirect**, with the **inferior epigastric vessels** serving as the differentiating landmark.

- **Direct hernia.** Results when the small intestine protrudes into the canal *medial* to the inferior epigastric vessels.
- **Indirect hernia.** Results when the small intestine protrudes into the canal *lateral* to the inferior epigastric vessels into the inguinal canal. ▽

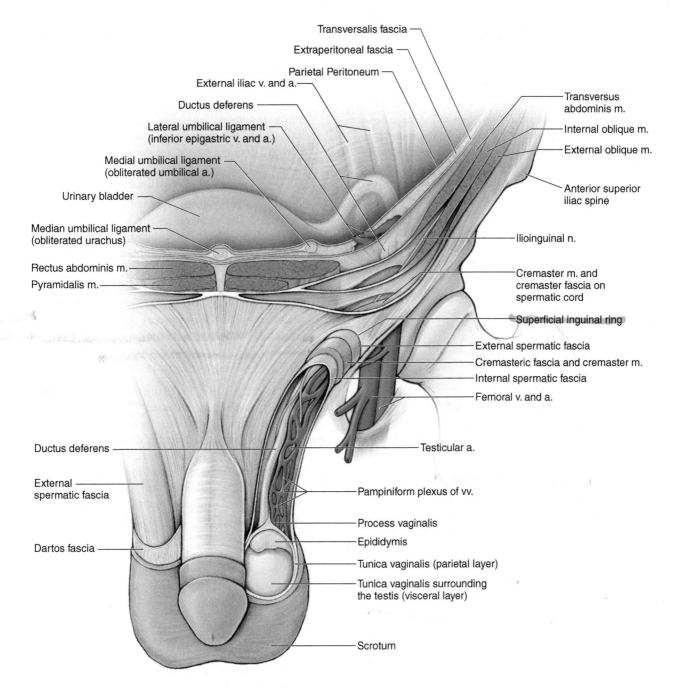

Figure 7-4: Schematic of the inguinal canal, spermatic cord, and scrotum.

SCROTUM AND SPERMATIC CORD

BIG PICTURE

The primary sex organs of the males are the **testes** because they produce **sperm**. The testes are housed in a layered sac of muscle and fascia called the **scrotum**, which is connected to the anterior abdominal wall via the **spermatic cord**. Unlike the spinal cord, which is protected by the vertebral column, bones do not protect the testes. A temperature of 34°C is required for the testes to produce sperm, a temperature that is 3°C lower than core body temperature; therefore, the testes must be housed outside the body. During embryonic development, the testes begin development in the region of the kidneys and descend throughout development until they traverse the inguinal canal, protruding through the inferior portion of the anterior abdominal wall. This developmental migration of the testes through the anterior abdominal wall is the basis of formation of the spermatic cord and scrotum from the muscle and fascial layers of the anterior abdominal wall. This developmental migration also is the basis of congenital (indirect) inguinal hernia in newborn male infants.

SCROTUM

The layers of the scrotal sac are as follows (it may be helpful to review the muscle and fascial layers of the anterior abdominal wall) (Figure 7-4):

- **Skin.**
- **Dartos fascia.** A continuation of Scarpa's fascia. Consists of a thin layer of loose connective tissue with some smooth muscle (**dartos muscle**). The smooth muscle contracts and thus wrinkles the skin of the scrotum, enhancing the radiation of heat when the temperature of the testes increases excessively (e.g., a hot bath).
- **External spermatic fascia.** Extension of the external oblique aponeurosis.
- **Cremasteric muscle.** Extension of the internal oblique muscle. If the temperature of the testes drops, contraction of the cremasteric muscle moves the testes closer to the body wall, thus helping to maintain appropriate temperature.

▽ The **sensory branch of the genitofemoral nerve** provides cutaneous innervation of the skin over the medial aspect of the thigh, whereas the **motor division** innervates the **cremasteric muscle**. Light touch of the medial region of the thigh elicits a motor reflex, causing contraction of the cremaster muscle to pull the testis on the same side of the body closer to the body. This is called the **cremaster reflex.** ▽

- **Internal spermatic fascia.** Extension of the transversalis fascia (the transverse abdominis aponeurosis has no contributions to the scrotal sac).
- **Tunica vaginalis.** Extensions of the parietal peritoneum. However, the tunica vaginalis is only found surrounding the testes and does not have a counterpart in the spermatic cord.

SPERMATIC CORD

Similar to the scrotal sac, the spermatic cord contains the same muscular and fascial layers as the anterior abdominal wall because the testis descends through the abdominal wall to the scrotum during embryonic development. The spermatic cord contains the following structures:

- **Ductus deferens.** During ejaculation, functions to transport sperm from the testes, through the spermatic cord and inguinal canal to the ejaculatory duct in the prostate. Much of the ductus deferens is composed of smooth muscle and, as a result, feels rigid and hard to the touch. Therefore, the ductus deferens is easy to palpate in the spermatic cord.

▽ The ductus deferens is also called the **vas deferens**, a misnomer because the word "vas" means vessel. The ductus deferens is not a vessel. However, this term persists frequently so that when each ductus deferens is cut as a form of male birth control, the procedure is called a "**vasectomy**," which means "cutting of the vas." A vasectomy is a simple procedure in that it only requires an incision through the skin and layers of each spermatic cord to cut the ductus deferens. ▽

- **Testicular (gonadal) artery.** The paired gonadal arteries are bilaterally symmetrical. Each gonadal artery originates on the abdominal aorta, inferior to the renal arteries. Each artery traverses the deep inguinal ring and courses through the inguinal canal to provide the vascular supply to the testes and ductus deferens on the corresponding side of the body.
- **Testicular (gonadal) vein.** The paired gonadal veins are not bilaterally symmetrical. Each vein drains blood from the testis and courses from the scrotal sac, through the spermatic cord, and traverses the superficial inguinal ring through the inguinal canal before entering the abdominal cavity. The right gonadal vein drains into the inferior vena cava, whereas the left gonadal vein drains into the left renal vein.
- **Autonomic neurons.** Sympathetic and visceral sensory neurons.

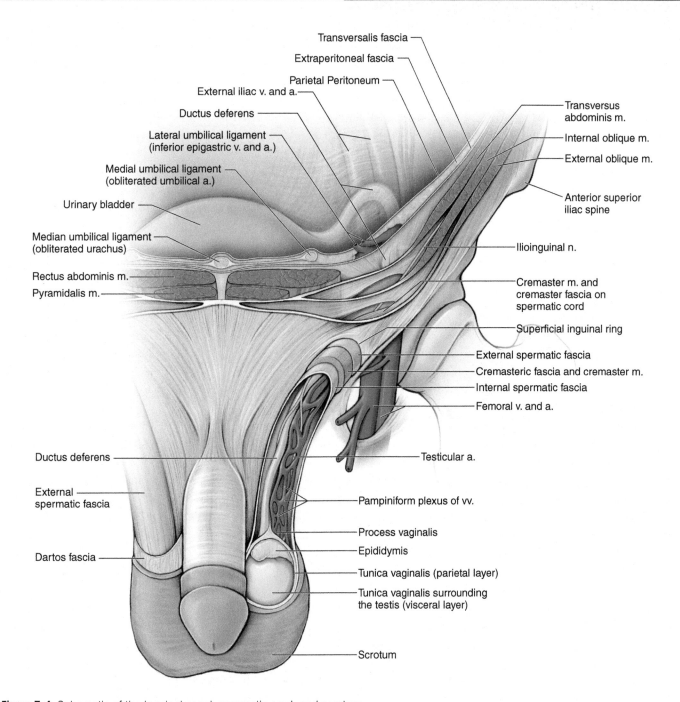

Figure 7-4: Schematic of the inguinal canal, spermatic cord, and scrotum.

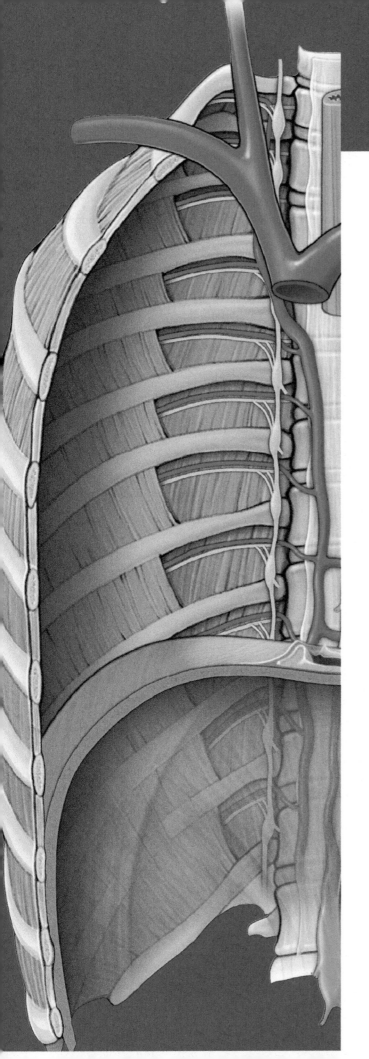

CHAPTER 8

SEROUS MEMBRANES OF THE ABDOMINAL CAVITY

PERITONEUM

BIG PICTURE

The abdominopelvic cavity is lined with a serous membrane called the **peritoneum**. This membrane expands from the internal surface of the abdominal wall to completely or partially surround organs of the abdominopelvic cavities.

PERITONEAL CAVITY

The peritoneum is a serous membrane that consists of two layers: parietal peritoneum and visceral peritoneum (Figure 8-1A–C). The **parietal peritoneum** lines the internal walls of the abdominal cavity, forming a closed sac known as the **peritoneal cavity**. The peritoneal cavity is completely closed in males. In females, the peritoneal cavity has two openings where the uterine tubes, uterus, and vagina provide a passage to the outside. The parietal peritoneum reflects off of the posterior abdominal wall, forming a fused, double layer of peritoneum surrounding the blood vessels, nerves, and lymphatics to abdominal organs. This double layer of peritoneum, known as the **mesentery**, suspends the jejunum and ileum from the posterior abdominal wall. The peritoneum that surrounds the gut tube is called the **visceral peritoneum**. The peritoneal membranes produce a serous fluid that lubricates the peritoneal surfaces, enabling the intraperitoneal organs to slide across one another with minimal friction.

OMENTUM

The omentum refers to modified mesenteries associated with the stomach and liver (Figure 8-1A).

- **Greater omentum.** An apron-like fold of mesentery that attaches between the transverse colon to the greater curvature of the stomach.
- **Lesser omentum.** Mesentery that attaches between the liver, stomach, and proximal portion of the duodenum. As a result, the lesser omentum is also referred to as the **hepatogastric ligament** and **hepatoduodenal ligament**. The lesser omentum forms a sac known as the **omental bursa**, which forms a subdivision of the peritoneal cavity known as the **lesser sac**. The **greater sac** is the remaining part of the peritoneal cavity. The greater and lesser sacs communicate with each other through the **epiploic foramen (of Winslow)**.

INNERVATION AND VASCULAR SUPPLY OF THE PERITONEUM

The neurovascular and lymphatic supply of the peritoneum course to and from the posterior abdominal wall and gut tube through the two-layered mesentery (Figure 8-1B). The vascular supply to the parietal peritoneum is through the same vessels that supply the abdominal body wall, mainly the **intercostal, lumbar, and epigastric vessels**. The vascular supply to the visceral peritoneum is through vessels that arise from the abdominal aorta. These vessels also supply the organs in the abdominal cavity.

The nerves supplying the parietal peritoneum are the same that supply the body wall (**intercostal nerves**). The parietal peritoneum receives somatic sensory innervation. **Somatic pain** is sharp, focused, and specific. The visceral peritoneum and abdominal organs receive sensory innervation by the visceral afferents that accompany the autonomic nerves (sympathetic and parasympathetic). **Visceral pain** is dull, diffuse, and nonspecific.

▽ The parietal and visceral peritoneum are innervated by different modalities of sensory neurons; that is, parietal peritoneum via somatic innervation and visceral peritoneum via visceral innervation. Therefore, pain experienced in the parietal peritoneum is sharp, focused, and specific. In contrast, pain experienced in the visceral peritoneum is dull, diffuse, and nonspecific. ▼

ORGANIZATION OF THE ABDOMINAL VISCERA

Abdominal viscera are classified as either intraperitoneal or retroperitoneal (Figure 8-1A and C).

- **Intraperitoneal.** Viscera that are suspended from the abdominal wall by mesenteries. Intraperitoneal organs are surrounded by visceral peritoneum (e.g., stomach).
- **Retroperitoneal.** Viscera that are not suspended from the abdominal wall by mesenteries. Retroperitoneal organs are covered on one of their surfaces by parietal peritoneum (e.g., kidney).

▽ **Surgical procedures** involving organs located in the **retroperitoneal space** can be accessed through the body wall, superficial to the parietal peritoneum. For example, to access organs in the retroperitoneal space, such as the kidney, a lateral incision may be made through the muscles of the body wall, leaving the parietal peritoneum intact. This approach reduces the risk of infection and peritonitis because the peritoneal cavity is not entered. ▼

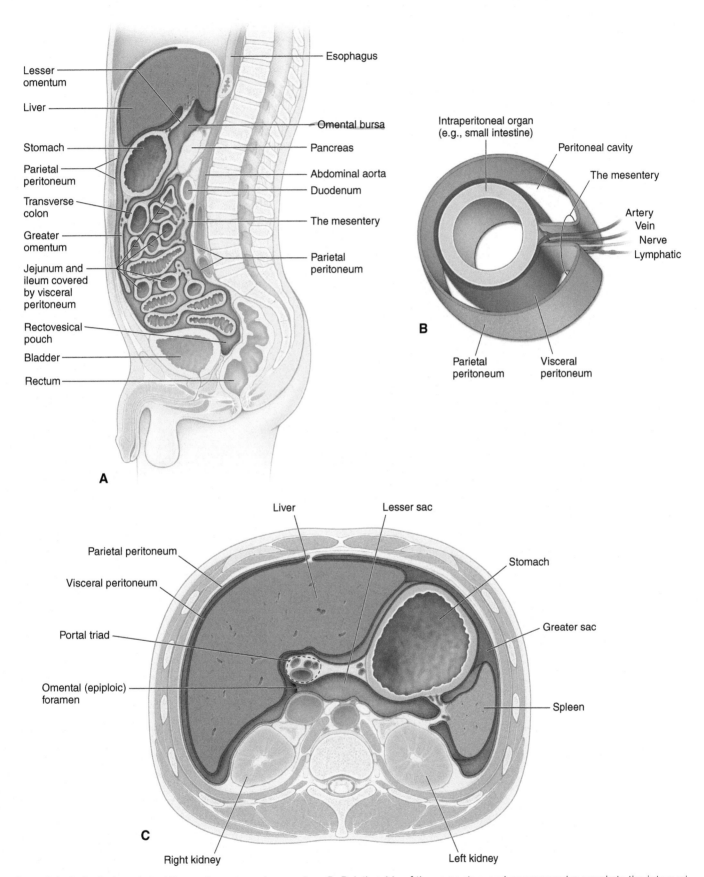

Figure 8-1: A. Sagittal section of the peritoneum and mesentery. **B**. Relationship of the mesentery and neurovascular supply to the intraperitoneal organs. **C**. Axial (cross-section) of the peritoneum and mesentery.

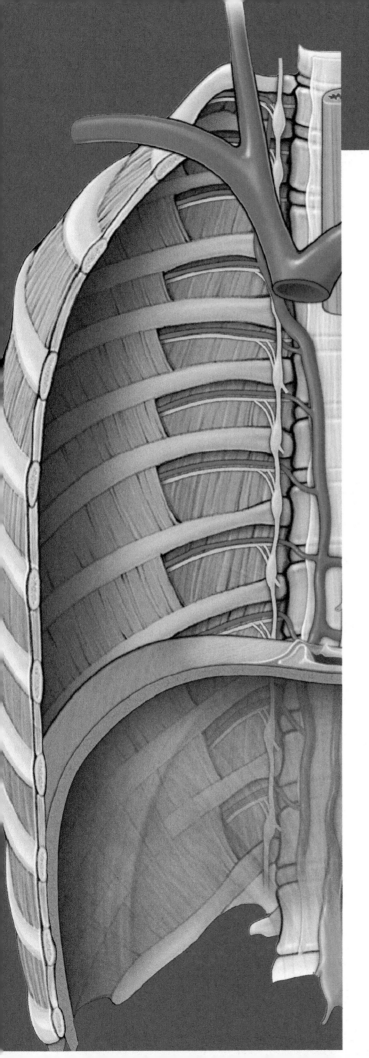

CHAPTER 9

FOREGUT

BIG PICTURE

The foregut consists of the distal end of the esophagus, the stomach, and a portion of the duodenum. In addition, the pancreas, liver, and gallbladder form embryologically from the foregut and thus also are included in this discussion. The celiac trunk is the principal (but not exclusive) artery supplying the foregut. The celiac trunk arises from the abdominal aorta.

ESOPHAGUS

The distal end of the **esophagus** enters the abdominal cavity in the upper left quadrant by traversing the diaphragm at the **T10 vertebral level**. The esophagus immediately transitions into the stomach, with the **cardiac sphincter** serving as the transition boundary. Coursing parallel to the esophagus are the **anterior and posterior vagal trunks**. The vagal trunks consist of visceral motor and sensory **parasympathetic fibers** from the **left and right vagus nerves**, respectively.

▼ A function of the stomach is the production of hydrochloric acid. If the cardiac sphincter fails to contain the acidic chyme produced by the stomach, the acid moves into the esophagus, irritating its mucosal lining and causing **gastroesophageal reflux disease (GERD)**. The irritation presents as an uncomfortable, perhaps burning sensation in the region of the esophagus, deep to the heart. As a result, this condition is also referred to as "**heart burn**." ▼

STOMACH

The stomach is a dilated, J-shaped portion of the foregut, juxtaposed between the esophagus and the duodenum (Figure 9-1A and B). The stomach is located in the upper left quadrant of the abdomen, with the spleen, pancreas, and aorta located deep to the stomach body.

Gastric secretions are churned in the stomach, with food, into a semifluid mixture (**chyme**) that is eventually transported from the stomach into the duodenum.

The stomach is partitioned into the following four regions:

- **Cardia.** Surrounds the gastroesophageal opening.
- **Fundus.** Dome-shaped region superior to the cardia.
- **Body.** Largest region of the stomach that consists of a **lesser curvature** and a **greater curvature**, where the **lesser omentum** and **greater omentum** attach, respectively.
- **Pylorus.** Distal end of the stomach containing the **pyloric sphincter**, which is located in the transpyloric plane at the L1 vertebral level.

▼ The distal end of the esophagus and gastric fundus can herniate through the esophageal hiatus of the diaphragm into the thoracic cavity. This is known as a **hiatal hernia** and results from conditions such as strain on the diaphragm due to childbirth or to congenital defects in the diaphragm. ▼

When food enters the stomach, it begins to expand and stretch, resulting in a **vagovagal reflex**. Visceral sensory neurons from the **vagus nerve [cranial nerve (CN) X]** relay the stretching of the stomach to the brainstem. In response, the brainstem relays impulses via the vagus nerve, inhibiting the tone of the muscularis externa (smooth muscle of the stomach). In this way, the wall of the stomach progressively expands to accommodate greater quantities of food. The stomach can hold up to 1.5 L of food before pressure within the stomach lumen increases.

DUODENUM

The duodenum is approximately 25-cm long and curves around the pancreatic head. It is divided into the following four parts (Figure 9-1A and B):

- **Part one (superior).** The stomach is an intraperitoneal organ as is the first part of the duodenum. The first part of the duodenum contains the duodenal cap, a dilation of the proximal duodenum that is easily identifiable on radiographs. In contrast, the second part of the duodenum courses deep to the parietal peritoneum and thus is retroperitoneal, as are the third and fourth parts of the duodenum. However, the distal portion of the fourth part is transitional, from retroperitoneal to intraperitoneal in the region of the **duodenojejunal junction**.

- **Part two (descending).** Courses deep to the transverse colon and anterior to the right kidney. The **common bile duct** enters its posterior wall. Within the duodenal wall, the common bile duct receives the **main pancreatic duct (of Wirsung)**. Immediately after the junction, there is an enlargement called the **major duodenal papilla (ampulla of Vater)**. The papilla is surrounded by smooth muscle called the **sphincter of Oddi**. An **accessory pancreatic duct (of Santorini)** may enter the duodenum proximal to the main pancreatic duct.

- **Part three (horizontal).** Turns left and courses horizontally across the inferior vena cava, the aorta, and the vertebral column. In addition, the **superior mesenteric artery and vein** course anteriorly to the third part of the duodenum.

- **Part four (ascending).** Ascends anterior to the aorta at the L2 vertebral level.

The duodenum receives its blood supply from branches of both the **celiac trunk** (superior pancreaticoduodenal arteries) and the **superior mesenteric trunk** (inferior pancreaticoduodenal arteries).

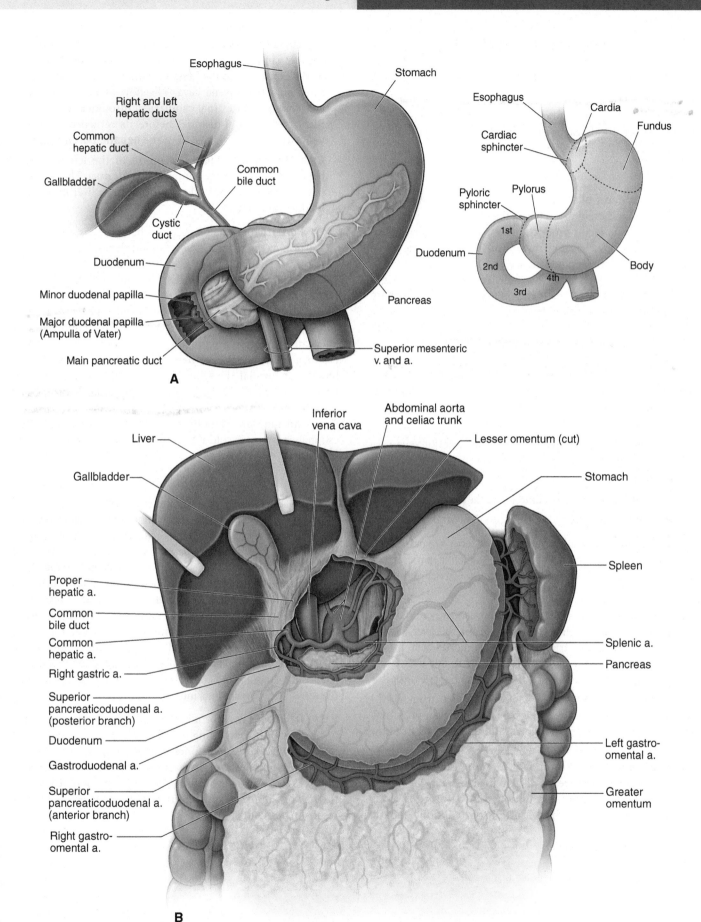

Figure 9-1: A. Parts of the stomach and duodenum. **B**. Anterior view of the foregut; the lesser omentum is partially removed.

LIVER AND GALLBLADDER

BIG PICTURE

In addition to its numerous metabolic activities, the liver secretes bile. Bile is transported to the gallbladder, where it is stored. When food reaches the duodenum, the gallbladder releases bile, which emulsifies fat in the duodenum.

LIVER

The liver produces **bile**, which emulsifies fat. The liver is also involved in cholesterol metabolism, the urea cycle, protein production, clotting factor production, detoxification, phagocytosis via the Kupffer cells lining the sinusoids, and receiving blood from the portal vein and hepatic artery.

The liver is attached to the inferior surface of the right dome of the **diaphragm** via the **coronary ligaments**. The **bare area of the liver** is a region devoid of peritoneum between the coronary ligaments and, therefore, lies in direct contact with the diaphragm. The **falciform ligament** is a peritoneal structure that courses between the left and right lobes of the liver and the anterior abdominal wall. The **ligamentum teres** is within the falciform ligament and is the embryonic remnant of the ductus venosus of the umbilical cord. The four lobes of the liver are as follows (Figure 9-2A and C):

- **Right lobe.** Positioned to the right of the inferior vena cava and gallbladder.
- **Left lobe.** Positioned to the left of the ligamentum teres.
- **Quadrate lobe.** Positioned anterior to the portal triad.
- **Caudate lobe.** Positioned posterior to the portal triad.

Functionally, the quadrate and caudate lobes are part of the left lobe because they are supplied by the left hepatic artery, drained by the left branch of the portal vein, and deliver bile via the left bile duct.

PORTAL TRIAD The portal triad lies between the caudate and quadrate lobes and is the structural unit of the liver (Figure 9-2B and C). The portal triad consists of the **portal vein, proper hepatic artery**, and the **common hepatic duct**. The portal vein is deep to the hepatic artery and the common hepatic duct.

- **Proper hepatic artery.** Branches from the celiac trunk via the common hepatic artery. The hepatic artery supplies oxygenated blood to the liver. The **cystic artery** arises from the hepatic artery to supply the gallbladder.
- **Portal vein.** Formed through the union of the splenic and superior mesenteric veins, deep to the pancreas. The portal vein collects nutrient-rich venous blood from the small and large intestines, where it is transported to the **hepatic**

sinusoids of the liver for filtration and detoxification. The hepatic sinusoids empty into the **common central vein**, which empties into the hepatic veins and ultimately drains into the inferior vena cava. The flow of blood from one capillary bed (intestinal capillaries) through a second capillary bed (liver sinusoids) before its return by systemic veins to the heart is defined as the **hepatic portal system**.

- **Common hepatic duct.** The union of the left and right hepatic ducts forms the common hepatic duct. The common hepatic duct transmits bile produced in the liver to the gallbladder for storage.

▽ **Portal hypertension** results when there is an obstruction to the regular flow of blood through the sinusoids of the liver. There are many possible causes, including cirrhosis of the hepatocytes (liver cells). Signs of portal hypertension include hemorrhoids and gastroesophageal bleeding, which result from the obstruction of the portal venous blood flow through the liver and the increased flow of blood through alternate routes to reach the inferior vena cava (e.g., rectal and esophageal veins). When these alternate paths receive more blood than normal, the veins dilate, distend, and become more prone to hemorrhage. For example, **esophageal varices** are distended esophageal veins, resulting from portal hypertension, and may precipitate life-threatening bleeding in the esophagus if hot or cold fluids are ingested or violent coughing occurs. ▽

GALLBLADDER

The gallbladder lies on the visceral surface of the liver, to the right of the quadrate lobe, and stores and concentrates **bile** secreted by the liver (Figure 9-2A–C). Bile is released into the duodenum when the gallbladder is stimulated after eating a fatty meal. Bile enters the **cystic duct**, which joins the **common hepatic duct**, becoming the **common bile duct**. The common bile duct courses within the **hepatoduodenal ligament** of the **lesser omentum**, deep to the first part of the duodenum, where it joins the **main pancreatic duct**. Together, the common bile duct and the main pancreatic duct enter the second part of the duodenum at the **hepatopancreatic ampulla (of Vater)**. The **sphincter of Oddi** surrounds the ampulla and controls the flow of bile and pancreatic digestive enzyme secretions into the duodenum.

▽ **Gallstones** may form in the gallbladder and obstruct the flow of bile, resulting in inflammation and enlargement of the gallbladder. These stones may be composed of bilirubin metabolites, cholesterol, or various calcium salts. They frequently obstruct the gallbladder, causing retention of bile and the risk of rupture into the peritoneal cavity, which ultimately results in **peritonitis**. ▽

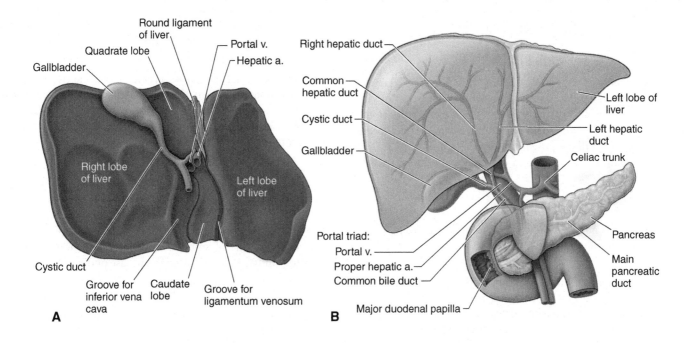

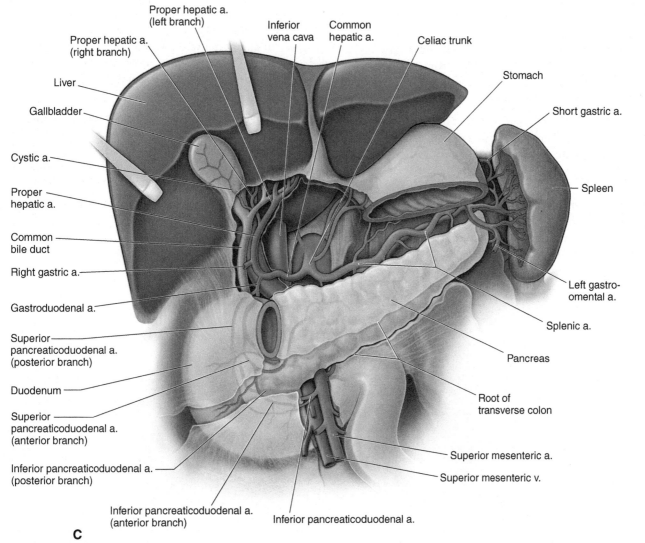

Figure 9-2: A. Visceral (inferior) view of the liver. **B**. Portal triad. **C**. Anterior view of the foregut with the body and pylorus of the stomach removed; the lesser omentum is also removed.

PANCREAS AND SPLEEN

BIG PICTURE

The pancreas is an organ of dual function: exocrine secretion for digestion and endocrine function for the regulation of glucose metabolism. By comparison, the spleen contributes to the formation of blood cells during fetal and early postnatal life and is involved in the development of immune cells (lymphocytes) and the clearance of red blood cells from the blood (tissue macrophages).

PANCREAS

The pancreas is a retroperitoneal organ located at the L2 vertebral level and contains the following structures (Figure 9-3A and B):

- **Head and neck.** Both parts are nestled within the concavity of the duodenum. The head and neck receive their blood supply from the **superior pancreaticoduodenal arteries** (celiac trunk) and the **inferior pancreaticoduodenal arteries** (superior mesenteric artery).
- **Body and tail.** Both parts are located anterior to the left kidney, with the tail touching the spleen. The body and tail receive their blood supply from the **splenic artery** (celiac trunk) and **pancreatic branches** from the superior mesenteric artery. The end of the pancreatic tale, where it touches the spleen, is the only portion of the pancreas that is not retroperitoneal.

The pancreas is a gland that consists of both **exocrine tissue** (a gland that produces and secretes products locally) and **endocrine tissue** (a gland that produces and secretes hormones into the blood).

- **Pancreas as an exocrine organ.** The pancreas produces enzymes that chemically digest carbohydrates, proteins, and fats. These enzymes are produced in the pancreatic exocrine glands and are secreted into the **main pancreatic duct**, which joins the **common bile duct** within the wall of the duodenum. An **accessory pancreatic duct** may open separately into the duodenum, proximal to the common bile duct.
- **Pancreas as an endocrine organ.** Endocrine tissue islands, called **pancreatic islets (of Langerhans)**, are found within the pancreas and produce the hormones **insulin and glucagon**. These hormones are secreted into the blood stream via the **pancreatic vein**, where they are transported by the blood to distant cellular targets.

Innervation most likely plays a limited role in the digestive process. Parasympathetic stimulation of the vagus nerve may increase exocrine digestive secretions. Sympathetic input (T5–T9 spinal cord levels via the greater splanchnic nerve) increases the tone of smooth muscle cells on the neck of the secretory units, thus inhibiting the release of digestive enzymes.

SPLEEN

The spleen is located in the left upper quadrant of the abdomen, between the stomach and the diaphragm (Figure 9-3B). The spleen is the size of a fist and stores blood, phagocytizes foreign blood particles, and produces mononuclear leukocytes. It also maintains "quality control" over erythrocytes by the removal of senescent and defective red blood cells. The spleen receives its blood supply from the splenic artery (via the celiac trunk) and venous drainage through the portal vein.

▽ An increase in the number of red blood cells may result in an enlarged spleen (**splenomegaly**). Splenomegaly can occur in patients who are diagnosed with diseases that change the shape of red blood cells (e.g., malaria); as a result, in these patients, the spleen filters an abnormally high number of red blood cells, which results in enlargement of the spleen. ▽

▽ Although protected by the ribs, the **spleen is frequently injured in trauma**. In many cases, the force applied to the ribcage pushes the ribs inward, fracturing the ribs, and the free edges puncture the spleen. ▽

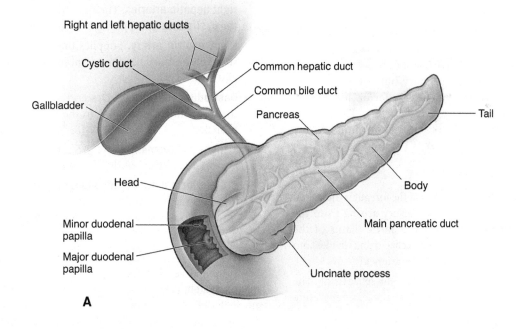

A

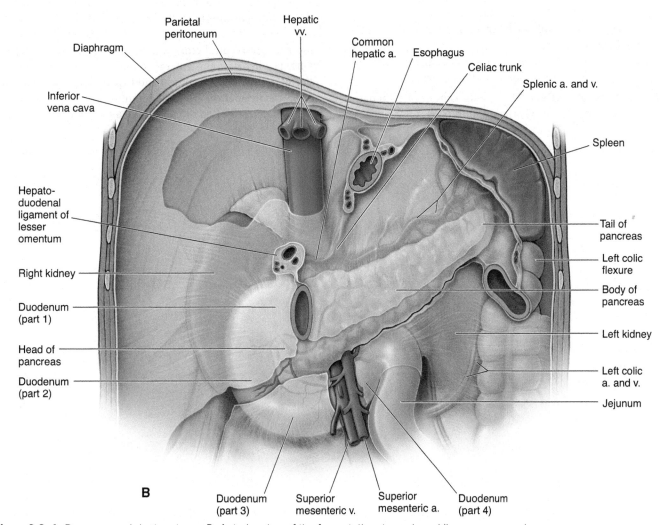

B

Figure 9-3: A. Pancreas and duct systems. **B**. Anterior view of the foregut; the stomach and liver are removed.

VASCULAR SUPPLY OF THE FOREGUT

BIG PICTURE

Knowledge of the three unpaired arterial trunks that arise from the anterior surface of the abdominal aorta will facilitate in the understanding of the three subdivisions of the gastrointestinal tract. Each subdivision receives a primary, but not sole, arterial supply: foregut–celiac trunk, midgut–superior mesenteric artery, and hindgut–inferior mesenteric artery. The foregut organs are the stomach, the first half of the duodenum, and the liver, gallbladder, pancreas, and spleen.

ARTERIAL SUPPLY

The principal blood supply to the organs of the foregut is the **celiac trunk**. The celiac trunk is an unpaired artery arising from the abdominal aorta, immediately below the aortic hiatus of the diaphragm at the T12 vertebral level. The celiac trunk divides into the left gastric, splenic, and common hepatic arteries (Figure 9-4A).

- **Left gastric artery.** The smallest branch of the celiac trunk. The left gastric artery provides blood supply to the lesser curvature of the stomach and the inferior portion of the esophagus.

- **Splenic artery.** The largest branch of the celiac trunk. The splenic artery provides blood to the spleen, the greater curvature of the stomach (**left gastroomental branch**), the fundus (**short gastric branches**), and the pancreas (**pancreatic branches**). The splenic artery follows a highly tortuous course along the superior border of the pancreas.

- **Common hepatic artery.** Divides into the proper hepatic artery, the gastroduodenal artery, and the right gastric artery.

- **Proper hepatic artery.** Ascends in the free edge of the lesser omentum and divides near the portal triad into the **left and right hepatic arteries**. The right hepatic artery gives rise to the **cystic artery**, supplying the gallbladder.

- **Right gastric artery.** Supplies the lesser curvature of the stomach and usually forms an anastomosis with the left gastric artery.

- **Gastroduodenal artery.** Descends deep to the first part of the duodenum, giving rise to the following two principal branches:

  - **Right gastroomental artery.** Supplies the right half of the greater curvature of the stomach.

  - **Superior pancreaticoduodenal artery.** Courses between the duodenum and the head of the pancreas to further split into anterior and posterior divisions.

VENOUS DRAINAGE

The principal venous drainage of the foregut is to the **portal venous system** (Figure 9-4B). The portal venous system drains nutrient-rich venous blood from the gastrointestinal tract and the spleen to the liver. Three collecting veins converge to form the portal vein: the **splenic vein** and the **superior mesenteric** and **inferior mesenteric veins**. Portal venous blood flows to the liver, where nutrients are metabolized. The metabolic products are collected in central veins, which are tributaries of the **hepatic veins**. The hepatic veins emerge from the liver to drain into the **inferior vena cava**. The portal and hepatic veins are connected within the liver, at the liver lobules; thus, the combined venous system is called the **hepatic portal system of veins**.

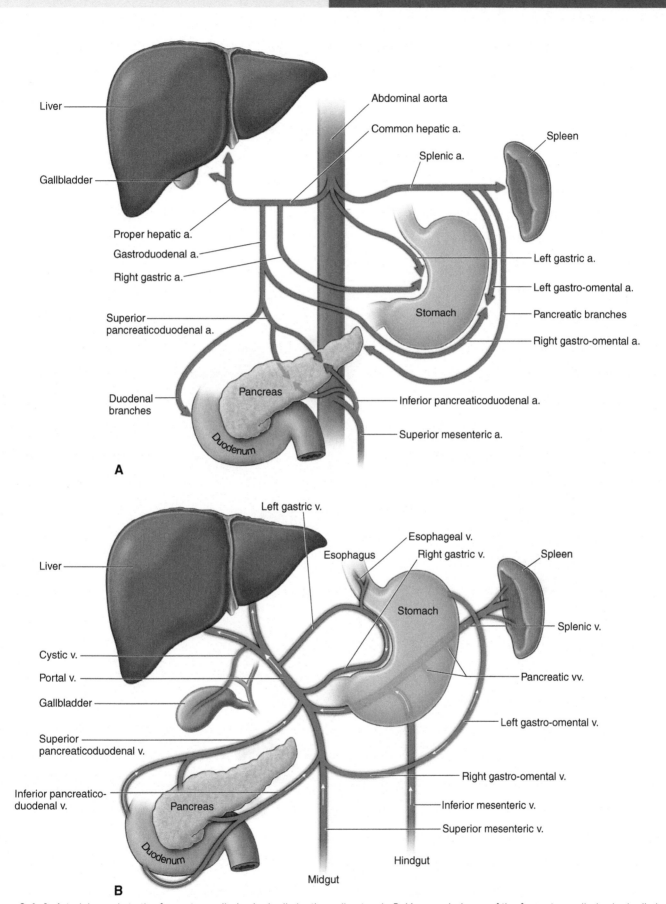

Figure 9-4: A. Arterial supply to the foregut supplied principally by the celiac trunk. **B**. Venous drainage of the foregut supplied principally by the portal vein.

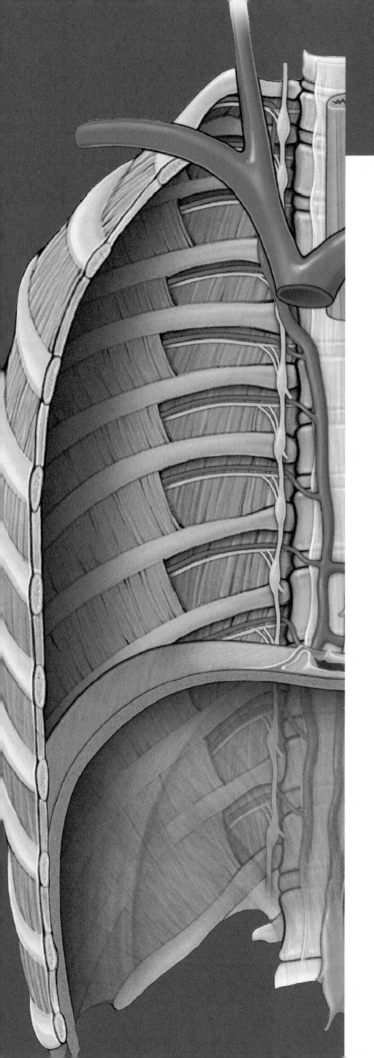

MIDGUT AND HINDGUT

MIDGUT

BIG PICTURE

The midgut consists of the distal half of the duodenum, jejunum, ileum, cecum, ascending colon, and the proximal half of the transverse colon (Figure 10-1A). Branches of the superior mesenteric arteries and veins provide the primary (but not exclusive) vascular supply for the midgut (Figure 10-1B).

DISTAL HALF OF THE DUODENUM

The duodenum is the first part of the small intestine. The chemical digestion of food (i.e., carbohydrates to simple sugars; fats to fatty acids and glycerol; proteins to amino acids) primarily occurs in the duodenum because of the secretion of pancreatic enzymes. The remainder of the small intestine (i.e., jejunum and ileum) primarily functions in absorption of these nutrients into the blood stream.

The duodenum is part of the **foregut (supplied by branches of the celiac artery)** and the **midgut (supplied by branches of the superior mesenteric artery)**, as noted by its dual vascular supply (Figure 10-1B). The junction between the duodenum and the jejunum is marked by the **suspensory ligament of the duodenum (ligament of Treitz)**. The suspensory ligament consists of connective tissue and smooth muscle and courses from the left crus of the diaphragm to the fourth part of the duodenum. Contraction of the smooth muscle within the ligament helps to open the **duodenojejunal flexure**, enabling the flow of chyme.

▽ The submucosal layer of the duodenum contains **Brunner's glands**, which protect the duodenum against the acidic chyme from the stomach. Despite this protection, the duodenum is a relatively common site of **ulcer formation.** ▽

JEJUNUM AND ILEUM

The **jejunum** is the second part of the small intestine and has the most highly developed **circular folds** lining the lumen, thereby increasing the surface area of the mucosal lining for absorption. In contrast to the ileum, the jejunum also has a greater number of **vasa recti**. A histologic section of the jejunum is usually identified negatively: it lacks Brunner's glands (like the duodenum) or Peyer's patches (like the ileum).

The **ileum** is the third part of the small intestine and contains large lymphatic aggregates known as **Peyer's patches**. In contrast to the jejunum, the ileum has fewer circular folds lining the lumen and more **vascular arcades**.

The **terminal end of the ileum** has a thickened smooth muscle layer known as the **ileocecal valve** (sphincter), which prevents feces from the cecum to move backward from the large intestine into the small intestine.

The jejunum and ileum receive their blood supply primarily via **jejunal and ileal branches** of the **superior mesenteric artery**.

CECUM

The cecum is the blind-ended sac at the beginning of the large intestine (Figure 10-1A and B). The cecum is inferior to the **ileocecal valve** and is located in the right lower quadrant of the abdomen, within the iliac fossa. Attached to the cecum is the **vermiform appendix** (unknown function in humans). The **taenia coli** (longitudinal smooth muscle bands) of the ascending colon lead directly to the base of the appendix.

▽ By identifying the taenia coli during surgery, surgeons locate the origin of the appendix on the cecum. The position of the remainder of the appendix varies because it is intraperitoneal (mobile). The surface projection of the appendix (**McBurney's point**) is most often located one-third of the distance between the right anterior superior iliac spine and the umbilicus. ▽

The **ileocolic artery**, a branch of the superior mesenteric artery, supplies the cecum. In addition, a small branch of the ileocolic artery, the **appendicular artery**, supplies the appendix (Figure 10-1B).

▽ The appendix may become inflamed, resulting in **appendicitis**. Sensory neurons from the visceral peritoneum of the appendix signal the central nervous system that the appendix is inflamed. These signals are transmitted via **visceral sensory neurons** in the **lesser splanchnic nerve**, which enters the **T10 vertebral level** of the spinal cord. However, somatic sensory neurons from the skin around the umbilicus also enter at the T10 vertebral level of the spinal cord. Because both visceral and somatic neurons enter the spinal cord at the same level and synapse in the same region, the brain interprets the inflammation from the appendix as if the pain originated in the region of the umbilicus. This phenomenon is known as **referred pain.** ▽

ASCENDING AND TRANSVERSE COLON

The **ascending colon** arises from the cecum and courses vertically to the liver, where the colon bends at the **right colic (hepatic) flexure** (Figure 10-1B). The parietal peritoneum covers its anterior surface, and thus the ascending colon is considered a retroperitoneal organ. A depression between the lateral surface of the ascending colon and the abdominal wall is known as the **right paracolic gutter**. Branches of the right colic artery that supply the ascending colon enter the bowel on its medial surface. It is possible during surgery to mobilize the ascending colon by cutting the peritoneum along the right paracolic gutter without injuring its major vessels or lymphatics.

The colon continues horizontally as the **transverse colon** to the spleen on the opposite side of the abdomen, inferior to the liver, gallbladder, and the greater curvature of the stomach. The transverse colon is connected to the greater curvature of the stomach via the **gastrocolic ligament**, which is part of the greater omentum. The duodenum, pancreas, duodenojejunal flexure, and parts of the small intestine are all located deep to the transverse colon.

The transverse colon is surrounded by visceral peritoneum and thus is considered an intraperitoneal structure.

The **right colic artery** supplies the ascending colon, whereas the **middle colic artery** supplies the transverse colon (both are branches of the superior mesenteric artery). The right and left colic arteries contribute to the blood supply of the transverse colon.

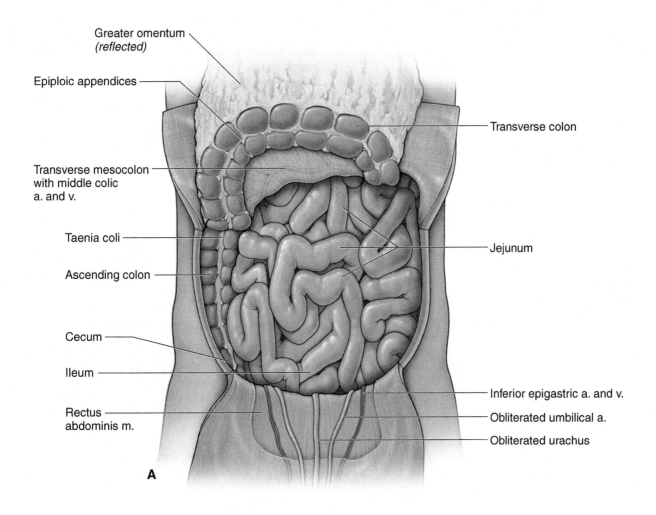

Greater omentum
(reflected)

Epiploic appendices

Transverse colon

Transverse mesocolon
with middle colic
a. and v.

Taenia coli

Ascending colon

Jejunum

Cecum

Ileum

Inferior epigastric a. and v.

Obliterated umbilical a.

Rectus
abdominis m.

Obliterated urachus

A

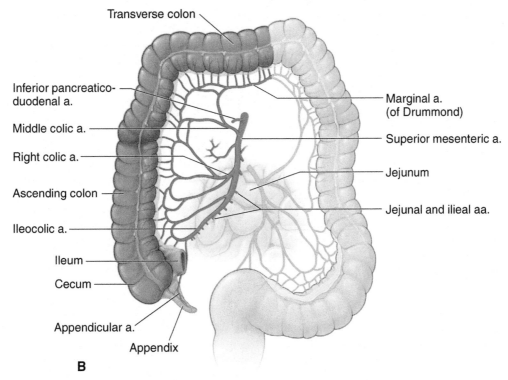

Transverse colon

Inferior pancreatico-
duodenal a.

Middle colic a.

Right colic a.

Marginal a.
(of Drummond)

Superior mesenteric a.

Ascending colon

Jejunum

Ileocolic a.

Jejunal and ilieal aa.

Ileum

Cecum

Appendicular a.

Appendix

B

Figure 10-1: A. Midgut with the greater omentum reflected superiorly and the anterior abdominal wall reflected inferiorly. **B**. Primary blood supply to the midgut is through the superior mesenteric artery.

HINDGUT

BIG PICTURE

The hindgut consists of the distal half of the transverse colon, descending colon, sigmoid colon, and the proximal third of the rectum. Branches of the inferior mesenteric artery and vein provide vascular supply to the hindgut.

TRANSVERSE, DESCENDING, AND SIGMOID COLON

The **transverse colon** ends at the spleen, where the colon bends as the **left colic (splenic) flexure** (Figure 10-2A). The colon continues vertically down the left wall of the abdomen as the **descending colon**. The parietal peritoneum covers only the anterior surface of the descending colon, and thus the descending colon is considered a retroperitoneal structure.

A depression between the lateral surface of the descending colon and the abdominal wall is known as the **left paracolic gutter**. Branches of the left colic artery supplying the descending colon enter the bowel on its medial surface. As such, it is possible to surgically mobilize the descending colon by cutting the peritoneum along the left paracolic gutter without injuring major vessels or lymphatics.

The transverse colon is part of both the **midgut** (supplied by left and middle colic arteries, branches of the superior mesenteric artery) and the **hindgut** (supplied by left colic arteries, branches of the inferior mesenteric artery). The **descending colon** also receives its vascular supply mainly from the **left colic artery** (branch of the inferior mesenteric artery) (Figure 10-2B).

▽ The **marginal artery (of Drummond)** is an arterial anastomosis between the superior and inferior mesenteric arteries. The marginal artery courses within the mesentery and parallels the ascending, transverse, and descending colon. The anastomosis is so complete that during the repair of an aortic aneurism, the inferior mesenteric artery often will not be regrafted to the aorta because the marginal artery will supply blood to the hindgut via the superior mesenteric artery. ▽

Near the left iliac fossa, the descending colon continues as the **sigmoid colon**, which is an intraperitoneal structure. The vascular supply of the sigmoid colon is via **sigmoid arteries** from the inferior mesenteric artery.

RECTUM AND ANUS

In the pelvic cavity, the **rectum** is the terminal, straight portion of the colon and ultimately terminates as the **anus**. The transition from sigmoid colon to rectum occurs at approximately the **S3 vertebral level**, where the rectum is covered anteriorly by parietal peritoneum and thus is a retroperitoneal structure.

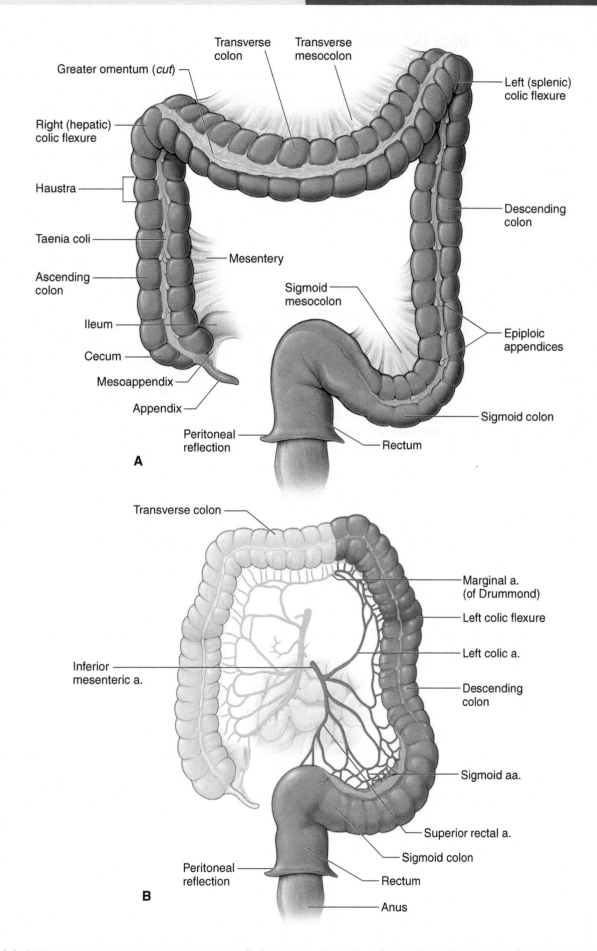

Figure 10-2: A. Hindgut with the small intestine removed. **B**. Primary blood supply to the hindgut is through the inferior mesenteric artery.

INNERVATION OF THE MIDGUT AND HINDGUT

BIG PICTURE

The midgut and hindgut receive sympathetic innervation from the midthoracic to the upper lumbar spinal nerves, parasympathetic innervation from the vagus nerves and pelvic splanchnic nerves (S2–S4) (Figure 10-3).

SYMPATHETIC MOTOR INNERVATION

Sympathetic motor innervation to the gastrointestinal tract decreases motility, peristalsis, sphincter muscle contraction, absorption, and glandular secretions, in addition to causing vasoconstriction. Sympathetic innervation to the gastrointestinal tract is accomplished via the following nerves:

- **Greater splanchnic nerve.** Carries preganglionic sympathetics originating from the T5–T9 level of the spinal cord that most likely synapse in the prevertebral plexus (celiac and superior mesenteric ganglia or plexuses).

- **Lesser splanchnic nerve.** Carries preganglionic sympathetics originating from the T10–T11 level of the spinal cord that most likely synapse in the prevertebral plexus (celiac and superior mesenteric ganglia or plexuses).

- **Least splanchnic nerve.** Carries preganglionic sympathetics from the T12 level of the spinal cord that most likely synapse in the prevertebral plexus (aorticorenal and inferior mesenteric ganglia or plexuses).

- **Lumbar splanchnic nerve.** Carries preganglionic sympathetics from the L1–L2 level of the spinal cord that most likely synapse in the prevertebral plexus (inferior mesenteric and inferior hypogastric ganglia or plexuses).

PARASYMPATHETIC MOTOR INNERVATION

Parasympathetic motor innervation increases motility, absorption, smooth muscle contraction, and glandular secretions. In addition, parasympathetic motor innervation relaxes the sphincter muscles. Parasympathetic innervation to the gastrointestinal tract is accomplished as follows:

- **Midgut.** Preganglionic parasympathetic fibers originating in the brainstem course in the **vagus nerve (CN X)** to the prevertebral plexus and accompany sympathetic fibers to regions of the midgut.

- **Hindgut.** Preganglionic parasympathetic fibers originating at the **S2–S4 levels** of the spinal cord are transported via the **pelvic splanchnic nerves** to the prevertebral plexus (inferior hypogastric plexus). Here, they accompany sympathetic fibers to regions of the hindgut and urogenital systems.

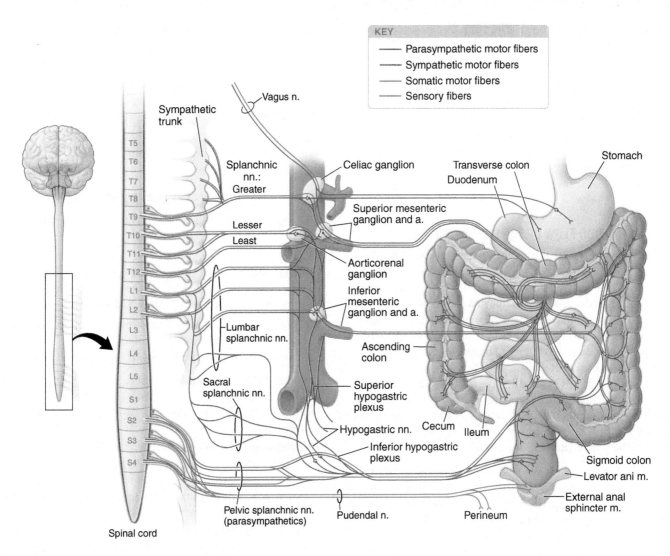

Figure 10-3: Innervation of the foregut, midgut, and hindgut.

PORTAL SYSTEM

BIG PICTURE

The portal system is responsible for transporting blood from most of the gastrointestinal tract to the liver for metabolic processing before the blood returns to the heart. The portal system drains venous blood from the distal end of the esophagus, stomach, small and large intestines, proximal portion of the rectum, pancreas, and spleen. The portal system is the venous counterpart to areas supplied by the celiac trunk and the superior and inferior mesenteric arteries.

BLOOD FLOW OF THE PORTAL SYSTEM

The liver is unique in that it receives both nutrient-rich deoxygenated blood (**portal vein**) and oxygenated blood (**hepatic arteries**). The portal vein branches as it enters the liver, where its blood percolates around **hepatocytes** in tiny vascular channels known as **sinusoids**. Hepatocytes detoxify the blood, metabolize fats, carbohydrates, and drugs, and produce bile. The sinusoids receive deoxygenated blood from the portal veins (provide blood for metabolism and detoxification) and oxygenated blood from the hepatic arteries (provide oxygen for hepatocytes). Blood exits the sinusoids into a **central vein**, which empties into the **hepatic veins** and ultimately into the **inferior vena cava**, which passes through the diaphragm before entering the right atrium of the heart.

▽ Oral drugs travel throughout the gastrointestinal tract, where they are absorbed by the small intestine. These drugs then travel to the liver via the hepatic portal system, where they are metabolized before entering the systemic circulation. Because of hepatic metabolism, the concentration of oral drugs is reduced before entering the systemic circulation. This is known as the **first-pass effect**. Therefore, drugs that are inactivated by the liver (e.g., nitroglycerin) must be administered by a different method. For example, nitroglycerin is administered sublingually (absorption under the tongue) because, if swallowed, the liver inactivates the drug before it can enter the systemic circulation. ▼

VEINS OF THE PORTAL SYSTEM

Veins of the portal system generally mirror the arterial branches of the celiac trunk and the superior and inferior mesenteric arteries (Figure 10-4A). The major veins of the portal system are as follows:

- **Splenic vein.** Drains blood from the foregut, including the spleen, pancreas, and part of the stomach. The splenic vein courses deep to the pancreas.
- **Superior mesenteric vein.** Drains blood from the midgut and part of the foregut. The superior mesenteric vein is located to the right of the superior mesenteric artery as it courses over the third part of the duodenum.
  - **Gastro-omental veins.** Drain blood from the greater curvature of the stomach into the superior mesenteric vein.
- **Inferior mesenteric vein.** Drains blood from the hindgut, including the proximal third of the rectum. The inferior

mesenteric vein usually drains into the superior mesenteric vein, inferior to its union with the portal vein.

- **Portal vein.** Collects blood from the foregut, midgut, and hindgut. The portal vein is located deep to the hepatic artery and cystic duct and is formed by the union of the superior mesenteric vein and splenic vein, deep to the neck of the pancreas.
  - **Gastric veins.** Drain blood from the lesser curvature of the stomach into the portal vein.

PORTAL–CAVAL ANASTOMOSES

To better understand the portal–caval anastomoses, recall that veins in the abdomen return blood to the heart via two routes (Figure 10-4B):

- **Portal system.** Veins from the foregut, midgut, and hindgut drain blood to the liver before the blood enters the inferior vena cava and ultimately returns to the heart.
- **Caval system.** Veins from the lower limbs, pelvis, and posterior abdominal wall transport blood directly to the inferior vena cava before the blood returns to the heart.

Portal–caval anastomoses are at regions of the gastrointestinal tract that are drained by both the portal and systemic (-caval) systems. The principal portal–caval anastomoses are as follows:

1. **Distal portion of the esophagus.** The left gastric vein of the hepatic portal system drains blood from the distal portion of the esophagus. However, most of the blood drained from the esophagus is through the esophageal veins, which drain into the azygos (caval) vein.

2. **Anterior abdominal wall.** The paraumbilical veins drain the tissue surrounding the umbilicus: Embryologically, these veins communicated with the umbilical veins. These connections may reopen during chronic portal hypertension. Normally in the adult, most of the venous drainage is from the inferior epigastric veins.

3. **Rectum.** The proximal portion of the rectum is drained via the superior rectal vein, which drains into the inferior mesenteric vein of the hepatic portal system. However, the remainder of the rectum is drained by the middle rectal vein (branch of the internal iliac vein) and inferior rectal vein (branch of the internal pudendal vein).

▽ When hepatocytes are damaged (e.g., due to disease, alcohol, or drugs), the liver cells are replaced by fibrous tissue, which impedes the flow of blood through the liver (cirrhosis). When the hepatic portal system is blocked, the return of blood from the intestines and spleen through the liver is impeded, resulting in **portal hypertension**. Therefore, veins that usually flow into the liver are blocked. Consequently, blood pressure in the blocked veins increases, causing them to dilate and gradually reopen previously closed connections with the caval system. Veins in the distal portion of the esophagus begin to enlarge (**esophageal varices**); veins in the rectum begin to enlarge (**internal hemorrhoids**); and in chronic cases, the veins of the paraumbilical region enlarge (**caput medusa**). ▼

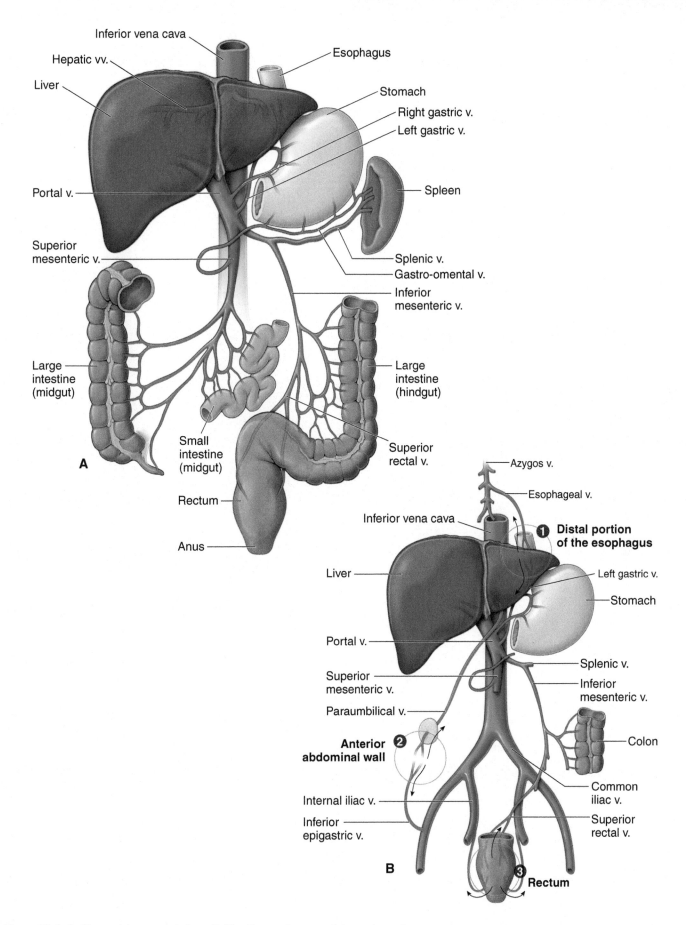

Figure 10-4: A. The portal venous system. **B**. The three primary portal–caval anastomoses.

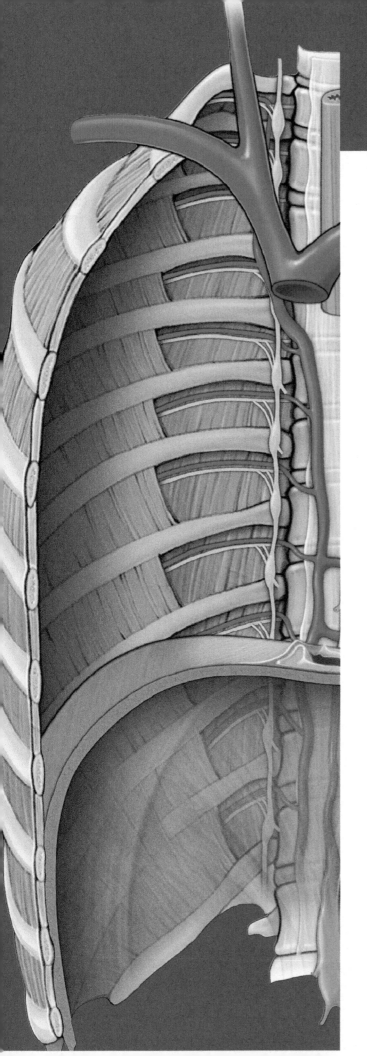

POSTERIOR ABDOMINAL WALL

MUSCLES

BIG PICTURE

The diaphragm forms the superior and much of the posterior border of the posterior abdominal wall. In addition, the psoas major, iliacus, and quadratus lumborum muscles form the posterior abdominal wall. These muscles function in respiration (diaphragm) as well as trunk and lower limb motion.

MUSCLES AND FASCIA

The muscles that form much of the structure of the posterosuperior abdominal wall are as follows (Figure 11-1):

- **Diaphragm.** A dome-shaped muscle that separates the abdominal cavity from the thoracic cavity. The origin of the muscles of the diaphragm is along the internal circumference of the ribcage, sternum, and lumbar vertebrae. This portion of muscle consists of the **esophageal hiatus**, at the T10 vertebral level, and the **aortic hiatus**, at the T12 vertebral level. The muscle fibers are directed to the center of the diaphragm, to the central tendon. The central tendon consists of collagen tissue and the **venal caval hiatus**, which is at the T8 vertebral level. When the muscle is stimulated to contract (phrenic nerve, C1–C4), the muscle fibers shorten, causing the central tendon to move inferiorly and flatten. This action results in inspiration. The structural components of the diaphragm are as follows:
  - **Right crus.** Forms part of the aortic hiatus. The right crus also loops around the esophagus to form the esophageal hiatus and contributes to the **suspensory ligament of the duodenum (ligament of Treitz).**
  - **Left crus.** Forms part of the aortic hiatus.
- **Quadratus lumborum muscle.** Attaches to the iliac crest, lumbar transverse processes, and the 12th rib. The quadratus lumborum muscle is the bed on which the kidneys lie. The muscle laterally flexes the vertebral column and stabilizes the 12th rib during breathing. The subcostal and lumbar intercostal nerves provide innervation.
- **Psoas major muscle.** Attaches to the lumbar vertebrae superiorly and to the lesser trochanter of the femur inferiorly. Between these attachments, the psoas major muscle courses deep to the inguinal ligament and lateral to the femoral nerve. The psoas major muscle flexes the hip joint (when the vertebrae are stabilized) and the lumbar vertebrae. The psoas major muscle is innervated by the L1–L3 spinal nerves.

▽ **The appendix** is in close relationship with the parietal peritoneum, including that covering the **right psoas muscle**. When the appendix is inflamed, the inflammation irritates the parietal peritoneum. The parietal peritoneum is innervated by somatic nerves, including pain fibers, and thus inflammation of the parietal peritoneum results in a "shooting" abdominal pain. To diagnose acute appendicitis, the physician will push on McBurney's point when the patient is in a supine position. If the patient suffers from acute appendicitis, the pain reflex will flex the hip joint. ▽

- **Psoas minor muscle.** Attaches to the L1 vertebra and pubic crest. The psoas minor muscle helps to tilt the pelvis and is innervated by the L1–L2 spinal nerves. *This muscle is not present in everyone.*
- **Iliacus muscle.** Attaches within the iliac fossa and lesser trochanter of the femur. Between its attachments, the iliacus muscle courses deep to the inguinal ligament and joins with the psoas major muscle to attach to the lesser trochanter of the femur. The combination of these two muscles in the thigh is often referred to as the **iliopsoas muscle**. The iliacus muscle flexes the hip. This muscle is innervated by the femoral nerve (L2–L3 spinal nerves).

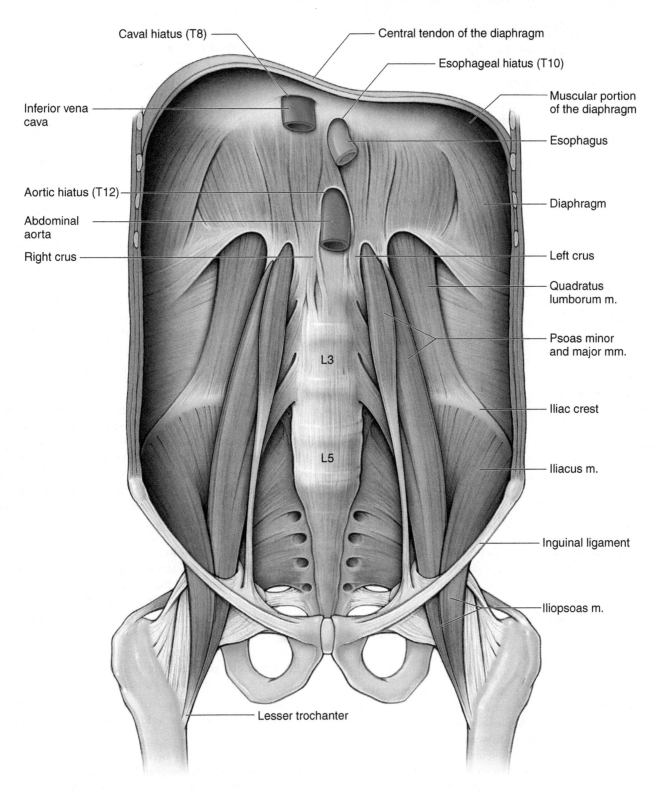

Caval hiatus (T8)

Central tendon of the diaphragm

Esophageal hiatus (T10)

Muscular portion of the diaphragm

Esophagus

Inferior vena cava

Aortic hiatus (T12)

Diaphragm

Abdominal aorta

Right crus

Left crus

Quadratus lumborum m.

Psoas minor and major mm.

L3

Iliac crest

L5

Iliacus m.

Inguinal ligament

Iliopsoas m.

Lesser trochanter

Figure 11-1: Muscles of the posterior abdominal wall.

VESSELS OF THE POSTERIOR ABDOMINAL WALL

BIG PICTURE

The abdominal aorta and inferior vena cava course vertically in the retroperitoneal space, providing the vascular supply for the abdomen, pelvis, and perineum.

ABDOMINAL AORTA

The aorta enters the abdomen from the thorax by traversing the aortic hiatus of the diaphragm at the **T12 vertebral level**. The aorta courses along the midline, on the anterior surface of vertebral bodies to the left of the inferior vena cava. The abdominal aorta has the following branches, from superior to inferior (Figure 11-2):

- **Inferior phrenic arteries.** The first paired branches of the aorta in the abdominal cavity. The inferior phrenic arteries supply the inferior surface of the diaphragm.
- **Middle suprarenal arteries.** One of three pairs of arteries supplying the adrenal glands.
- **Gonadal arteries.** Paired arteries that supply the gonads.
- **Lumbar arteries.** Usually, four pairs of arteries that supply the abdominal wall, similar to the intercostal arteries of the thorax.
- **Celiac trunk.** Unpaired artery that is located approximately at the L1 vertebral level. Supplies the foregut and organs associated with the foregut.
- **Superior mesenteric artery.** Unpaired artery that is located immediately below the celiac trunk. The superior mesenteric artery supplies the midgut.
- **Inferior mesenteric artery.** Unpaired artery that is located 4 to 5 cm superior to the bifurcation of the abdominal aorta into the common iliac arteries. The inferior mesenteric artery supplies the hindgut.
- **Common iliac arteries.** At the L4 vertebral level, the abdominal aorta bifurcates into the left and right common iliac arteries.

▽ The aorta is the largest artery in the body and as such channels blood under high pressure. An **abdominal aortic aneurysm** (clinically referred to as an AAA) is a condition in which a section of the abdominal aorta expands or bulges, much like a balloon, in response to weakening of the vessel wall. An AAA can occur anywhere within the thoracic and abdominal aorta, but most occur inferior to the renal arteries. An AAA is a serious health condition because rupture of the abdominal aorta results in severe abdominal bleeding and is fatal within minutes. To prevent rupture, the weakened part of the aorta is often replaced with a tube-like replacement (aortic graft). ▼

INFERIOR VENA CAVA

The inferior vena cava is located to the right of the abdominal aorta. The union of the left and right common iliac veins forms the inferior vena cava. The inferior vena cava ascends along the right side of the vertebral bodies. Before entering the thoracic cavity, the inferior vena cava courses within a groove on the posterior surface of the liver. This portion of the inferior vena cava receives the **hepatic veins**. Along its course in the abdomen, the inferior vena cava receives the following tributaries:

- **Right gonadal vein.** Drains the right testis or ovary by entering the inferior vena cava, inferior to the right renal vein.
- **Renal veins.** Drain the kidneys. The gonadal veins are not bilaterally symmetrical. The **left gonadal vein** drains into the left renal vein, in contrast to the right gonadal vein.
  - **Adrenal veins.** Drain the adrenal glands, typically by entering the left and right renal veins.
- **Inferior phrenic veins.** Drain the inferior surface of the diaphragm.

The **lumbar veins** drain into a pair of **ascending lumbar veins**, which ascend posterior to the diaphragm to empty into the **azygos system of veins** in the thoracic cavity. Connections exist between the ascending lumbar veins and the inferior vena cava. Therefore, blood in the posterior abdominal wall may drain through the azygos vein or through the inferior vena cava. Remember, blood from the abdominal viscera drains through the hepatic portal system.

▽ Blood from the lower limbs and the retroperitoneal organs drains into the inferior vena cava and ascends through the abdomen and into the thorax before entering the right atrium. In **chronic thrombosis of the inferior vena cava**, a blood clot decreases or obstructs blood flow. As a result, venous blood must flow via a different route on its return to the heart.

- Blood from the lower limbs and pelvis drains into the superficial epigastric veins in Camper's fascia. These veins anastomose with other epigastric veins.
- Blood flows from the epigastric veins to the internal thoracic veins, the brachiocephalic veins, the superior vena cava, and then the right atrium.
- The epigastric veins may swell because they accommodate so much blood that they appear in the skin as irregular veins coursing vertically on the abdominal wall. ▼

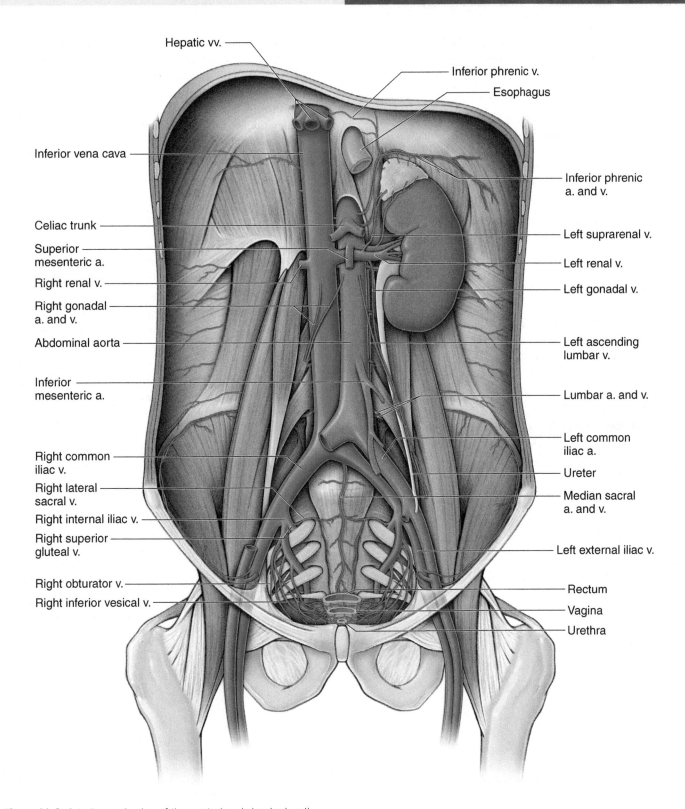

Figure 11-2: Arteries and veins of the posterior abdominal wall.

SOMATIC NERVES OF THE POSTERIOR ABDOMINAL WALL

BIG PICTURE

The ventral rami of the lower thoracic and lumbar spinal nerves provide somatic innervation to the abdominal wall muscles and skin.

SOMATIC NERVES

The somatic nerves of the posterior abdominal wall are the ventral rami of the subcostal and lumbar spinal nerves (Figure 11-3).

- **Subcostal (T12), iliohypogastric (L1), and ilioinguinal (L1) nerves.** These three nerves emerge along the lateral surface of the psoas major muscle and course between the internal oblique and transverse abdominis muscles, innervating these abdominal wall muscles.

  - **Iliohypogastric nerve (L1)** In addition the: provides cutaneous innervation to the skin in the hypogastric region.

  - **Ilioinguinal nerve (L1)** traverses the superficial inguinal ring, enters the spermatic cord, and provides cutaneous innervation to the anterior surface of the scrotal sac and labia majora.

- **Genitofemoral nerve (L1–L2).** Pierces through the anterior surface of the psoas major muscle and courses along its surface before bifurcating into the **femoral branch** (cutaneous innervation of anterior thigh) and the **genital branch** (courses through the superficial inguinal ring). In males, the genital branch courses within the spermatic cord and innervates the cremaster muscle and the skin of the scrotum. In females, the genital branch innervates the skin of the mons pubis and labia majora.

- **Lateral cutaneous nerve of the thigh (L2–L3).** Emerges along the lateral border of the psoas major muscle and crosses anteriorly to the iliacus muscle, deep to the inguinal ring. This nerve provides cutaneous innervation to the skin over the lateral aspect of the thigh.

- **Femoral nerve (L2–L4).** Emerges from the lateral surface of the psoas major muscle and provides innervation to the anterior compartment muscles of the thigh responsible for knee extension (also known as the quadriceps).

- **Obturator nerve (L2–L4).** Emerges from the medial surface of the psoas major muscle and provides innervation to the medial compartment muscles of the thigh responsible for hip adduction.

- **Lumbosacral trunk (L4–L5).** Branches of the L4 and L5 ventral rami that unite and course inferiorly over the pelvic brim into the pelvic cavity and contribute to the sciatic nerve (L4–S3).

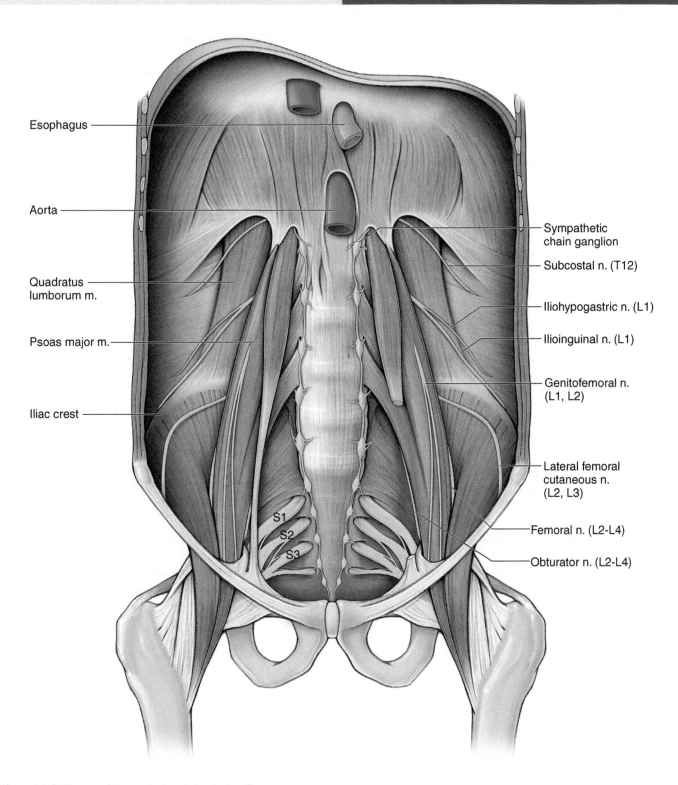

Esophagus

Aorta

Quadratus lumborum m.

Psoas major m.

Iliac crest

S1
S2
S3

Sympathetic chain ganglion

Subcostal n. (T12)

Iliohypogastric n. (L1)

Ilioinguinal n. (L1)

Genitofemoral n. (L1, L2)

Lateral femoral cutaneous n. (L2, L3)

Femoral n. (L2-L4)

Obturator n. (L2-L4)

Figure 11-3: Nerves of the posterior abdominal wall.

AUTONOMICS OF THE POSTERIOR ABDOMINAL WALL

BIG PICTURE

The prevertebral plexus is a network of sympathetic and parasympathetic fibers that innervate the digestive, urinary, and reproductive systems. Sympathetic nerves contribute to the prevertebral plexus via splanchnic nerves from the sympathetic trunk. Parasympathetic nerves contribute to the prevertebral plexus via the vagus nerve [cranial nerve (CN X)] and pelvic splanchnics from the S2–S4 spinal nerves.

SYMPATHETIC CONTRIBUTIONS TO THE PREVERTEBRAL PLEXUS

The sympathetic trunk in the abdomen is continuous with the sympathetic trunk in the thorax (Figure 11-4A). The sympathetic trunk is located along the anterolateral surface of the vertebrae. The trunks are bilateral and descend over the sacral promontory to enter the pelvic cavity. There are approximately four sympathetic ganglia in the posterior abdominal wall on each side of the body, and each ganglion houses the cell bodies of postganglionic sympathetic neurons. The following splanchnic nerves carry preganglionic sympathetic fibers and visceral afferent fibers to and from the prevertebral plexus (Figure 11-4B):

- **Greater splanchnic nerve.** Courses from the sympathetic trunks in the thorax at the T5–T9 spinal nerve levels and contributes primarily to the celiac ganglion and to a lesser extent to the superior mesenteric ganglion.
- **Lesser splanchnic nerve.** Courses from the sympathetic trunks in the thorax at the T10–T11 spinal nerve levels and contributes primarily to the superior mesenteric and aorticorenal ganglia and to a lesser extent to the celiac ganglion.

- **Least splanchnic nerve.** Courses from the T12 sympathetic trunk and contributes primarily to the renal plexus.
- **Lumbar splanchnic nerves.** Course from the sympathetic trunks at the lumbar spinal nerve level and contribute primarily to the inferior mesenteric plexus.
- **Sacral splanchnic nerves.** Course from sympathetic trunks in the sacral spinal nerve level and contribute primarily to the inferior hypogastric plexus.

PARASYMPATHETIC CONTRIBUTIONS TO THE PREVERTEBRAL PLEXUS

Parasympathetic fibers from the vagus nerve (CN X) innervate all of the abdominal viscera, distally to the transverse colon; the remainder of the gastrointestinal tract is innervated by pelvic splanchnics (Figure 11-4B).

- **Vagus nerve (CN X).** The right and left vagus nerves form the esophageal plexus in the thoracic cavity and enter the abdomen through the esophageal hiatus as the anterior and posterior vagal trunks. The **anterior vagal trunk** is derived from the left vagus nerve and primarily innervates the stomach, liver, and gallbladder. The **posterior vagal trunk** is derived from the right vagus nerve and primarily contributes to the celiac plexus of the prevertebral plexus.
- **Pelvic splanchnics (S2–S4).** Preganglionic parasympathetic neurons originate in the S2–S4 levels of the spinal cord and course in the ventral root and rami and into the pelvic splanchnic nerves, which contribute to the inferior hypogastric plexus en route to innervate the distal part of the transverse colon, descending colon, sigmoid colon, and rectum, as well as the urinary and reproductive systems.

All preganglionic parasympathetic fibers synapse within the wall of the end organ. In the gastrointestinal system, the synapse with the **postganglionic parasympathetic fibers** occurs in the myenteric (Auerbach's) plexus and the submucosal (Meissner's) plexuses.

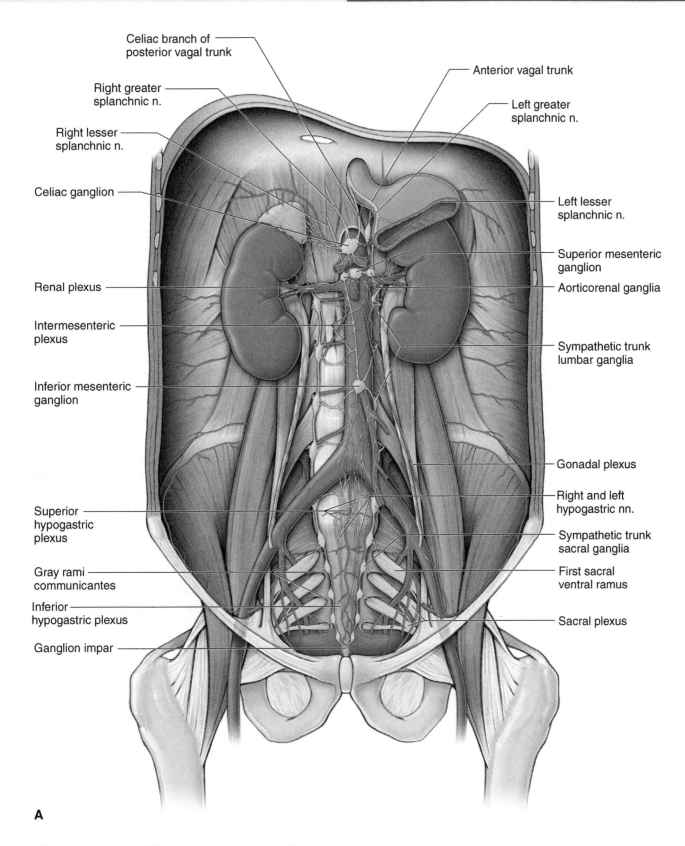

Figure 11-4: A. Autonomics of the posterior abdominal wall.

PREVERTEBRAL GANGLIA AND PLEXUS

The **prevertebral (preaortic) plexus** is a network of autonomic nerve fibers covering the abdominal aorta and extending into the pelvic cavity between the common iliac arteries. This plexus serves as a common pathway for the following autonomics (Figure 11-4A and B):

- **Preganglionic sympathetic nerves.** From greater, lesser, least, and lumbar splanchnic nerves.
- **Preganglionic parasympathetic nerves.** From the vagus nerve (CN X) and pelvic splanchnics (S2–S4).
- **Visceral afferents.** From both sympathetic and parasympathetic pathways.

Sympathetic (prevertebral) ganglia that are named for the associated branch of the abdominal aorta are located within the prevertebral plexus. These ganglia are collections of postganglionic sympathetic neurons. Neurons within the prevertebral plexus connect autonomics to and from the digestive, urinary, and reproductive organs.

The prevertebral plexus is subdivided into smaller plexuses and ganglia. Many of these plexuses and ganglia are located very close together and are variable and interrelated. Their principal features can be described as follows:

- **Celiac ganglia and plexus.** Located around the origin of the celiac trunk and distributed along its branches. The celiac plexus receives preganglionic sympathetic nerves from the greater splanchnic nerve. Parasympathetics from the vagus nerve course through the celiac ganglion en route to the viscera without synapsing. Nerves exiting the celiac ganglion supply postganglionic sympathetic nerves to the liver, gallbladder, stomach, spleen, pancreas, and proximal part of the duodenum.
- **Superior mesenteric ganglia and plexus.** Located around the origin of the superior mesenteric artery and distributed along its branches. The ganglia receive preganglionic contributions from the lesser splanchnic nerve. The superior mesenteric plexus supplies postganglionic sympathetic fibers to the head of the pancreas; the distal part of the duodenum; the jejunum, ileum, and cecum; the ascending colon; and the transverse colon.
- **Aorticorenal ganglia and plexus.** Located at the origin of the renal arteries and distributed along their branches. The ganglia receive contributions from the lesser splanchnic nerves and supply the adrenal glands, the kidneys, and the proximal part of the ureters.
- **Inferior mesenteric ganglia and plexus.** Located at the origin of the inferior mesenteric artery and distributed along its branches. The inferior mesenteric ganglion receives contributions from the superior mesenteric plexus, the first and second lumbar splanchnic nerves (sympathetics), and fibers from the superior hypogastric plexus (sympathetics and parasympathetics). The inferior mesenteric plexus supplies the descending colon, the sigmoid colon, and the upper portion of the rectum.
- **Superior hypogastric plexus.** Located inferior to the bifurcation of the aorta, between the common iliac arteries. The superior hypogastric plexus is formed by contributions of the inferior mesenteric plexus, the third and fourth lumbar splanchnic nerves (sympathetics), and the parasympathetic nerves that ascend from the inferior hypogastric plexus. The superior hypogastric plexus continues caudally into the pelvis via the hypogastric nerves.
- **Inferior hypogastric plexus and pelvic ganglia.** In males, the inferior hypogastric plexus is posterolateral to the bladder, seminal vesicles, and prostate. In females, it is posterolateral to the bladder and cervix. The inferior hypogastric plexus and pelvic ganglia are formed primarily from the following fibers:
  - **Sympathetics.** Preganglionic sympathetic fibers enter the inferior hypogastric plexus through the **sacral splanchnic nerves.** Additionally, some sympathetic fibers arising from the lumbar splanchnic nerves descend from the superior hypogastric plexus into the inferior hypogastric plexus. Preganglionic sympathetic neurons usually synapse with postganglionic sympathetic neurons somewhere in the prevertebral plexus (i.e. inferior hypogastric plexus).
  - **Parasympathetics.** Preganglionic parasympathetic fibers enter the inferior hypogastric plexus through the pelvic splanchnic nerves. Once most of the parasympathetic fibers enter the inferior hypogastric plexus, they ascend out of the pelvis and into the superior hypogastric plexus to innervate the hindgut. However, some nerves exit the pelvis to innervate the urinary and reproductive systems. Preganglionic parasympathetic neurons usually synapse with postganglionic parasympathetic neurons somewhere in the wall of the target organ (i.e., Aurbachis plexus).

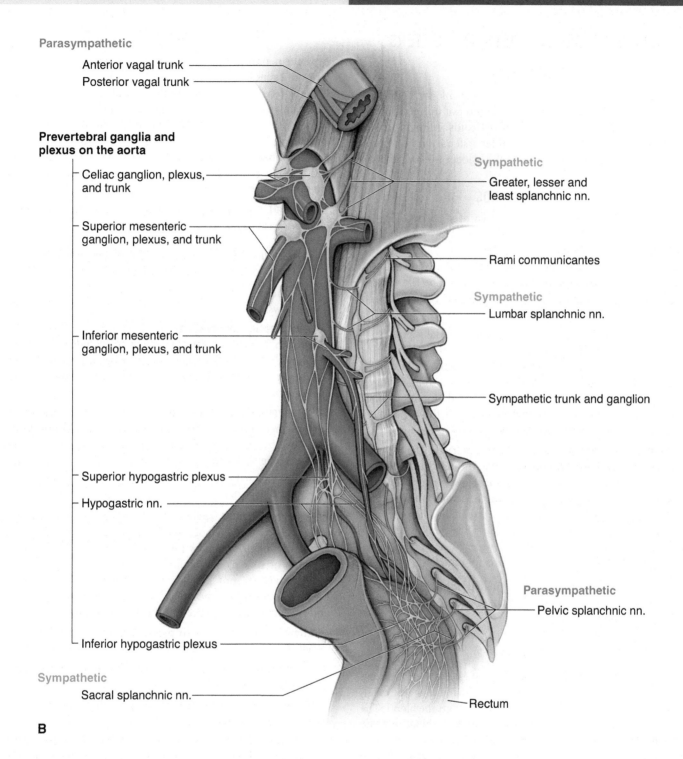

Parasympathetic

Anterior vagal trunk

Posterior vagal trunk

Prevertebral ganglia and plexus on the aorta

Celiac ganglion, plexus, and trunk

Superior mesenteric ganglion, plexus, and trunk

Inferior mesenteric ganglion, plexus, and trunk

Superior hypogastric plexus

Hypogastric nn.

Inferior hypogastric plexus

Sympathetic

Sacral splanchnic nn.

Sympathetic

Greater, lesser and least splanchnic nn.

Rami communicantes

Sympathetic

Lumbar splanchnic nn.

Sympathetic trunk and ganglion

Parasympathetic

Pelvic splanchnic nn.

Rectum

B

Figure 11-4: (*continued*) **B**. Anterolateral view of the autonomics of the posterior abdominal wall.

ADRENAL GLANDS, KIDNEYS, AND URETERS

BIG PICTURE

The adrenal (suprarenal) glands are responsible for regulating stress through the production and secretion of hormones such as adrenalin (epinephrine), glucocorticoids, mineralocorticoids, and androgens. The kidneys filter systemic blood to produce urine, which is transported to the bladder by the ureters.

ADRENAL GLANDS

The adrenal (suprarenal) glands consist of a **cortex** (secretes mineralocorticoids, glucocorticoids, and androgens) and a **medulla** (secretes adrenalin). The adrenal glands lie at the T12 vertebral level, on the superior pole of each kidney, and are separated from the kidneys by the renal capsule. The left adrenal gland is positioned more medially than the right adrenal gland (Figure 11-5).

- **Vascular supply.** Each gland receives an arterial supply from three arteries: **superior adrenal artery** (branch of the inferior phrenic artery), **middle adrenal artery** (branch of the aorta), and **inferior adrenal artery** (branch of the renal artery). Venous drainage of the right adrenal gland empties directly into the inferior vena cava, whereas that of the left adrenal gland empties into the left renal vein.

- **Innervation. Sympathetic nerves** innervate the adrenal medulla. The **least splanchnic nerve**, through the **aorticorenal ganglion**, provides innervation. However, unlike other organs innervated by the sympathetic nerves, only **preganglionic neurons** enter the adrenal medulla. The adrenalin (epinephrine) hormones released into the blood stream act like a systemic postganglionic neuron to stimulate tissues around the body that have receptors for adrenalin.

The adrenal cortex is primarily regulated within the endocrine system by hormones from the pituitary gland (e.g., adrenocorticotropic hormone).

KIDNEYS

The kidneys are the functional unit of the urinary system. Within the kidneys, millions of nephrons maintain body water, salt, and pH balance and eliminate solutes from the blood. The kidneys accomplish these functions by **filtering blood plasma into urine**, which is eliminated from the body via the ureters, bladder, and urethra. The kidneys are a pair of **bean-shaped** organs located adjacent to the vertebrae at the T12–L2 vertebral level. The right kidney is slightly lower than the left rib because of the presence of the liver. The ureter, bladder, and kidney are all retroperitoneally positioned. **Perirenal fat** surrounds and thus cushions the kidneys (Figure 11-5).

- **Vascular supply. Renal arteries** supply blood to the kidney; **renal veins** drain blood from the kidneys.

URETERS

After the nephrons filter the blood plasma and produce urine, a series of tubes collect the urine and transport it to the bladder. Urine is received by **collecting ducts**, which in turn drain into **minor calyces** and then into **major calyces**. The major calyces in the kidney form the **renal pelvis**, which exits the kidney as the **ureter**. The ureter descends anterior to the transverse processes of the lumbar vertebrae and course over the sacral promontory and into the pelvis. In the female, the ureters course **inferior to the uterine arteries** before entering the base of the bladder.

- **Vascular supply.** Is provided by branches from the renal arteries and veins.

INNERVATION OF THE KIDNEYS AND THE PROXIMAL PART OF THE URETERS

The autonomic innervation of the kidneys and proximal ureters is as follows:

- **Sympathetic innervation.** Preganglionic neurons course through the **lesser, least, and lumbar splanchnic nerves**. They synapse in the aorticorenal and superior mesenteric ganglia, where postganglionic neurons course in a renal plexus to the kidneys.

- **Parasympathetic innervation.** Preganglionic parasympathetic neurons originate in the brainstem and course in the vagus nerve to enter the abdomen. The vagus nerve branches to become the posterior vagal trunk, which enters the celiac plexus. The neurons join the renal plexus and synapse in the wall of the kidney.

▽ **Kidney stones (renal calculi)** are accumulations of crystals that are condensed from the urine. The size of a stone, or calculi, is not as much of a concern as is its shape (smooth or spiked) or site of lodging. Stones that form in the renal pelvis are not problematic. However, if (and when) the stone moves into the ureter, the stone may serve as a dam and prevent or partially prevent the flow of urine.

- The kidney continues to produce urine, which backs up behind the stone and may cause the kidney to swell.

- The pressure buildup causes much of the pain associated with kidney stones.

- However, the pressure also helps to push the stone along until it enters the bladder.

- Depending on the level of the location of the stone, which will vary as pressure moves it along, pain will most likely be referred from the **lumbar region or to the hypogastric region to the groin.**

- The pain is referred to the cutaneous areas supplied by the same spinal cord segments that supply the ureter (i.e., lesser splanchnic, T10–T11; least splanchnic, T12; lumbar splanchnic nerves, L1–L2). The urethra has a larger lumen than the ureter, and therefore, the symptoms are usually relieved at this point. ▽

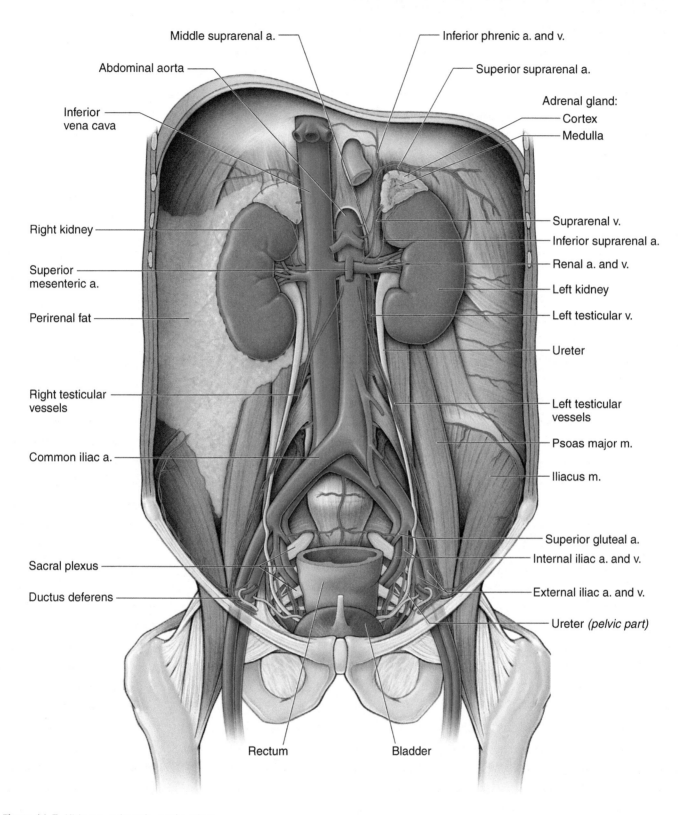

Middle suprarenal a.

Abdominal aorta

Inferior
vena cava

Right kidney

Superior
mesenteric a.

Perirenal fat

Right testicular
vessels

Common iliac a.

Sacral plexus

Ductus deferens

Inferior phrenic a. and v.

Superior suprarenal a.

Adrenal gland:
Cortex
Medulla

Suprarenal v.

Inferior suprarenal a.

Renal a. and v.

Left kidney

Left testicular v.

Ureter

Left testicular
vessels

Psoas major m.

Iliacus m.

Superior gluteal a.

Internal iliac a. and v.

External iliac a. and v.

Ureter (pelvic part)

Rectum

Bladder

Figure 11-5: Kidneys, adrenals, and ureters.

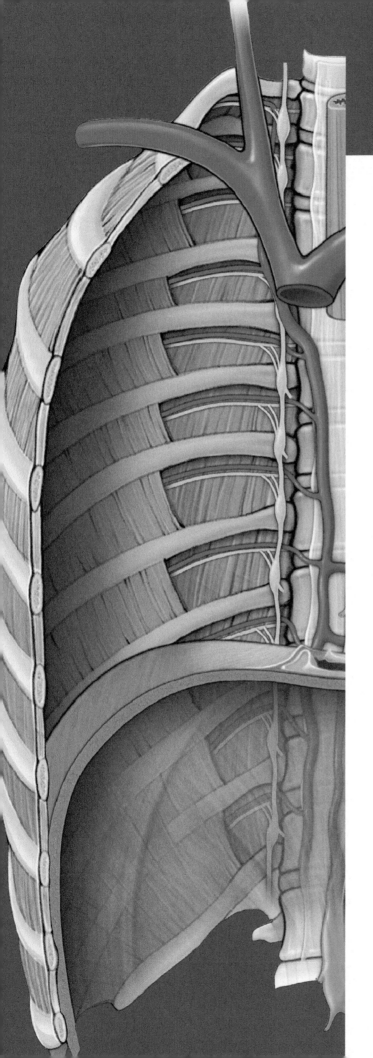

CHAPTER 12

PELVIS AND PERINEUM

PELVIC FLOOR

BIG PICTURE

The pelvic diaphragm forms the floor of the pelvis and serves as a bed for the pelvic organs.

PELVIC DIAPHRAGM

The pelvic diaphragm is formed by the union of the **levator ani** and the **coccygeus muscles**. A layer of fascia lines the superior and inferior aspects of the pelvic diaphragm (Figures 12-1A). The levator ani muscle consists of three separate muscles: pubococcygeus, puborectalis, and iliococcygeus.

The pelvic diaphragm circumferentially attaches along the pubis, lateral pelvic walls, and coccyx. The rectum pierces the center of the pelvic diaphragm, giving the appearance of a funnel suspended within the pelvis. In addition to the rectum, the urethra and the vagina (in females) and the urethra (in males) pierce the pelvic diaphragm.

Functions of the pelvic diaphragm are as follows:

- Closes the pelvic outlet.
- Supports the abdominopelvic viscera.
- Resists increases in intra-abdominal pressure.
- Controls the openings of the rectum, urethra, and vagina (e.g., helps retain or release feces during a bowel movement).
- Marks the boundary between the rectum and the anal canal.

▽ The complex organization of overlapping muscles and fascia cause the pelvic diaphragm to be susceptible to injury and damage, especially in women. Repetitive stresses, such as those that occur during labor and delivery, can stretch and damage the levator ani muscles and cause **pelvic floor insufficiency** and its associated clinical problems (e.g., uterine prolapse; urinary incontinence). ▽

OTHER MUSCLES OF THE PELVIC FLOOR

- **Obturator internus muscle.** Covers and lines most of the lateral wall of the pelvis. The obturator nerves and vessels and other branches of the internal iliac vessels course along the medial surface of the obturator internus muscle. The obturator internus muscle exits the pelvis through the lesser sciatic foramen and inserts on the greater trochanter of the femur and performs external hip rotation.
- **Piriformis muscle.** Covers most of the posterior wall of the pelvis. The piriformis muscle exits the pelvis through the greater sciatic foramen and inserts on the greater trochanter of the femur and performs external hip rotation. The sacral plexus of nerves is medial to the piriformis muscle.

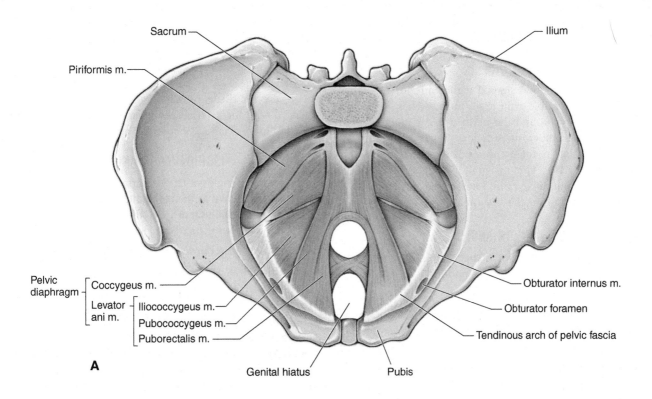

A

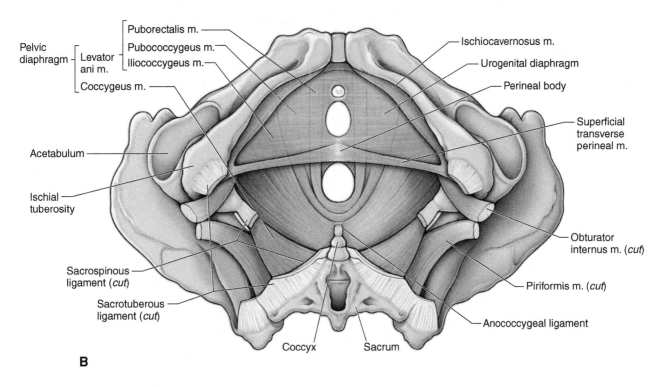

B

Figure 12-1: Superior (**A**) and inferior (**B**) views of the pelvic diaphragm muscles.

PERINEUM

BIG PICTURE

The perineum is the diamond-shaped region inferior to the pelvic diaphragm. The pubic symphysis, pubic arches, ischial tuberosities, and coccyx bound the perineum. An imaginary line between the ischial tuberosities divides the perineum into an anterior (urogenital) triangle and a posterior (ischioanal) triangle (Figure 12-2A). The urogenital triangle extends between the paired ischiopubic rami. The ischioanal triangle is the fat-filled area surrounding the anal canal.

UROGENITAL TRIANGLE

The urogenital triangle consists of a **superficial** and a **deep layer of fascia** (Figure 12-2A–D), with a middle layer of skeletal muscle called the **urogenital diaphragm**.

DEEP PERINEAL SPACE The deep perineal space is filled by the muscles that comprise the urogenital diaphragm. The deep perineal space contains the following structures (Figure 12-2B and D):

- **External urethral sphincter.** This skeletal muscle voluntarily prevents urine from spilling from the bladder.
- **Deep transverse perineal muscle.**
- **Bulbourethral (Cowper's) gland.** In males, this gland produces and secretes mucus, which lines the lumen of the urethra during ejaculation.

SUPERFICIAL PERINEAL SPACE The superficial perineal space is the region inferior to the urogenital diaphragm and its inferior fascia, and is enclosed by the **superficial perineal (Colles') fascia** (Figure 12-2A and C). The superficial perineal space contains erectile tissue, muscles, and neurovascular structures associated with the external genitalia. These structures will be covered in more detail in Chapters 13 and 14.

ISCHIOANAL TRIANGLE

The ischioanal triangle contains a horseshoe-shaped fossa filled with fat and is known as the **ischioanal fossa**. The **anal canal** and its associated **sphincters** are in the center of the triangle.

The boundaries of the ischioanal fossa are as follows:

- **Anterior.** Posterior border of the urogenital triangle.
- **Posterior.** Gluteus maximus muscle and the sacrotuberous ligament.
- **Lateral.** Deep fascia of the obturator internus muscle, which condenses around the **pudendal nerve (S2–S4)** and the **internal pudendal artery and vein**, forming the **pudendal (Alcock's) canal**. The **inferior rectal vessels and nerves** cross the fossa from the pudendal canal and supply the inferior portion of the rectum. The pudendal nerve and internal pudendal artery and vein exit the pudendal canal and course into the urogenital triangle to supply the structures in the triangle and the external genitalia.
- **Floor.** Deep fascia of the levator ani muscle.

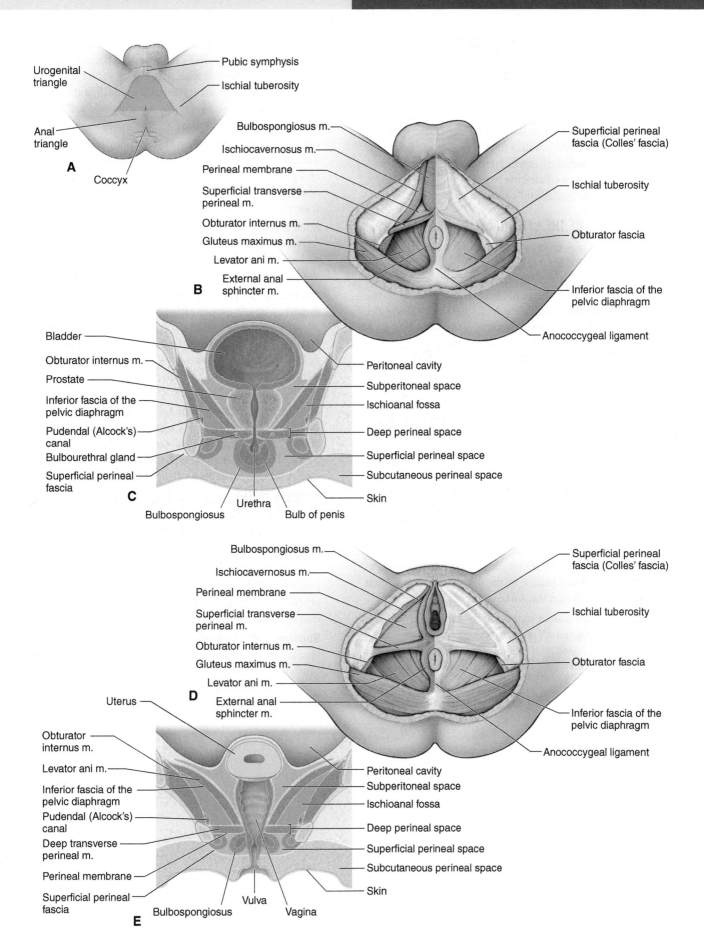

Figure 12-2: A. Urogenital and anal triangles. **B.** Male perineum. **C.** Coronal section of the male perineum. **D.** Female perineum. **E.** Coronal section of the female perineum.

PELVIC VASCULATURE

BIG PICTURE

The common iliac arteries divide at the sacroiliac joints and become the external and internal iliac arteries. The external iliac arteries mainly serve the lower limb. The internal iliac arteries distribute blood to the pelvic walls and viscera (i.e., rectum, bladder, prostate, ductus deferens, uterus, uterine tubes, and ovaries). Additionally, these arteries distribute blood to the gluteal region, the perineum, and the medial compartment of the thigh.

BRANCHES OF THE EXTERNAL ILIAC ARTERY

The external iliac artery courses deep to the inguinal ligament, where it becomes the **femoral artery** (Figure 12-3A). The femoral artery is the principal blood supplier to the lower limb. Before it exits the pelvis, the external iliac artery gives rise to the **inferior epigastric artery**, which ascends vertically along the internal surface of the anterior abdominal wall.

BRANCHES OF THE INTERNAL ILIAC ARTERY

The internal iliac artery divides into an anterior and a posterior trunk near the greater sciatic foramen (Figure 12-3A and B).

ANTERIOR TRUNK Branches from the anterior trunk supply the pelvic viscera, perineum, gluteal region, and medial compartment of the thigh. The branches off the anterior trunk of the internal iliac artery are as follows:

- **Umbilical artery.** Is usually the first branch off the anterior trunk and ascends out of the pelvis along the internal surface of the anterior abdominal wall to terminate at the umbilicus. In the fetus, the umbilical artery transports blood from the fetus through the umbilical cord to the placenta. After birth, the vessel collapses after the umbilical cord is cut. The **superior vesical artery** is a branch of the umbilical artery, which supplies the bladder and the ductus deferens.

- **Obturator artery.** Courses along the medial surface of the obturator internus muscle and exits the pelvis through the obturator canal, along with the obturator nerve and vein. The obturator artery supplies the medial compartment of the thigh.

- **Uterine artery.** In females, the uterine artery courses within the base of the broad ligament, superior to the ureter before reaching the cervix. At the cervix, the uterine artery ascends along the lateral margin of the uterus to the uterine tube and forms an anastomosis with the ovarian artery. During pregnancy, the uterine artery enlarges significantly to supply blood to the uterus, ovaries, and vaginal walls. *This artery is not present in males.*

- **Inferior vesical and vaginal artery.** In males, the **inferior vesical artery** supplies the bladder, ureter, seminal vesical, and prostate. In females, the **vaginal artery** is the equivalent of the inferior vesical artery in males and, as its name implies, it supplies the vagina as well as parts of the bladder.

- **Middle rectal artery.** Supplies the rectum and forms anastomoses with the superior rectal artery (branch of inferior mesenteric artery) and the inferior rectal artery (branch of the internal pudendal artery).

- **Internal pudendal artery.** Exits the pelvis through the greater sciatic foramen, inferior to the piriformis muscle. Along with the pudendal nerve, the internal pudendal artery courses along the pelvis lateral to the ischial spine and then passes through the lesser sciatic foramen to enter the urogenital triangle. The internal pudendal artery supplies the perineum, including the erectile tissues of the penis and clitoris.

- **Inferior gluteal artery.** The terminal branch of the anterior division of the internal iliac artery. The inferior gluteal artery courses between the ventral rami of S1 and S2 of the sacral plexus and exits the pelvis through the greater sciatic foramen. The inferior gluteal artery supplies the gluteal region.

POSTERIOR TRUNK Branches from the posterior trunk of the internal iliac artery serve the lower portion of the posterior abdominal and pelvic walls and the gluteal region.

- **Iliolumbar artery.** Gives risk to branches that supply the posterior portion of the abdominal and pelvic walls.

- **Lateral sacral artery.** Gives rise to branches that traverse the anterior sacral foramina to supply the posterior sacrum and overlying muscles.

- **Superior gluteal artery.** The terminal branch of the posterior trunk. The superior gluteal artery courses between the lumbosacral trunk and the ventral ramus of the S1 spinal nerve. This artery supplies the gluteal region, with anastomotic overlap with the inferior gluteal artery.

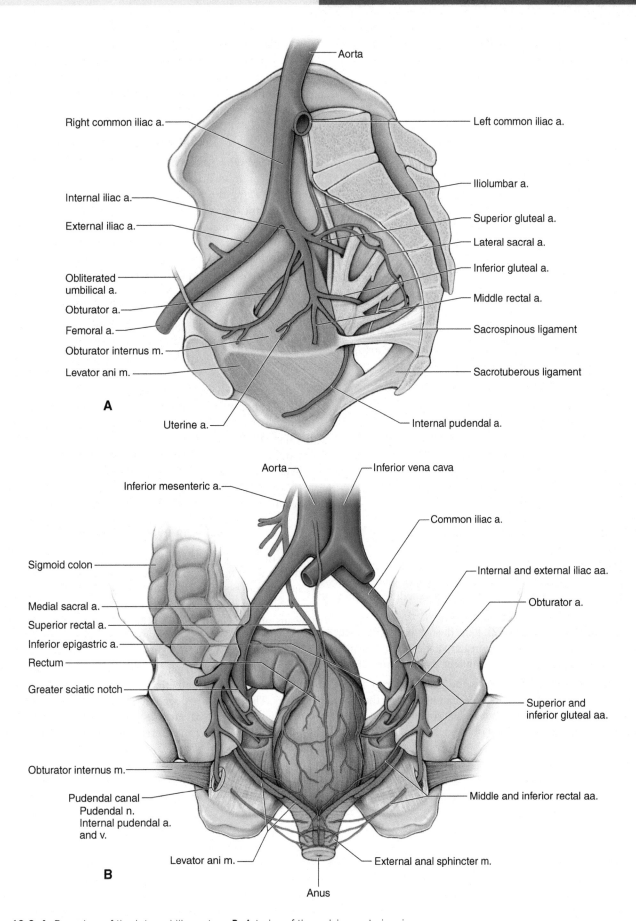

Figure 12-3: A. Branches of the internal iliac artery. **B**. Arteries of the pelvis, posterior view.

PELVIC INNERVATION

BIG PICTURE

Somatic and autonomic nerves contribute to pelvic innervation. The obturator nerve and the sacral plexus provide innervation to skeletal muscles and skin in the pelvis and lower limbs. All autonomics of the pelvis and perineum pass through the inferior hypogastric plexus. Sympathetic and parasympathetic nerves contribute to the inferior hypogastric plexus through the sacral splanchnics and pelvic splanchnics, respectively.

OBTURATOR NERVE

The obturator nerve originates from the ventral rami of spinal nerves L2–L4 and courses along the obturator internus muscle, superior to the obturator artery and vein. The neurovascular bundle exits the pelvis through the obturator canal and supplies the medial compartment of the thigh.

▼ The course of the obturator nerve is near the ovary. Therefore, the obturator nerve is at risk during an **oophorectomy** (surgical removal of an ovary). If the obturator nerve is damaged, the adductor muscles of the medial compartment of the thigh may spasm or lose function. In addition, loss or change of cutaneous sensation will occur over the medial surface of the thigh. ▼

SACRAL PLEXUS

The sacral plexus is formed by the lumbosacral trunk (ventral rami of spinal nerves L4–L5) and the ventral rami of spinal nerves S1–S4 (Figure 12-4A and B). The sacral plexus lies on the anterior surface of the piriformis muscle. The following nerves branch from the sacral plexus:

- **Superior gluteal nerve (L4–S1).** Exits the pelvis superior to the piriformis muscle, through the greater sciatic notch. The superior gluteal nerve supplies the gluteus medius and minimus muscles and the tensor fascia lata muscle.
- **Inferior gluteal nerve (L5–S2).** Exits the pelvis inferior to the piriformis muscle, through the greater sciatic notch. The inferior gluteal nerve innervates the gluteus maximus muscle.
- **Pudendal nerve (S2–S4).** Exits the pelvis inferior to the piriformis muscle and enters the perineum through the lesser sciatic foramen, where the pudendal nerve enters the pudendal canal along the lateral wall of the ischioanal fossa.

Sciatic nerve (L4–S3). The largest peripheral nerve in the body. The sciatic nerve is comprised of the **tibial** and **common peroneal nerve** and exits the pelvis inferior to the piriformis muscle, between the ischial tuberosity and the greater trochanter of the femur.

SACRAL SYMPATHETIC TRUNK

The sacral sympathetic trunk crosses the pelvic brim posterior to the common iliac vessels (Figure 12-4A). Four ganglia are present along the trunk. The trunks of the two sides unite in front of the coccyx at a small swelling called the **ganglion impar**. The sacral sympathetic trunk contributes sympathetic nerves to the somatic branches of the sacral nerves (targeting the skin) and contributes visceral branches to the inferior hypogastric plexus (targeting the pelvic viscera and perineum).

PELVIC SPLANCHNIC NERVES

The pelvic splanchnic nerves are the only splanchnic nerves that carry parasympathetic fibers (Figure 12-4A). All other splanchnic nerves, such as the greater splanchnic nerve, carry only sympathetic fibers.

Preganglionic parasympathetic fibers originate from the **S2–S4 spinal cord levels**. The fibers course through the ventral rami of **spinal nerves S2–S4** and then through the **pelvic splanchnic nerves**. Pelvic splanchnic nerves carry the parasympathetic and visceral afferent fibers to and from the **inferior hypogastric plexus**. These nerves supply the distal portion of the transverse colon, descending colon, sigmoid colon, and rectum, as well as organs of the pelvis and perineum.

INFERIOR HYPOGASTRIC PLEXUS

The inferior hypogastric plexus is formed by the union of nerves from the superior hypogastric plexus, sacral splanchnic nerves, and pelvic splanchnic nerves (Figure 12-4A). The inferior hypogastric plexus is located diffusely around the lateral walls of the rectum, bladder, and vagina. The plexus contains ganglia in which both sympathetic and parasympathetic preganglionic fibers synapse. Therefore, the inferior hypogastric plexus consists of preganglionic and postganglionic sympathetic and parasympathetic fibers, as well as visceral sensory fibers. The inferior hypogastric plexus gives rise to many other smaller plexuses that provide innervation to organs involved with urination, defecation, erection, ejaculation, and orgasm.

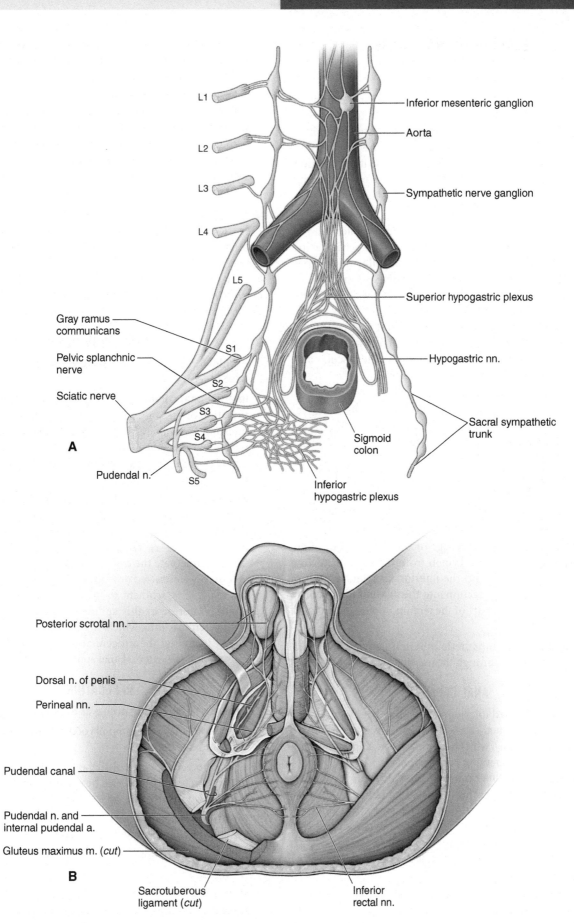

Figure 12-4: A. Prevertebral and sacral plexuses. **B**. Innervation of the perineum.

RECTUM AND ANAL CANAL

BIG PICTURE

The colon terminates at the rectum, which in turn terminates at the anus. The rectum is located superior to the pelvic diaphragm in the pelvic cavity. The anus is located inferior to the pelvic diaphragm in the ischioanal fossa. The terminal end of the gastrointestinal tract is significant because it is a junction of two embryologic origins of tissue: endoderm and ectoderm.

RECTUM

The rectum is a continuation of the sigmoid colon and is located superior to the pelvic diaphragm in the pelvic cavity. The rectum has a dilated region called the **ampulla**, which stores feces. As feces collect in the rectum, increasing pressure causes the walls to bulge. Stretch receptors in the rectal wall relay messages to the brain for the need to defecate.

Peritoneum covers the anterior and lateral surfaces of the proximal portion of the rectum; however, there is no peritoneal covering for the distal portion of the rectum. The vascular and lymphatic supply of the rectum is as follows:

- **Arteries.** The rectum receives its blood supply from the superior rectal artery (a branch of the inferior mesenteric artery), the middle rectal artery (a branch of the internal iliac artery), and the inferior rectal artery (a branch of the internal pudendal artery) (Figure 12-5A).
- **Veins.** Venous return is through the superior rectal vein (a branch of the inferior mesenteric vein of the portal system), the middle rectal vein (the internal iliac vein), and the inferior rectal vein (the internal pudendal vein) (Figure 12-5B).
- **Lymphatics.** Lymphatic drainage of the rectum is through three principal directions. Lymph from the superior region of the rectum drains into the **inferior mesenteric nodes**, from the middle of the rectum into the **internal iliac nodes**, and from the inferior part of the rectum into the **superficial inguinal nodes**.
- **Innervation.** Parasympathetic innervation via the pelvic splanchnic nerves is through the inferior hypogastric plexus.

ANAL CANAL

The anal canal is located inferior to the pelvic diaphragm in the ischioanal fossa. The anal canal is divided into an upper two-thirds (visceral portion), which is part of the large intestine, and a lower one-third (somatic portion), which is part of the perineum. The **pectinate line** (anorectal junction) is an important landmark in that it divides the anal canal into upper and lower portions. Developmentally, the pectinate line is the junction between the development of the hindgut (gut tube) and the proctodeum (body wall). The pectineal line is an important anatomic landmark in that it distinguishes the vascular, nerve, and lymphatic supplies as follows:

- **Superior to the pectinate line.** The vascular supply is from the superior and middle rectal arteries and veins. Visceral motor and sensory innervation is via the inferior hypogastric plexus. Lymph drainage is to the internal iliac, inferior mesenteric, and pararectal lymph nodes. The epithelium is simple columnar, as is the remainder of the small and large intestines, which reflects the endodermal origin of this part of the anal canal.
- **Inferior to the pectineal line.** The vascular supply is from the inferior rectal arteries and veins. Somatic motor and sensory innervation is via the inferior rectal nerves. Lymph drainage is to the superficial inguinal lymph nodes. The epithelium is stratified squamous keratinized epithelium, similar to the skin, which reflects the ectodermal origin of this part of the anal canal.

▽ **Hemorrhoids** are dilated and inflamed venous plexuses within the anorectal canal. Hemorrhoids are classified as **internal** (superior to the pectineal line) or **external** (inferior to the pectineal line). Internal hemorrhoids usually are not painful because visceral sensory nerves lack pain receptors. In contrast, external hemorrhoids are usually painful because their innervation is from somatic sensory nerves, which detect pain. ▽

ANAL SPHINCTERS

The following two sphincters are responsible for regulating passage of fecal matter:

- **Internal anal sphincter.** The internal anal sphincter is a continuation of the smooth muscle layer of the remainder of the intestine. Sympathetic innervation (lower lumbar splanchnic nerves) causes contraction of the internal anal sphincter, whereas parasympathetic innervation (pelvic splanchnic nerves) causes relaxation.
- **External anal sphincter.** The external anal sphincter is a voluntary skeletal muscle that encircles the distal portion of the anus and enables voluntary control of defecation. The paired inferior rectal nerves from the pudendal nerve (S2–S4) innervate the external anal sphincter. The muscle is voluntarily contracted or relaxed. The external anal sphincter has a fascial attachment to the coccyx (**anococcygeal ligament**). Anterior to this sphincter is the **perineal body**, a strong tendon into which many of the perineal muscles insert, including the urogenital diaphragm, the levator ani muscle, and the external anal sphincter.

DEFECATION

The process of defecation occurs as follows:

1. Distension of the rectal ampulla occurs from feces passing from the sigmoid colon.
2. The puborectalis portion of the levator ani muscle relaxes, thereby decreasing the angle between the rectal ampulla and the upper portion of the anal canal.
3. Intra-abdominal pressure increases when the diaphragm and the abdominal body wall muscles contract.
4. The internal anal sphincter relaxes, as does the external anal sphincter.
5. Feces pass out of the rectum and anus.
6. After defecation has occurred, the puborectalis muscle contracts once again, increasing the angle between the rectal ampulla and the upper portion of the anal canal, as do the anal sphincters to close the anus.

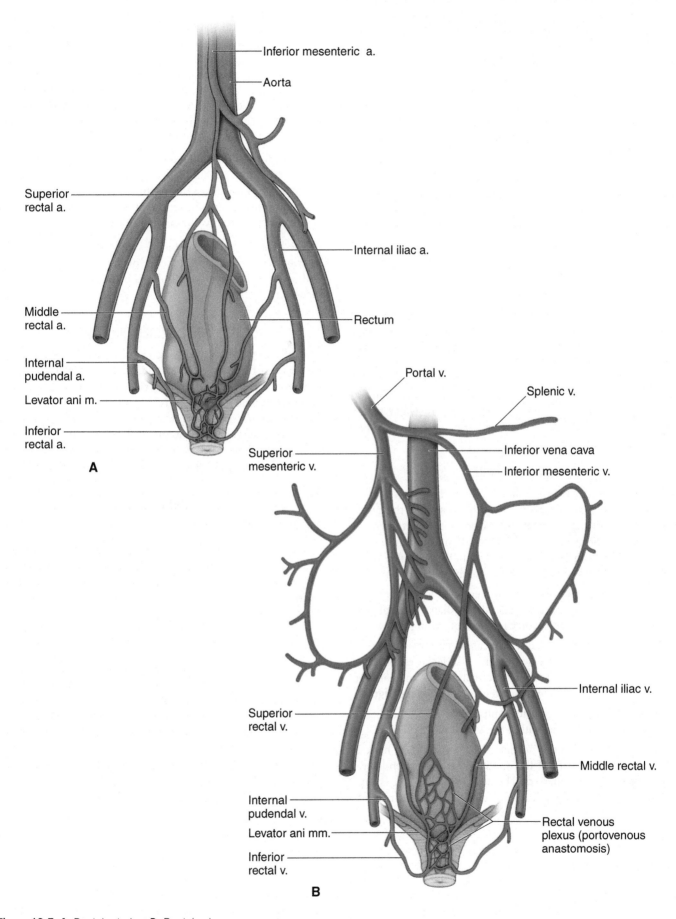

Figure 12-5: A. Rectal arteries. **B**. Rectal veins.

URETERS AND URINARY BLADDER

BIG PICTURE

The ureters connect the kidneys to the urinary bladder, which stores urine until urination occurs. The urethra voids urine from the urinary bladder.

URETER

The ureter is a retroperitoneal organ that transports urine that is propelled by peristaltic waves from the kidney to the urinary bladder (Figure 12-6A). **Three constrictions** occur along the length of the urethra: (1) where the renal pelvis becomes the ureter; (2) where the ureter crosses the pelvic brim; and (3) where the ureter enters the bladder (Figure 12-6C). In males, the ureter courses posterior and medial to the ductus deferens and anterior to the seminal vesicle. In females, the ureter courses lateral to the cervix, where the ureter courses inferior to the uterine artery.

▽ An important anatomic relationship that pelvic surgeons rely on is that the ureter courses inferior to the uterine artery. When performing a hysterectomy, the surgeon clamps the uterine artery to prevent bleeding. If the surgeon is not careful, the ureter may be clamped and cut by accident. A mnemonic for remembering this relationship is "**water under the bridge**," where "water" represents "urine" and the "bridge" represents the "uterine artery" (Figure 12-6C). ▽

URINARY BLADDER

The urinary bladder is retroperitoneal, as are the ureters (Figure 12-6A). The urinary bladder is positioned against the pubis between the pelvic diaphragm and the obturator internus muscles. The superior surface of the bladder is dome shaped when the bladder is empty and swells superiorly into the abdomen when the bladder is full. The bladder consists of the following parts:

- **Apex.** The top of the bladder is located at the top of the pubic symphysis. The apex continues as the median umbilical ligament, which is the embryonic remnant of the urachus.
- **Base.** The base of the bladder is located inferiorly and posteriorly. The paired ureters enter the bladder at each of the superior corners of the base. Internally, the triangular area between the openings of the ureters is known as the **trigone**.
- **Detrusor muscle.** The detrusor muscle consists of bundles of smooth muscle located within the wall of the bladder.
- **Neck.** The most inferior portion of the bladder is the neck. It surrounds the origin of the urethra where the inferolateral surfaces and base intersect. The neck of the bladder is supported by the pubovesical ligament, fibromuscular bands that attach between the neck and pubic bones.

- **Internal urethral sphincter.** The internal urethral sphincter is smooth muscle that involuntarily contracts or relaxes, thereby regulating the emptying of the bladder (Figure 12-6B). The constant muscular tone of the internal urethral sphincter is through sympathetic innervation. However, to urinate, the muscle is relaxed via parasympathetic innervation. This is the primary muscle for preventing the release of urine.
- **External urethral sphincter.** This sphincter is composed of skeletal muscle within the urogenital diaphragm that voluntarily opens and closes the urethra to void urine. The external urethral sphincter is innervated by the pudendal nerve and, therefore, is under voluntary control.

The bladder receives its blood supply via the superior and inferior vesical arteries (branches of the internal iliac artery) and is drained by the vesical plexus of veins (into the internal iliac vein).

URETER AND BLADDER INNERVATION

The inferior portions of the ureter and urinary bladder receive innervation from the **vesical and prostatic plexuses** (extensions of the inferior hypogastric plexus) (Figure 12-6B). Parasympathetic innervation is from the S2–S4 spinal cord levels, which enter the inferior hypogastric plexus, as do sacral splanchnic nerves for sympathetic innervation. The inferior hypogastric plexus then gives rise to the vesical and prostatic plexuses, which innervate the ureter and urinary bladder.

URINATION

The **process of urination** occurs as follows:

- The bladder fills with urine, causing the bladder to distend.
- Visceral sensory fibers relay to the spinal cord (S2–S4), via the pelvic splanchnic nerves, that the bladder wall is stretched.
- Preganglionic parasympathetic fibers from the S2–S4 spinal cord segments enter the spinal nerves and pelvic splanchnic nerves. The pelvic splanchnic nerves enter the inferior hypogastric plexus; the preganglionic parasympathetic fibers course from the inferior hypogastric plexus to the bladder, where they synapse with postganglionic parasympathetic fibers. Stimulation of these parasympathetic nerves causes the detrusor muscle to contract and the internal urethral sphincter to relax.
- Somatic motor neurons in the pudendal nerve cause relaxation of the external urethral sphincter and contraction of the bulbospongiosus muscles, which expel the last drops of urine from the urethra.

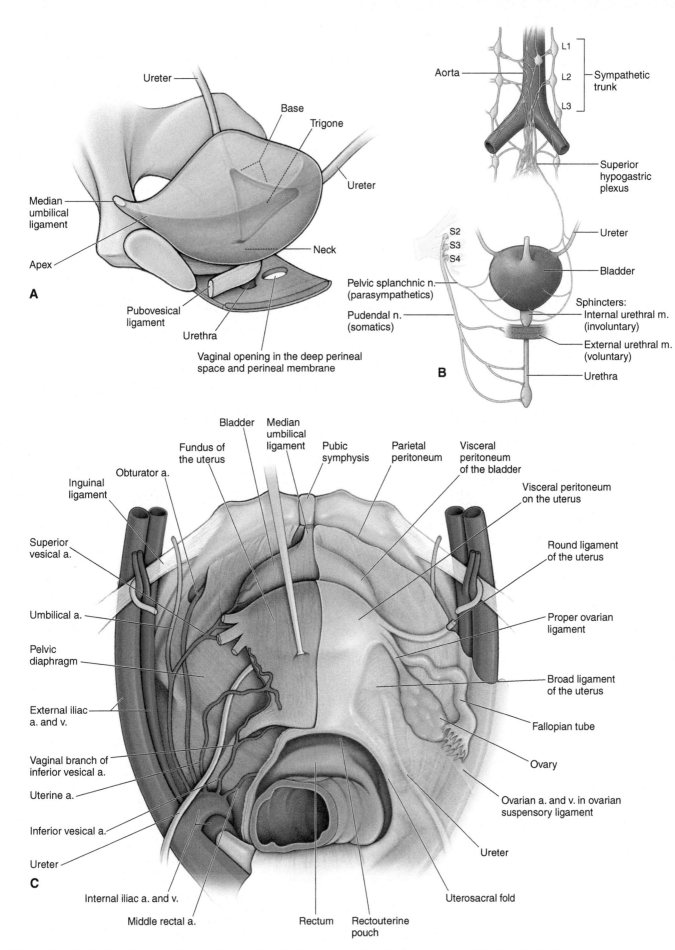

Figure 12-6: A. Bladder. **B**. Innervation of distal ureter, bladder, and urethra. **C**. Superior view of the bladder in situ.

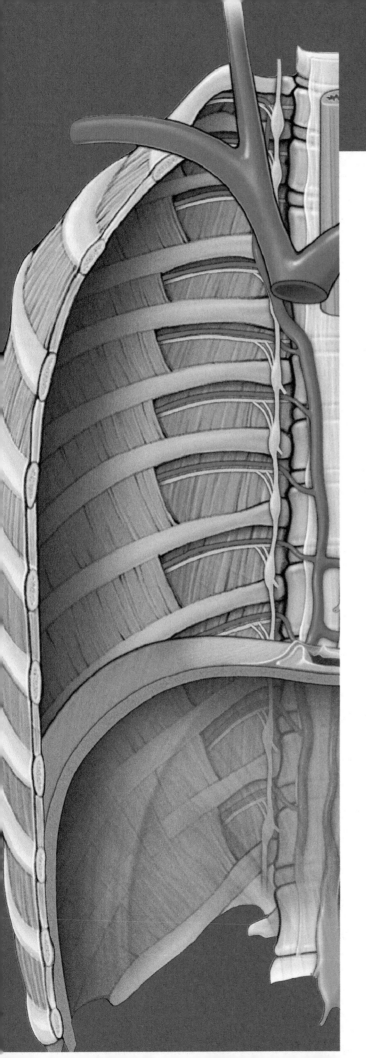

CHAPTER 13

MALE REPRODUCTIVE SYSTEM

MALE REPRODUCTIVE SYSTEM

BIG PICTURE

The male reproductive system primarily consists of the paired testes and the penis. In addition, accessory sex glands contribute to seminal fluid. The male reproductive system matures during adolescence and remains active for the remainder of the lifespan of the male.

MALE GENITAL ORGANS AND GLANDS

The following genital organs and glands comprise the male reproductive system (Figure 13-1A):

- **Testes.** The primary male sex organ. The testes produce sperm and sex hormones (e.g., testosterone) and are located within the scrotum.
- **Epididymis.** A convoluted duct that sits on the superior pole of each testis. Sperm are stored in the epididymis during the maturation process.
- **Ductus deferens.** A thick-walled tube in the spermatic cord that transports sperm from the epididymis to the ejaculatory ducts in the prostate gland. The ductus deferens traverses the superficial inguinal ring, coursing through the inguinal canal, and enters the pelvis through the deep inguinal ring lateral to the inferior epigastric artery. En route to the ejaculatory duct, the ductus deferens crosses the medial side of the umbilical artery and the obturator neurovascular structures. **Sympathetic nerves** from the inferior hypogastric plexus cause peristaltic contractions in the thick smooth muscle wall and propel sperm during ejaculation.
- **Ejaculatory ducts.** Formed by the union of the ductus deferens and ducts from the seminal vesicles. The ejaculatory ducts open into the prostatic urethra.
- **Seminal vesicles.** Lobular glands located on the base of the bladder. During emission and ejaculation, the seminal vesicles empty their secretions (e.g., fructose, citric acid, prostaglandins, and fibrinogen) into the ejaculatory duct, along with sperm from the ductus deferens. Seminal vesicle secretions add substantially to the volume of semen.
- **Prostate gland.** Composed of five lobes, all surrounding the **prostatic urethra**. The prostate gland is located superior to the pelvic diaphragm and anterior to the rectum. The prostate gland secretes a milky fluid that contributes to the bulk of the semen.

▽ The prostate gland may hypertrophy as men age. As a result of an **enlarged prostate gland**, affected men may have difficulty urinating because the gland surrounds the urethra. Because of its proximal anterior location to the rectum, the prostate gland is relatively easy to palpate. A **digital rectal examination** is performed to determine the size of the prostate gland. During a digital rectal examination, the physician may also palpate the seminal vesicles and the ductus deferens. ▽

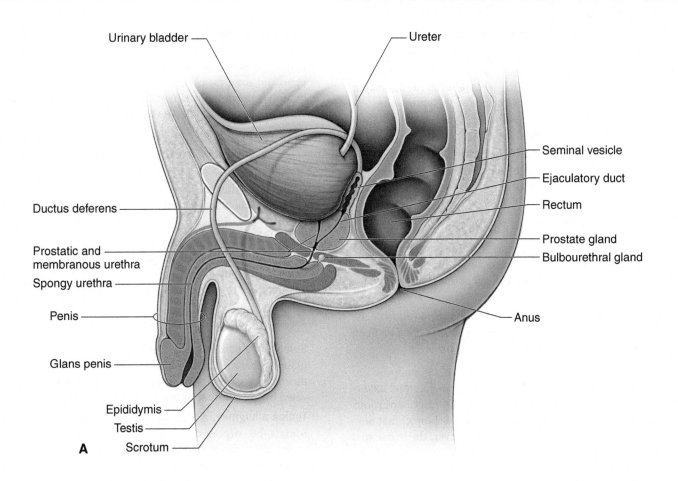

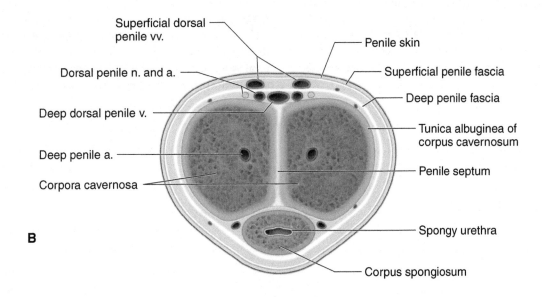

Figure 13-1: A. Male reproductive system. **B.** Cross-section of the penis.

DEEP PERINEAL SPACE

The deep perineal space (pouch) is the region within the **urogenital diaphragm** (Figure 13-1D). In males, this space contains the **bulbourethral glands** and the **internal urethral sphincter**. The bulbourethral glands secrete mucus that flows through the urethra during sexual intercourse to aid in lubrication. Parasympathetic nerves from the inferior hypogastric plexus innervate the bulbourethral glands.

- **Membranous urethra.** The portion of the urethra that courses through the deep perineal space.

SUPERFICIAL PERINEAL SPACE

The superficial perineal space (pouch) is the region inferior to the urogenital diaphragm. It is enclosed by the superficial perineal (Colles') fascia. The superficial perineal space contains (Figure 13-1A–D):

- **Deep (Buck's) fascia of the penis.** Continuous with the external spermatic fascia and deep perineal fascia. The **deep dorsal vein of the penis** is inside this fascial layer.

CRURA OF THE PENIS Each crus of the penis is composed of erectile tissue that is continuous with the paired **corpus cavernosa** of the penis. Erectile tissue consists of large, cavernous venous sinusoids (spaces like those found in a sponge) that normally are somewhat void of blood. However, during sexual arousal, the corpus cavernosa fill with blood, causing the penis to become erect (Figure 13-1B and C).

- **Corpus cavernosa.** The crura of the penis are separate along the ischiopubic ramus. However, at the pubic symphysis, the corpora cavernosa course along the dorsum of the penis, adjacent to each other. The **deep arteries of the penis** (branches of the internal pudendal artery) course within the center of the corpus cavernosa, providing blood that is necessary for an erection.

- **Ischiocavernosus muscle.** The ischiocavernosus muscles are voluntary skeletal muscles that surround the crura of the penis. The ischiocavernosus muscle helps to stabilize an erect penis and compresses the crus of the penis to impede venous blood return to maintain an erection. This muscle is innervated by the **perineal nerve** (branch of the pudendal nerve).

BULB OF THE PENIS   The bulb of the penis is the origin of the corpus spongiosum. The bulb of the penis is composed of erectile tissue and is continuous with the **corpus spongiosum** of the penis.

- **Corpus spongiosum.** The ventrally located erectile tissue that surrounds the **spongy urethra**, which transports urine and semen.

- **Glans penis.** The terminal part of the corpus spongiosum. A fold of skin called the **prepuce** (foreskin) covers the glans penis. The **frenulum** of the prepuce is a median ventral fold passing from the deep surface of the prepuce.

▽   **Circumcision** is the surgical removal of the prepuce.   ▼

- **Bulbospongiosus muscle.** Contributes to erection and ejaculation. This muscle also helps expel the final drops of urine during micturition. The bulbospongiosus muscle is innervated by the **perineal nerve** (branch of the pudendal nerve).

- **Tunica albuginea.** A thin layer of connective tissue that surrounds the corpora cavernosa and corpus spongiosum. The tunica albuginea is denser around the corpora cavernosa and inhibits blood return during an erection. The tunica albuginea is more elastic around the corpus spongiosum and enables semen to pass through the urethra during ejaculation.

▽   The outline of the superficial perineal space can be observed when the **spongy urethra is ruptured** inferior to the urogenital diaphragm. Urine flows into the superficial perineal space (extravasated urine) by spreading into the scrotum, around the penis, and superiorly into the abdominal wall.   ▼

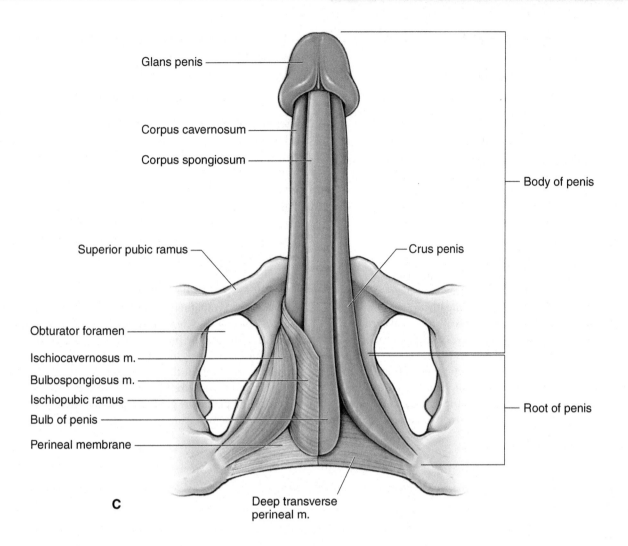

C

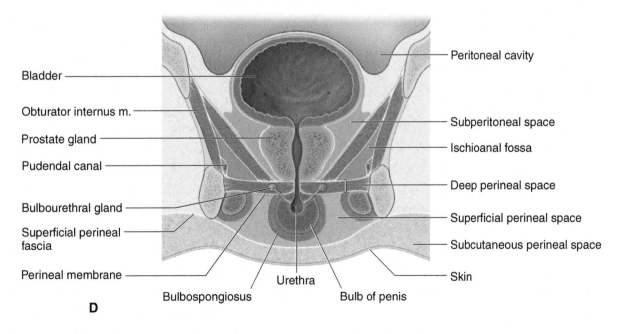

D

Figure 13-1: (*continued*) **C**. Erectile muscles and tissues. **D**. Coronal section of the male perineum.

MALE SEX ACT

BIG PICTURE

The male sex act begins with sexual stimulation. Somatic sensory nerves relay this information to the central nervous system. Parasympathetic impulses from the S2–S4 levels of the spinal cord cause blood to flow into the erectile tissue of the penis, resulting in penile erection. Sympathetic impulses from the T10–L2 spinal cord levels cause seminal fluids to mix with the sperm in the urethra in a process called emission. Ejaculation is the expulsion of the semen from the penis, which is caused by sympathetic innervation as well.

ERECTION

The corpora cavernosa and corpus spongiosum are bodies of erectile tissue. Erectile tissue is made of cavernous sinusoids.

Penile erection is initiated by parasympathetic innervation. The impulses originate in the S2–S4 levels of the spinal cord and travel through the pelvic splanchnic nerves to the central arteries of the corpora cavernosa and corpus spongiosum (erectile tissue) (Figure 13-2). The parasympathetic innervation relaxes the smooth muscle in arteries. As a result, blood fills the spaces of the erectile tissue, causing the erectile bodies to enlarge. Because the erectile tissue is surrounded by the deep penile (Buck's) fascia: pressure within the sinusoids increases, causing expansion of the erectile tissue so that the penis becomes elongated and rigid, forming an **erection**.

EMISSION AND EJACULATION

Emission and ejaculation are the climax of the male sex act; both processes result through sympathetic innervation. Impulses originating in the T10–L2 spinal cord levels travel through the lumbar splanchnic nerves and through the hypogastric plexus to the male reproductive organs and glands.

- **Emission.** Begins with the contraction of the ductus deferens by moving sperm into the urethra, where contractions of the seminal vesicles and the prostate and bulbourethral glands secrete seminal fluid into the urethra and force the sperm forward. All of the fluids mix in the urethra, forming semen. This process is known as emission.

- **Ejaculation.** Sympathetic impulses cause rhythmic contraction of smooth muscle within the ductus deferens and urethra. In addition, the pudendal nerves cause contraction of the ischiocavernosus and bulbocavernosus muscles, which compress the bases of the penile erectile tissue. These effects together cause wavelike increases in pressure in the erectile tissue of the penis, the genital ducts, and the urethra, which ejaculate the semen from the urethra to the exterior.

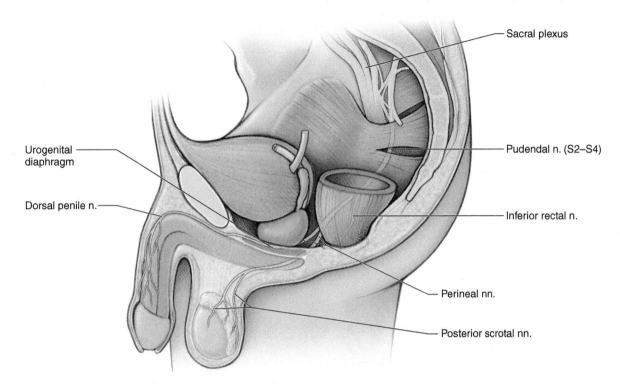

Figure 13-2: Innervation of the penis.

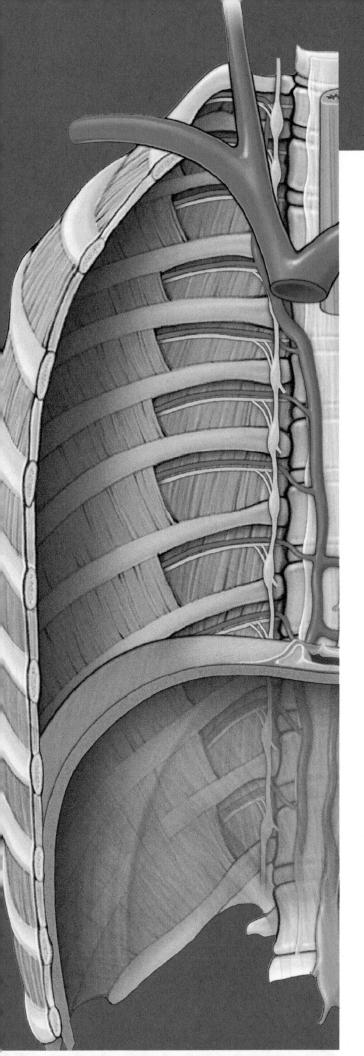

CHAPTER 14

FEMALE REPRODUCTIVE SYSTEM

FEMALE REPRODUCTIVE SYSTEM

BIG PICTURE

The female reproductive system consists of the ovaries, uterine tubes, uterus, vagina, and external genitalia. These organs remain underdeveloped for about the first 10 years of life. During adolescence, sexual development occurs and menses first occur (menarche). Cyclic changes occur throughout the reproductive period, with an average cycle length of approximately 28 days. These cycles cease at about the fifth decade of life (menopause), at which time the reproductive organs become atrophic.

OVARIES

The ovaries are the primary female sex organ because they produce eggs (ovum or oocytes) and sex hormones (e.g., estrogen). The ovaries are located within the pelvic cavity.

UTERINE TUBES

The paired uterine tubes are also called fallopian tubes, or oviducts, and extend from the ovaries to the uterus (Figure 14-1A and B). The luminal diameter of the uterine tubes is very narrow and, in fact, is only as wide as a human hair.

- **Infundibulum and fimbriae.** The infundibulum is the funnel-shaped, peripheral end of the uterine tube. The infundibulum has fingerlike projections called fimbriae. The fimbriated end of the uterine tube is not covered by peritoneum, which provides open communication between the uterine tube and the peritoneal (pelvic) cavity. In contrast to the male reproductive system, where the tubules are continuous with the testes, the uterine tubes are separate from the ovaries. Oocytes are released (ovulation) into the peritoneal cavity. The beating of the fimbriae may create currents in the peritoneal fluid, which carry oocytes into the uterine tube lumen.
- **Ampulla.** The ampulla is a region of the uterine tube where fertilization usually occurs.
- **Isthmus.** The isthmus is the constricted region of the uterine tube where each tube attaches to the superolateral wall of the uterus.

▼ An **ectopic pregnancy** occurs when a fertilized egg implants in the uterine tube or peritoneal cavity. The uterine tubes are not continuous with the ovaries, and therefore, there is also a risk that fertilization and implantation will occur outside of the uterine tubes in the peritoneal cavity. Ectopic pregnancies usually result in loss of the fertilized ovum and in hemorrhage, putting the health of the woman at risk. ▼

UTERUS

The uterus, known as the womb, resembles an inverted pear and is located in the pelvic cavity between the rectum and the urinary bladder (Figure 14-1A and B). The uterus is a hollow organ that functions to receive and nourish a fertilized oocyte until birth. Normally, the uterus is flexed anteriorly, where it joins the vagina; however, the uterus may also be retroverted (flexed posteriorly). The pelvic and urogenital diaphragms support the uterus. The uterus consists of the following subdivisions:

- **Fundus.** The rounded superior surface between the uterine tubes.
- **Body.** The main part of the uterus located between the uterine tubes and isthmus. The lumen of the uterus is triangular in the coronal section and continuous with the uterine tubes and vaginal canal.
- **Isthmus.** The narrow region between the body and the cervix.
- **Cervix.** The outlet that projects into the vagina. The **internal os** is the junction of the cervical canal and the uterine body; the **external os** communicates with the vaginal canal.
- **Arterial supply.** From branches of the internal iliac artery (uterine arteries) and aorta (ovarian arteries).

▼ **Cervical cancer** is a common form of cancer that occurs in women aged 30 to 55 years. This cancer usually arises from the epithelium that covers the cervix. The most effective method of detecting cervical cancer is by a **Papanicolaou (Pap) smear**, in which cervical epithelial cells are scraped from the cervix and examined to determine if the cells are abnormal. ▼

BROAD LIGAMENT The uterine tubes and uterus are covered by a layer of peritoneum on the anterior, superior, and posterior surfaces (Figure 14-1C). Inferior to the uterine tube and lateral to the uterus, the peritoneal membrane is fused into a double layer called the broad ligament.

- **Mesosalpinx.** The uterine tubes course along the upper edge of the broad ligaments. This portion of the broad ligament is called the mesosalpinx.
- **Suspensory ligament of the ovary.** The ovarian arteries and veins course between the double layers of the broad ligament. This portion of the broad ligament is called the suspensory ligament.
- **Round ligament.** The round ligament is a fibrous cord that courses from the uterus through the deep inguinal ring and inguinal canal, exits the superficial inguinal ring, and attaches to the labia majora. The round ligament courses between the double layers of the broad ligament.
- **Ovarian ligament.** This ligament is a fibrous cord that connects the ovary to the uterine body.
- **Mesovarium.** The ovary is partially covered by a separate posterior fold of the broad ligament called the mesovarium.

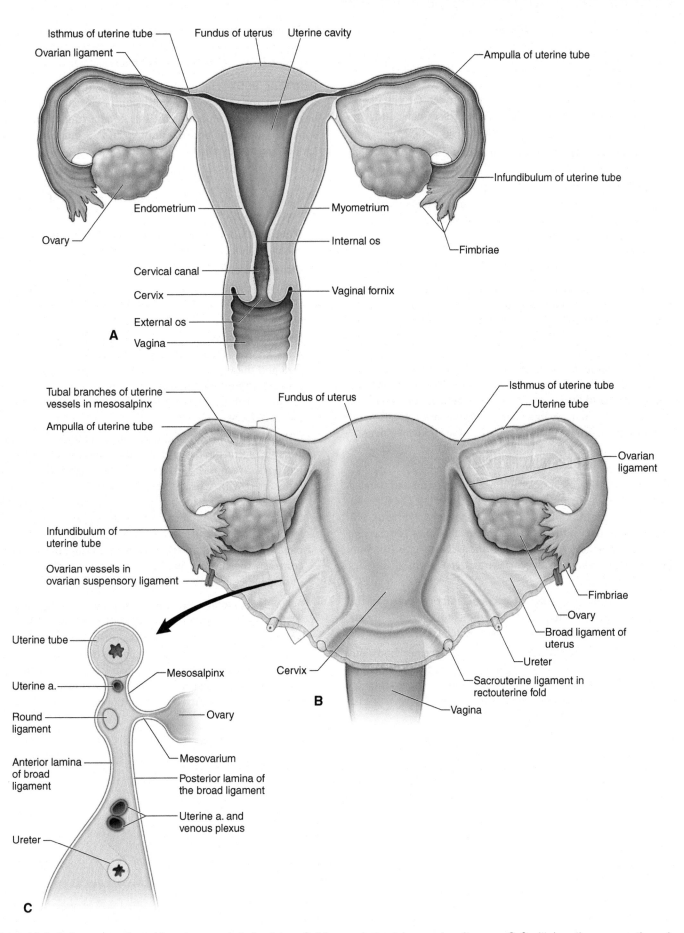

Figure 14-1: A. Coronal section of the uterus and uterine tubes. **B**. Uterus, uterine tubes, and peritoneum. **C**. Sagittal section as seen through the broad ligament of the uterus.

VAGINA

The vagina serves as the inferior region of the birth canal. It also serves as the passageway for the sloughed endometrium that results from menstruation and is the receptacle for the penis during sexual intercourse. The recesses between the cervix and the vaginal wall are known as the fornices. The vascular and lymphatic supply for the vagina is as follows (Figure 14-2A and B):

- **Vascular supply.** Receives blood supply from the vaginal branches of the uterine artery and the internal iliac artery.
- **Lymphatic drainage.** Lymph drains in two directions. The lymphatics from the upper region drain into the internal iliac nodes. Lymphatics from the lower region of the vagina drain into the superficial inguinal nodes.

DEEP PERINEAL SPACE

The deep perineal space is located between the superior and inferior fascia of the **urogenital diaphragm** (Figure 14-2C and D). The urogenital diaphragm stretches between the paired pubic rami and the ischial rami. The urethra and the vagina pierce the urogenital diaphragm. The urogenital diaphragm consists of the **deep transverse perineal muscle** and the **internal urethral sphincter, both of which are covered by fascia (superior and inferior fascia of the UG diaphram).** The **perineal branch** of the internal pudendal nerve innervates both muscles. The space also contains the internal pudendal vessels and the pudendal nerve.

SUPERFICIAL PERINEAL SPACE

The superficial perineal space is the region inferior and parallel to the urogenital diaphragm (Figure 14-2C and D). The superior boundary is the inferior fascia of the urogenital diaphragm. The inferior boundary is the superficial perineal (Colles') fascia. The superficial perineal space contains the ischiocavernosus, bulbospongiosus, and the superficial transverse perineal muscles (Figure 14-2C).

- **Ischiocavernosus muscle.** Covers the inferior surface of the corpus cavernosum (the crus of the clitoris) and is innervated by the perineal branch of the pudendal nerve.
- **Bulbospongiosus muscle.** Arises from the perineal body and inserts into the corpus spongiosum. The perineal branch of the pudendal nerve innervates the bulbospongiosus muscle. This muscle compresses the erectile tissue of the vestibular bulbs and constricts the vaginal orifice.

- **Superficial transverse perineal muscle.** Attaches to and supports the perineal body. The perineal branch of the pudendal nerve innervates the superficial perineal muscle.

PERINEAL BODY The perineal body is a fibromuscular mass located at the center of the perineum, between the anus and vagina. The perineal body serves as an attachment site for the superficial and deep perineal muscles and the bulbospongiosus, levator ani, and external anal sphincter muscles.

EXTERNAL GENITALIA

The follow structures form the external genitalia of the female reproductive system (Figure 14-2A):

- **Mons pubis.** A rounded area superficial to the pubic symphysis containing adipose tissue.
- **Labia majora.** Paired longitudinal ridges of skin that are inferior and posterior to the mons pubis. The outer surfaces are covered with pubic hair. The round ligaments of the uterus are attached to the labia majora.
- **Labia minora.** Paired hairless skin ridges flanking a midline space known as the vestibule. The vestibule contains the urethra and the vagina. Flanking the vaginal opening are small **greater vestibular glands**, which release mucus into the vestibule that helps to keep the vagina lubricated during sexual intercourse.
- **Clitoris.** The junction of the labia minora folds. The clitoris consists of erectile tissue that is continuous with two crura (corpora cavernosa), and a glans. The clitoris, which is richly innervated with sensory nerve endings sensitive to touch, become engorged with blood and erect with blood during tactile stimulation, contributing to sexual arousal in the female. The female urinary and reproductive tracts are completely separate, with neither running through the clitoris. In contrast, the urethra in the male carries both urine and semen and courses through the penis.

▽ An **episiotomy** is a surgical incision made between the posterior edge of the vagina and the perineal body to enlarge the superficial opening of the birth canal. The incision can be midline or at an angle. An episiotomy is performed to prevent random tearing of the perineal structures, particularly across the external anal sphincter during childbirth. If the external and sphincter is torn, rectal incontinence can occur. ▼

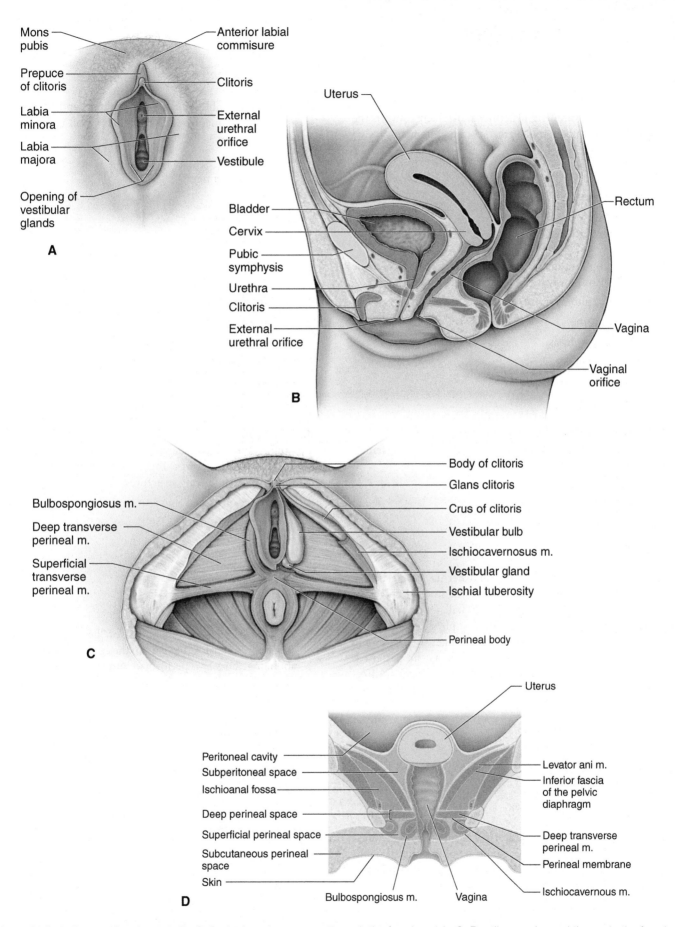

Figure 14-2: A. External female genitalia. **B**. Sagittal section as seen through the female pelvis. **C**. Erectile muscles and tissues in the female. **D**. Coronal section of the female perineum.

STUDY QUESTIONS

Directions: Each of the numbered items or incomplete statements is followed by lettered options. Select the **one** lettered option that is **best** in each case.

1. Which of the following structures most likely converts the greater sciatic notch to the greater sciatic foramen?
   A. Obturator membrane
   B. Obturator internus muscle
   C. Piriformis muscle
   D. Sacrospinous ligament
   E. Sacrotuberous ligament

2. The ischiopubic or conjoint ramus is formed when the ischial ramus joins which of the following structures?
   A. Inferior pubic ramus
   B. Ischial spine
   C. Pubic symphysis
   D. Pubic tubercle
   E. Superior pubic ramus
   F. Sciatic notch

3. The female pubic arch differs from the male pubic arch to facilitate childbirth. When compared to the male, the female pubic arch can best be described as
   A. narrower
   B. shorter
   C. taller
   D. wider

4. Diagnosis of an indirect inguinal hernia is determined when intestine protrudes lateral to the inferior epigastric artery through the abdominal body wall. During the physical examination of a male patient, a physician will assess for an indirect hernia by inserting a finger in the scrotum and feeling for bowel that protrudes, as the patient is instructed to turn his head and cough. If an indirect inguinal hernia is present, the physician will most likely feel bowel at which of the following sites?
   A. Anterior superior iliac spine
   B. Deep inguinal ring
   C. McBurney's point
   D. Pubic symphysis
   E. Superficial inguinal ring

5. During the initial examination of a 3.6 kg (8 lb) male infant delivered at term, urine is found to be leaking from the umbilicus. This infant most likely has an abnormality of which of the following fetal structures?
   A. Umbilical arteries
   B. Umbilical vein

C. Urachus
D. Urogenital sinus
E. Urorectal septum

6. The external oblique, internal oblique, and transversus abdominis aponeuroses all have a common insertion into which structure?
   A. Arcuate line
   B. Inguinal ligament
   C. Linea alba
   D. Pectineal line
   E. Pubic tubercle

7. Collateral circulation between the subclavian and external iliac arteries is created by an anastomosis between which of the following structures?
   A. Epigastric arteries
   B. Lumbar arteries
   C. Posterior intercostal arteries
   D. Round ligament of the liver
   E. Superficial epigastric arteries

8. When performing gastric bypass surgery on a 36-year-old woman, the surgeon identifies the hepatogastric and hepatoduodenal ligaments. Together, both ligaments create which of the following structures?
   A. Greater omentum
   B. Lesser omentum
   C. Mesentery
   D. Parietal peritoneum
   E. Omental bursa
   F. Visceral peritoneum

9. A 38-year-old man with a history of "heartburn" suddenly experiences excruciating pain in the epigastric region of his abdomen. Surgery is performed immediately, and evidence of a perforated ulcer in the posterior wall of the stomach is noted. Stomach contents that have seeped out will most likely be found in which of the following structures?
   A. Between the parietal peritoneum and the posterior body wall
   B. Greater peritoneal sac
   C. Ischioanal fossa
   D. Lesser peritoneal sac
   E. Paracolic gutter

10. A 20-year-old woman is involved in a vehicular accident and struck on the driver's side of the automobile she is driving. She is taken to the emergency department, where physical examination shows low blood pressure and tenderness on the left midaxillary line. Upon further examination of the patient, the physician also notes a large swelling that protrudes downward and medially below the left costal margin. Which of the following abdominal organs in this patient was most likely injured?

A. Descending colon

B. Left kidney

C. Liver

D. Pancreas

E. Spleen

F. Stomach

11. A 55-year-old man who has alcoholic cirrhosis of the liver is brought to the emergency department because he has been vomiting blood for the past 2 hours. He has a 2-month history of abdominal distention, dilated veins over the anterior abdominal wall, and internal hemorrhoids. Which of the following veins is the most likely origin of the hematemesis?

A. Esophageal veins

B. Inferior mesenteric veins

C. Paraumbilical veins

D. Superior mesenteric vein

E. Superior vena cava

12. A 70-year-old-man has a blockage at the origin of the inferior mesenteric artery. He does not have ischemic pain because of collateral arterial supply. Which of the following arteries is the most likely additional source of blood to the descending colon?

A. Left gastroepiploic

B. Middle colic

C. Sigmoid

D. Splenic

E. Superior rectal

13. A 65-year-old man is admitted to hospital with symptoms of an upper bowel obstruction. A CT scan reveals that a large vessel is compressing the third (transverse) portion of the duodenum. Which of the following vessels is most likely involved in the obstruction?

A. Gastroduodenal artery

B. Inferior mesenteric artery

C. Portal vein

D. Splenic artery

E. Superior mesenteric artery

14. A 25-year-old medical student in good health develops severe pain in the area around her umbilicus. She complains of nausea and is taken to the emergency department. While there, the pain becomes more localized in the lower right quadrant of her abdomen and the physician diagnoses appendicitis. Which of the following nerves perceived pain in the area around the umbilicus and most likely carried the pain sensations to the central nervous system?

A. Inferior hypogastric nerves

B. Lesser splanchnic nerves

C. Pudendal nerves

D. Superior hypogastric nerves

E. Vagus nerves

15. A 52-year-old man undergoes surgery to biopsy iliac lymph nodes. The physician tells the patient that it is important to identify the peripheral spinal nerves to protect them from being damaged during the surgery. The most likely location to find the genitofemoral nerve is coursing along which of the following surfaces?

A. Anterior surface of the psoas major muscle

B. Anterior surface of the quadratus lumborum muscle

C. Inferior surface of the iliacus muscle

D. Inferior surface of rib 12

E. Medial surface of the quadratus lumborum muscle

F. Medial surface of psoas major muscle

16. Three days after giving birth, a 32-year-old woman develops a fever and right lower abdominal pain. Ultrasonography shows a right ovarian vein thrombosis extending proximally. The thrombus most likely extends into the

A. ascending lumbar vein

B. hepatic portal vein

C. inferior vena cava

D. renal vein

E. right internal iliac vein

17. Parasympathetic innervation to the hindgut originates in the S2–S4 spinal cord segments. Parasympathetic neurons travel to the prevertebral plexus via which of the following nerves?

A. Greater splanchnic nerves

B. Least splanchnic nerves

C. Lesser splanchnic nerves

D. Lumbar splanchnic nerves

E. Pelvic splanchnic nerves

F. Sacral splanchnic nerves

18. Sweat glands within the S2 dermatome along the posterior region of the thigh most likely receive innervation via preganglionic sympathetic neurons originating from which of the following central nervous system levels?

    A. Brainstem

    B. C2 spinal cord level

    C. L2 spinal cord level

    D. S2 spinal cord level

    E. T2 spinal cord level

19. A 56-year-old man who is diagnosed with rectal cancer is undergoing biopsy of several lymph nodes. The nodes most likely to be sampled from this patient will be from the inferior mesenteric nodes, inguinal nodes, and the

    A. gonadal nodes

    B. internal iliac nodes

    C. portal vein nodes

    D. renal nodes

    E. superior mesenteric nodes

20. A potential complication of multiple term gestational births and vaginal deliveries is a prolapsed uterus. To prevent this condition, Kegel exercises may be advised for supporting the uterus. Which pelvic floor muscle is most likely targeted in Kegel exercises?

    A. External anal sphincter

    B. Bulbospongiosus muscle

    C. Obturator internus muscle

    D. Pelvic diaphragm

    E. Superficial transverse perineal muscle

21. A 30-year-old woman sustains a stage 4 tear in the perineum during a difficult delivery. In preparation to repair the tear, an anesthetic nerve block is administered to the pudendal nerve as it courses around the sacrospinous ligament. Which of the following areas is most likely blocked by the anesthetic?

    A. L2–L4 cutaneous field

    B. S1 cutaneous field

    C. S2–S4 cutaneous field

    D. L2–L4 dermatomes

    E. S1 dermatome

    F. S2–S4 dermatomes

22. During sexual arousal, an erection is caused by a dilation of arteries filling the erectile tissue of the penis. Innervation of the penile arteries is provided by which of the following nerves?

    A. Genitofemoral nerves

    B. Ilioinguinal nerves

    C. Pelvic splanchnic nerves

    D. Pudendal nerves

    E. Sacral splanchnic nerves

23. A 42-year-old man has a vasectomy. The physician explains to him that 3 to 4 months after the procedure, when he has an orgasm during sexual intercourse, most likely he will

    A. no longer produce an ejaculate

    B. still produce an ejaculate and the ejaculate will contain sperm

    C. still produce an ejaculate but the ejaculate will not contain sperm

24. Which structure can be palpated anterior to the cervix during a pelvic examination?

    A. Cardinal ligament

    B. Ovary

    C. Pelvic diaphragm

    D. Bladder

    E. Uterine tube

25. During the radical hysterectomy of a 52-year-old woman, the surgeon is careful to avoid damaging the ureters when removing the uterus. The landmark relationship that the surgeon should look for adjacent to the uterus to ensure preservation of each ureter is the ureter coursing

    A. inferior to the ovarian artery

    B. superior to the ovarian artery

    C. inferior to the uterine artery

    D. superior to the uterine artery

    E. inferior to the uterine tube

    F. superior to the uterine tube

26. A 17-year-old girl is brought to a refugee camp and has significant blood loss. She recently underwent a form of genital mutilation called excision, where the clitoris and labia minora were removed. Direct branches of which of the following arteries are most likely responsible for the blood loss?

    A. External iliac

    B. Inferior rectal

    C. Internal pudendal

    D. Ovarian

    E. Uterine

ANSWERS

1—D: The sacrospinous ligament courses from the sacrum to the ischial spine and encloses the greater sciatic notch to form a foramen. The obturator membrane and the internus muscle cover the obturator foramen. The piriformis muscle courses through the greater sciatic foramen, but does not form it.

2—A: The inferior pubic ramus joins the ischial ramus along the inferior aspect of the os coxa to form the ischiopubic ramus.

3—D: The female pubic arch is wider than the male pubic arch. The female pubic arch is about 85 degrees compared to 60 degrees in the male.

4—E: An indirect hernia results from bowel protruding through the deep inguinal ring, and through the inguinal canal and into the spermatic cord via the superficial inguinal ring. Therefore, during the physical examination, the physician will attempt to feel for herniation by digitally palpating the superficial inguinal ring through the scrotal sac.

5—C: The obliterated urachus is a fetal structure that functions by draining urine from the bladder through the umbilicus into the amniotic sac. If the urachus remains patent, it is possible that urine may be leaking out of the umbilicus.

6—C: The aponeuroses from the external oblique, internal oblique, and transversus abdominis muscles create the rectus sheath and then insert on the linea alba between the two rectus abdominis muscles.

7—A: The inferior epigastric artery branches off the external iliac artery and forms an anastomosis with the superior epigastric artery on the posterior surface of the rectus abdominis muscle. The superior epigastric artery branches off the internal thoracic artery, a branch of the subclavian artery.

8—B: The hepatogastric and hepatoduodenal ligaments are the two components of the lesser omentum. They are named for their attachments to the liver, stomach, and duodenum.

9—D: The ulcer in this patient is located on the deep surface of the stomach. Therefore, gastric contents that have seeped out will most likely be found in the lesser peritoneal sac. Recall how the greater peritoneal sac occupies the entire peritoneal cavity, with the exception of the region deep to stomach that is accessed via the epiploic foramen.

10—E: The spleen is located in the upper left quadrant of the abdomen, deep to the left costal margin. The descending colon is posterior to the midaxillary line in the retroperitoneal position, as is the left kidney. The liver is located in the upper right quadrant. The pancreas is in the retroperitoneal position in the midline. The stomach is in the upper left quadrant, but it would not create a swelling as would the damaged spleen.

11—A: A cirrhotic liver prevents all portal blood from flowing through the liver sinusoids. Therefore, portal hypertension occurs with blood backing up to the sites of portocaval anastomoses, including the esophageal veins. Chronic portal hypertension will result in swelling of the esophageal veins and potential hemorrhaging, causing hematemesis.

12—B: The marginal artery of Drummond consists of contributions from the inferior mesenteric artery as well as branches from the superior mesenteric artery via the right and middle colic arteries. Therefore, if the inferior mesenteric artery is blocked, blood flowing from the middle colic artery would provide the additional source of blood to the descending colon.

13—E: The superior mesenteric artery (and vein) course over the third portion of the duodenum.

14—B: The T10 dermatome is associated with the umbilical region. Sensory neurons course from the umbilical skin to the T10 spinal cord level. Visceral sensory neurons course from the appendix to the T10 spinal cord level as well as via the lesser splanchnic nerves. Therefore, the referred pain comes from the lesser splanchnic nerves.

15—A: The genitofemoral nerve courses along the anterior surface of the psoas major muscle.

16—C: The right ovarian vein courses from the right ovary to the inferior vena cava. Therefore, if the thrombosis extends proximally, it will course into the inferior vena cava. If the thrombosis were in the left ovarian vein, it would extend into the left renal vein.

17—E: Pelvic splanchnic nerves exit the ventral rami of spinal nerves S2–S4 and contain preganglionic parasympathetic neurons to the prevertebral plexus, such as the inferior hypogastric plexus. The other splanchnic nerves listed in the choices (i.e., greater, least, lesser, lumbar, and sacral splanchnic nerves) contain only sympathetic neurons.

18—C: Preganglionic sympathetic neurons originate between the T1 and L2 spinal cord levels. Dermatomes within the sacral region, such as the S2 dermatome described in this question, are supplied by sympathetics from the L2 spinal cord level, the lowest of all sympathetic innervation origin.

19—B: Lymphatics in the abdomen generally follow its associated arteries. Clusters of lymph nodes, which are important in monitoring the immune system, are found along the course of the regional arteries. The rectum is supplied by the following:

- Superior rectal artery—branch off the inferior mesenteric artery
- Middle rectal artery—branch off the internal iliac artery
- Inferior rectal artery—branch off the internal pudendal artery

Therefore, if the rectal cancer spreads, it can potentially do so parallel to all three arterial origins. The lymph nodes for the superior and inferior rectal arteries are provided in the stem of the question. The only nodes not mentioned are the internal iliac nodes for the origin of the middle rectal artery.

20—D: The pelvic diaphragm, consisting of the levator ani and coccygeus muscles, forms a hammock-like support to the pelvic floor. In females, it supports the bladder, uterus, and rectum. As such, Kegel exercises, which contract and relax pelvic floor muscles, give strength to the pelvic diaphragm in hopes of preventing tears during childbirth.

21—C: The pudendal nerve carries sensory axons from the genital region to the S2–S4 spinal cord levels. Therefore, it supplies sensory distribution for a region of the S2, S3, and S4 dermatomes, but not all of the parts of each dermatome. Therefore, the anesthetic blocked a cutaneous field, not a dermatome.

22—C: Dilation of penile arteries resulting in blood filling erectile tissue is under parasympathetic innervation. Therefore, the pelvic splanchnic nerves are responsible for transporting parasympathetic nerves to the penile arteries. The genitofemoral, ilioinguinal, and pudendal nerves are all somatic and do not cause an erection. The sacral splanchnics are responsible for transporting the sympathetics and will result in ejaculation. Remember, "point" and "shoot" ("p" parasympathetic; "s" sympathetic).

23—C: A vasectomy (a surgical procedure in which the ductus deferens is cut for the purpose of sterilization) will eventually sterilize the male by inhibiting sperm from entering the ejaculate. However, seminal contributions from the seminal vesicles, prostate, and bulbourethral gland will continue. Therefore, ejaculation will result in an ejaculate but without any sperm.

24—D: The bladder is anterior to the vagina.

25—C: The uterine artery courses superiorly over the ureter; in other words, "the water (ureter) courses under the bridge (uterine artery)."

26—C: The internal pudendal artery supplies all of the perineum, including the clitoris and labia minora.

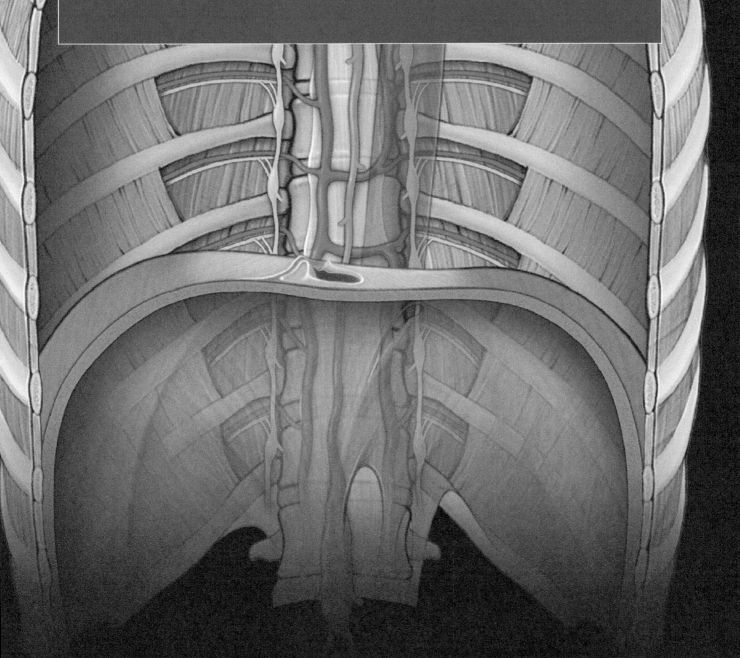

SECTION 4

HEAD

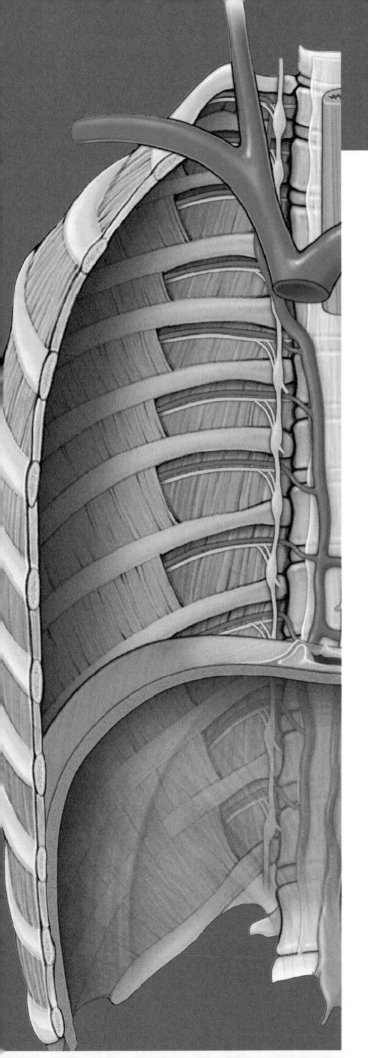

CHAPTER 15

SCALP, SKULL, AND MENINGES

ANATOMY OF THE SCALP

BIG PICTURE

The scalp consists of five layers of tissue. The five layers, from superficial to deep, are skin, subcutaneous connective tissue, a muscular aponeurotic layer, a loose connective tissue layer, and the pericranium.

LAYERS OF THE SCALP

The layers of the scalp can best be remembered by the acronym "**SCALP**," with each letter of the word representing the tissue layer associated with it Figure 15-1A .

- **S**kin. The skin of the scalp contains sweat and sebaceous glands and usually numerous hair follicles.
- **C**onnective tissue. The tissue between the skin and the aponeu-rotic layers is composed of dense collagenous connective tissue and contains the arteries, veins and nerves supplying the scalp.
- **A**poneurosis. The **superficial musculoaponeurotic system** of the scalp consists of the **occipitofrontalis muscle** and its investing fascia. This fascia is specialized to form a tendinous epicranial aponeurosis known as the **galea aponeurotica**. The galea continues into the temples, investing the auricular muscles, and terminates by attaching to the mastoid processes and the zygomatic arch. The frontalis muscle is instrumental in movements of the eyebrows and forehead and is an important muscle of facial expression are inner-vated by the facial nerve, cranial nerve (CN) VII.
- **L**oose connective tissue. A sponge-like layer of loose connec-tive tissue forms a subaponeurotic compartment that enables free movement of the top three scalp layers across the peri-cranium. It also contains the emissary veins.
- **P**ericranium. The pericranium is the **periosteum** over the external surface of the skull where the fibrous tissue knits into the sutures.

▽ Understanding the structure of the scalp is important when treating patients with **scalp wounds**. Superficial scalp wounds do not gape because of the strength of the under-lying aponeurosis, which holds the margins of the wound together. However, if the aponeurosis is lacerated in the coronal plane, deep scalp wounds gape because of the contraction of the frontalis and occipitalis muscles, which contract in opposite directions. This aponeurotic layer of the scalp is often tightened during cosmetic surgery (e.g., "facelifts") to help reduce wrin-kles in the face and forehead. ▽

▽ Injury to the fourth layer of the scalp (loose connective tissue) is dangerous because infection can potentially spread from the scalp through emissary veins into the cranial cavity. In addition, an infection or fluid can enter the eyelids because the frontalis muscle inserts into the skin and subcuta-neous tissue (not to bone), resulting in **ecchymosis**, or "**black-eyes.**" ▽

INNERVATION OF THE SCALP

The scalp receives its cutaneous innervation as follows (Figure 15-1B):

- **Posterior region of the scalp.** Innervated by branches of the cervical plexus, which are principally derived from C2 and C3 (**lesser occipital** and **greater occipital nerves**).
- **Anterior region of the scalp.** Innervated by the **supraorbital** and **supratrochlear nerves**, which are derived from the oph-thalmic division of the trigeminal nerve (CN V-1).
- **Lateral region of the scalp.** Innervated by branches of the maxillary (CN V-2) and mandibular (CN V-3) divisions of the trigeminal nerve (the **zygomaticotemporal** and **auricu-lotemporal nerves**, respectively).

VASCULAR SUPPLY OF THE SCALP

The scalp is highly vascularized via branches of the external and internal carotid arteries (Figure 15-1C).

- **External carotid artery.** Branches include the **occipital, pos-terior auricular**, and **superficial temporal arteries**.
- **Internal carotid artery.** Branches include the **supraorbital** and **supratrochlear arteries**.

▽ **Scalp lacerations** usually bleed profusely, primarily because arteries enter the scalp and bleed from both ends of the artery as a result of abundant anastamoses. In addition, the severed arteries do not contract when they are cut because the vessel lumens are held open by the dense connective tissue in the second layer of the scalp. As a result, bleeding from the scalp can be profuse. ▽

The supraorbital and supratrochlear veins unite to form the facial vein. The superficial temporal vein joins with the maxil-lary vein to form the retromandibular vein in the parotid sali-vary gland. The posterior auricular vein unites with the poste-rior division of the retromandibular vein to form the external jugular vein.

Scalp veins connect with the **diploic and emissary veins**.

- **Diploic veins.** Diploic veins course in the diploe cranial bones of the skull and connect with the dural venous sinuses via emissary veins.
- **Emissary veins.** Small veins connect veins of the scalp and skull with the dural venous sinuses.

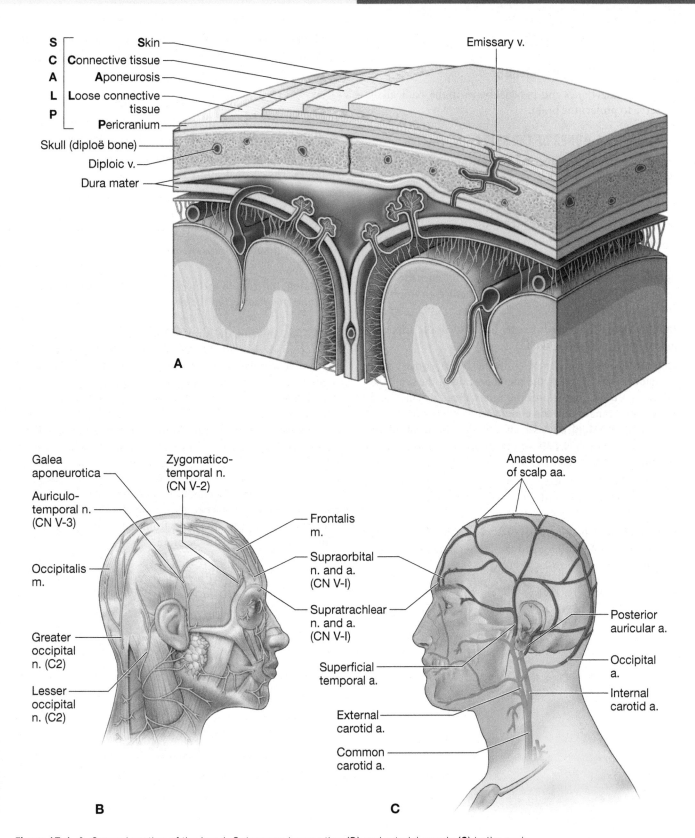

S — **S**kin
C — **C**onnective tissue
A — **A**poneurosis
L — **L**oose connective tissue
P — **P**ericranium

Emissary v.

Skull (diploë bone)
Diploic v.
Dura mater

Galea aponeurotica
Auriculo-temporal n. (CN V-3)
Occipitalis m.
Greater occipital n. (C2)
Lesser occipital n. (C2)

Zygomatico-temporal n. (CN V-2)
Frontalis m.
Supraorbital n. and a. (CN V-I)
Supratrachlear n. and a. (CN V-I)

Anastomoses of scalp aa.
Posterior auricular a.
Occipital a.
Internal carotid a.

Superficial temporal a.
External carotid a.
Common carotid a.

B   **C**

Figure 15-1: A. Coronal section of the head. Cutaneous innervation (**B**) and arterial supply (**C**) to the scalp.

SKULL

BIG PICTURE

There are 8 cranial bones and 14 facial bones in the skull, all of which serve to protect the brain.

BONES OF THE SKULL

The skull protects the brain and its surrounding meninges (Figure 15-2A). The outer and inner surfaces of the skull, are covered by periosteum, known respectively as the pericranium and the endocranium. The periosteum is continuous at the sutures of the skull. The cranial bones consist of spongy bone "sandwiched" between two layers of compact bone. The bones of the skull are as follows (Figure 15-2B–E):

- **Frontal bone.** The unpaired frontal bone underlies the forehead, roof of the orbit, and a smooth median prominence called the **glabella.**
- **Parietal bone.** The paired parietal bones form the superior and lateral aspects of the skull.
- **Temporal bone.** The paired temporal bones consist of a **squamous part** that forms the lateral portion of the skull; the **petrous part**, which encloses the internal ear (cochlea and semicircular canals) and the middle ear (malleus, incus, and stapes); the **mastoid part**, which contains the mastoid air cells; and the **tympanic part**, which houses the external auditory meatus and tympanic cavity.
- **Occipital bone.** The occipital bone encloses the foramen magnum, which transmits the spinal cord and vertebral arteries.
- **Sphenoid bone.** The sphenoid bone consists of a body, which houses the sphenoid sinus, and the greater and lesser wings and the pterygoid processes.

- **Ethmoid bone.** The ethmoid bone is located between the orbits and consists of the cribriform plate, the perpendicular plate, and the ethmoid air cells.

SUTURES OF THE SKULL

Sutures are immovable fibrous joints between the bones of the skull. The principal sutures of the skull are as follows:

- **Coronal suture.** Joins the frontal bone and the two parietal bones. The coronal suture courses in the coronal plane.
- **Sagittal suture.** Joins the paired parietal bones. The sagittal suture courses in the sagittal plane.
- **Squamous suture.** Joins the parietal and temporal bones.
- **Lambdoid suture.** Joins the parietal bones with the occipital bone.
- **Pterion.** Junction of the frontal, parietal, and temporal bones in the lateral aspect of the skull.

▼ The anterior division of the middle meningeal artery courses deep to the pterion on the inner surface of the skull. A blow to the pterion, therefore, could rupture the middle meningeal artery and result in an **epidural (extradural) hematoma**, which is a buildup of blood between the dura mater and the skull. On a radiograph, an epidural hematoma appears as a convex shape because sutures at the sites where the periosteal dura is more firmly attached to the skull stop the hematoma from expanding. As a result, epidural hematomas expand inward toward the brain instead of along the side of the skull, which occurs in a subdural hematoma. Unconsciousness and death occur rapidly because of the bleeding that dissects a wide space as it strips the dura from the inner surface of the skull, causing pressure on the brain. ▼

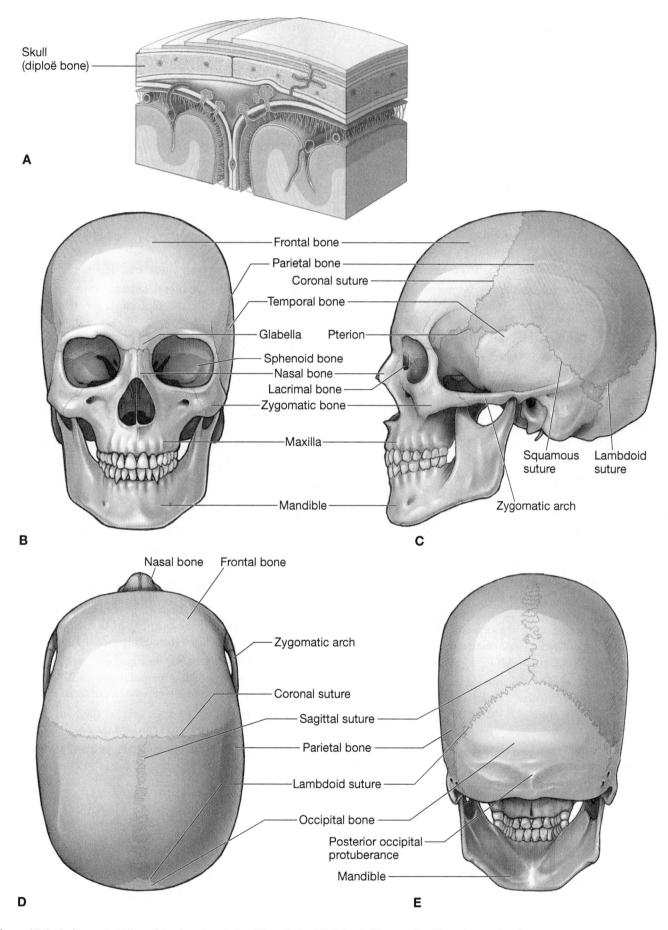

Skull
(diploë bone)

A

Frontal bone
Parietal bone
Coronal suture
Temporal bone
Glabella   Pterion
Sphenoid bone
Nasal bone
Lacrimal bone
Zygomatic bone
Maxilla
Squamous   Lambdoid
suture   suture
Zygomatic arch
Mandible

B   C

Nasal bone   Frontal bone
Zygomatic arch
Coronal suture
Sagittal suture
Parietal bone
Lambdoid suture
Occipital bone
Posterior occipital
protuberance
Mandible

D   E

Figure 15-2: A. Coronal section of the head; anterior (**B**), anterior (**C**), lateral (**D**), superior (**E**) and posterior views.

CRANIAL FOSSAE

BIG PICTURE

The base of the skull forms the floor on which the brain lies and consists of three large depressions that lie on different levels known as the anterior middle, and posterior cranial fossae (Figure 15-3A).

ANTERIOR CRANIAL FOSSA

The anterior cranial fossa houses the frontal lobes of the cerebrum and has the following bony landmarks (Figure 15-3B and C):

- **Cribriform plate.** Transmits the olfactory nerves (CN I) from the nasal cavity to the olfactory bulbs.
- **Crista galli.** Projects superiorly from the cribriform plate and serves as an attachment for the falx cerebri.
- **Lesser wing of the sphenoid bone.** Forms ridge separating the anterior and the middle cranial fossae.

MIDDLE CRANIAL FOSSA

The middle cranial fossa is deeper than the anterior cranial fossa and is separated from the posterior cranial fossa by the clivus. The middle cranial fossa supports the temporal lobes of the cerebrum and has the following bony landmarks (Figure 15-3B and C):

- **Sella turcica.** A deep central depression in the sphenoid bone that houses the pituitary gland. It is located superior to the sphenoid sinus.
- **Optic canal.** A passage that transmits the optic nerve (CN II) and the ophthalmic artery.
- **Superior orbital fissure.** A longitudinal fissure in the orbit that transmits the oculomotor nerve (CN III), trochlear nerve (CN IV), ophthalmic division of the trigeminal nerve (CN V-1), abducens nerve (CN VI), and the superior ophthalmic veins.
- **Carotid canal.** Transmits the internal carotid artery from the neck into the cranium. In addition, the carotid plexus of sympathetic nerves accompanies the internal carotid artery to provide innervation to the superior tarsal muscle, the dilator pupil muscle, the sweat glands of the face and scalp, and the blood vessels in the head.
- **Foramen rotundum.** Located posterior to the medial end of the superior orbital fissure. The foramen rotundum transmits the maxillary nerve (CN V-2) en route to the pterygopalatine fossa. CN V-2 supplies the skin, teeth, and mucosa associated with the maxillary bone.
- **Foramen ovale.** A large ovale foramen posterolateral to the foramen rotundum. The foramen ovale communicates with the infratemporal fossa and transmits the mandibular nerve (CN V-3).
- **Foramen spinosum.** Transmits the middle meningeal artery.
- **Foramen lacerum.** A triangular hole located on the sides of the sphenoid body. In a dried skull (e.g., such as that found in an anatomy laboratory), this foramen is patent; however, in vivo, the foramen is occluded by cartilage. After exiting the carotid canal, the internal carotid artery travels over the roof of the foramen lacerum. The greater petrosal nerve courses through the foramen lacerum en route to the pterygoid canal.

POSTERIOR CRANIAL FOSSA

The posterior cranial fossa is the lowest and deepest of the fossae and houses the cerebellum, pons, and medulla oblongata and has the following bony landmarks (Figure 15-3B and C):

- **Internal acoustic (auditory) meatus.** Transmits the facial nerve (CN VII) and the vestibulocochlear nerve (CN VIII) along with the labyrinthine artery.
- **Jugular foramen.** Transmits the glossopharyngeal (CN IX), vagus (CN X), and spinal accessory (CN XI) nerves as well as the internal jugular vein. The temporal and occipital bones form the jugular foramen.
- **Hypoglossal canal.** Transmits the hypoglossal nerve (CN XII).
- **Foramen magnum.** The largest foramen of the skull. The medulla oblongata, an extension of the spinal cord, enters and exits the cranial vault, along with the vertebral arteries, through the foramen magnum.
- **Mastoid foramen.** A branch of the occipital artery to the dura mater and mastoid emissary vein that traverses the mastoid foramen.

FORAMINA IN THE BASE OF THE SKULL

- **Petrotympanic fissure.** Located posterior to the mandibular fossa and serves as the opening of the chordae tympani nerve, a branch of CN VII, entering the infratemporal fossa.
- **Stylomastoid foramen.** An opening between the styloid and mastoid processes of the temporal bone. CN VII exits the foramen en route to innervate muscles of facial expression.
- **Incisive canal.** Located anteriorly on the hard palate. The nasopalatine nerve traverses through the canal.
- **Greater and lesser palatine canals.** Located posteriorly on the hard palate. The greater and lesser palatine nerves traverse the canals.

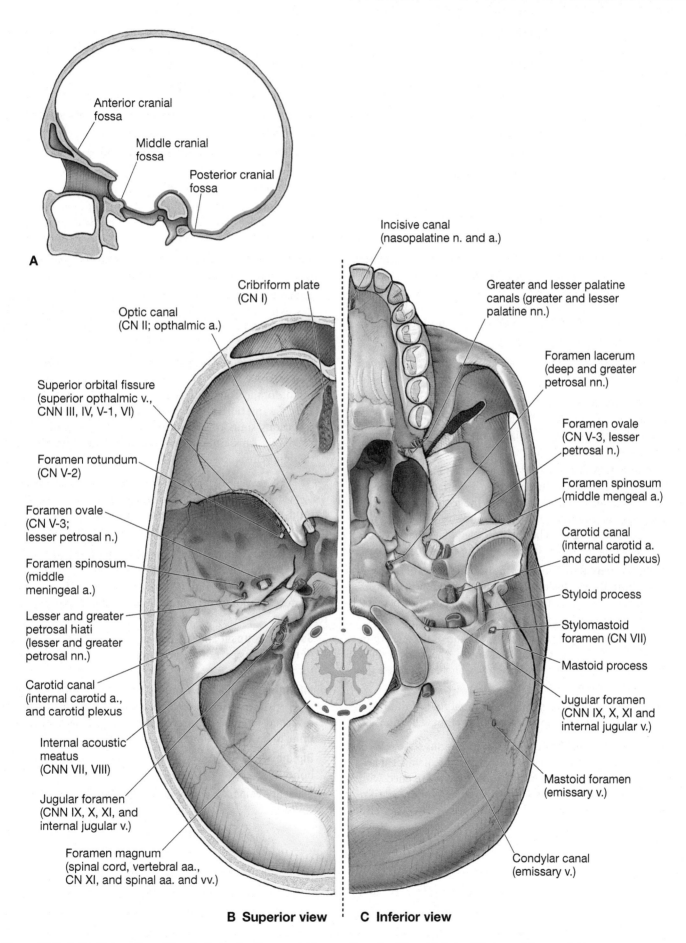

Figure 15-3: A. Sagittal section of the skull showing the cranial fossae. Superior (**B**) and inferior (**C**) views of the cranial base.

MENINGES

BIG PICTURE

The brain is surrounded and protected by three connective tissue layers called meninges. These meninges, from superficial to deep, are the dura mater, arachnoid mater, and pia mater.

DURA MATER

The dura mater is composed of the following two layers (Figure 15-4A):

Periosteal layer. The periosteal layer of the dura mater is the periosteum lining the internal surface of the skull and, as such, is intimately attached to the cranial bones and sutures.

Meningeal layer. The meningeal layer is the dura mater proper and is composed of dense collagenous connective tissue that is continuous with the dura mater of the spinal cord. The dura mater envelopes the cranial nerves like a sleeve, which then fuse with the epineurium of the nerves outside the skull.

The two layers of dura mater are bound together. However, the layers separate at numerous locations to form dural septae or dural venous sinuses.

DURAL SEPTAE Much like a seat belt assists in protecting a passenger from hitting the inside of the vehicle during an accident, four dural septae restrict displacement of the brain during everyday movements (Figure 15-4B).

Falx cerebri. A sickle-shaped layer of dura mater between the cerebral hemispheres. The falx cerebri is attached anteriorly to the crista galli and posteriorly to the tentorium cerebelli. The superior and inferior sagittal sinuses form the superior and inferior margins. The inferior edge of the falx cerebri courses along the superior surface of the corpus callosum.

Tentorium cerebelli. Separates the occipital lobes of the cerebrum from the cerebellum. The tentorium cerebelli encloses the transverse sinuses posteriorly and the superior petrosal sinuses anteriorly (Figure 15-4C).

Falx cerebelli. A small triangular extension of the tentorium cerebelli containing the occipital sinus. The falx cerebelli descends, separating the cerebellar lobes, and terminates at the foramen magnum.

Diaphragma sellae. A circular horizontal fold of dura that covers the pituitary by forming a roof over the sella turcica.

DURAL VENOUS SINUSES The dural venous sinuses are venous channels located between the periosteal and the meningeal layers of the dura mater. The dural venous sinuses are lined with endothelium and lack valves. They serve as a receptacle for blood from the cerebral, diploic, and emissary veins. They also receive the cerebrospinal fluid (CSF), drained by the arachnoid granulations. Blood in the dural venous sinuses primarily drains into the internal jugular veins.

Superior and inferior sagittal sinuses. Course in the midline along the superior and inferior borders of the falx cerebri. The inferior sagittal sinus joins with the **great cerebral vein (of Galen)** to form the straight sinus.

Straight sinus. Courses along the line of attachment of the falx cerebri to the tentorium cerebelli.

Occipital sinus. Courses within the falx cerebelli.

Confluence of sinuses. Forms the union of the superior sagittal straight and occipital sinuses and is located deep to the internal occipital protuberance.

Transverse sinus. Courses laterally from the confluence of sinuses within the tentorium cerebelli.

Sigmoid sinus. A continuation of the transverse sinus and descends in an S-shaped groove to join with the **inferior petrosal sinus** at the jugular foramen, forming the **internal jugular vein**.

Cavernous sinus. Located on each side of the sella turcica. The internal carotid artery and CN VI course through the middle of the sinus. Cranial nerves (CNN) III, IV, V-1, and V-2 course anteriorly through the lateral walls of the sinus. The cavernous sinus communicates with the pterygoid venous plexus via emissary veins and the superior and inferior ophthalmic veins.

▽ The cavernous sinus is the only location in the body where an artery courses through a venous structure. A **carotid-cavernous sinus fistula** forms when the internal carotid artery ruptures within the cavernous sinus. ▽

▽ **Pituitary tumors** can expand in the direction of least resistance and compress the cavernous sinus structures (CNN III, IV, V-1, V-2 and VI), causing paralysis of the extraocular muscles and sensory loss in the forehead and maxillary region. ▽

Superior petrosal sinus. Courses along the superior portion of the petrous part of the temporal bone and drains into the transverse sinus.

Inferior petrosal sinus. Courses along the inferior portion of the petrous part of the temporal bone and drains into the cavernous sinus. The inferior petrosal and sigmoid sinuses unite to become the internal jugular vein.

Diploic veins. Course within the spongy portion of cranial bones.

Emissary veins. Course between the scalp and dural venous sinuses.

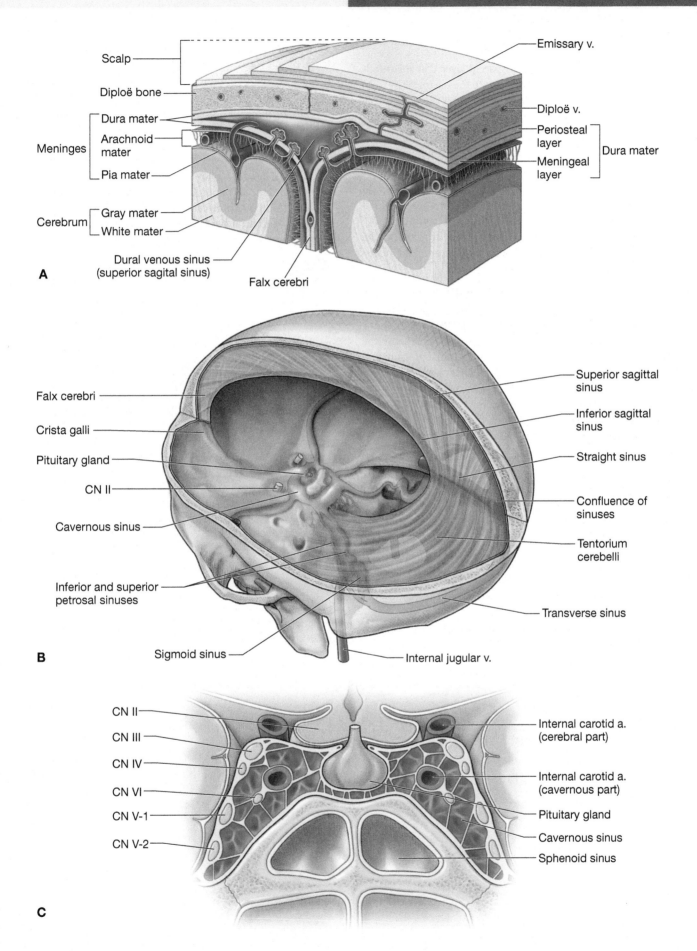

Figure 15-4: A. Coronal section of the head. **B**. Posterosuperior view of the dural septae and dural venous sinuses. **C**. Coronal section through the sphenoid bone highlighting the cavernous sinuses.

ARACHNOID MATER

The arachnoid mater is a thin, transparent layer that surrounds the brain and spinal cord. It is connected to the pia mater by web-like filaments, hence the name "arachnoid" mater (Figure 15-4D).

Subarachnoid space. The space between the arachnoid mater and the pia mater in which CSF circulates. Many cerebral vessels course around the surface of the brain within the subarachnoid space.

Arachnoid villi (granulations). Highly folded arachnoid mater that projects into the superior sagittal sinus and lateral lacunae (lateral extensions of the superior sagittal sinus). Arachnoid villi serve as sites where CSF diffuses into the superior sagittal sinus. Arachnoid villi often produce indentations in the inner surface of the calvarium.

▽ Generally, there is no space between the dura mater and the arachnoid mater. However, trauma to the head may stretch and rupture a bridging (cerebral) vein, resulting in bleeding into the subdural space (**subdural hematoma**). Because the damaged vessel is a vein, the increase in intracranial pressure and the effect of compressing the brain is much slower when compared to an epidural hematoma, which is caused by tearing of an artery. As a result, a subdural hematoma may develop over a period of days or even a week. Enlarging the subdural space is one factor that increases the risk of a subdural hematoma. As the subdural space enlarges, the bridging veins that traverse the space travel over a wider distance, causing them to be more vulnerable to tears. As a result, infants (who have smaller brains), the elderly (whose brains atrophy with age), and alcoholics (whose brains atrophy from alcohol use) are at increased risk of developing a subdural hematoma because of the tension of traversing vessels from the shrinking brain to the dural venous sinus. Subdural hematomas spread along the internal surface of the skull, creating a concave shape that follows the curve of the brain. The spread of blood is limited to one side of the brain due to dural reflections such as the tentorium cerebelli and falx cerebri. Contrast the spread of subdural hematomas to that of epidural hematomas that are limited in their spread due to the sutures. ▼

▽ A **subarachnoid hemorrhage** is defined as bleeding into the subarachnoid space because of a ruptured cerebral aneurysm. ▼

PIA MATER

The pia mater is the most internal and delicate of the meninges surrounding the brain and spinal cord (Figure 15-4D). The pia mater forms a sheath around blood vessels as they course into the fissures and sulci and penetrate the brain. The pia mater joins with the ependymal cells that line the ventricles of the brain to form choroid plexuses that produce CSF.

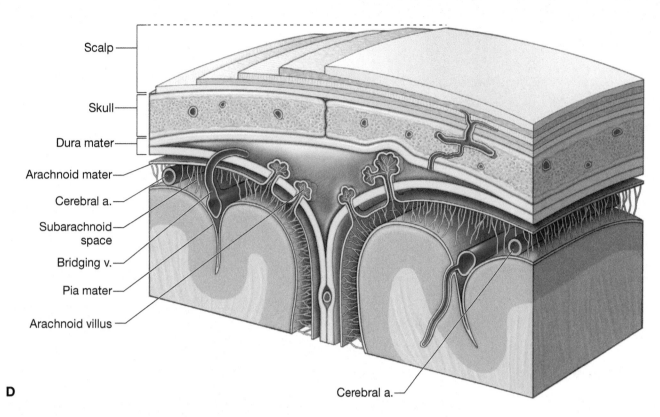

Scalp

Skull

Dura mater

Arachnoid mater

Cerebral a.

Subarachnoid space

Bridging v.

Pia mater

Arachnoid villus

Cerebral a.

D

Figure 15-4: (*continued*) **D**. Coronal section of the head highlighting the arachnoid and pia mater.

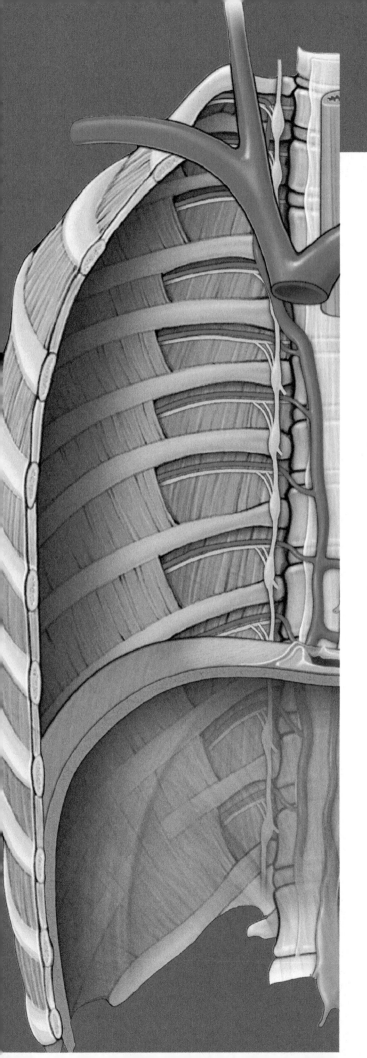

CHAPTER 16

BRAIN

ANATOMY OF THE BRAIN

BIG PICTURE

The brain contains millions of neurons arranged in a vast array of synaptic connections that provide seemingly unfathomable circuitry. Through that circuitry, the brain integrates and processes sensory information and provides motor output.

DIVISIONS OF THE BRAIN

The brain is divided into the cerebrum, diencephalon, brainstem, and cerebellum (Figure 16-1A and B).

CEREBRUM The cerebrum is the organ of thought and serves as the control site of the nervous system, enabling us to possess the qualities associated with consciousness such as perception, communication, understanding, and memory (Figure 16-1A). The **cerebral hemispheres** consist of elevations (**gyri**) and valleys (**sulci**), with a **longitudinal cerebral fissure** separating the hemispheres. Each cerebral hemisphere is divided into lobes, which correspond roughly to the overlying bones of the skull.

- **Frontal lobe.** The frontal lobe is located in the anterior cranial fossa. The **central sulcus** divides the frontal lobe from the parietal lobe in a coronal plane. The gyrus anterior to the central sulcus is called the **precentral sulcus** and serves as the primary motor area of the brain. The remainder of the frontal lobe is used in modifying motor actions.

- **Parietal lobe.** The parietal lobe interprets sensations from the body. The gyrus posterior to the central sulcus, the **postcentral sulcus**, is the primary area for receipt of these sensations.

- **Occipital lobe.** The occipital lobe is located superior to the tentorium cerebelli, in the posterior cranial fossa, and is primarily concerned with vision.

- **Temporal lobe.** The temporal lobe is located in the middle cranial fossa and is primarily concerned with hearing.

DIENCEPHALON The diencephalon consists of the **thalamus, hypothalamus, epithalamus**, and **subthalamus**, and is situated between the cerebrum and the brainstem (Figure 16-1B). The diencephalon serves as the main processing center for information destined to reach the cerebral cortex from the ascending pathways.

BRAINSTEM The brainstem consists of the midbrain, pons, and medulla oblongata (Figure 16-1B).

- **Midbrain.** The midbrain (mesencephalon) contains the nuclei for the oculomotor nerve and the trochlear nerve, cranial nerves (CNN) III and IV, respectively. The **cerebral aqueduct** is a portion of the ventricular system and courses through the center of the midbrain to connect the third and fourth ventricles.

- **Pons.** The pons is situated against the clivus and the dorsum sellae and contains the nuclei for the trigeminal, abducens, facial, and vestibulocochlear nerves (CNN V, VI, VII, and VIII, respectively).

- **Medulla oblongata.** The medulla oblongata, commonly called the medulla, is located at the level of the foramen magnum. It serves as the major autonomic reflex center that relays visceral motor control to the heart, blood vessels, respiratory system, and gastrointestinal tract. It possesses the nuclei for the glossopharyngeal, vagal, accessory, and hypoglossal nerves (CNN IX, X, XI, and XII, respectively).

CEREBELLUM The cerebellum lies in the posterior cranial fossa and assists in the coordination of skeletal muscle contraction (Figure 16-1A and B). It functions at a subconscious level and provides skeletal muscles with precise timing and appropriate patterns of contraction needed for smooth, coordinated movements.

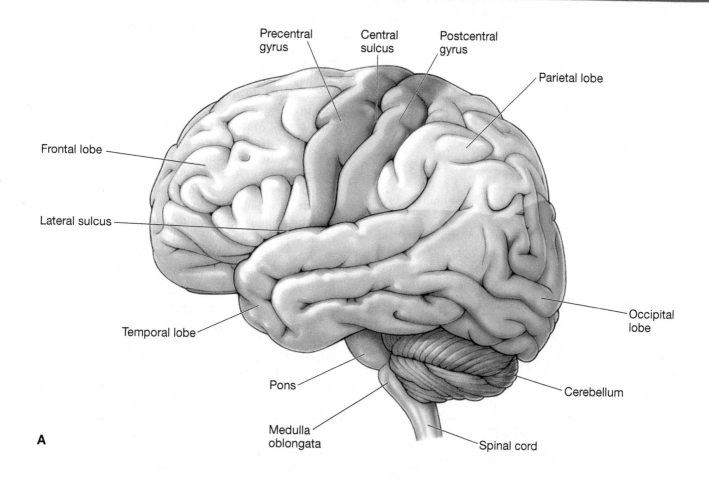

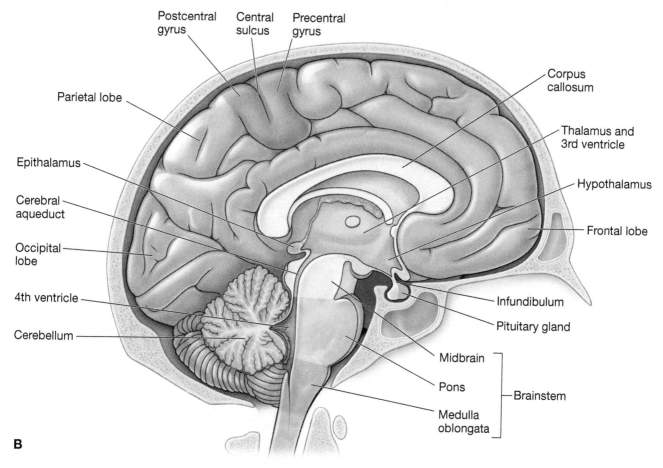

Figure 16-1: A. Lateral view of the brain. **B**. Medial view of the sagittal section of the brain.

VENTRICULAR SYSTEM OF THE BRAIN

BIG PICTURE

The ventricular system of the brain is a set of four chambers within the brain and is continuous with the central canal of the spinal cord. Cerebrospinal fluid (CSF) flows within these chambers and serves as a liquid cushion, providing buoyancy to the brain and spinal cord.

VENTRICLES

The four connected ventricles form chambers within the brain and are filled with CSF (Figure 16-2A and B).

- **Lateral ventricles.** These paired, C-shaped chambers are located deep within each cerebral hemisphere. Each lateral ventricle communicates via the **interventricular foramen (of Monro)** with the third ventricle.

- **Third ventricle.** The third ventricle is narrow and is located midline and inferior to the lateral ventricles. The third ventricle is positioned between the left and right diencephalon. The third ventricle communicates with the fourth ventricle via the **cerebral aqueduct (of Sylvius).**

- **Fourth ventricle.** The fourth ventricle is located superior to the pons and the medulla oblongata.

CEREBROSPINAL FLUID

CSF is produced by the choroid plexuses, located within each of the ventricles, and flows from the **lateral** and **third ventricles** to the fourth ventricle via the **cerebral aqueduct**. From the fourth ventricle, CSF enters an enlarged part of the subarachnoid space (**cisterna magna**) via the central **median aperture (of Magendie)** and the **lateral apertures (of Luschka).** The CSF circulates around the spinal cord and brain in the subarachnoid space to empty into the superior sagittal sinus via the **arachnoid granulations**. Arachnoid granulations are projections of the arachnoid mater along the superior sagittal sinus. The brain and spinal cord are very delicate and, as such, are susceptible to damage. Therefore, the brain and spinal cord are well protected through the bony encasement (skull and vertebral column) as well as by a second protective layer of connective tissue coverings (the meninges). The third protective mechanism is CSF, which supports the brain and spinal cord (Figure 16-2C). Despite all of these protections, trauma to the brain and spinal cord can still occur and can result in devastating injuries and deficits.

▽ The lumen of the cerebral aqueduct or the fourth ventricular apertures may become obstructed. When either of these conditions occurs, CSF continues to be secreted, producing excessive pressure within the ventricles. In children, this results in **hydrocephalus**, a condition in which the head enlarges because the skull bones have not yet fused. In adults, however, hydrocephalus is a different challenge because the skull is rigid. Therefore, the accumulating CSF compresses the brain tissue. In most cases, hydrocephalus is treated by inserting a shunt into the ventricles to drain the excess CSF into either a jugular vein or the peritoneal cavity. ▼

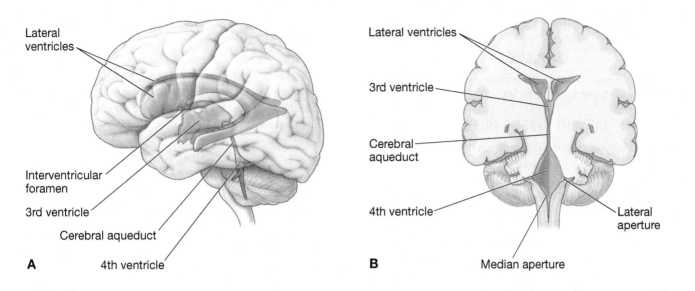

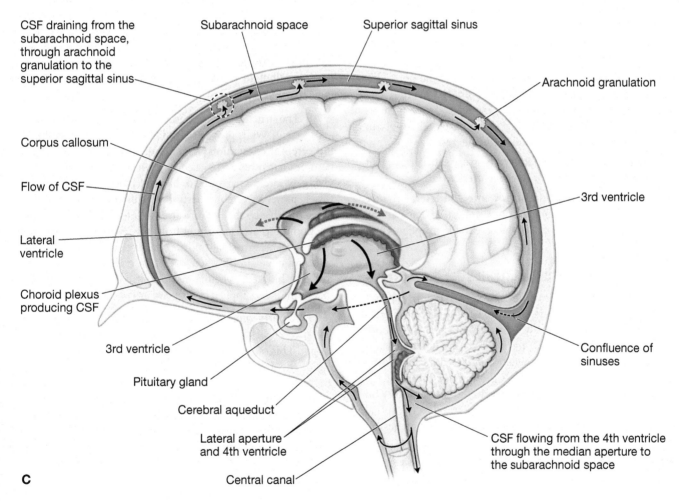

Figure 16-2: A. Three-dimensional lateral view of the ventricles of the brain. **B**. Coronal section of the brain showing the ventricles. **C**. Formation, location, and circulation of cerebrospinal fluid (CSF).

BLOOD SUPPLY TO THE BRAIN

BIG PICTURE

The brain receives its arterial supply from two sources, the **internal carotid** and the **vertebral arteries**.

INTERNAL CAROTID ARTERY

The common carotid arteries originate from the aortic arch on the left and the brachiocephalic trunk on the right. The common carotid arteries bifurcate at the level of the thyroid cartilage into the external and internal carotid arteries. The external carotid artery sends branches to the neck and face, whereas the internal carotid artery ascends to the base of the skull, entering the **carotid canal**. Upon exiting the carotid canal, the internal carotid artery courses horizontally over the foramen lacerum and enters the cavernous sinus and, after turning superiorly, divides into its terminal branches. The terminal branches of the internal carotid are as follows (Figure 16-3A–C):

Ophthalmic artery. Courses through the optic canal to supply the retina, orbit and part of the scalp.

Posterior communicating artery. Joins the posterior cerebral artery with the internal carotid artery.

Anterior cerebral artery. Courses superior to the optic chiasma and enters the **longitudinal cerebral fissures**. The anterior cerebral artery courses superiorly and then posteriorly along the corpus callosum, providing blood supply to the medial sides of both cerebral hemispheres.

• **Anterior communicating artery.** Is a very short artery that connects the two anterior cerebral arteries.

Middle cerebral artery. Courses into the lateral fissure between the parietal and temporal lobes. It sends many branches to the lateral sides of the cerebral hemispheres and central branches into the brain.

VERTEBRAL ARTERIES

Each vertebral artery arises from the subclavian artery and ascends through the transverse foramina of C1–C6. The vertebral artery courses horizontally across C1 through the suboccipital triangle before entering the skull via the **foramen magnum** (Figure 16-3A–C). After penetrating the dura mater, the vertebral arteries then course along the inferior aspect of the medulla oblongata before converging into the **basilar artery** on the pons. Branches of the vertebral arteries travel to the spinal cord, the meninges, and the brainstem. The major branches are as follows:

Posterior inferior cerebellar arteries. Course between the origins of cranial nerve (CN) X and CN XI en route to the inferior surface of the cerebellum. Often referred to by the acronym PICA.

Posterior cerebral arteries. The terminal branches of the basilar artery provide vascular supply to that part of the brain base that is superior to the tentorium cerebelli. CN III and CN IV exit the brain between the superior cerebellar and the posterior cerebral arteries.

BASILAR ARTERY

The basilar artery ascends along the ventral surface of the pons and gives rise to the following branches:

Anterior inferior cerebellar artery. Courses along the inferior surface of the cerebellum.

Superior cerebellar artery. Courses along the superior surface of the cerebellum.

CIRCLE OF WILLIS

The anterior communicating artery connects the two anterior cerebral arteries, and the posterior communicating arteries connect the internal carotid and posterior cerebral arteries. As a result of these connections, an arterial circle, known as the **cerebral arterial circle (of Willis)**, is formed around the **infundibulum** (stalk connecting the pituitary gland to the hypothalamus).

The structure of the arterial supply of the brain into the circle of Willis provides a **collateral circulatory pathway** in the cerebral circulation. As a result, if one part of the circle becomes narrowed or blocked or if one of the four arteries supplying the circle is narrowed or blocked, blood flow from the other vessels preserves cerebral perfusion. This collateral circulation is made possible by the lack of valves in arteries. However, some of the numerous arteries arising from this circle that penetrate the brain substance are small and are considered end arteries (without collateral circulation). Therefore, if an end artery becomes narrowed or blocked, ischemia may occur in the region of the brain that is uniquely supplied by that end artery. ▼

A **berry aneurysm** is a balloon-like outpouching of a cerebral arterial wall that is berry shaped (hence, the name). This outpouching most often reflects a gradual weakening of the arterial wall as a result of chronic hypertension or arteriosclerosis and places the artery at risk to rupture, causing a **stroke**. Some cerebral vessels are inherently weak and susceptible to berry aneurysms, such as the arteries associated with the circle of Willis, where small communicating arteries connect larger cerebral arteries (internal carotid, vertebral, and basilar arteries). A ruptured berry aneurysm bleeds into the subarachoid space. ▼

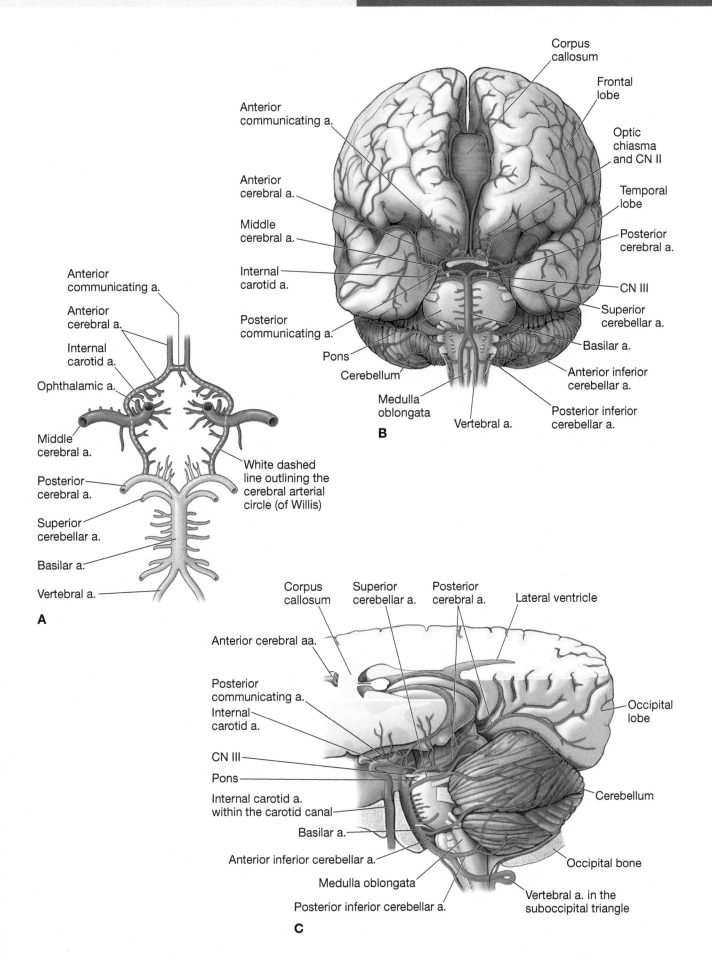

Figure 16-3: A. Cerebral arterial circle (of Willis). **B**. Anterior view of the brain showing the arteries. **C**. Lateral view of the brain showing the arteries (CN II, optic nerve; CN III, oculomotor nerve).

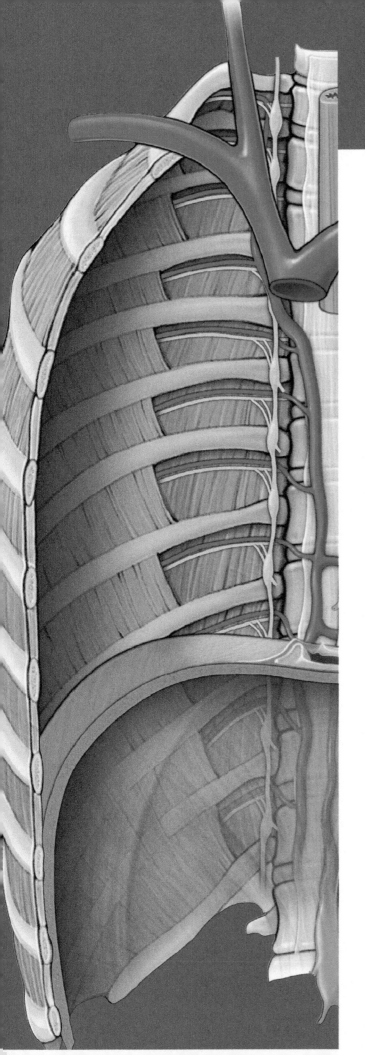

CHAPTER 17

CRANIAL NERVES

OVERVIEW OF THE CRANIAL NERVES

BIG PICTURE

Cranial nerves (CNN) emerge through openings in the skull and are covered by tubular sheaths of connective tissue derived from the cranial meninges. There are 12 pairs of cranial nerves, numbered I to XII, from rostral to caudal, according to their attachment to the brain. The names of the cranial nerves reflect their general distribution and function. Like spinal nerves, cranial nerves are bundles of sensory and motor neurons that conduct impulses from sensory receptors and innervate muscles or glands.

RAPID REVIEW OF THE NERVOUS SYSTEM

To best understand the cranial nerves, it is helpful to remember the following information:

Neuron versus nerve. A neuron is a single sensory or motor nerve cell, whereas a nerve is a bundle of neuronal fibers (axons). Cranial nerves have three types of sensory and three types of motor neurons, known as modalities. Therefore, a nerve may be composed of a combination of sensory or motor neurons (e.g., the facial nerve possesses sensory and motor neurons).

Ganglion. A ganglion is a collection of nerve cell bodies in the peripheral nervous system.

Nucleus. A nucleus is a collection of nerve cell bodies in the central nervous system (CNS).

CRANIAL NERVE MODALITIES

The 12 pairs of cranial nerves may possess one or a combination of the following **sensory** and **motor modalities** (Figure 17-1; Table 17-1):

Sensory (afferent) neurons. Conduct information from the body tissues to the CNS.

- **General sensory (general somatic afferent).** Transmit sensory information (e.g., touch, pain, and temperature), conducted mainly by CN V but also by CNN VII, IX, and X.
- **Special sensory (special visceral afferent).** Include special sensory neurons (e.g., smell, vision, taste, hearing, and equilibrium), mainly conducted by the olfactory, optic, and vestibulocochlear nerves (CNN I, II, and VIII, respectively) as well as by CN VII and CN X.
- **Visceral sensory (general visceral afferent).** Convey sensory information from the viscera, including the gastrointestinal tract, trachea, bronchi, lungs, and heart, as well as the carotid body and sinus. Visceral sensory neurons course within CN IX and CN X.

Motor (efferent) neurons. Conduct information from the CNS to body tissues.

- **Somatic motor (general somatic efferent) neurons.** Innervate skeletal muscles derived from somites, including the extraocular and tongue muscles. Innervation is accomplished via the oculomotor, trochlear, abducens, and hypoglossal nerves (CNN III, IV, VI, and XII, respectively).
- **Branchial motor (special visceral efferent) neurons.** Innervate skeletal muscles derived from the branchial arches, including the muscles of mastication and facial expression and the palatal, pharyngeal, laryngeal, trapezius and sternocleidomastoid muscles. Innervation is accomplished via the trigeminal, facial, glossopharyngeal, vagus, and spinal accessory nerves (CNN V, VII, IX, X, and XI, respectively).
- **Visceral motor (general visceral efferent) neurons.** Innervate involuntary (smooth) muscles or glands, including visceral motor neurons that constitute the cranial outflow of the parasympathetic division of the autonomic nervous system. The preganglionic neurons originate in the brainstem and synapse outside the brain in parasympathetic ganglia. The postganglionic neurons innervate smooth muscles and glands via CNN III, VII, IX, and X.

The nuclei of the cranial nerves (where motor neurons originate or sensory neurons terminate) are located in the brainstem, with the exception of CN I and CN II, which are extensions of the forebrain.

CRANIAL NERVE TARGETS

The specific functions of cranial nerves depend on the nature of the anatomic targets of the cranial nerves (Table 17-2).

CNN V, VII, IX, and X. Innervate almost all of the structures of the head and neck, such as the skin, mucous membranes, muscle, and glands derived from the pharyngeal arches. Of these four nerves, CN V and CN VII innervate most of these structures, whereas CN IX innervates only a few structures of the head, oral cavity, pharynx, and neck. Almost all targets of CN X are in the trunk.

CNN III, IV, and VI. Innervate only structures in the orbit.

CNN I, II, and VIII. Possess only special sensory neurons for smell, sight, balance, and hearing.

CN XII. Innervates only the tongue muscles.

CNN III, VII, IX, and X. The cranial nerves that carry parasympathetic neurons.

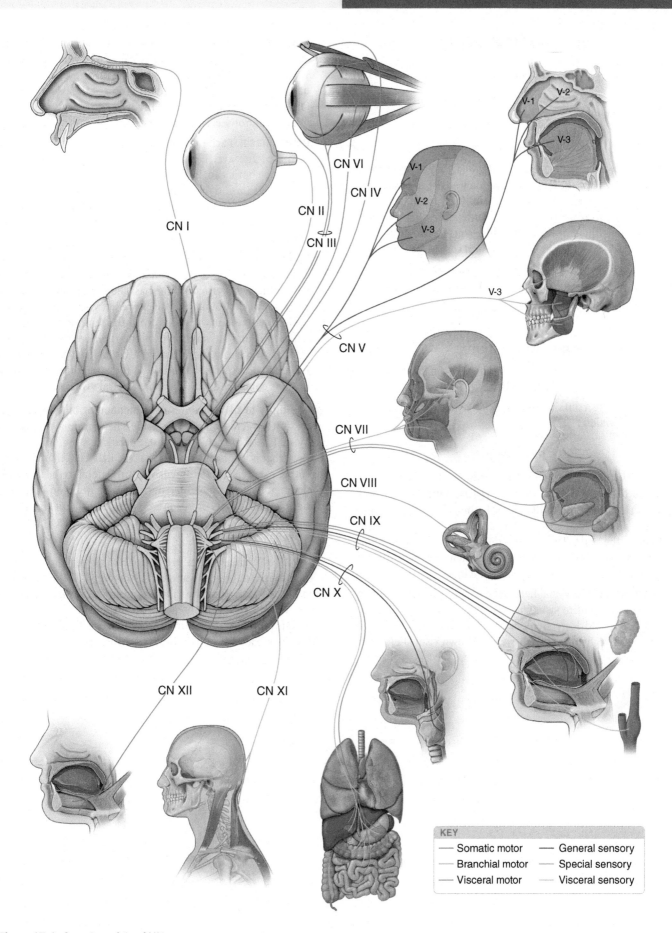

Figure 17-1: Overview of the CNN.

CN I: OLFACTORY NERVE

BIG PICTURE

The olfactory nerves contain only special sensory neurons concerned with **smell** (Figure 17-2A).

PATHWAYS

The olfactory neurons originate in the olfactory epithelium in the superior part of the lateral and septal walls of the nasal cavity. The nerves ascend through the **cribriform foramina** of the ethmoid bone to reach the **olfactory bulbs**. The olfactory neurons synapse with neurons in the bulbs, which course to the primary and association areas of the cerebral cortex.

▽   **Injury to CN I** (e.g., a fracture of the cranial base) can result in **anosmia** (loss of smell), tearing of the meninges, or cerebrospinal fluid rhinorrhea. ▼

CN II: OPTIC NERVE

BIG PICTURE

The optic nerve contains only special sensory neurons concerned with **vision** (Figure 17-2B).

PATHWAYS

Optic nerve fibers arise from the retina and all converge at the **optic disc**. CN II exits the orbit via the **optic canals**. Both optic nerves form the **optic chiasm**, the site where neurons from the nasal side of either retina cross over to the contralateral side of the brain. The neurons then pass via the **optic tracts** to the thalamus, where they synapse with neurons that course to the primary visual cortex of the occipital lobe.

- **Visual fields.** The visual field is the part of the world seen by the eyes. The entire visual field is divided into right and left and upper and lower regions, defined when the patient is looking straight ahead. Each eye has its own visual field or, in other words, the part of the world seen by each eye alone. The lateral visual field of an eye is called the **temporal** field, whereas the medial visual field of the same eye is called the **nasal** field. Although the fields of vision of the two eyes overlap greatly, the right eye sees things far to the right that the left eye cannot see, and vice versa.

- **Optic chiasm.** Each optic nerve carries axons from the entire retina of the eye. However, after coursing through their respective optic canals, the right and left optic nerves engage in a redistribution of axons at the optic chiasma, located just anterior to the pituitary stalk. The optic chiasma is created by neurons from the nasal half of each retina crossing over to the opposite side.

- **Optic tracts.** The two optic tracts emerge from the optic chiasma. The right optic tract contains axons from the temporal half of the right retina and the nasal half of the left retina and carries information about the entire left visual field. The left optic tract contains neurons from the temporal half of the left retina and the nasal half of the right retina and carries information about the entire right visual field. An optic tract is named according to the side of the body on which it lies, but it is concerned with the contralateral visual field.

▽   **Injury to CN II** may result in **monocular blindness**. If an optic tract is injured, the result is hemianopia; that is, half the visual field of each eye is lost. Specifically, the half-field of one eye that is lost is the same side as the half-field that is lost by the other eye. The lost fields will be contralateral to the damaged optic tract. For example, interruption of the function in the left optic tract causes loss of vision in the right visual fields of both eyes. In addition, injury to CN II may result in a **loss of the papillary reflex** of CN III. ▼

▽   Most cranial nerves are peripheral nerves and therefore myelinated by Schwann cells. However, CN II is an extension of the forebrain and as such is myelinated by oligodendrocytes. **Multiple sclerosis** is an autoimmune disorder that attacks myelin in oligodendrocytes. Therefore, CN II is the only cranial nerve affected by multiple sclerosis. ▼

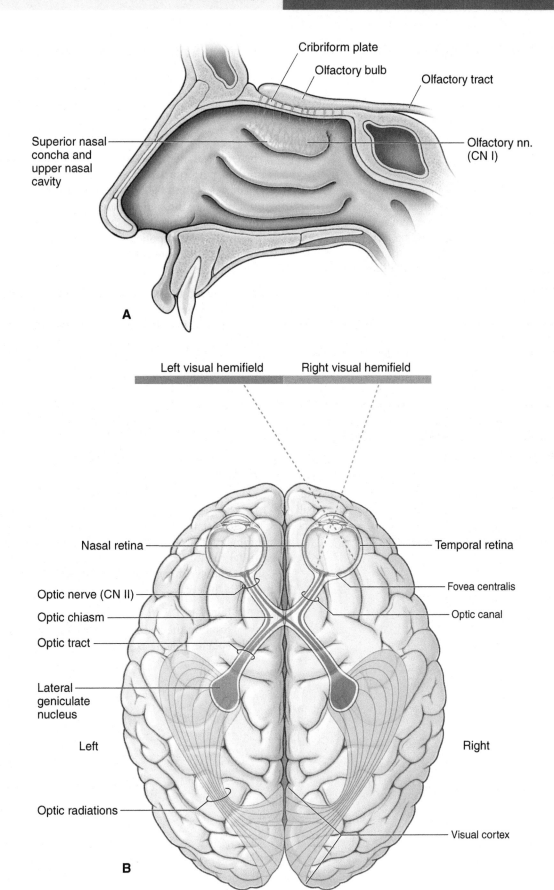

Figure 17-2: A. Special sensory innervation from the olfactory nerve (CN I). **B**. Special sensory innervation from the optic nerve (CN II) and visual fields.

CN III: OCULOMOTOR NERVE

BIG PICTURE

The oculomotor nerve innervates the levator palpebrae superioris muscle, four of the six extraocular muscles as well as the pupillary constrictor and ciliary muscles.

PATHWAYS

Upon exiting the midbrain, the oculomotor nerve courses between the posterior cerebral and superior cerebellar arteries and runs along the lateral wall of the cavernous sinus superior to CN IV (Figure 17-3A). CN III enters the orbit through the **superior orbital fissure**, where it divides into superior and inferior divisions. CN III has the following **modalities**:

- **Somatic motor neurons.** CN III innervates four of the **extraocular muscles** (superior rectus, medial rectus, inferior rectus, and inferior oblique) and the **levator palpebrae superioris muscle**. The somatic motor component of CN III plays a major role in controlling the muscles responsible for the precise movements of the eyes.

- **Visceral motor neurons.** Preganglionic parasympathetic neurons of CN III originate in the **Edinger–Westphal nucleus** and synapse in the **ciliary ganglion**, providing innervation to the **ciliary body** (lens accommodation) and **sphincter pupillae muscle** (pupil constriction) (Figure 17-3B).

▽ **Injury to CN III** (e.g., cavernous sinus thrombosis, pressure on the nerve, or an aneurysm) can result in **mydriasis** (dilated pupil), **ptosis** (droopy eyelid), turning the eyeball down and out, loss of the **pupillary reflex** on the side of the lesion, and loss of the **accommodation reflex**. ▽

CN IV: TROCHLEAR NERVE

BIG PICTURE

The trochlear nerve innervates the superior oblique muscle.

PATHWAYS

The trochlear nerve is the only cranial nerve that originates from the posterior aspect of the brainstem (Figure 17-3A). CN IV courses around the brainstem in the free edge of the tentorium cerebelli, through the lateral wall of the cavernous sinus, and enters the orbit via the superior orbital fissure. The trochlear nerve supplies somatic motor innervation to the **superior oblique muscle**, which causes the eyeball to move down and out.

▽ **Injury to CN IV** (e.g., a **fracture of the orbit** or **stretching of the nerve** during its course around the brainstem) can result in the **inability to look down** when the eyeball is adducted. The patient may also tilt the head away from the side of the lesion to accomodate for the extorsion caused by the loss of the superior oblique muscle. ▽

CN VI: ABDUCENS NERVE

BIG PICTURE

The abducens nerve innervates the lateral rectus muscle.

PATHWAYS

The abducens nerve originates from the pons and courses through the cavernous sinus, entering the orbit via the superior orbital fissure (Figure 17-3A). The abducens nerve supplies somatic motor innervation to the **lateral rectus muscle**, which abducts the eye.

▽ **Injury to CN VI,** the cavernous sinus, or the orbit may result in the inability to move the eye laterally, resulting in **diplopia** on lateral gaze. ▽

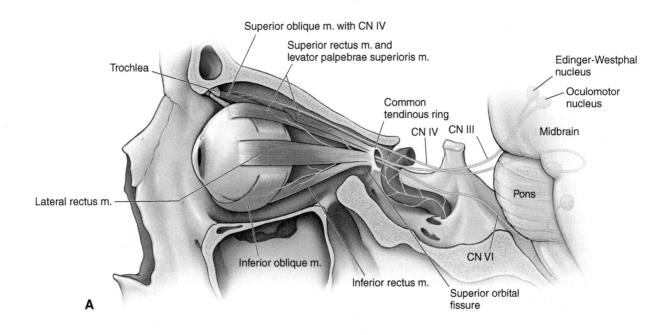

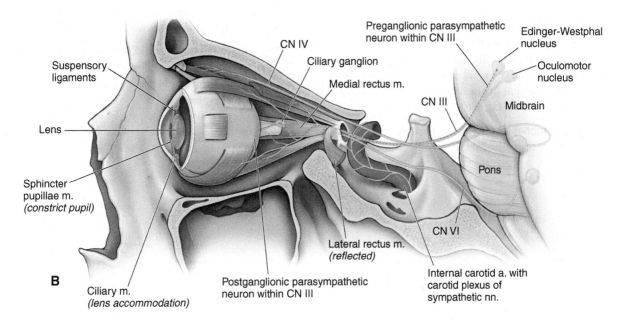

Figure 17-3: A. Somatic motor innervation from the oculomotor, trochlear, and abducens nerves (CNN III, IV, and VI, respectively). **B**. Visceral motor parasympathetic component of CN III.

CN V: TRIGEMINAL NERVE

BIG PICTURE

The trigeminal nerve is the principal **general sensory supply to the head** (Figure 17-4A).

PATHWAYS

CN V originates from the lateral surface of the pons as a large sensory root and a smaller motor root. These roots enter the **trigeminal (Meckel's) cave** of the dura, lateral to the body of the sphenoid bone and the cavernous sinus. The sensory root leads to the **trigeminal (semilunar) ganglion**, which houses the cell bodies for the general sensory neurons. The motor root runs parallel to the sensory root, bypassing the ganglion and becoming part of the mandibular nerve (CN V-3). As well as being the primary general sensory distribution to the head, CN V also aids in distributing postganglionic parasympathetic neurons of the head to their destinations for CNN III, VII, and IX. The trigeminal ganglion gives rise to three divisions, named for the cranial location to the eyes (**ophthalmic**), the maxilla (**maxillary**), and the mandible (**mandibular**).

CN V-1: OPHTHALMIC DIVISION

CN V-1 courses along the lateral wall of the cavernous sinus and enters the **orbit** via the superior orbital fissure (Figure 17-4A and B). CN V-1 provides general sensory innervation to the orbit, cornea, and the skin of the bridge of the nose, scalp and forehead (above the lateral corners of the eye).

 Injury to branches of CN V-1 can result in a loss of sensation in the skin of the forehead and scalp. CN V-1 also innervates the cornea; therefore, mediation of the sensory limb of the **corneal reflex** is via the nasociliary branch. ▼

CN V-2: MAXILLARY DIVISION

CN V-2 passes through the lateral wall of the cavernous sinus and through the foramen rotundum into the **pterygopalatine fossa** (Figure 17-4A and B). This nerve provides general sensory innervation to the maxillary face (between the lateral corners of the eye and the corners of the mouth), including the palate, nasal cavity, paranasal sinuses, and the maxillary teeth.

 Injury to branches of CN V-2 can result in a loss of sensation in the skin over the maxilla and maxillary teeth. ▼

CN V-3: MANDIBULAR DIVISION

CN V-3 courses through the foramen ovale into the **infratemporal fossa** (Figure 17-4A–C). CN V-3 provides general sensory innervation to the lower part of the face (below the lateral corners of the mouth), including the anterior two-thirds of the tongue, the mandibular teeth, the mandibular face, and even part of the scalp. CN V-3 also has branchial motor neurons, which innervate muscles derived from the first branchial arch which include the muscles of mastication (temporalis, masseter, and medial and lateral pterygoid muscles), mylohyoid, anterior belly of the digastricus, tensor tympani, and tensor veli palatine muscles.

▽ **Injury to branches of CN V-3** can result in a **loss of sensation** in the mandibular skin and teeth as well as the anterior two-thirds of the tongue. Because the motor division of CN V-3 innervates the muscles of mastication (e.g., temporalis and masseter muscles), the patient may experience weakness in chewing and deviation of the mandible on the side of the lesion when the mouth is opened. ▼

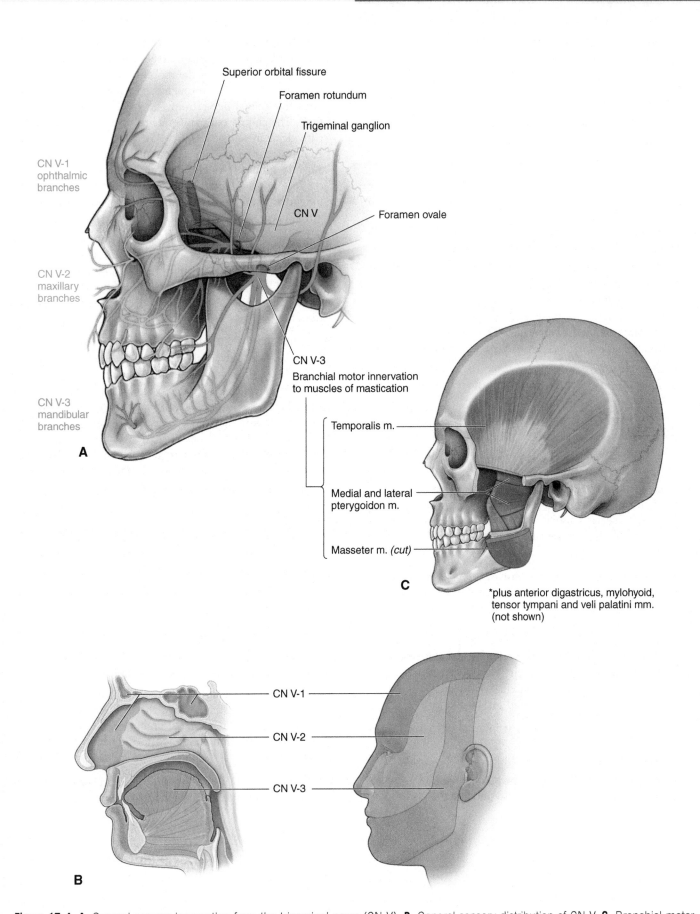

Figure 17-4: A. General sensory innervation from the trigeminal nerve (CN V). **B**. General sensory distribution of CN V. **C**. Branchial motor distribution of the mandibular division of the trigeminal nerve (CN V-3) to muscles of mastication.

CN VII: FACIAL NERVE

BIG PICTURE

The facial nerve provides motor innervation to the muscles of facial expression, lacrimal gland, and submandibular and sublingual salivary glands, as well as taste to the anterior two-thirds of the tongue.

CN VII MODALITIES

The facial nerve traverses the internal acoustic meatus carrying **four modalities** (Figure 17-5A).

- **Branchial motor neurons.** Supply muscles derived from the second branchial arch including the muscles of facial expression as well as the stapedius, posterior belly of the digastricus, and stylohyoid muscles.
- **Visceral motor neurons.** Provide parasympathetic innervation to almost all glands of the head (e.g., lacrimal, submandibular, sublingual, nasal, and palatal). The only exception is the parotid gland, which receives its visceral motor innervation from CN IX.
- **Special sensory neurons.** Transmit taste sensation from the anterior two-thirds of the tongue.
- **General sensory neurons.** Transmit general sensation from a portion of the external acoustic meatus and auricle.

CN VII BRANCHES

Two distinct fascial sheaths package the four modalities carried by CN VII, with branchial motor neurons in one sheath and visceral motor, special sensory, and general sensory neurons in another sheath called the **nervus intermedius.**

BRANCHIAL MOTOR NERVE TRUNK The branchial motor components constitute the largest portion of CN VII. After entering the temporal bone via the internal acoustic meatus, a small branch of CN VII courses to the stapedius muscle, where branchial motor neurons course through the facial canal to exit the skull via the stylomastoid foramen. In the parotid gland, five terminal branches (i.e., **temporal, zygomatic, buccal, mandibular, and cervical**) provide voluntary control of the muscles of facial expression, including the buccinator, occipitalis, platysma, posterior digastricus, and stylohyoid muscles.

NERVUS INTERMEDIUS The nervus intermedius gives rise to the following nerves:

- **Greater petrosal nerve.** Contains preganglionic parasympathetic neurons that synapse in the **pterygopalatine ganglion** en route to the **lacrimal, nasal,** and **palatine glands.**

- **Chorda tympani.** Arises in the descending part of the facial canal and crosses the medial aspect of the tympanic membrane, passing between the malleus and incus. The chorda tympani exits the skull through the petrotympanic fissure and joins the lingual nerve from CN V-3 in the infratemporal fossa. The chorda tympani contains preganglionic parasympathetic neurons that synapse in the **submandibular ganglion** en route to innervate the **submandibular** and the **sublingual salivary glands**. The chorda tympani also contains special sensory neurons (taste) from the anterior two-thirds of the tongue, with cell bodies located in the geniculate ganglion.
- **Auricular branches.** Arise from the external acoustic meatus and auricle and carry general sensory neurons through the geniculate ganglion to the brainstem. The **geniculate ganglion** is a knee-shaped bend in CN VII, located within the temporal bone and housing sensory cell bodies for the special sensory neurons for taste and general sensory neurons from the ear.

▽ **Injury to CN VII,** after it exits the brainstem, results in paralysis of the facial muscles (**Bell's palsy**) on the ipsilateral side. Fracture of the temporal bone can result in the abnormalities just described, plus increased sensitivity to noise (**hyperacusis**) due to the lack of innervation to the stapedius muscle, dry mouth due to a decrease in salivation, dry corneas due to the lack of lacrimal gland activity, and a loss of taste on the anterior two-thirds of the tongue. It should be noted that a brainstem injury to CN VII results in paralysis of the contralateral facial muscles below the eye. ▽

CN VIII: VESTIBULOCOCHLEAR NERVE

BIG PICTURE

The vestibulocochlear nerve provides sensory innervation for hearing and equilibrium.

PATHWAYS

The vestibulocochlear nerve traverses the internal acoustic meatus with CN VII and has the following **modalities** (Figure 17-5B):

- **Special sensory neurons.** CN VIII originates from the grooves between the pons and the medulla oblongata. CN VIII divides into the **cochlear branch** to the **cochlea** (hearing) and the **vestibular branch** to the **semicircular canal (equilibrium).**

 Injury to CN VIII can result in ipsilateral deafness, tinnitus (ringing in the ear), and vertigo (loss of balance). ▽

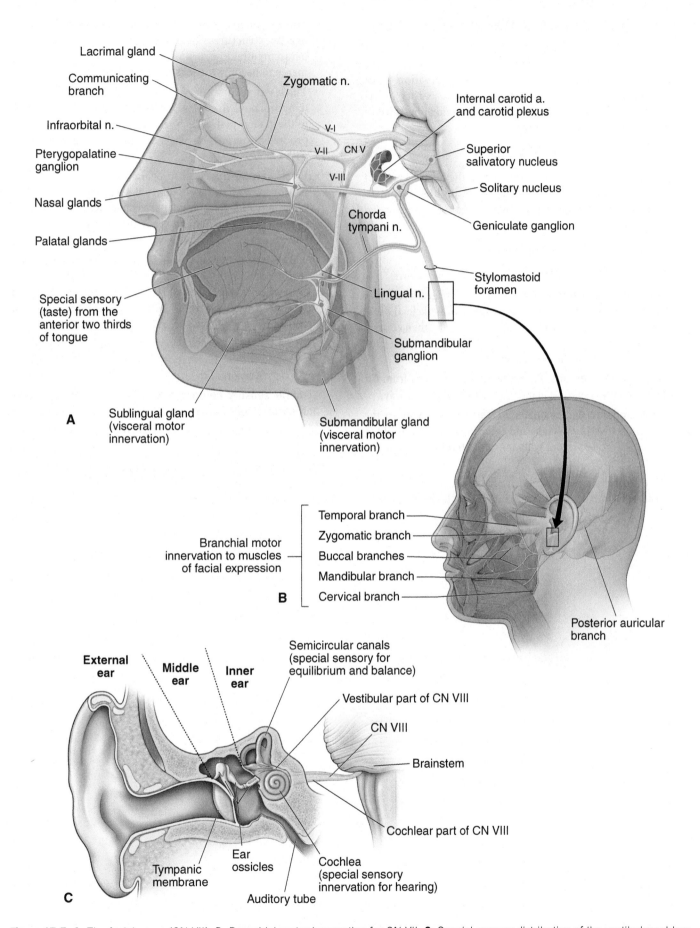

Figure 17-5: A. The facial nerve (CN VII). **B.** Branchial motor innervation for CN VII. **C.** Special sensory distribution of the vestibulocochlear nerve (CN VIII).

CN IX: GLOSSOPHARYNGEAL NERVE

BIG PICTURE

The glossopharyngeal nerve provides motor innervation to the stylopharyngeus muscle and parotid gland and sensory innervation from the carotid body and sinus, posterior one-third of the tongue, and the auditory tube.

BRAINSTEM ORIGIN

The glossopharyngeal nerve emerges from the lateral aspect of the medulla oblongata and traverses the **jugular foramen**, where its superior and inferior sensory ganglia are located (Figure 17-6).

MODALITIES

CN IX consists of the **five modalities** described as follows.

Branchial motor neurons. Course in a pharyngeal branch to innervate the stylopharyngeus muscle the only skeletal muscle derived from the third branchial arch. CN IX then courses between the superior and middle pharyngeal constrictor muscles en route to the oropharynx and the tongue.

Visceral motor neurons. Parasympathetic neurons from CN IX innervate the parotid gland via the following pathway:

- *Pre*ganglionic parasympathetic neurons originate in the inferior salivatory nucleus and enter the petrous part of the temporal bone via the **tympanic canaliculus** to the tympanic cavity.
- Within the tympanic cavity, the tympanic nerve forms the **tympanic plexus** on the promontory of the middle ear to provide general sensation to the tympanic membrane and nearby structures. The visceral motor preganglionic parasympathetic neurons course through the plexus and merge to become the **lesser petrosal nerve**.

- The lesser petrosal nerve reenters and travels through the temporal bone to emerge in the middle cranial fossa, lateral to the greater petrosal nerve. The lesser petrosal nerve then proceeds anteriorly to exit the skull via the **foramen ovale** with CN V-3.

- Upon exiting the skull, preganglionic parasympathetic neurons within the lesser petrosal nerve synapse in the **otic ganglion.** *Post*ganglionic parasympathetic neurons exit the otic ganglion and travel with the auriculotemporal branch of CN V-3 to innervate the parotid gland.

General sensory neurons. Course within the tympanic plexus, providing innervation to the internal surface of the tympanic membrane, the middle ear, and the auditory tube. General sensory neurons also course in the pharyngeal branch from the posterior third of the tongue and the oropharynx.

Special sensory neurons. Course from the posterior third of the tongue within the pharyngeal branch of CN IX, providing taste sensation to the brain.

Visceral sensory neurons. Course from the **carotid sinus and carotid body** and ascend in the sinus nerve, joining CN IX. The cell bodies of the visceral sensory neurons reside in the inferior ganglion.

▽ **Injury to CN IX** may result in **loss of the gag reflex**, alteration in taste to the posterior third of the tongue, and altered **vasovagal reflex**. ▼

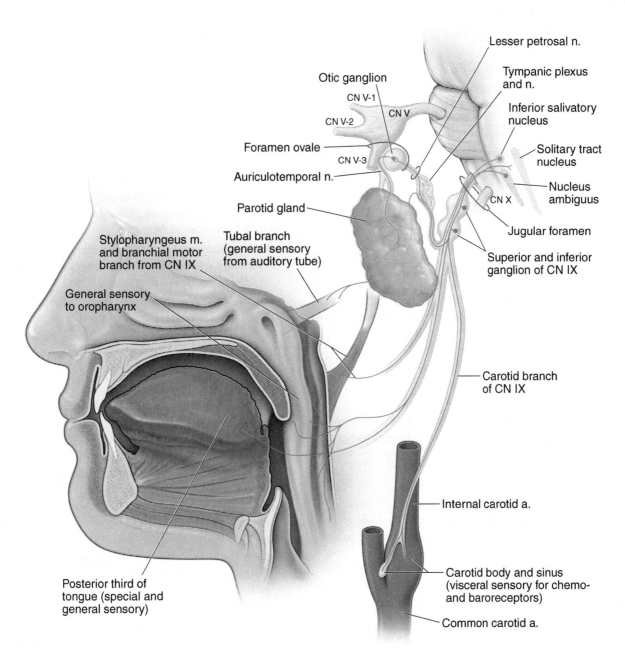

Figure 17-6: The glossopharyngeal nerve (CN IX).

CN X: VAGUS NERVE

BIG PICTURE

The vagus nerve innervates muscles of the larynx, pharynx, palate in addition to the gut tube, heart and lungs.

BRAINSTEM ORIGIN

The vagus nerve emerges from the lateral aspect of the medulla oblongata and traverses the **jugular foramen,** where the superior and inferior sensory ganglia are located.

CN X MODALITIES

The vagus nerve exits the medulla oblongata and travels with CNN IX and XI into the jugular foramen. CN X consists primarily of the following **four modalities:**

Visceral sensory neurons. Provide visceral sensory information from the larynx (below the vocal folds), trachea, and esophagus, and the thoracic and abdominal viscera as well as the stretch receptors of the aortic arch and the chemoreceptors of the aortic and carotid bodies (Figure 17-7A and B).

General sensory neurons. Provide general sensory information from part of the external acoustic meatus, the pinna, and the laryngopharynx (Figure 17-7A).

Branchial motor neurons. Supply the palatoglossus, laryngeal, and pharyngeal muscles (except for the stylopharyngeus muscle, which is supplied by CN IX) and the palatal muscles (except for the tensor veli palatini, which is supplied by CN V-3) (Figure 17-7A).

Visceral motor neurons. Provide parasympathetic innervation to the smooth muscle and the glands of the respiratory system and gastrointestinal tract to the transverse colon (Figure 17-7B). In general, CN X increases secretion from glands and smooth muscle contraction. CN X slows the heart rate, stimulates bronchiolar secretions, bronchoconstriction, and peristalsis, and increases secretions.

Upon traversing the jugular foramen, CN X travels between the internal jugular vein and the internal carotid artery within the carotid sheath.

BRANCHES OF CN X

The **branchial motor fibers** exit the vagus nerve as the following branches:

Pharyngeal branch. The pharyngeal branch is the principal motor innervation of the pharyngeal muscles. It branches just below the inferior vagal ganglion and courses between the internal and external carotid arteries. The pharyngeal branch then enters the middle pharyngeal constrictor muscle, where it forms the pharyngeal plexus. The pharyngeal plexus innervates all of the pharyngeal muscles (i.e., superior, middle, and inferior pharyngeal constrictors; salpingopharyngeus, palatopharyngeus, palatoglossus, and levator veli palatini), with the exception of the stylopharyngeus muscle by CN IX and the tensor veli palatini by CN V-3.

Superior laryngeal branch. The superior laryngeal nerve branches immediately below the pharyngeal nerve. The nerve descends in the neck adjacent to the pharynx and **splits to form the following nerves:**

- **External laryngeal nerve.** Provides branchial motor innervation to part of the inferior pharyngeal constrictor muscle and the cricothyroideus muscle.

- **Internal laryngeal nerve.** Pierces the thyrohyoid membrane and provides general sensory innervation of the larynx above the vocal folds.

Recurrent laryngeal branch. The path of the recurrent laryngeal nerve differs on the right and left sides of the body.

- **Left recurrent laryngeal nerve.** Branches from the vagus nerve at the level of the aortic arch. The nerve loops posteriorly around the aortic arch by the ligamentum arteriosus and ascends through the superior mediastinum to enter the groove between the esophagus and the trachea.

- **Right recurrent laryngeal nerve.** Branches from the vagus nerve before entering the superior mediastinum at the level of the right subclavian artery. The nerve hooks posteriorly around the subclavian artery and also ascends in the groove between the esophagus and trachea.

- Both recurrent laryngeal nerves pass deep to the lower margin of the inferior constrictor muscle to innervate the intrinsic laryngeal muscles and visceral sensory innervation below the vocal folds.

▽ **Injury to CN X** may result in **hoarseness** (due to paralysis of the intrinsic laryngeal muscles) and **difficulty swallowing** (due to paralysis of pharyngeal muscles). On examination, the soft palate droops on the affected side, and the uvula deviates opposite the affected side as a result of the unopposed action of the intact levator veli palatini muscle. There also may be loss of the gag reflex, where CN IX provides the sensory limb and CN XI provides the motor limb through innervation of the pharyngeal muscles. ▽

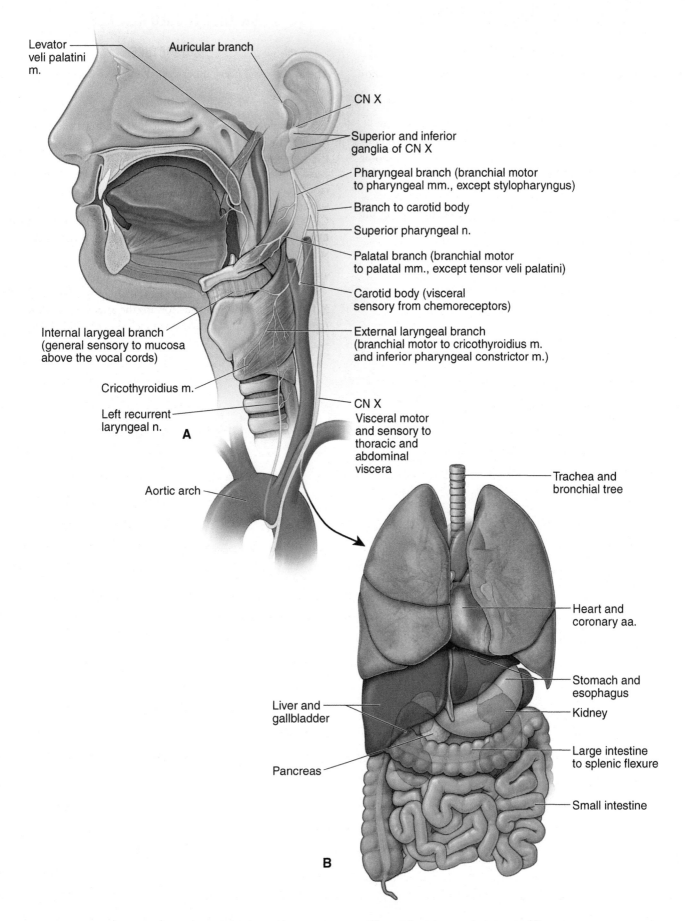

Figure 17-7: Distribution of the vagus nerve (CN X) to the head and neck (**A**) and the thorax and abdomen (**B**).

CN XI: SPINAL ACCESSORY NERVE

BIG PICTURE

The spinal accessory nerve innervates the trapezius and sterno-cleidomastoid muscles.

PATHWAYS

The spinal accessory nerve originates from the medulla oblongata, exits the jugular foramen, and provides branchial motor innervation to the **trapezius** and **sternocleidomastoid muscles** (Figure 17-8).

 Injury to CN XI may result in weakness in turning the head to the opposite side and in **shoulder drop.** ▼

CN XII: HYPOGLOSSAL NERVE

BIG PICTURE

The hypoglossal nerve innervates the tongue muscles.

PATHWAYS

The hypoglossal nerve exits the medulla oblongata in the groove between the pyramid and the olive (Figure 17-8). Upon exiting the hypoglossal canal, CN XII courses between the internal carotid artery and the internal jugular vein, deep to the posterior digastricus muscle. CN XII then courses along the lateral surface of the hyoglossus muscle deep to the mylohyoid muscle. CN XII provides somatic motor innervation to all **intrinsic** and **extrinsic tongue muscles** (with the exception of the palatoglossus, supplied by CN X).

▽ **Injury to CN XII** can be assessed by having the patient stick the tongue straight out. If there is a lesion on CN XII, the tongue will protrude toward the affected side. A way to remember this is that a patient with a lesion of CN XII will "lick his wound." ▼

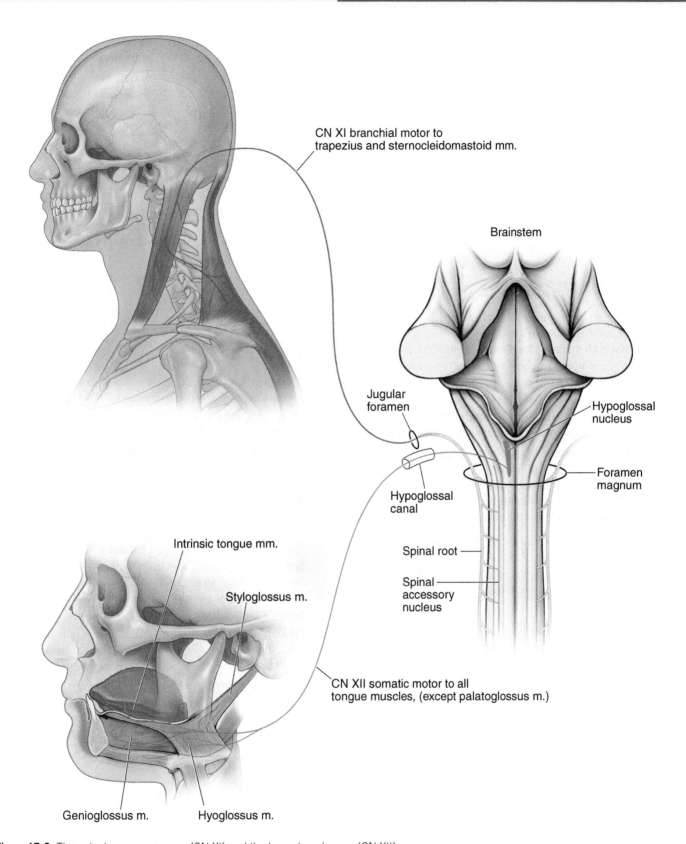

CN XI branchial motor to
trapezius and sternocleidomastoid mm.

Brainstem

Jugular
foramen

Hypoglossal
nucleus

Foramen
magnum

Hypoglossal
canal

Intrinsic tongue mm.

Styloglossus m.

Spinal root

Spinal
accessory
nucleus

CN XII somatic motor to all
tongue muscles, (except palatoglossus m.)

Genioglossus m.   Hyoglossus m.

Figure 17-8: The spinal accessory nerve (CN XI) and the hypoglossal nerve (CN XII).

AUTONOMIC INNERVATION OF THE HEAD

BIG PICTURE

All preganglionic sympathetic neurons destined for the head originate at the T1 level of the spinal cord and synapse in the superior cervical ganglion. Postganglionic sympathetic neurons course along cranial arteries to the end organs such as the sweat glands, the superior tarsal muscle, and the dilator pupillae muscle.

Preganglionic parasympathetic neurons originate in the brainstem, course in CNN III, VII, IX, or X, and synapse in one of four ganglia (i.e., ciliary, pterygopalatine, submandibular, and otic). Postganglionic parasympathetic neurons then course along nerves to their end organs (e.g., salivary glands and pupillary sphincter muscle).

SYMPATHETIC INNERVATION

Remember that all preganglionic sympathetic neurons in the body originate at spinal cord levels T1–L2 (Figure 17-9). Therefore, all **visceral motor sympathetic innervation** to the head originates in the upper levels of the thoracic spinal cord, with most originating at the T1 level.

- *Pre*ganglionic sympathetic fibers. Originate from T1 of the spinal cord, ascend in the **sympathetic trunk**, and synapse with postganglionic fibers in the **superior cervical ganglion**.
- *Post*ganglionic sympathetic fibers. Originate in the superior cervical ganglion and follow the arteries throughout the head to innervate the **blood vessels, sweat glands, superior tarsal muscle** (to elevate the upper eyelid), and the **dilatator pupillae muscle** (to dilate the pupil).

▽ Lesions in the sympathetic neurons to the head result in **Horner's syndrome**. Patients with Horner's syndrome usually present with **ptosis** (eyelid droop), **miosis** (constricted pupil), **anhydrosis** (loss of sweating), and flushed face. ▽

PARASYMPATHETIC INNERVATION

There are **four parasympathetic ganglia** associated with the cranial nerves (Figure 17-9; Table 17-3).

- **Ciliary ganglion.** Located posterior to the eyeball between the optic nerve and the lateral rectus muscle. Preganglionic parasympathetic neurons from the inferior division of CN III synapse in the ciliary ganglion and send postganglionic parasympathetic neurons to the sphincter pupillae and the ciliary muscles via the **short ciliary nerves**.
- **Pterygopalatine ganglion.** Located in the pterygopalatine fossa, just inferior to CN V-2 and lateral to the sphenopalatine foramen. Preganglionic parasympathetic neurons from CN VII synapse in the pterygopalatine ganglion and send postganglionic parasympathetic neurons to the **lacrimal, nasal,** and **palatal glands**.
- **Submandibular ganglion.** Suspended from the lingual nerve of CN V-3, deep to the mylohyoid muscle. Preganglionic parasympathetic neurons from CN VII synapse in the submandibular ganglion and send postganglionic parasympathetic neurons to supply the **submandibular** and **sublingual salivary glands**.
- **Otic ganglion.** Located in the infratemporal fossa inferior to the foramen ovale between CN V-3 and the tensor veli palatini muscle. Preganglionic parasympathetic neurons from CN IX synapse in the otic ganglion and send postganglionic neurons in the auriculotemporal nerve of CN V-3 to supply the parotid gland.

There are four cranial nerves that carry visceral motor parasympathetic innervation to the head: CNN III, VII, IX, and X.

- **CN III (oculomotor nerve).** Preganglionic parasympathetic neurons synapse in the **ciliary ganglion**, with postganglionic parasympathetic neurons serving the **ciliary muscles** and **sphincter pupillae** for light accommodation and constriction of the pupil.
- **CN VII (facial nerve).** Preganglionic neurons traveling in the greater petrosal nerve synapse in the **pterygopalatine ganglion**, with postganglionic parasympathetic neurons coursing to serve the **lacrimal, nasal**, and **palatal glands**. Preganglionic neurons from CN VII also travel within the **chorda tympani nerve** to synapse in the **submandibular ganglion**, with postganglionic neurons serving the **submandibular** and **sublingual salivary glands**.
- **CN IX (glossopharyngeal nerve).** Preganglionic parasympathetic neurons synapse in the **otic ganglion**, with postganglionic parasympathetic neurons serving the **parotid gland**.
- **CN X (vagus nerve).** Preganglionic parasympathetic neurons synapse at or near the target organ, with postganglionic parasympathetic neurons serving smooth muscle and glands of the **gastrointestinal tract to the transverse colon**.

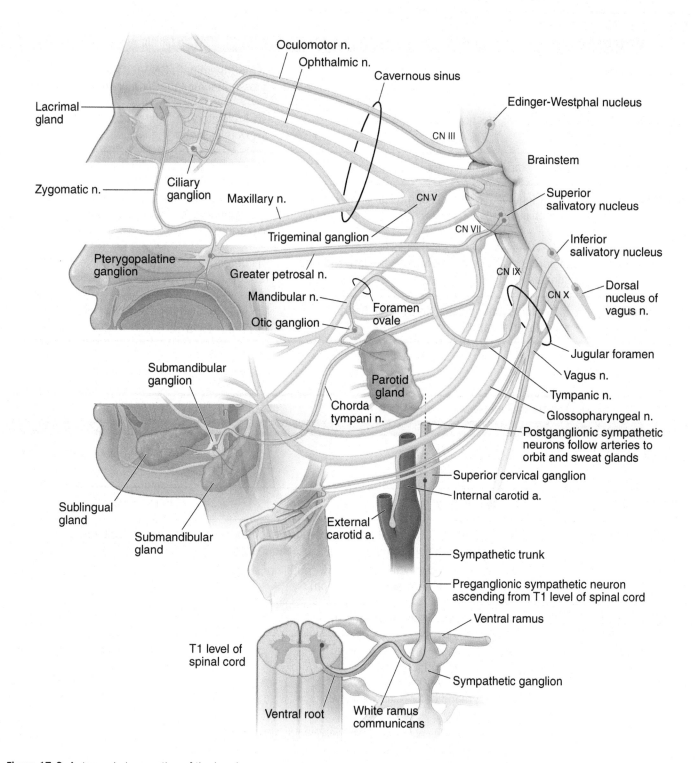

Figure 17-9: Autonomic innervation of the head.

TABLE 17-1. Modalities of the Cranial Nerves

Modality	General Function	CNN Containing the Modality
General sensory	Perception of touch, pain, temperature	CN V (trigeminal), CN VII (facial), CN IX (glossopharyngeal), CN X (vagus)
Special sensory	Vision, smell, hearing, balance, taste	CN I (olfactory), CN II (optic), CN VII (facial), CN IX (glossopharyngeal)
Visceral sensory	Sensory input from viscera	CN IX (glossopharyngeal), CN X (vagus)
Branchial motor	Motor innervation to skeletal muscle derived from branchial arches	CN V-3 (mandibular), CN VII (facial), CN IX (glossopharyngeal), CN X (vagus), CN XI (spinal accessory)
Somatic motor	Motor innervation of skeletal muscle derived from somites	CN III (oculomotor), CN IV (trochlear), CN VI (abducens), CN XII (hypoglossal)
Visceral motor	Motor innervation to smooth muscle, heart muscle, and glands	CN III (oculomotor), CN VII (facial), CN IX (glossopharyngeal), CN X (vagus)

TABLE 17-2. Overview of the Cranial Nerves

CN	Modalities and Function	Exit from Skull
CN I (olfactory)	Special sensory: smell	Cribriform plate of the ethmoid bone
CN II (optic)	Special sensory: sight	Optic canal
CN III (oculomotor)	Somatic motor: levator palpebrae superioris m.; superior, medial, and inferior rectus mm.; inferior oblique mm. Visceral motor: sphincter pupillae m. (pupil constriction) and ciliary mm. (lens accommodation)	Superior orbital fissure
CN IV (trochlear)	Somatic motor: superior oblique m.	Superior orbital fissure
CN V (trigeminal)	General sensory: CN V-1: orbit and forehead CN V-2: maxillary region CN V-3: mandibular region, tongue Branchial motor: CN V-3: muscles of mastication, mylohyoid, anterior digastricus, tensor tympani, and tensor veli palatine mm.	CN V-1: superior orbital fissure CN V-2: foramen rotundum CN V-3: foramen ovale
CN VI (abducens)	Somatic motor: lateral rectus m.	Superior orbital fissure
CN VII (facial)	General sensory: external acoustic meatus and auricle Special sensory: anterior two-thirds of tongue Branchial motor: muscles of facial expression and stylohyoid, posterior digastricus, stapedius mm. Visceral motor: all glands of the head (lacrimal, submandibular, sublingual, palatal, nasal) except the one it courses through (does not innervate the parotid)	Internal acoustic meatus

TABLE 17-2. Overview of the Cranial Nerves *(continued)*

CN	Modalities and Function	Exit from Skull
CN VIII (vestibulocochlear)	Special sensory: hearing, balance, and equilibrium	Internal acoustic meatus
CN IX (glossopharyngeal)	General sensory: posterior third of tongue, oropharynx, tympanic membrane, middle ear, and auditory tube Special sensory: taste from posterior one-third of tongue Visceral sensory: carotid sinus (baroreceptor) and carotid body (chemoreceptor) Branchial motor: stylopharyngeus m. Visceral motor: parotid gland	Jugular foramen
CN X (vagus)	General sensory: skin of the posterior ear and external acoustic meatus Visceral sensory: aortic and carotid bodies (chemoreceptors) and aortic arch (baroreceptor) Branchial motor: all palatal muscles (except tensor tympani); all pharyngeal muscles (except stylopharyngeus m.) and all laryngeal mm. Visceral motor: heart, smooth muscle, and glands of the respiratory tract, gastrointestinal tube, and viscera of the foregut and midgut	Jugular foramen
CN XI (spinal accessory)	Branchial motor: trapezius and sternocleidomastoid mm.	Jugular foramen
CN XII (hypoglossal)	Somatic motor: tongue mm. (except palatoglossus m.)	Hypoglossal canal

mm, muscles; m, muscle.

TABLE 17-3. Parasympathetic Innervation of the Head

CN	Preganglionic Parasympathetic Cell Body Origin	Postganglionic Parasympathetic Cell Body Origin	Function
CN III (oculomotor)	Edinger–Westphal nucleus	Ciliary ganglion	Sphincter pupillae m. (constricts pupil) and ciliary m. (lens accommodation for near vision)
CN VII (facial)	Superior salivatory nucleus	Pterygopalatine ganglion Submandibular ganglion	Lacrimal, nasal, and palatal glands Submandibular and sublingual salivary glands
CN IX (glossopharyngeal)	Inferior salivatory nucleus	Otic ganglion	Parotid gland
CN X (vagus)	Posterior vagal nucleus	Intramural ganglia	Innervates heart, smooth muscle of respiratory tract, gastrointestinal tract (up to splenic flexor), and viscera associated with foregut and midgut

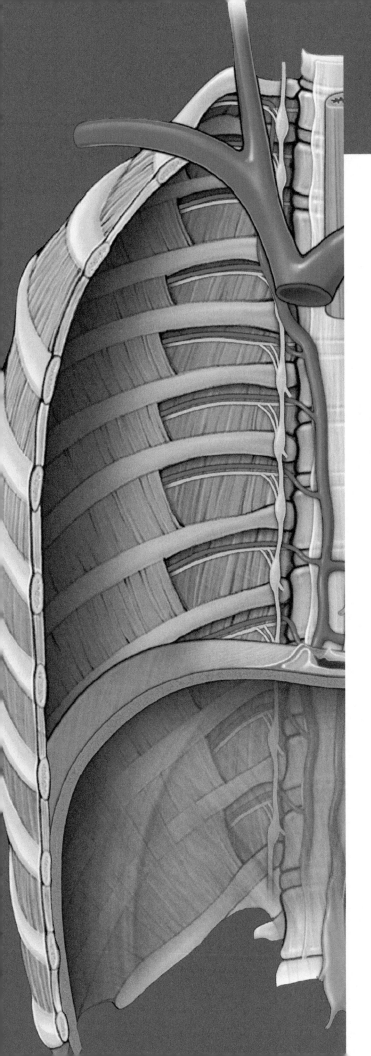

CHAPTER 18

ORBIT

Orbital Region 212

The Eye 214

Extraocular Muscle Movement 216

Innervation of the Orbit 220

Vascular Supply of the Orbit 222

ORBITAL REGION

BIG PICTURE

The primary organ responsible for vision is the eye. The eyeball is located within a bony orbital encasement, which protects it. The lacrimal apparatus keeps the eye moist and free of dust and other irritating particles through the production and drainage of tears. Eyelids protect the eye from external stimuli such as dust, wind, and excessive light.

BONY ORBIT

The bony orbit is the region of the skull that surrounds the eye and is composed of the following structures (Figure 18-1A):

- **Superior wall.** Formed by the frontal bone and the lesser wing of the sphenoid bone.
  - **Supraorbital foramen.** Transmits the supraorbital nerve [cranial nerve (CN) V-1] and vessels to the scalp.
- **Lateral wall.** Formed by the zygomatic bone and the greater wing of the sphenoid bone.
- **Inferior wall.** Formed by the maxillary, zygomatic, and palatine bones.
  - **Infraorbital foramen.** Transmits the infraorbital nerve (CN V-2) and vessels to the maxillary region of the face.
- **Medial wall.** Formed by the ethmoid, frontal, lacrimal, and sphenoid bones.
  - **Anterior and posterior ethmoidal foramina.** Transmits the anterior and posterior ethmoidal nerves and vessels, to the nasal cavity and the sphenoid and ethmoid sinuses.
  - **Nasolacrimal canal.** Formed by the maxillary, lacrimal, and inferior nasal concha bones. Drains tears from the eye to the inferior meatus in the nasal cavity.
- **Optic canal.** Transmits the optic nerve (CN II) and the ophthalmic artery.
- **Superior orbital fissure.** An opening between the greater and lesser wings of the sphenoid bone that transmits the oculomotor, trochlear, ophthalmic, and abducens nerves (CNN III, IV, V-1, and VI, respectively), and the ophthalmic veins.
- **Inferior orbital fissure.** Communicates with the infratemporal and pterygopalatine fossae. The inferior orbital fissure transmits CN V-2 and the infraorbital artery and vein.

LACRIMAL APPARATUS

The **lacrimal gland** lies in the superolateral corner of the orbit (Figure 18-1B).

- **Tears.** The lacrimal gland **secretes tears** that spread evenly over the eyeball through blinking and cleanse the eye of dust and foreign particles.
- **Drainage.** Tears drain from the eyeball, via the nasolacrimal duct, into the inferior nasal meatus of the nasal cavity.

- **Innervation.** The lacrimal gland is innervated by visceral motor parasympathetic neurons from CN VII (Figure 18-4C).

▽ Because the mucosa of the nasal cavity is continuous with the mucosa of the nasolacrimal duct system, a cold or **"stuffy" nose** often causes the lacrimal mucosa to become inflamed and swollen. Swelling constricts the ducts and prevents tears from draining from the eye surface, causing **"watery" eyes**. ▼

EYELIDS

Eyelids protect the eye from foreign particles and from bright light (Figure 18-1C). The external surface of the eyelids is covered by skin, whereas the conjunctiva covers the internal surface.

CONJUNCTIVA The conjunctiva is a mucous membrane that lines the internal surface of the eyelids. The conjunctiva is reflected at the superior and inferior fornices onto the anterior surface of the eyeball and forms the conjunctival sac when the eyes are closed.

EYELID MUSCLES Tears produced by the lacrimal gland are secreted continually and are spread over the conjunctiva and cornea by movement of the eyelids (blinking). The following muscles are involved in movement of the eyelids:

- **Orbicular oculi muscle.** Considered a muscle of facial expression. Its circular fibers attach to the anterior surface of the bony orbit as well as in the eyelid. When the facial nerve (CN VII) activates the orbicular oculi muscle, the eyes close.

▽ The **corneal, or blink, reflex** is elicited by stimulation of the cornea. The nasociliary nerve (CN V-1) mediates the sensory portion of this reflex, with CN VII initiating the motor response by innervating the orbicularis oculi muscle. ▼

- **Levator palpebrae superioris muscle.** The levator palpebrae superioris muscle elevates the upper eyelid and is innervated by CN III.
- **Superior tarsal (Müller's) muscle.** Composed of smooth muscle and attaches to the inferior surface of the levator palpebrae superioris muscle and the upper eyelid. These two muscles work synergistically to keep the eyelid elevated. The superior tarsal muscle is innervated by postganglionic sympathetic neurons that originate in the superior cervical ganglion and are carried to the orbit by the internal carotid and ophthalmic arteries.

▽ **Horner's syndrome** is caused by damage to the sympathetic innervation to the head (Figure 18-1D). Signs of Horner's syndrome found on the ipsilateral side of the injury include the following:

- **Ptosis** (drooping upper eyelid). Results from the loss of sympathetic innervation to the superior tarsal muscle.
- **Anhidrosis** (decreased sweating). Results from the loss of sympathetic innervation to sweat glands.
- **Miosis** (constricted pupil). Results from the loss of sympathetic innervation to the dilator pupillary muscle. ▼

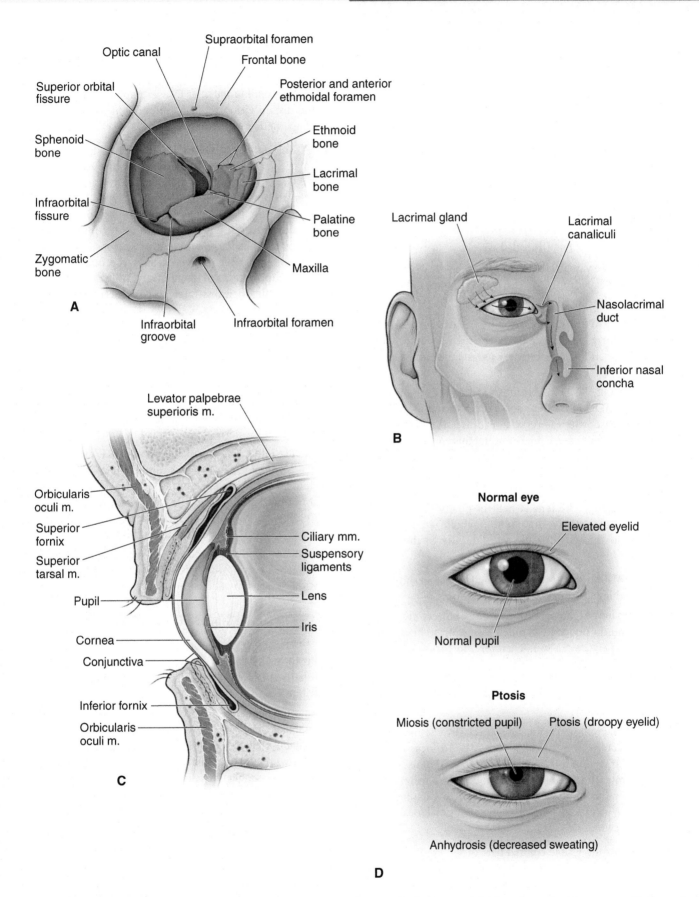

Figure 18-1: A. Bony orbit. **B**. Lacrimal apparatus. **C**. Sagittal section of the eyelid. **D**. A normal right eye in contrast to an eye with Horner's syndrome.

THE EYE

BIG PICTURE

The eyeball consists of the sclera, choroid, and retina.

SCLERA

The sclera is the white, fibrous covering of the eye into which muscles insert (Figure 18-2A).

- **Cornea.** The sclera is continuous anteriorly as the **cornea**, which forms a bulging, transparent region specialized for refracting light as it enters the eye.

CHOROID

The choroid is the vascular, middle layer of the eye where blood vessels course (Figure 18-2A). The choroid layer is also composed of the following structures:

CILIARY APPARATUS The anterior region of the choroid is thickened to form the **ciliary muscle**, which is composed of smooth muscle and is innervated by parasympathetic neurons from CN III (Figure 18-2B).

- **Ciliary muscle.** This circular muscle surrounds the periphery of the lens. Contraction causes the diameter of the muscle to decrease, whereas relaxation causes the diameter of the muscle to increase. Understanding this concept is essential to understanding how the lens focuses on images near and far.
- **Lens.** A transparent biconvex structure enclosed in an elastic covering. The lens is held in position by radially arranged **suspensory ligaments**, which are attached medially to the lens capsule and laterally to the ciliary muscles.

ACCOMMODATION Process by which the lens charges to maintain a focused image from distant or near objects.

- **Distant vision.** Light rays from distant objects are nearly parallel and do not need as much refraction to bring them into focus (Figure 18-2C). Therefore, to focus on distant objects, the ciliary muscle relaxes, which stretches the suspensory ligaments and flattens the lens. When the lens is flat (least rounded), it is at optimal focal length for distant viewing.
- **Near vision.** Light rays from close objects diverge and require more refraction to focus (Figure 18-2D). To focus on near objects, parasympathetic neurons in CN III cause ciliary muscle contraction and thus relaxation of the tension on the suspensory ligaments, allowing the lens to become more rounded. When the lens is round, it is at optimal focal length for near viewing.

IRIS The iris is the visible colored part of the eye (Figure 18-2B). The round central opening of the iris is the **pupil**, which allows light to enter the eye. The iris consists of smooth muscle under autonomic control, which contract reflexively and vary the size of the pupil.

- **Sphincter pupillae muscle.** Causes pupil constriction in bright light and close vision via visceral motor (parasympathetic) innervation from CN III.
- **Dilator pupillae muscle.** Causes pupil dilation in dim light and distant vision to enable more light to enter the eye via sympathetic innervation.

▽ The **pupillary reflex** regulates the intensity of light entering the eye by controlling the diameter of the pupil. Greater intensity of light causes the pupil to become smaller, thus allowing less light to reach the retina. CN II is responsible for the sensory limb of the pupillary reflex by sensing the incoming light. Visceral motor innervation from CN III is responsible for the motor response by constricting the pupil. Light entering one eye should produce constriction of both pupils. **Lack of a pupillary reflex or an abnormal pupillary reflex** demonstrates potential lesions in CN II, CN III, or the brainstem. ▼

RETINA

The retina is the innermost layer of the eyeball (Figure 18-2A).

- **Optic disc.** The optic disc is located medial to the **macula lutea**. The optic disc has no photoreceptors and is insensitive to light, hence the nickname "the blind spot." The optic disc consists of the nerve fibers formed by CN II.
- **Macula lutea.** The macula lutea is a yellow area of the retina that is lateral to the optic disc and contains a central depression called the **fovea centralis**. This region contains only cones and functions for detailed vision (maximal visual acuity). The eye is positioned such that light rays from the object in the center of the view fall upon this region.
- **Rods and cones.** Rods are specialized cells for vision in dim light (black, gray, white vision), whereas cones are associated with visual acuity and color vision.

CHAMBERS OF THE EYE

The eye is divided into the following segments (Figure 18-2B).

- **Posterior segment.** Known as the **vitreous chamber**, it is filled with a clear gel called **vitreous humor**, which contributes to the intraocular pressure.
- **Anterior segment.** Subdivided by the iris into the **anterior chamber** (between the cornea and iris) and the **posterior chamber** (between the iris and lens). The anterior segment is filled with **aqueous humor**, which has a composition similar to plasma. Unlike vitreous humor, which is constant and is never replaced, aqueous humor forms and drains continually into the venous system via the **canal of Schlemm.**

▽ **Aqueous humor** normally is produced and drained at a constant rate, and as a result, a constant intraocular pressure is maintained. If drainage of aqueous humor is impaired, the pressure within the eye may increase and cause compression of the retina and damage the optic nerve, resulting in a condition known as **glaucoma.** ▼

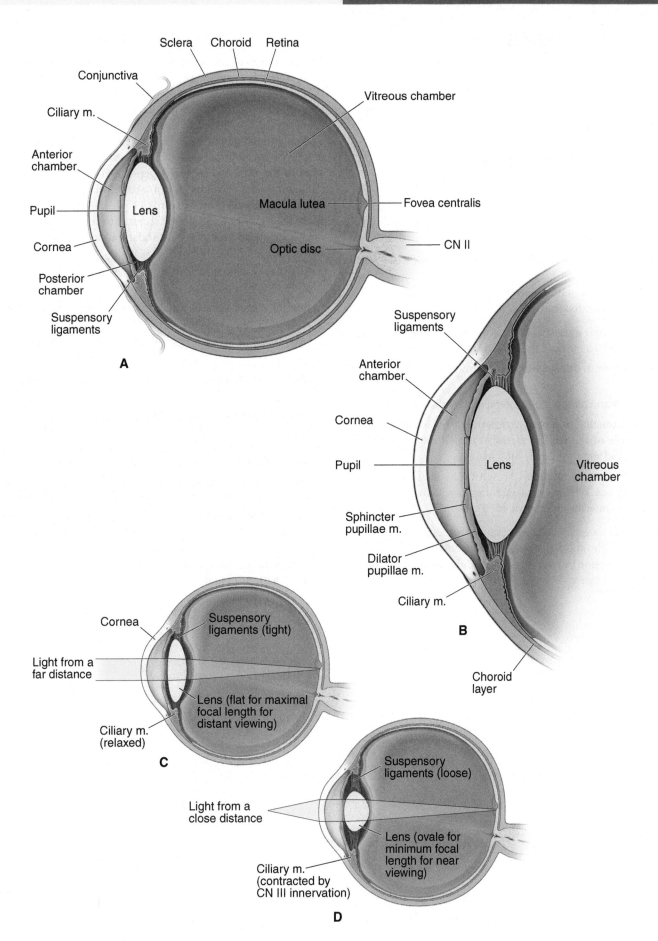

Figure 18-2: A. Axial section of the eye. **B**. Close-up of the axial section of the anterior portion of the eye. **C**. Light from a distance is bent by the stretched lens to strike the retina. **D**. Light from a source nearby is bent even more sharply by the relaxed lens to strike the retina.

EXTRAOCULAR MUSCLE MOVEMENT

BIG PICTURE

Six strap-like extraocular muscles (four rectus and two oblique) control the movement of the eye (elevation, depression, abduction, adduction, intorsion, and extorsion).

EYE MOVEMENT

The eye moves in the following **three axes of rotation** (Figure 18-3A):

- **X-axis.** A horizontal axis in the axial plane enabling abduction and adduction.
- **Y-axis.** A vertical axis in the sagittal plane resulting in elevation and depression.
- **Z-axis.** A horizontal axis in the sagittal plane resulting in intorsion and extorsion.

The extraocular muscles act upon the eye in these three axes to cause the eye to move in the following dimensions:

- **Elevation** and **depression.** An upward and downward gaze.
- **Abduction** and **adduction.** A lateral and medial gaze.
- **Intorsion (medial rotation).** An inward medial rotation of the upper pole of the vertical meridian caused by the superior oblique and superior rectus muscles.
- **Extorsion (lateral rotation).** An outward lateral rotation of the upper pole of the vertical meridian caused by the inferior rectus and inferior oblique muscles.

Each of the six extraocular muscles exerts rotational forces in all three axes (elevation–depression, abduction–adduction, and intorsion–extorsion) to varying degrees, providing fine motor control to direct gaze.

EXTRAOCULAR MUSCLES

The movement of each eyeball is controlled by the six extraocular muscles (four rectus and two oblique) innervated by CNN III, IV, and VI. The extraocular muscles are named according to their position on the eyeball (Figure 18-3B and C).

All of the muscles except the inferior oblique muscle arise from a **common tendinous ring** in the posterior part of the orbit. To determine the action of each muscle, it should be noted that the **apex of the orbit** (where the superior and inferior rectus and oblique muscles originate) is not parallel with the **optical axis** (when looking directly forward) (Figure 18-3D). Therefore, when the superior and inferior rectus muscles and the oblique muscles contract, there are secondary actions on the eyeball movement, as follows (Figure 18-3E):

- **Superior rectus muscle (CN III).** Elevation of the eye, with adduction and intorsion.
- **Inferior rectus muscle (CN III).** Depression of the eye, with adduction and extorsion.
- **Lateral rectus muscle (CN VI).** Only action is abduction of the eye.
- **Medial rectus muscle (CN III).** Only action is adduction of the eye.
- **Superior oblique muscle (CN IV).** Originates in common with the rectus muscles, courses along the medial wall of the orbit, and then turns and loops through a fibrous sling, called the **trochlea**, before inserting on the superolateral region of the eyeball. The superior oblique muscle moves the eyeball down and out (depression and abduction), with intorsion.
- **Inferior oblique muscle (CN III).** Originates on the medial wall of the bony orbit and courses laterally and obliquely to insert on the inferolateral surface of the eyeball. The inferior oblique moves the eyeball up and out (elevation and abduction), with extorsion.

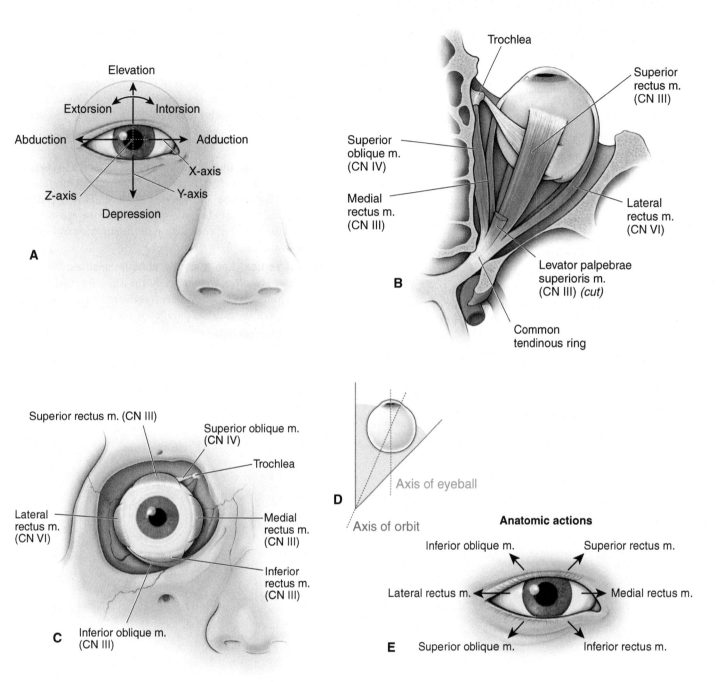

Figure 18-3: A. Movements of the eyeball. Extraocular muscles of the right eye; **(B)** superior and **(C)** anterior views. **D**. Axes of the eyeball and orbit. **E**. Anatomic actions of the right extraocular muscles.

CLINICAL EXAMINATION OF THE EXTRAOCULAR MUSCLES

When performing a physical examination of the eye, a physician will test each of the extraocular muscles and their associated cranial nerves by drawing an "H" pattern in the air in front of the patient's face (Figure 18-3F). The patient is instructed to follow the physician's finger with her eyes only.

Horizontal line of the "H." The horizontal line of the "H" will test the medial and lateral rectus muscles. The medial rectus muscle adducts the eye, whereas the lateral rectus muscle abducts the eye. The medial and lateral rectus muscles are the only muscles that move the eye in the horizontal plane and are therefore easy to test.

Vertical lines of the "H." The vertical motion of the eye is a little more complex than the horizontal motion. The superior and inferior rectus muscles and the superior and inferior oblique muscles control the vertical motion of the eyeball. When a patient's gaze is straight up, it occurs as a result of the combined action of the superior rectus and inferior oblique muscles. When the gaze is straight down, it is from the combined action of the inferior rectus and superior oblique muscles. Therefore, it is essential in testing the extraocular muscles to ensure that only one muscle is tested at a time, without influence from another extraocular muscle. To test any of these four muscles, each muscle must be isolated from the others.

Lateral vertical line tests the rectus muscles. When the right eye is fully abducted, only the superior and inferior rectus muscles elevate and depress the eye. This is purely a mechanical property because of the axis of the eye lining up parallel to the line of contraction of the superior and inferior rectus muscles.

Medial vertical line tests the oblique muscles. When the right eye is fully adducted, only the superior and inferior oblique muscles elevate and depress the eye. Again, this is due to the axis of the muscles paralleling the axis of the eye.

Clinical testing

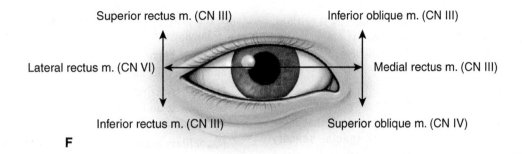

Superior rectus m. (CN III)   Inferior oblique m. (CN III)

Lateral rectus m. (CN VI)   Medial rectus m. (CN III)

Inferior rectus m. (CN III)   Superior oblique m. (CN IV)

F

Superior rectus m.

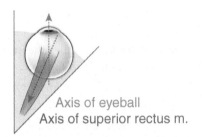

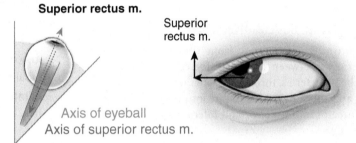

Axis of eyeball
Axis of superior rectus m.

Superior
rectus m.

Axis of eyeball
Axis of superior rectus m.

The axis of the eyeball and superior rectus m. are **NOT PARALLEL**; therefore the superior rectus m. cannot be isolated

Once the eyeball has been abducted, the eyeball now is **PARALLEL** with the superior rectus m.; therefore only the superior rectus m. can elevate the eye

Superior oblique m.

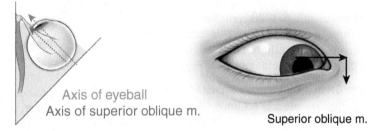

Axis of eyeball
Axis of superior oblique m.

Axis of eyeball
Axis of superior oblique m.

Superior oblique m.

The axis of the eyeball and superior oblique m. are **NOT PARALLEL**; therefore the superior oblique m. cannot be isolated

Once the eyeball has been adducted, the eyeball now is **PARALLEL** with the superior oblique m.; therefore only the superior oblique m. can depress the eye

Figure 18-3: (*continued*) **F**. Clinical examination of the extraocular muscles and associated CNs. The superior rectus and superior oblique muscles are highlighted.

INNERVATION OF THE ORBIT

BIG PICTURE

The cranial nerves associated with the orbit are CN II (vision), CN III (eye movement), CN IV (eye movement), CN V-1 (general sensory to eye and scalp), CN VI (eye movement), and CN VII (crying and closing the eyes).

CN II: OPTIC NERVE

The optic nerve enters the orbit through the optic canal along with the ophthalmic artery and provides **special sensory innervation (vision)** from the **retina** to the brain. CN II from one eye joins CN II from the corresponding eye to form the **optic chiasma**.

CN III: OCULOMOTOR NERVE

The oculomotor nerve emerges from the midbrain and courses anteriorly between the posterior cerebral and the superior cerebellar arteries, through the lateral wall of the cavernous sinus and superior orbital fissure into the orbit, where the nerve bifurcates into superior and inferior divisions (Figure 18-4B and C).

- **Superior division.** Provides **somatic motor innervation** to the **levator palpebrae superioris** and **superior rectus muscles**.

- **Inferior division.** Provides **somatic motor innervation** to the **medial rectus, inferior rectus**, and **inferior oblique muscles**. CN III also provides **visceral motor innervation**. Preganglionic parasympathetic neurons originate in the **Edinger–Westphal nucleus** and course in the inferior division of CN III and synapse in the **ciliary ganglion** (between CN II and the lateral rectus muscle). Postganglionic parasympathetic neurons course in the **short ciliary nerves** to enter the eyeball and innervate the **ciliary muscles** (lens accommodation) and **pupillary sphincter muscles** (pupil constriction).

▽ **Injury to CN III** results in a fixed, **dilated pupil** because the dilator pupillae muscle no longer has an antagonistic muscle counteracting its force. The **lens does not accommodate** (due to loss of innervation to the ciliary muscles) and **ptosis** develops (due to loss of innervation to the levator palpebrae superioris muscle). Because of the unopposed action of the lateral rectus and superior oblique muscles, the eye deviates to the abducted and down and out position.

CN IV: TROCHLEAR NERVE

The trochlear nerve arises in the midbrain and is the only cranial nerve to exit from the posterior surface of the brainstem. After curving around the midbrain, CN IV courses through the lateral wall of the cavernous sinus to enter the orbit via the superior orbital fissure (Figure 18-4A). CN IV provides **somatic motor innervation** to the **superior oblique muscle**.

▽ Injury to CN IV results in the inability to view an object inferiorly when the eye is adducted. ▼

CN V-1: OPHTHALMIC BRANCH OF THE TRIGEMINAL NERVE

The ophthalmic nerve courses through the superior orbital fissure and provides **general sensory innervation to the orbit** via the following three branches (Figure 18-4A–C):

- **Lacrimal nerve.** Innervates the lacrimal gland, conjunctiva, and skin of the upper eyelid.

- **Frontal nerve.** Courses superior to the levator palpebrae superioris muscle and bifurcates into the **supraorbital** and **supratrochlear nerves** to innervate the skin of the forehead and scalp.

- **Nasociliary nerve.** Provides **general sensory innervation** via the following branches:
  - **Long ciliary nerves.** Supplies the cornea (sensory limb of the **corneal reflex**).
  - **Posterior ethmoidal nerve.** Supplies the sphenoid and ethmoidal sinuses.
  - **Anterior ethmoidal nerve.** Supplies the ethmoid sinus, nasal cavity, and skin on the tip of the nose.
  - **Infratrochlear nerve.** Supplies the lower eyelid, conjunctiva, and skin of the nose.

CN VI: ABDUCENS NERVE

The abducens nerve enters the orbit via the superior orbital fissure and provides **somatic motor innervation** to the **lateral rectus muscle** (Figure 18-4B).

 Injury to CN VI results in the inability to abduct the eye, resulting in **diplopia and loss of parallel gaze**. ▼

CN VII: FACIAL NERVE

In the orbit, the facial nerve "makes you cry" (visceral motor) and "close your eyes" (branchial motor) (Figure 18-4C).

- **Preganglionic parasympathetic neurons from CN VII** originate in the **superior salivatory nucleus** (pons) and exit the cranial cavity through the internal acoustic meatus. At the geniculate ganglion, CN VII gives off the **greater petrosal nerve**, which joins with the deep petrosal nerve (sympathetics from the **carotid plexus**), becoming the **nerve of the pterygoid canal (Vidian nerve)**. The nerve of the pterygoid canal enters the pterygopalatine fossa, where preganglionic parasympathetic neurons synapse in the **pterygopalatine ganglion**. **Postganglionic parasympathetic neurons** join with the **zygomatic nerve** (CN V-2) and then with the **lacrimal nerve** (CN V-1) to innervate the **lacrimal gland**.

- CN VII provides **branchial motor innervation** to the **orbicularis oculi muscle**.

ORBITAL SYMPATHETICS

Sympathetics to the orbit course in the following pathways (Figure 18-4C):

- Postganglionic sympathetic neurons originating in the superior cervical ganglion ascend along the internal carotid artery up to the ophthalmic artery.

- Sympathetic nerves innervate the **superior tarsal muscle** (keeping the eyelid elevated) and the dilator pupillae muscles.

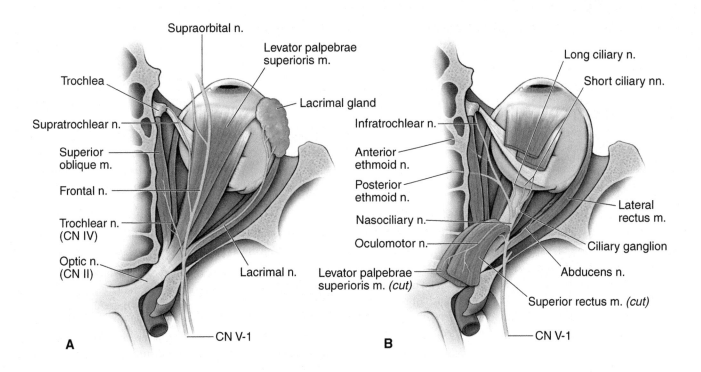

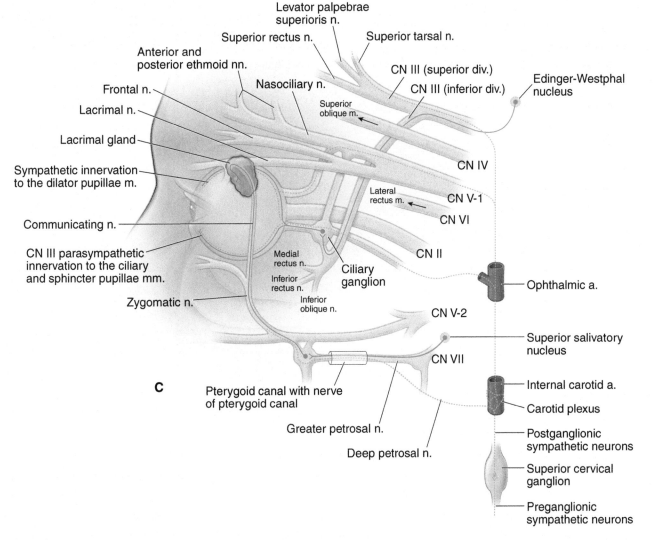

Figure 18-4: Superior view of the nerves of the orbit (**A**. superficial). (**B**. deep). **C**. Comprehensive innervation of the orbit highlighting autonomics.

VASCULAR SUPPLY OF THE ORBIT

BIG PICTURE

The ophthalmic artery provides the principal vascular supply to the orbit. The superior and inferior ophthalmic veins drain anteriorly to the facial vein, posteriorly to the cavernous sinus, and inferiorly to the pterygoid plexus.

OPHTHALMIC ARTERY

The ophthalmic artery enters the orbit through the optic canal with CN II (Figure 18-5A). Branches travel to the eye, lacrimal gland, eyelids, scalp, extraocular muscles, and the side of the nose. Major branches of the ophthalmic arteries are as follows:

- **Central retinal artery.** Travels within CN II and at the optic disc sends numerous branches to the retina. The central artery of the retina is the most important branch of the ophthalmic artery because it supplies all the nerve fibers of CN II. Therefore, if the central artery is occluded, there is complete loss of vision in that eye.
- **Posterior ethmoidal artery.** Courses through the posterior ethmoidal foramen and provides vascular supply to the ethmoid sinus.
- **Anterior ethmoidal artery.** Courses through the anterior ethmoidal foramen and provides vascular supply to the ethmoid sinus, frontal sinus, nasal cavity, and the external part of the nose.

OPHTHALMIC VEINS

The principal veins that drain the orbit are the superior and inferior ophthalmic veins (Figure 18-5B).

- **Superior ophthalmic vein.** Formed by the union of the supraorbital, supratrochlear, and **angular (facial) veins** and communicates posteriorly with the **cavernous sinus**. The superior ophthalmic vein may also receive tributaries from the inferior ophthalmic vein. Unlike most other veins, the superior ophthalmic vein lacks valves and, therefore, blood may drain anteriorly through the angular vein or posteriorly in the cavernous sinus.

▽ The **medial angle of the eye, nose, and lips** is a triangular area of potential danger in the face. The blood in this region usually drains inferiorly via the facial vein. However, blood can also drain superiorly through the facial vein to the superior ophthalmic vein to the cavernous sinus. Therefore, an **infection of the face** may spread to the cavernous sinus and pterygoid plexus. ▽

- **Inferior ophthalmic vein.** Communicates with the **pterygoid plexus of veins** through the infraorbital fissure, as well as with the infraorbital vein. The inferior ophthalmic vein terminates in the **cavernous sinus**.

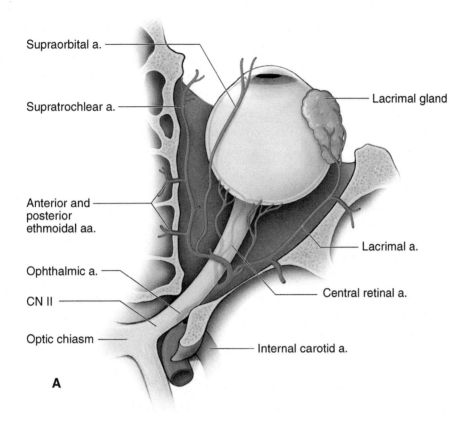

Supraorbital a.

Supratrochlear a.

Lacrimal gland

Anterior and posterior ethmoidal aa.

Ophthalmic a.

CN II

Optic chiasm

Lacrimal a.

Central retinal a.

Internal carotid a.

A

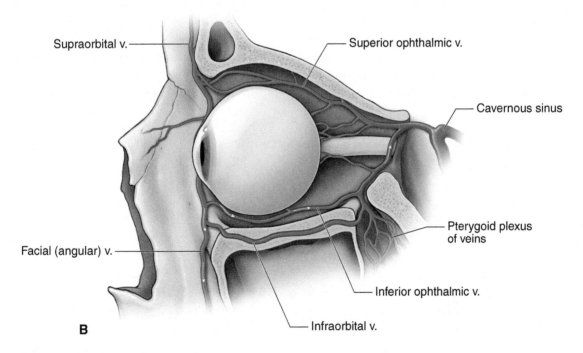

Supraorbital v.

Superior ophthalmic v.

Cavernous sinus

Pterygoid plexus of veins

Facial (angular) v.

Inferior ophthalmic v.

Infraorbital v.

B

Figure 18-5: A. Arteries of the orbit. **B**. Veins of the orbit.

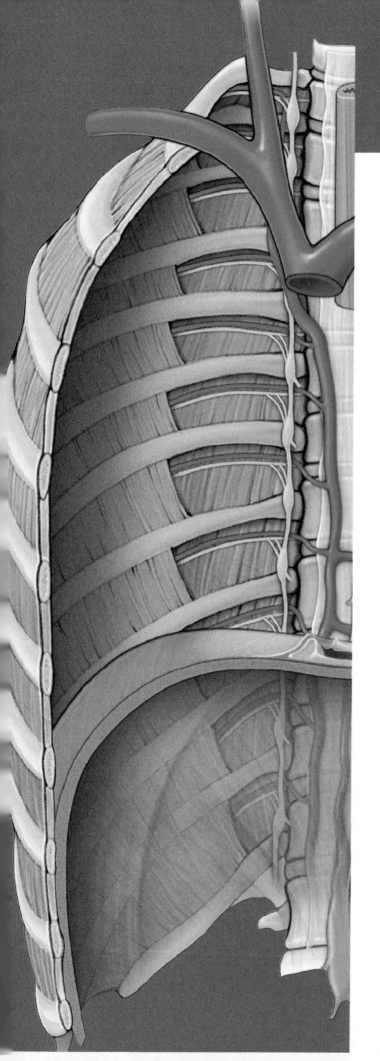

EAR

THE EAR

BIG PICTURE

The **external ear** collects sound waves and transports them through the external acoustic meatus to the tympanic membrane. The tympanic membrane vibrates, setting three tiny ear ossicles (malleus, incus, and stapes) in the **middle ear** into motion. The stapes attaches to the lateral wall of the **inner ear**, where the vibration is transduced into fluid movement. The fluid causes the basilar membrane in the cochlea to vibrate. The vestibulocochlear nerve [cranial nerve (CN) VIII] receives and conducts the impulses to the brain, where there is integration of sound and equilibrium.

EXTERNAL EAR

The external ear consists of the **auricle**, or pinna, which lies at the outer end of a short tube called the **external acoustic meatus** (Figure 19-1A). The auricle funnels sound waves through the external acoustic meatus to the **tympanic membrane**. The external ear receives general sensory innervation from the trigeminal, facial, and vagus nerves (**cranial nerves (CNN) V, VII,** and **X**, respectively) and from the **great auricular nerve** (cervical plexus C2–C3).

TYMPANIC MEMBRANE The **tympanic membrane**, or "eardrum," is a three-layered circular structure (Figure 19-1A–C).

- **Outer layer.** Composed of modified skin that is continuous with the external acoustic meatus.
- **Middle layer.** Composed of connective tissue through which the **chorda tympani nerve (CN VII)** passes.
- **Inner layer.** Lined with the mucosa of the middle ear, and receives general sensory innervation via the **tympanic plexus (CN IX)**.

▽ A physician uses an **otoscope** to view the health of a patient's external and middle ear. One of the structures seen on the tympanic membrane is the where the **handle of the malleus** attaches on its internal surface. When a physician shines the light of the otoscope onto a healthy tympanic membrane, the malleus causes a **cone of light** to appear in the anterior–inferior quadrant (Figure 19-1B). ▽

MIDDLE EAR

The middle ear is an air-filled chamber that transmits sound waves from air to the auditory ossicles and then to the fluid-filled inner ear (Figure 19-1A). The middle ear consists of the tympanic cavity proper, auditory tube, ear ossicles, and branches of CNN VII and IX.

TYMPANIC CAVITY PROPER The tympanic cavity proper is the space between the **tympanic membrane** and the **vestibular window**. Its mucosa receives general sensory innervation from the **tympanic nerve and the tympanic plexus (CN IX)** (Figure 19-1A–D). In addition, visceral motor preganglionic parasympathetic fibers from CN IX branch from the tympanic plexus to exit the middle ear as the **lesser petrosal nerve** on route to innervate the **parotid gland.**

- **Vestibular (oval) window.** A membrane-covered opening between the middle ear and the **vestibule** of the inner ear. The oval window is pushed back and forth by the footplate of the stapes and transmits the vibrations of the ossicles to the **perilymph** at the origin of the **scala vestibuli** in the inner ear.
- **Cochlear (round) window.** A membrane-covered opening that accommodates the pressure waves transmitted to the perilymph at the end of the **scala tympani**.

AUDITORY TUBE The **auditory (eustachian) tube** is an osseous–cartilaginous tube that connects the nasopharynx and the middle ear (Figure 19-1A). The auditory tube receives **general sensory** innervation from the **tympanic plexus** (CN IX) and also serves as an attachment point for the **tensor tympani muscle.**

▽ The **auditory tube** normally is closed, but yawning or swallowing can open the tube, allowing air to enter, which **equalizes the pressure between the middle ear and the atmosphere.** When air enters, which can occur when in an airplane or at a high elevation, a soft "pop" sound may be felt. ▽

▽ A patient with **otitis media** (middle ear infection) will present with a **red bulging tympanic membrane**, which is usually due to a buildup of fluid or mucus. This inflammation is often the result of a **pharyngeal infection** transmitted via the auditory tube to the middle ear. Because the **auditory tube is shorter and more horizontal in children**, it is easier for infection to spread from the nasopharynx to the middle ear, resulting in a higher prevalence of otitis media in children compared to adults. Hearing may be diminished because of the pressure on the eardrum, and taste may be altered due to the effect on the chorda tympani nerve. Infection can easily spread from the tympanic cavity to the mastoid air cells, causing **mastoiditis**. ▽

EAR OSSICLES The ear ossicles are three small bones known as the **malleus, incus, and stapes**, which transmit vibrations from the tympanic membrane to the inner ear (Figure 19-1A–D). The ossicles function as amplifiers to overcome the impedance mismatch at the air–fluid interface of the middle and inner ear.

- **Malleus.** The malleus is attached to the internal surface of the tympanic membrane and articulates with the incus (Figure 19-1B). The **tensor tympani muscle** attaches between the auditory tube and malleus and serves to reduce the movement of the tympanic membrane. It is innervated by **CN V-3.**
- **Incus.** The incus articulates with the stapes.
- **Stapes.** The **footplate of the stapes** attaches to the **vestibular window**, which separates the **air environment** of the middle ear from the **fluid environment** of the inner ear. The **stapedius muscle** is the smallest skeletal muscle in the body. The stapedius muscle prevents excess movement of the stapes and controls the amplitude of sound waves from the external environment to the middle ear. The stapedius is innervated by **CN VII.**

▽ **Paralysis of the stapedius muscle** is usually caused by a lesion of CN VII, resulting in wider oscillation of the stapes; consequentially, there is a heightened reaction of the auditory ossicles to sound vibration. This condition is known as **hyperacusis** and results in an increased sensitivity to loud sounds. ▽

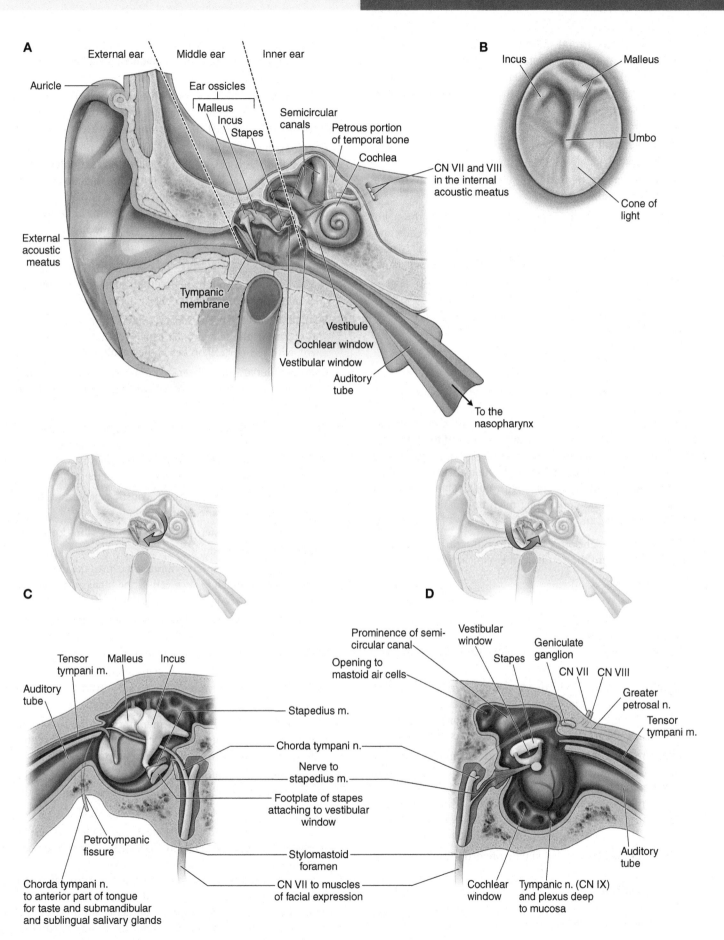

Figure 19-1: A. Coronal section of the temporal bone showing the hearing apparatus. **B**. Right tympanic membrane viewed through an otoscope. Lateral (**C**) and medial (**D**) wall of the middle ear.

BRANCHES OF THE FACIAL NERVE The facial nerve (CN VII) enters the **internal acoustic meatus** along with CN VIII. CN VII enters the facial canal and continues laterally between the internal and middle ear. It is at this point that the sensory **geniculate ganglion** forms a bulge in CN VII and gives rise to the following branches (Figure 19-1C and D):

- **Greater petrosal nerve.** Provides visceral motor innervation to the lacrimal, nasal, and palatal glands.
- **Branchial motor neurons.** Provides innervation **to the stapedius muscle**.
- **Chorda tympani nerve.** Arises before CN VII exits the stylomastoid foramen. The chorda tympani nerve ascends and courses through the posterior wall of the middle ear, passes through the middle layer of the tympanic membrane, continues between the malleus and stapes, and exits the skull at the petrotympanic fissure. The chorda tympani innervates the submandibular and sublingual salivary glands, and conveys taste sensation (special sensory) from the anterior two-thirds of the tongue.

INNER EAR

The inner ear contains the functional organs for hearing and equilibrium. It consists of a series of bony cavities (**bony labyrinth**), within which is a series of membranous ducts (**membranous labyrinth**), all within the **petrous part of the temporal bone**. The space between the bony and membranous labyrinths is filled with a fluid called **perilymph**. The tubular chambers of the membranous labyrinth are filled with **endolymph**. These two fluids provide a fluid-conducting medium for the vibrations involved in hearing and equilibrium.

BONY LABYRINTH The bony labyrinth is structurally and functionally divided into the vestibule, the semicircular canals, and the cochlea (Figure 19-2A).

Vestibule. The vestibule is the central portion of the bony labyrinth.

- **Vestibular window.** The vestibular window serves as a membranous interface between the stapes from the middle ear and the vestibule of the inner ear.
- **Utricle and saccule.** The membranous labyrinth within the vestibule consists of two connected sacs called the utricle and saccule. Both the utricle and saccule contain receptors that are sensitive to **gravity** and **linear movements of the head**.
- **Semicircular canals.** The three bony semicircular canals of the inner ear are at right angles to each other. The narrow **semicircular ducts** of the membranous labyrinth are located within the semicircular canals. Receptors within the semicircular ducts are sensitive to angular acceleration and deceleration of the head, as occurs in **rotational movement**.

Cochlea. The cochlea is a coiled tube divided into three chambers (Figure 19-2B).

- **Scala vestibuli.** Forms the upper chamber of the cochlea. The scala vestibuli begins at the vestibular window, where it is continuous with the vestibule, and contains perilymph.

- **Scala tympani.** Forms the lower chamber of the cochlea. The scala tympani terminates at the **cochlear window** and contains perilymph.
  - **Helicotrema.** The scala vestibuli and the scala tympani are separated completely, except at the narrow apex of the cochlea called the helicotrema, where they are continuous.
- **Cochlear duct.** Forms the middle chamber of the cochlea. The roof of the cochlear duct is called the **vestibular membrane**, and the floor is called the **basilar membrane**. The cochlear duct is filled with **endolymph** and ends at the helicotrema. The cochlear duct houses the **spiral organ (of Corti)**, where sound receptors transduce mechanical vibrations into nerve impulses.

VESTIBULOCOCHLEAR NERVE The vestibulocochlear nerve (CN VIII) courses through the **internal acoustic meatus** and divides into the vestibular and cochlear nerves (Figure 19-2A).

- **Vestibular nerve.** Special sensory innervation of the utricle and saccule of the semicircular canals (equilibrium and balance), with sensory cell bodies in the **vestibular ganglion**.
- **Cochlear nerve.** Special sensory innervation of the spiral organ (of Corti) in the cochlea (hearing), with sensory cell bodies in the **spiral ganglion**.

▽ **Sound waves** travel in all directions from their source, similar to ripples in water after a stone is dropped (Figure 19-2C). Sound waves are characterized by their pitch (high or low frequency) and intensity (loudness or quietness). ▼

1. A **sound wave** enters the external acoustic meatus and strikes the tympanic membrane.
2. The sound wave transfers its energy into the vibration of the tympanic membrane.
3. As the tympanic membrane vibrates, it causes the malleus to move medially, which in turn causes the incus and stapes to move sequentially, amplifying the sound wave.
4. The stapes is attached to the vestibular window; thus, the vestibular window also moves, resulting in a wave forming in the perilymph within the scala vestibuli of the cochlea.
5. The fluid wave in the perilymph progresses from the scala vestibuli of the cochlea, resulting in an outward bulging of the cochlear window at the end of the scala tympani.
6. This bulging causes the basilar membrane in the cochlea to vibrate, which in turn results in stimulation of the receptor cells in the spiral organ (of Corti).
7. The receptor cells conduct impulses to the brain through the cochlear division of CN VIII, where the brain interprets the wave as sound.

The difference between a **sound wave** and **sound** can best be explained by the age-old question, *"If a tree falls in a forest and no one is around to hear it, does it make a sound?"*

Sound, as we interpret it, results from transduction and perception of amplitude, frequency, and complexity of a sound wave by the brain. The falling tree produces sound waves, but there is no perception of sound without the brain interpreting the sound wave. Therefore, the tree does not make a sound unless someone's auditory apparatus is there to hear it. ▼

A

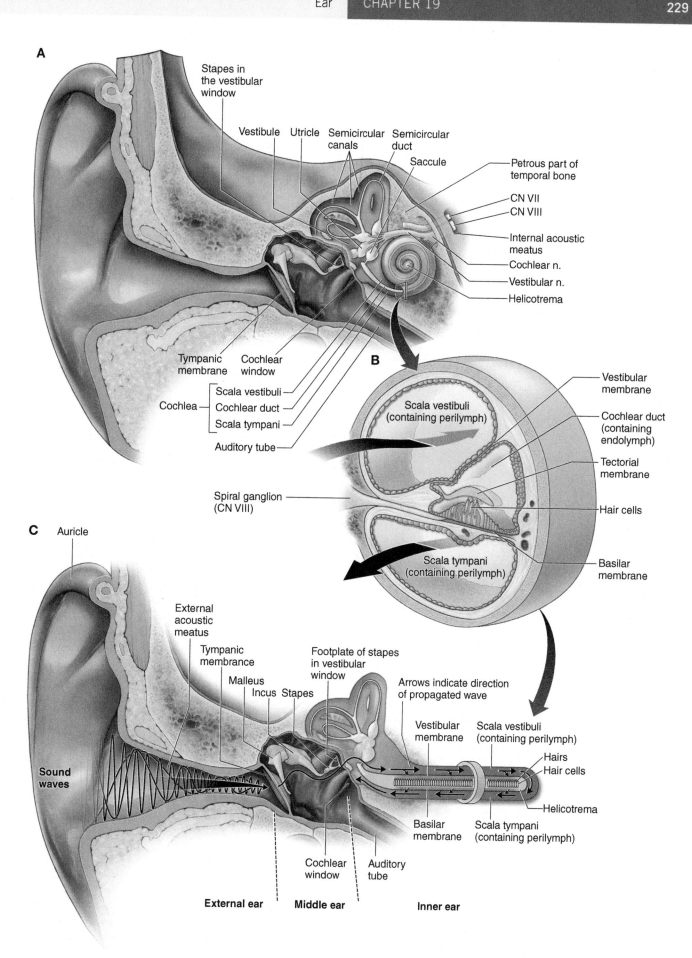

Figure 19-2: A. Coronal section of the internal ear. **B**. Cross-section of the cochlea. **C**. Pathway for the transmission of sound.

Figure 18-2. A

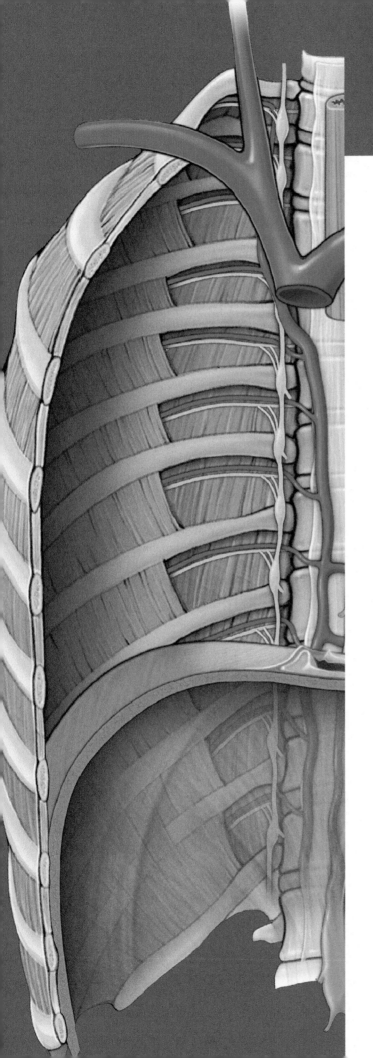

CHAPTER 20

SUPERFICIAL FACE

CUTANEOUS INNERVATION AND VASCULATURE OF THE FACE

BIG PICTURE

The sensory innervation of the face is provided by the three divisions of the trigeminal nerve [cranial nerve (CN) V], with each division supplying the upper, middle, and lower third of the face. The facial artery and the superficial temporal artery provide vascular supply.

CUTANEOUS INNERVATION OF THE FACE

The skin of the face and scalp is innervated by the cutaneous branches of the three divisions of CN V and by some nerves from the cervical plexus (Figure 20-1A and B).

- **CN V-1 (ophthalmic nerve).** Provides cutaneous innervation to the anterior region of the scalp via the **supraorbital nerve** and the **supratrochlear nerve**, the skin of the upper eyelid via the **lacrimal nerve**, and the bridge of the nose via the **external nasal nerve** and the **infratrochlear nerve**.

- **CN V-2 (maxillary nerve).** Provides cutaneous innervation to the skin along the zygomatic arch via the **zygomaticofacial** and the **zygomaticotemporal nerves** and the skin of the maxillary region, lower eyelid, and upper lip via the **branches of the infraorbital nerve**.

- **CN V-3 (mandibular nerve).** Provides cutaneous innervation to the skin of the lateral aspect of the scalp and the lateral part of the face, anterior to the external acoustic meatus, via the **auriculotemporal nerve**, the skin covering the mandible via the **buccal nerve**, and the skin of the lower lip via the **mental nerve**.

- **Cervical plexus.** The **great auricular nerve (C2–C3)** innervates the skin over the angle of the mandible just in front of the ear.

▽ **Trigeminal neuralgia (tic douloureux)** is a condition marked by paroxysmal pain along the course of CN V. Sectioning the sensory root of CN V at the trigeminal ganglion may alleviate the pain. ▼

VESSELS OF THE FACE

The face has a very rich blood supply provided primarily from the following vessels (Figure 20-1C):

- **Facial artery.** Branches off the external carotid artery and courses deep to the **submandibular gland**; winds around the inferior border of the mandible, anterior to the masseter muscle, to enter the face. As it ascends in the face, the facial artery supplies most of the face by way of the **inferior labial, superior labial, lateral nasal**, and **angular arteries**.

- **Supraorbital** and **supratrochlear arteries.** Terminal branches of the **ophthalmic artery**, a branch of the **internal carotid artery**, which supply the anterior portion of the scalp.

- **Superficial temporal artery.** A terminal branch of the external carotid artery provides arterial supply to the lateral surface of the face and scalp.

- **Facial vein.** Formed by the **union of the supraorbital and supratrochlear veins**. The facial vein descends in the face and receives tributaries corresponding to the branches of the facial artery. The facial vein drains into the **internal jugular vein**.

▽ In the skull, the **facial vein** connects with the **cavernous sinus** via the superior ophthalmic vein. Unlike other systemic veins, the facial and superior ophthalmic veins lack valves, thus providing a **potential pathway for the spread of infection** from the face to the cavernous sinus. ▼

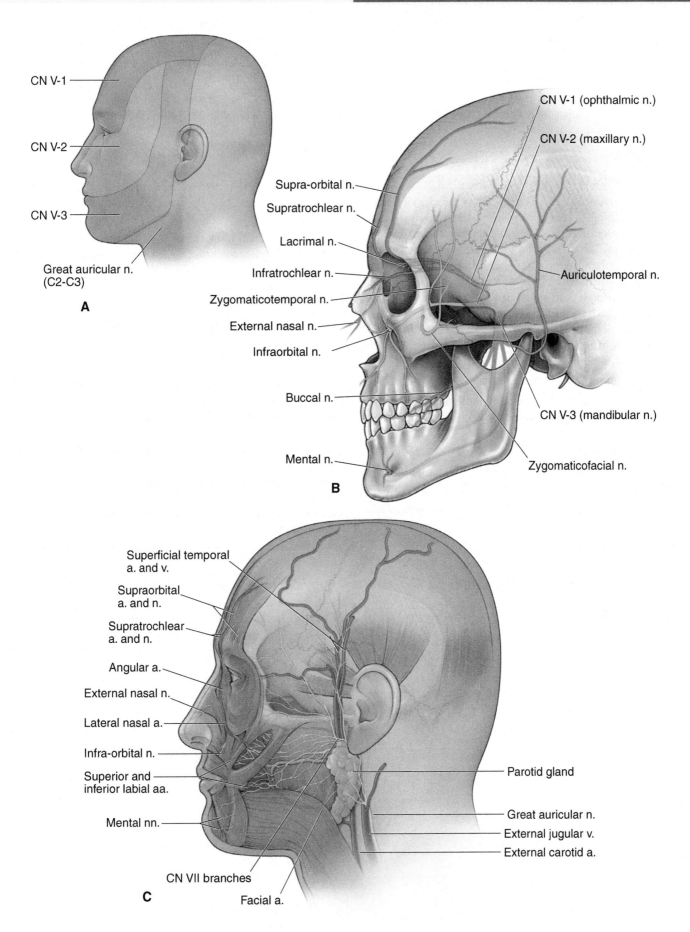

Figure 20-1: A. The trigeminal nerve (CN V) and its cutaneous fields of the face. **B**. Branches of CN V in the face. **C**. Vasculature of the face.

MUSCLES AND INNERVATION OF THE FACE

BIG PICTURE

Muscles of facial expression are located in the superficial fascia, and as their name implies, they control facial expression. The muscles of facial expression are innervated by the facial nerve (CN VII).

MUSCLES OF FACIAL EXPRESSION

The muscles of facial expression are voluntary muscles located in the superficial fascia. In general, they arise from bones or fascia of the skull and insert into the skin, which enables a wide array of facial expression. The muscles are sometimes indistinct at their borders because they develop embryologically from a continuous sheet of musculature derived from the second branchial arch. The muscles of facial expression are located superficially in the neck, face, and scalp. Each muscle is innervated by CN VII (except the muscles of mastication, which are innervated by CN V-3).

The muscles of facial expression can be organized into the following groups (Figure 20-2A):

Scalp and forehead (see Chapter 15)

- **Frontalis.** Connects with the **occipitalis muscle** by a cranial aponeurosis (**galea aponeurotica**); and wrinkles the forehead.

Muscles of the orbit

- **Orbicularis oculi.** Consists of **orbital** and **palpebral portions**, forming a sphincter muscle that closes the eyelids.
- **Corrugator supercilii.** Located deep to the orbicularis oculi; draws the eyebrows medially.

Muscles of the nose

- **Procerus.** Wrinkles the skin over the root of the nose.
- **Nasalis** and **levator labii superioris alaquae nasi.** Flare the nostrils.

Muscles of the mouth

- **Orbicularis oris.** Originates from the bones or fascia of the skull and inserts in the substance of the lips, forming an oral sphincter.
- **Levator labii superioris** and **levator anguli oris.** Raise the upper lip.
- **Zygomaticus major** and **minor**. Raise the corners of the mouth (smile).

- **Risorius.** Draws the corners of the lips laterally.
- **Depressor labii inferioris** and **depressor anguli oris.** Lower the bottom lip.
- **Buccinator.** Compresses the cheek when whistling, blowing, or sucking; holds food between the teeth during chewing.

Neck

- **Platysma.** Tenses the skin of the neck and lowers the mandible. Primarily located in the neck, although it does have attachments in the lower mandible and corners of the mouth.

▽ The **corneal blink reflex** is tested by touching the cornea with a piece of cotton, which should cause bilateral contraction of the orbicularis occuli muscles. The afferent limb is the nasociliary nerve of CN V-1, and the efferent limb of the reflex arch is CN VII. ▼

MOTOR NERVE SUPPLY TO THE FACE

CN VII provides motor innervation to the muscles of facial expression (Figure 20-2B). The facial nerve exits the skull through the **stylomastoid foramen** and immediately gives off the posterior auricular nerve and other branches that supply the occipitalis, stylohyoid, and posterior digastricus muscles and the posterior auricular muscle. CN VII courses superficial to the **external carotid artery** and the **retromandibular vein**, enters the **parotid gland**, and divides into the following five terminal branches: **temporal, zygomatic, buccal, mandibular,** and **cervical nerves**, which in turn supply the muscles of facial expression. Other muscles of the face include muscles of mastication (temporalis, masseter, and the medial pterygoid and lateral pterygoid muscles), which are innervated by the motor division of CN V-3.

▽ One of the most common problems involving CN VII occurs in the facial canal, just above the stylomastoid foramen. Here, an inflammatory disease of unknown etiology causes a condition known as **Bell's palsy**, where all of the facial muscles on one side of the face are paralyzed. Bell's palsy is characterized by facial drooping on the affected side, typified by the inability to close the eye, a sagging lower eyelid, and tearing. In addition, the patient has difficulty smiling, and saliva may dribble from the corner of the mouth. If the inflammation spreads, the chorda tympani and nerve to the stapedius muscle may be involved. ▼

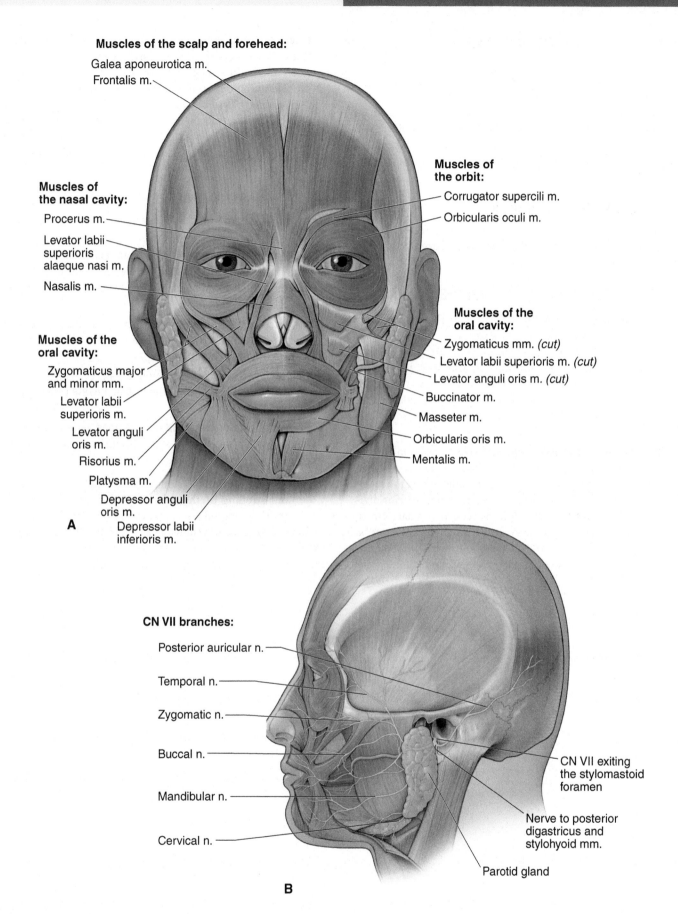

Muscles of the scalp and forehead:
Galea aponeurotica m.
Frontalis m.

Muscles of the orbit:
Corrugator supercili m.
Orbicularis oculi m.

Muscles of the nasal cavity:
Procerus m.
Levator labii superioris alaeque nasi m.
Nasalis m.

Muscles of the oral cavity:
Zygomaticus mm. *(cut)*
Levator labii superioris m. *(cut)*
Levator anguli oris m. *(cut)*
Buccinator m.
Masseter m.
Orbicularis oris m.
Mentalis m.

Muscles of the oral cavity:
Zygomaticus major and minor mm.
Levator labii superioris m.
Levator anguli oris m.
Risorius m.
Platysma m.
Depressor anguli oris m.
Depressor labii inferioris m.

A

CN VII branches:
Posterior auricular n.
Temporal n.
Zygomatic n.
Buccal n.
Mandibular n.
Cervical n.

CN VII exiting the stylomastoid foramen
Nerve to posterior digastricus and stylohyoid mm.
Parotid gland

B

Figure 20-2: A. Muscles of facial expression. **B**. Branches of the facial nerve (CN VII) in the face.

PAROTID GLAND

BIG PICTURE

The parotid gland is situated in the lateral part of the face on the surface of the masseter muscle, anterior to the sternocleidomastoid muscle. A dense fascia covers the gland. The parotid gland produces and secretes saliva into the oral cavity and is innervated by visceral motor parasympathetic fibers from the glossopharyngeal nerve (CN IX).

PAROTID DUCT

The parotid duct crosses the masseter muscle, pierces the buccinator muscle, and opens into the oral cavity adjacent to the second upper molar (Figure 20-3A).

STRUCTURES ASSOCIATED WITH THE PAROTID GLAND

Motor branches of CN VII. CN VII exits the stylomastoid foramen, enters the posterior surface of the parotid gland, and courses through the gland en route to innervate the muscles of facial expression. CN VII also provides visceral motor innervation to all glands of the head, with the only exception being the parotid gland, the gland that you would think CN VII would innervate (Figure 20-3A and B).

Retromandibular vein. The superficial temporal vein and the maxillary vein join to form the retromandibular vein within the parotid gland and exit its inferior border, where the retromandibular vein bifurcates into anterior and posterior divisions. The **anterior division** joins with the facial vein to become the common facial vein, which enters the internal jugular vein. The **posterior division** joins with the posterior auricular vein, forming the external jugular vein.

External carotid artery. The common carotid artery bifurcates at the level of the thyroid cartilage into the external and internal carotid arteries. The external carotid artery enters the inferior border of the parotid gland and within the gland divides into the superficial temporal artery and the maxillary artery. Both arteries emerge from the anterosuperior border and provide vascular supply to the temporal fossa and scalp (superficial temporal artery) and the deep face (maxillary artery).

INNERVATION OF THE PAROTID GLAND

The parotid gland is innervated by visceral motor neurons from **CN IX** (Figure 20-3C).

Preganglionic parasympathetic neurons from CN IX originate in the **inferior salivatory nucleus** of the medulla and exit the **jugular foramen** along with the vagus nerve (CN X) and the spinal accessory nerve (CN XI).

A branch of CN IX, the tympanic nerve, re-enters the skull via the **tympanic canaliculus**, enters the middle ear, and forms the **tympanic plexus**.

CN IX gives rise to the **lesser petrosal nerve**, which exits the middle ear and the skull via the **foramen ovale** to synapse in the otic ganglion.

Postganglionic parasympathetic neurons exit the **otic ganglion** and then "hitch-hike" along the auriculotemporal nerve en route to the parotid gland.

▽ **Mumps** is a viral infection characterized by inflammation and swelling of the parotid gland, resulting in pain within the tight parotid fascia that covers the gland. Symptoms include discomfort in swallowing and chewing. The disease will usually run its course, with analgesics given to treat the pain and fever that is associated with the infection. ▼

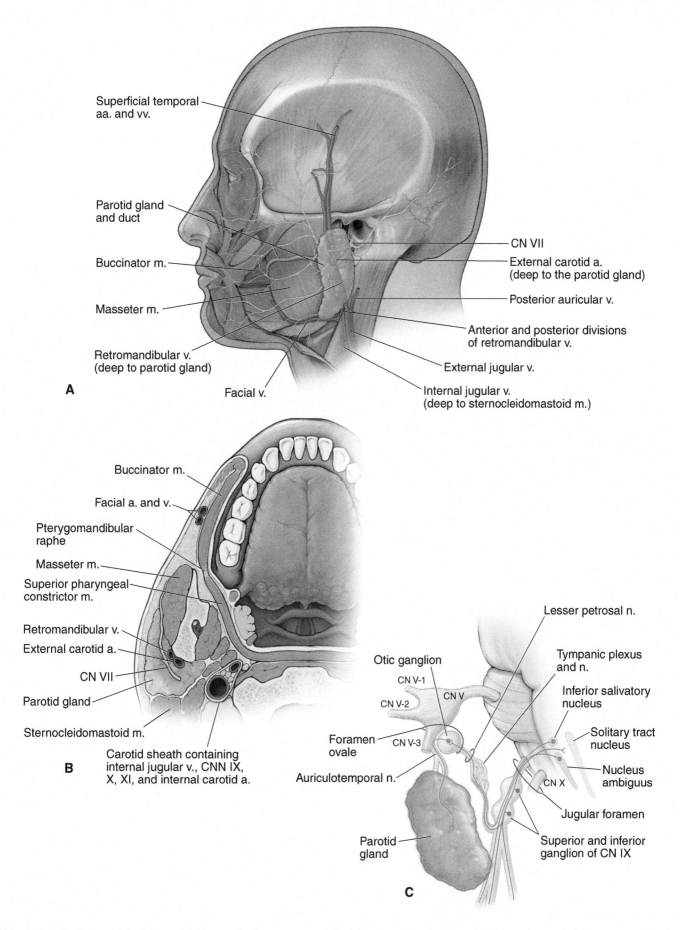

Figure 20-3: A. Topography of the parotid gland. **B**. Horizontal section through the parotid gland. **C**. Glossopharyngeal nerve (CN IX) and its innervation of the parotid gland.

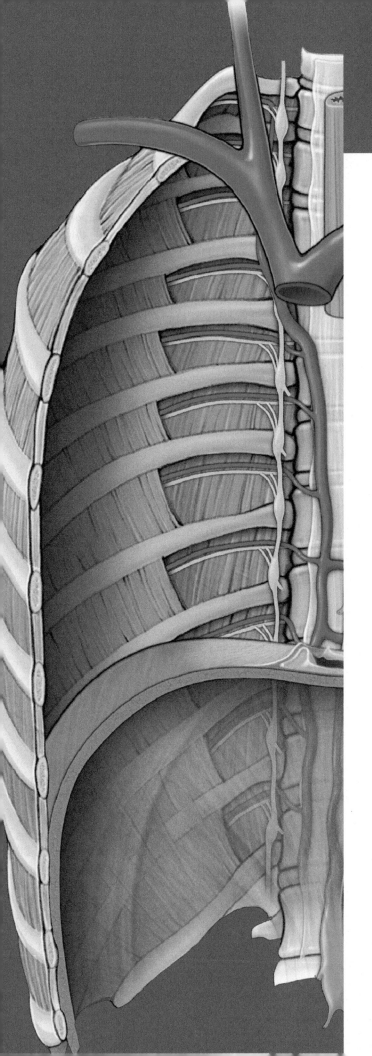

CHAPTER 21

INFRATEMPORAL FOSSA

OVERVIEW OF THE INFRATEMPORAL FOSSA

BIG PICTURE

The infratemporal fossa is the region deep to the ramus of the mandible. The infratemporal fossa accommodates the insertion of the temporalis muscle, medial and lateral pterygoid muscles, mandibular nerve [cranial nerve (CN) V-3], otic ganglion, chorda tympani nerve, maxillary artery, and the pterygoid plexus of the veins.

BOUNDARIES

The infratemporal fossa has the following boundaries (Figure 21-1A):

- Deep to the zygomatic arch.
- Anterior to the mastoid and styloid processes of the temporal bone.
- Posterior to the maxilla.
- Medial to the ramus of the mandible.
- Lateral to the pterygoid plate and the **pterygomaxillary fissure**, a communication between the infratemporal fossa and the pterygopalatine fossa.

TEMPOROMANDIBULAR JOINT

BIG PICTURE

The articulations between the **temporal bone (mandibular fossa)** and the **mandibular condyle** form a **synovial joint**, known as the temporomandibular joint (**TMJ**).

MOVEMENTS OF THE TMJ

The left and right TMJ's work together, enabling the **mandible to move** as follows:

- **Elevation (up).** Generated by the temporalis, masseter, and medial pterygoid muscles.
- **Depression (down).** Generated by the digastricus, geniohyoid, and mylohyoid muscles, and assisted by gravity.
- **Protraction.** Generated primarily through the lateral pterygoid muscle. Involves the anterior movement of the mandibular condyle and the articular disc.
- **Retraction.** Generated by the geniohyoid, digastricus, and temporalis muscles.
- **Side-to-side.** Generated by the pterygoid muscles.

STRUCTURE OF THE TMJ

A unique feature of the TMJ is the fibrocartilaginous **articular disc**, located within the joint capsule between the mandibular fossa and condyle. The disc divides the joint capsule into two distinct compartments.

- **Inferior compartment.** Enables the hinge-like rotation of the mandibular condyle, corresponding to the first 2 cm of opening the mouth (**depression**) (Figure 21-1B).
- **Superior compartment.** For the mouth to be opened more than 2 cm, the superior compartment within the joint capsule enables both the **mandibular condyle** and the **articular disc** to slide anteriorly (**protrusion**), incorporating a translational movement anteriorly when opening the mouth wider (Figure 21-1C).

▽ **TMJ disorder** is associated with painful and limited movement of the jaw. The disorder is poorly understood because of the complexity of the TMJ, which incorporates hinge-like movements along with movements that slide anteriorly and from side to side. Symptoms of TMJ disorder include pain and tenderness in and around the jaw, difficulty and painful chewing, headache, and clicking sounds when the jaw opens and closes. TMJ disorder can occur when the articular disc is damaged, eroded, or has slipped out of alignment. ▽

MUSCLE MOVEMENT OF THE TMJ

The muscles acting upon the TMJ are primarily the muscles that generate the various movements associated with chewing; hence, these muscles are often called the **muscles of mastication** (Table 21-1). The branchial motor division of **CN V-3** innervates each of the following muscles (Figure 21-1D and E):

- **Temporalis muscle.** Originates in the temporal fossa and courses deep to the zygomatic arch, inserting in the infratemporal fossa on the coronoid process of the mandible. The temporalis muscle elevates the mandible.
- **Masseter muscle.** Attaches between the zygomatic arch and the external surface of the ramus of the mandible. It elevates the mandible. The masseter muscle is not located in the infratemporal fossa; however, because it works functionally with the temporalis and pterygoid muscles to move the mandible at the TMJ, it is a muscle of mastication.
- **Lateral pterygoid muscle.** Attaches between the lateral pterygoid plate and the mandibular condyle. Contraction causes the mandibular condyle and the articular disc to move anteriorly, resulting in both protraction and depression of the mandible. The lateral pterygoid muscle works synergistically with the medial pterygoid muscle to move the mandible from side to side.
- **Medial pterygoid muscle.** Attaches between the medial and lateral pterygoid plates and the internal surface of the ramus of the mandible. The medial pterygoid muscle elevates the mandible and moves it from side to side.

▽ During a physical examination, the physician **palpates the muscles of mastication to test the function of CN V-3**. The physician will instruct the patient to clench his jaw as the temporalis and masseter muscles are palpated to determine if there is equal bilateral contraction in each muscle. The patient is then instructed to open and close his mouth against resistance. If there is muscle weakness of the medial or lateral pterygoid muscles, the jaw will deviate toward the weak side. ▽

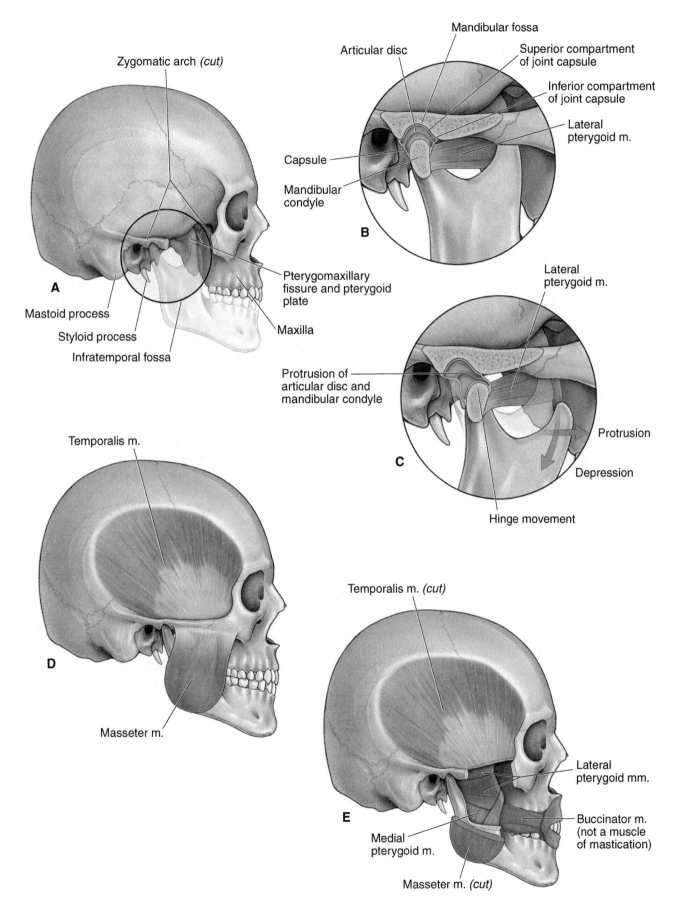

Figure 21-1: A. Boundaries of the infratemporal fossa. **B**. Compartments of the temporomandibular joint (TMJ). **C**. Opening of the TMJ. Superficial (**D**) and deep (**E**) views of muscles of mastication.

INNERVATION AND VASCULAR SUPPLY OF THE INFRATEMPORAL FOSSA

BIG PICTURE

The trigeminal nerve (CN V) provides most of the general sensory innervation to the head. Branches of the ophthalmic nerve (CN V-1) and the maxillary nerve (CN V-2) provide only general sensory innervation. In contrast, CN V-3 enters the infratemporal fossa and has both a general sensory and a branchial motor root. Branches of the maxillary artery provide the vascular supply to the infratemporal fossa.

INNERVATION OF THE INFRATEMPORAL FOSSA

The **sensory and motor roots** of **CN V-3** fuse together as a **trunk** and descend into the infratemporal fossa via the foramen ovale (Figure 21-2A). Once in the infratemporal fossa, CN V-3 bifurcates into **anterior** and **posterior divisions**.

- **Trunk.** Provides branchial motor innervation via branches to the medial pterygoid, tensor veli palatine, and tensor tympani muscles.
- **Anterior division**
  - **Motor branches.** Provides branchial motor innervation via branches to the temporalis, masseter, and lateral pterygoid muscles.
  - **Buccal nerve.** Provides general sensory innervation from the **skin of the cheek**, passing between the two heads of the lateral pterygoid muscle. The buccal nerve is also referred to as the **long buccal nerve** to distinguish it from the buccal branch of CN VII.
- **Posterior division**
  - **Auriculotemporal nerve.** Splits around the middle meningeal artery to provide general sensory innervation to the **temporal region of the face and scalp**.
  - **Otic ganglion.** Preganglionic parasympathetic neurons from the glossopharyngeal nerve (**CN IX**) exit the middle ear as the **lesser petrosal nerve** and enter the infratemporal fossa through the foramen ovale and synapse in the otic ganglion. Postganglionic parasympathetic neurons "hitchhike" along with the auriculotemporal nerve, providing visceral motor innervation to the **parotid gland**.

- **Inferior alveolar nerve.** Enters the mandibular foramen along with the inferior alveolar artery, providing general sensory innervation from the mandibular teeth the gingivae, lower lip, and chin.
- **Lingual nerve.** Provides general sensory innervation from the anterior two-thirds of the tongue (Figure 21-2A and B).
  - **Chorda tympani nerve (CN VII).** Enters the infratemporal fossa via the petrotympanic fissure, where it joins with the lingual nerve. The chorda tympani nerve provides **special sensory taste** from the **anterior two-thirds of the tongue**, as well as **visceral motor parasympathetic innervation** to the **submandibular** and **sublingual salivary glands**.
  - **Submandibular ganglion.** Suspended from the lingual nerve, where preganglionic and postganglionic parasympathetic neurons from the chorda tympani nerve (CN VII) synapse en route to providing visceral motor innervation to the submandibular and sublingual salivary glands.
- **Nerve to the mylohyoid.** Provides branchial motor innervation to the mylohyoid and anterior digastricus muscles.

VASCULAR SUPPLY OF THE INTRATEMPORAL FOSSA

The external carotid artery terminates as the maxillary artery and superficial temporal arteries, posterior to the neck of the mandible (Figure 21-2C). The **maxillary artery** courses medially and anteriorly to enter the infratemporal fossa. The maxillary artery disappears medially through the pterygomaxillary fissure into the pterygopalatine fossa as the sphenopalatine and infraorbital arteries in the nasal cavity and orbit, respectively. Although the maxillary artery gives off numerous branches, the two most important arteries are as follows:

- **Middle meningeal artery.** Ascends through the foramen spinosum, coursing between the roots of the auriculotemporal nerve. The middle meningeal artery provides the principal blood supply to the dura mater in the skull.
- **Inferior alveolar artery.** Joins the inferior alveolar nerve and enters the mandibular foramen.

The **pterygoid plexus of veins** is situated around the lateral pterygoid muscle. The pterygoid plexus communicates with veins in the orbit, cavernous sinus, and facial region.

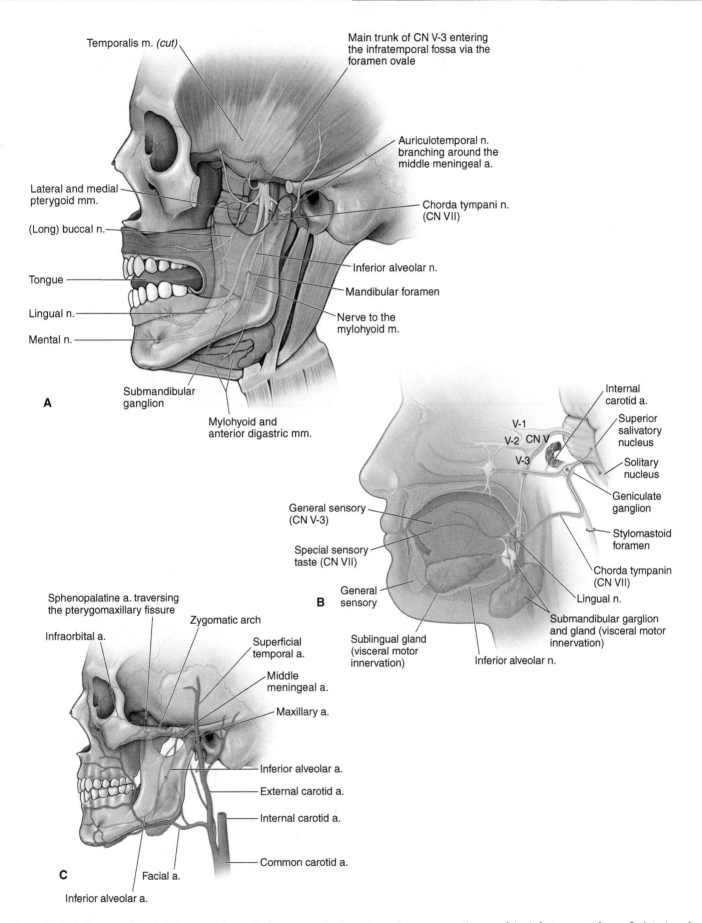

Figure 21-2: A. Nerves of the infratemporal fossa. **B**. Parasympathetic and special sensory pathways of the infratemporal fossa. **C**. Arteries of the infratemporal fossa.

TABLE 21-1. Muscles Derived from the First Branchial Arch and Innervated by CN V-3

Muscle	Origin	Insertion	Action	Innervation
Muscles of mastication				
• **Temporalis**	Temporal fossa, temporal and parietal bones	Coronoid process of mandible	Elevates and retracts mandible	
• **Masseter**	Zygomatic arch	Ramus of mandible (external surface)	Elevates mandible	
• **Lateral pterygoid**	Lateral pterygoid plate	Coronoid process of mandible	Protracts and laterally moves mandible	
• **Medial pterygoid**	Medial and lateral pterygoid plate	Ramus of mandible (internal surface)	Elevates and laterally moves mandible	CN V-3 (mandibular n.)
Anterior digastricus	Hyoid bone	Mandible	Depresses mandible	
Mylohyoid	Mandible	Hyoid bone	Depresses mandible	
Tensor veli palatini	Sphenoid bone	Soft palate	Tenses soft palate	
Tensor tympani	Auditory tube	Malleus	Dampens ossicles during chewing	

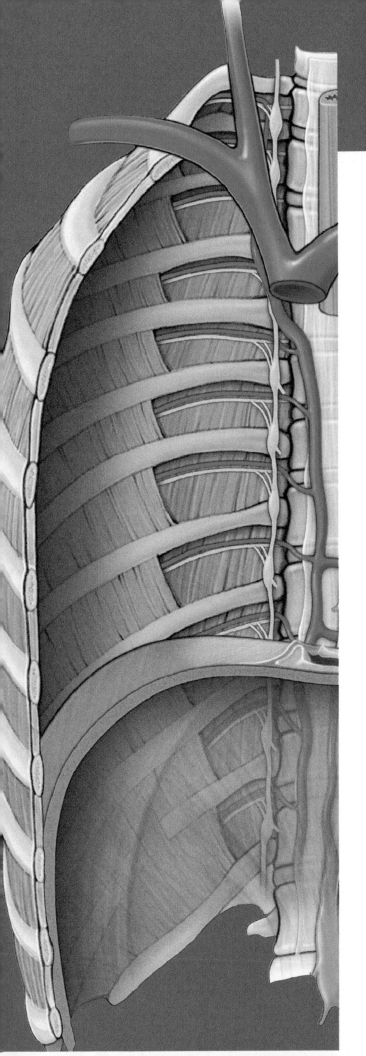

CHAPTER 22

PTERYGOPALATINE FOSSA

OVERVIEW OF THE PTERYGOPALATINE FOSSA

BIG PICTURE

The pterygopalatine fossa is the region between the pterygomaxillary fissure and the nasal cavity. The fossa accommodates branches of the maxillary nerve [cranial nerve (CN) V-2], the pterygopalatine ganglion, and the terminal branches of the maxillary artery.

BOUNDARIES OF THE PTERYGOPALATINE FOSSA

The pterygopalatine fossa is an irregular space where neurovascular structures course through to the nasal cavity, palate, pharynx, orbit and face (Figure 22-1A and B). The neurovascular structures enter and exit the fossa through the following boundaries:

Anterior boundary. Posterior surface of the maxilla.

Posterior boundary. Pterygoid processes and the greater wing of the sphenoid bone, with openings for the following structures:

- **Foramen rotundum** for CN V-2.
- **Pterygoid canal** for the nerve of the pterygoid canal (Vidian nerve).
- **Pharyngeal (palatovaginal) canal** for the pharyngeal branch of CN V-2.

Medial boundary. Perpendicular plate of the palatine bone containing the **sphenopalatine foramen**, which transmits the nasopalatine nerve (CN V-2 branch) and the sphenopalatine artery.

Lateral boundary. Pterygomaxillary fissure, which communicates with the infratemporal fossa.

Superior boundary. Greater wing and body of the sphenoid bone, **with the infraorbital fissure** transmitting the infraorbital nerve and the vessels in the orbit.

Inferior boundary. Palatine process of the maxilla and the pterygoid process of the sphenoid bone **with the greater and lesser palatine canals and foramina**, which transmit the greater and lesser palatine nerves and vessels.

INNERVATION OF THE PTERYGOPALATINE FOSSA

Branches of CN V-2 form most of the nerves that enter and exit the pterygopalatine fossa (Figure 22-1C and D). CN V-2 enters the fossa via the foramen rotundum and branches as follows:

Posterior superior alveolar nerves. Enters the posterior superior alveolar canals, providing general sensation to the maxillary molar teeth and gingivae.

Infraorbital nerve. Courses through the infraorbital fissure, groove, canal, and ultimately the foramen providing general sensation to the inferior eyelid, the lateral nose, and the superior lip. The infraorbital nerve also gives rise to the **middle** and **anterior superior alveolar nerves**, which supply the maxillary premolars, canines, and incisors, and the gingivae and mucosal lining of the maxillary sinus.

Zygomatic nerve. Enters the orbit via the infraorbital fissure, dividing into the **zygomaticotemporal** and **zygomaticofacial nerves**, which supply the skin over the zygomatic arch and the temporal region. In addition, the zygomatic nerve communicates with the lacrimal nerve in the orbit and carries parasympathetic neurons from the pterygopalatine ganglion to the lacrimal gland.

Pharyngeal nerve. Courses through the pharyngeal canal, supplying part of the nasopharynx.

Greater and lesser palatine nerves. Descend through the palatal canals, emerging through the greater and lesser palatine foramina to innervate the hard and soft palates.

Nasopalatine nerve. Traverses the sphenopalatine foramen, supplying the nasal septum before coursing through the incisive canal to supply part of the hard palate.

Pterygopalatine ganglion. Lies inferior to CN V-2 and receives preganglionic parasympathetic neurons from the facial nerve (CN VII) via the greater petrosal nerve, traversing the pterygoid canal. The pterygopalatine ganglion sends postganglionic parasympathetic neurons to the **lacrimal gland**, via communicating branches between the zygomatic nerve and the lacrimal nerve (CN V-1), and the **nasal and palatal glands**, via the nasopalatine, greater palatine, and lesser palatine nerves.

- **Sympathetics. Postganglionic sympathetic neurons** from the **superior cervical ganglion** course along the internal carotid artery and give rise to the **deep petrosal nerve**. The deep petrosal nerve joins with the **greater petrosal nerve** (parasympathetics from CN VII) at the foramen lacerum to become the **nerve of the pterygoid canal** (Vidian nerve). The postganglionic sympathetic neurons course through but do not synapse in the pterygopalatine ganglion and inhibit lacrimal and nasal gland secretion.

ARTERIES OF THE PTERYGOPALATINE FOSSA

The **maxillary artery** is a terminal branch of the external carotid artery, courses anteriorly through the infratemporal fossa, traverses the pterygomaxillary fissure, and enters the pterygopalatine fossa (Figure 22-1E and F). The maxillary artery supplies the maxilla, maxillary teeth, and palate before traversing the sphenopalatine foramen to terminate in the nasal cavity. The major **branches of the maxillary artery** in the pterygopalatine fossa are as follows:

Posterior superior alveolar artery. Supplies the maxillary molar teeth.

Descending palatine artery. Gives rise to the greater and lesser palatine arteries, which supply the soft and hard palates.

Infraorbital artery. Supplies the maxillary tooth and skin of the face.

Sphenopalatine artery. Traverses the sphenopalatine foramen to supply the nasal cavity.

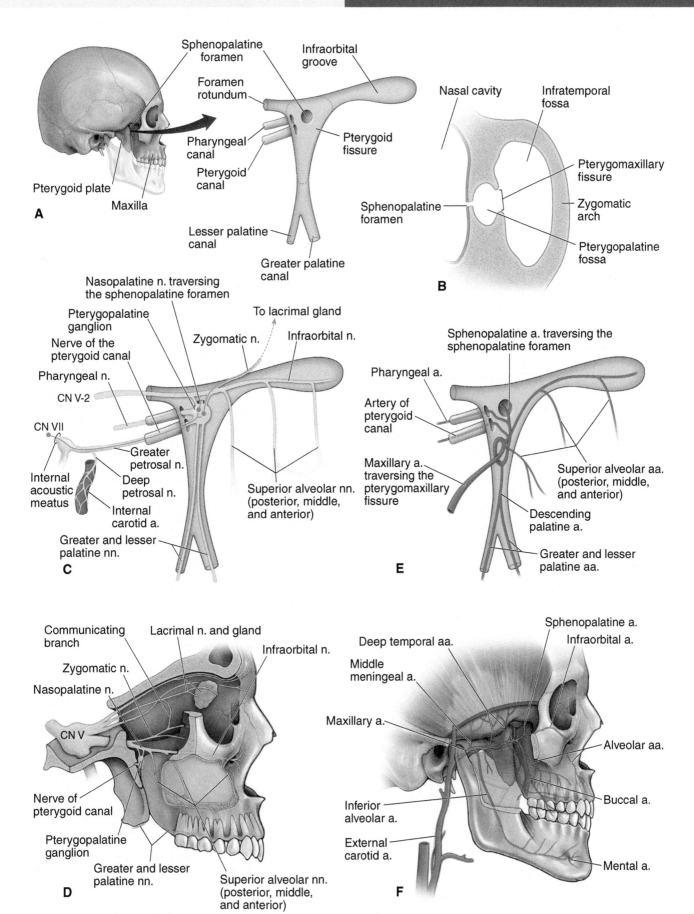

Figure 22-1: A. Outline of the pterygopalatine fossa. **B**. Axial section of the pterygopalatine fossa. **C. and D**. Nerves of the pterygopalatine fossa. **E. and F**. Arteries of the pterygopalatine fossa.

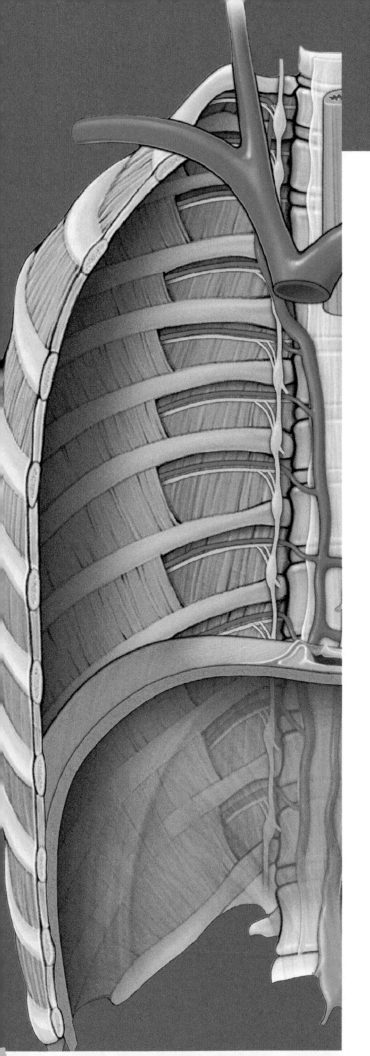

NASAL CAVITY

OVERVIEW OF THE NASAL CAVITY

BIG PICTURE

The **nasal cavity** is divided into two lateral compartments separated down the middle by the **nasal septum**. The nasal cavity communicates anteriorly through the **nostrils** and posteriorly with the nasopharynx through openings called **choanae**. The nasal cavities and septum are lined with a mucous membrane and are richly vascularized by branches of the maxillary, facial, and ophthalmic arteries. The nasal cavity receives innervation via branches of the olfactory [cranial nerve (CN) I], ophthalmic (CN V-1), and maxillary nerves (CN V-2).

BOUNDARIES OF THE NASAL CAVITY

The nasal cavity is bordered by the following structures (Figure 23-1A–C):

Roof. Formed by the nasal, frontal, sphenoid, and ethmoid bones (**cribriform foramina**, which transmits CN I for smell).

Floor. Formed by the maxilla and the palatine bones. The **incisive foramen** transmits branches of the sphenopalatine artery and the nasopalatine nerve for general sensation from the nasal cavity and palate.

Medial wall (nasal septum). Formed by the perpendicular plate of the ethmoid bone, the vomer bone, and the septal cartilage.

Lateral wall. Formed by the superior, middle and inferior nasal conchae. In addition, the maxillary, sphenoid, and palatine bones contribute to the lateral wall. The lateral wall contains the following **openings**:

- **Sphenoethmoidal recess.** The space between the superior nasal concha and the sphenoid bone, with openings from the sphenoid sinus.
- **Superior meatus.** The space inferior to the superior nasal concha, with openings from the posterior ethmoidal air cells.
- **Middle meatus.** The space inferior to the middle nasal concha, with openings for the frontal sinus via the **nasofrontal duct**, the middle ethmoidal air cells on the **ethmoidal bulla**, and the anterior ethmoidal air cells and maxillary sinus in the **hiatus semilunaris**.
- **Inferior meatus.** The space inferior to the inferior nasal concha, with an opening for the **nasolacrimal duct**, which drains tears from the eye into the nasal cavity.
- **Sphenopalatine foramen.** An opening posterior to the middle nasal concha receives the nasopalatine nerve and the sphenopalatine artery from the **pterygopalatine fossa** into the nasal cavity.

▽ **Rhinorrhea**, or "runny nose," is evident by the clear fluid that leaks out of the nostrils. A runny nose usually accompanies the **common cold**. Rhinorrhea usually results from overproduction of mucus resulting from conditions such as **sinusitis**, **hay fever**, and **allergic reactions**. However, rhinorrhea that occurs after an accident involving head trauma may indicate a basilar skull fracture, resulting in leakage of **cerebrospinal fluid** from the subarachnoid space through the fracture (often the ethmoid bone) into the nasal cavity and out of the nostrils. ▼

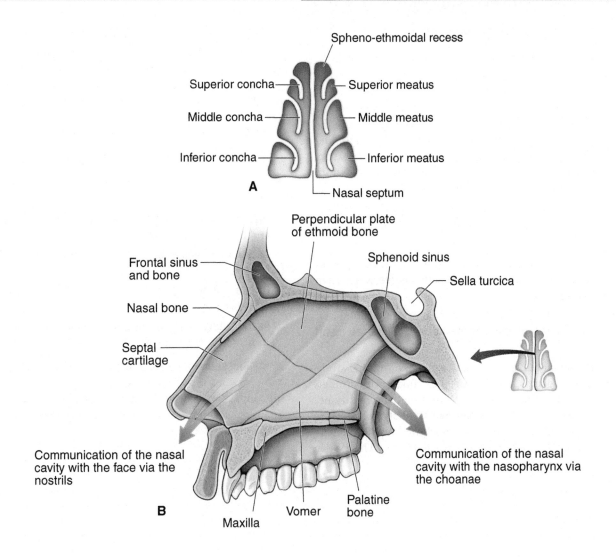

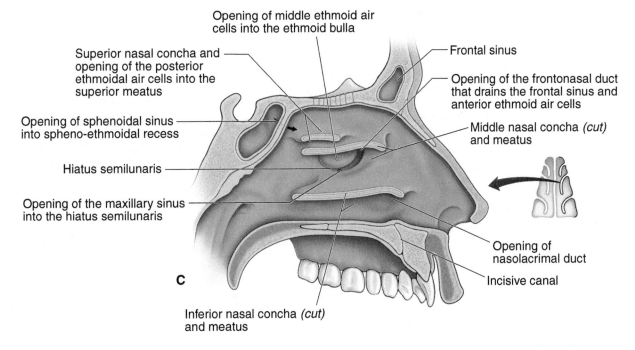

Figure 23-1: A. Coronal section through the nasal cavity. **B**. Nasal septum from the left side. **C**. Lateral nasal wall of the left nasal cavity.

NERVE SUPPLY OF THE NASAL CAVITY

The nasal cavity contains the following nerves (Figure 23-1D):

CN I. Originates in the mucosa lining the superior nasal concha and the superior septum, where the nerve provides special sensation for smell. Neurons from CN I course from the nasal cavity into the anterior cranial fossa and through the numerous foramina of the cribriform plate of the ethmoid bone. The neurons enter the olfactory bulb, where they synapse with interneurons that course along the olfactory tract, transporting information to the brain.

▽ **Anosmia** (lack of smell) can result from trauma to the ethmoid bone and the nasal region, where the optic nerve (CN II) endings become damaged. ▽

CN V-1. Provides general sensation to the superior aspect of the nasal cavity via the **anterior ethmoidal nerve**, a branch of the **nasociliary nerve**.

CN V-2. Provides general sensation to most of the nasal cavity via **branches of the nasopalatine** and **lateral nasal nerves**.

• **CN VII (facial nerve).** Provides visceral motor innervation to the nasal glands. CN VII exits the cranial cavity through the internal acoustic meatus. Within the temporal bone, CN VII gives rise to the **greater petrosal nerve**, which carries the visceral motor preganglionic parasympathetic neurons from CN VII en route to the nasal cavity. The greater petrosal nerve joins up with the deep petrosal nerve to form the **nerve of the pterygoid canal (Vidian nerve)**. The nerve of the pterygoid canal enters the pterygopalatine fossa, where parasympathetics from CN VII synapse. Postganglionic parasympathetic neurons exit the ganglion and "hitch-hike" along CN V-2 branches to the nasal mucosa, where the mucosal glands are innervated.

VASCULAR SUPPLY OF THE NASAL CAVITY

The nasal cavity receives its **vascular supply via the following arteries** (the nasal veins parallel the arteries) (Figure 23-1E):

Sphenopalatine artery. Principal blood supply to the septum and the lateral nasal wall.

Anterior and **posterior ethmoidal arteries**. Supply the superior portion of the nasal cavity.

Greater palatine artery. Supplies the inferior nasal septum via the incisive canal.

Facial artery. Supplies the anterior portion of the nasal septum and the lateral nasal wall.

▽ **Kiesselbach's area (plexus)** is a region in the anteroinferior region of the nasal septum where branches of the sphenopalatine, anterior ethmoidal, greater palatine, and facial arteries anastamose. Most nosebleeds (**epistaxis**) usually occur in this area. ▽

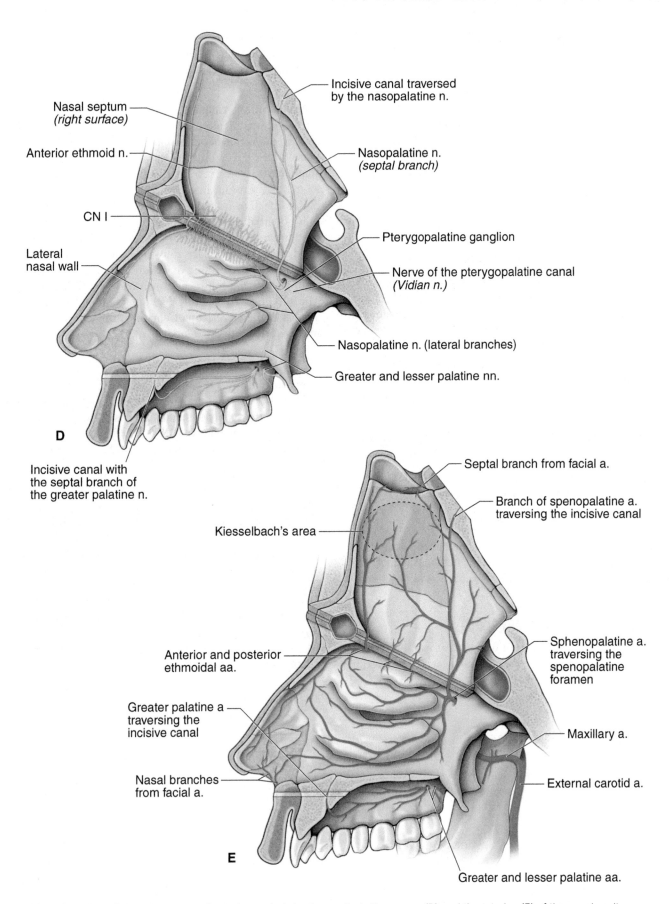

Incisive canal traversed
by the nasopalatine n.

Nasal septum
(right surface)

Anterior ethmoid n.

Nasopalatine n.
(septal branch)

CN I

Pterygopalatine ganglion

Nerve of the pterygopalatine canal
(Vidian n.)

Lateral
nasal wall

Nasopalatine n. (lateral branches)

Greater and lesser palatine nn.

D

Incisive canal with
the septal branch of
the greater palatine n.

Septal branch from facial a.

Branch of spenopalatine a.
traversing the incisive canal

Kiesselbach's area

Anterior and posterior
ethmoidal aa.

Sphenopalatine a.
traversing the
spenopalatine
foramen

Greater palatine a
traversing the
incisive canal

Maxillary a.

Nasal branches
from facial a.

External carotid a.

E

Greater and lesser palatine aa.

Figure 23-1: (continued) Nasal septum reflected superiorly to demonstrate the nerves (**D**) and the arteries (**E**) of the nasal cavity.

PARANASAL SINUSES

BIG PICTURE

The paranasal sinuses are hollow cavities within the ethmoid, frontal, maxillary, and sphenoid bones (Figure 23-2A and B). They help to decrease the weight of the skull, resonate sound produced through speech, and produce mucus. The paranasal cavities communicate with the nasal cavity, where mucus is drained. Branches of CN V provide general sensory innervation.

▼ The **paranasal sinuses** are easily recognizable on an x-ray because the sinuses are filled with air and thus appear as darker shadows on the radiograph. ▼

ETHMOIDAL SINUS

Unlike the frontal, maxillary, and sphenoid paranasal sinuses, the ethmoidal sinus consists of numerous small cavities (air cells) within the bone, as opposed to one or two large sinuses. The **subdivisions of the ethmoidal air cells** (anterior, middle, and posterior) **communicate with the nasal cavity**.

- **Anterior ethmoidal air cells.** Drain through tiny openings in the hiatus semilunaris of the middle meatus.
- **Middle ethmoidal air cells.** Drain through the ethmoidal bulla of the middle meatus.
- **Posterior ethmoidal air cells.** Drain through openings in the superior meatus.

The **posterior ethmoidal nerve (CN V-1)** provides general sensory innervation for the ethmoidal air cells.

FRONTAL SINUS

The frontal sinus is located in the frontal bone and opens into the anterior part of the middle meatus via the **frontonasal duct**. The **supraorbital nerve (CN V-1)** provides general sensory innervation for the frontal sinus.

MAXILLARY SINUS

The maxillary sinus is the largest of the paranasal sinuses and is located in the maxilla, lateral to the nasal cavity and inferior to the orbit. The maxillary sinus opens into the posterior aspect of the hiatus semilunaris in the middle meatus. The **infraorbital nerve (CN V-2)** primarily innervates the maxillary sinus.

▼ **Maxillary sinusitis** results from inflammation of the mucous membrane lining the maxillary sinus and is a common infection because of its pattern of drainage. The maxillary sinus drains into the nasal cavity through the **hiatus semilunaris**, which is located superiorly in the sinus (Figure 23-2C). As a result, infection has to move against gravity to drain. Infection from the frontal sinus and the ethmoidal air cells potentially can pass into the maxillary sinus, compounding the problem. In addition, the maxillary molars are separated from the maxillary sinus only by a thin layer of bone. Therefore, if an infecting organism erodes the bone, infection from an infected tooth can potentially spread into the sinus. The **infraorbital nerve (CN V-2)** innervates the maxillary teeth and sinus; therefore, pain originating from a tooth or the sinus may be difficult to differentiate. ▼

SPHENOID SINUS

The sphenoid sinus is contained within the body of the sphenoid bone and is inferior to the sella turcica. The sphenoid sinus opens into the **sphenoethmoidal recess** of the nasal cavity. The **posterior ethmoidal nerve (CN V-1)** and branches from CN V-2 provide general sensory innervation of the sphenoid sinus.

▼ The **pituitary gland** is located in the roof of the sphenoid bone. The gland is important for the production and release of hormones targeting the gonads, adrenals, thyroid, kidney, uterus, and the mammary glands. **Tumors of the pituitary gland** can cause an overproduction of these hormones or may affect vision by compressing CN II. Surgery may be necessary to remove the tumor. The sphenoid sinus is separated from the nasal cavity by a thin layer of bone. Therefore, the pituitary gland can be approached surgically by going through the nasal cavity into the sphenoid sinus and finally through the superior aspect of the sphenoid sinus into the sella turcica, where the pituitary gland is located (**transsphenoidal hypophysectomy**). ▼

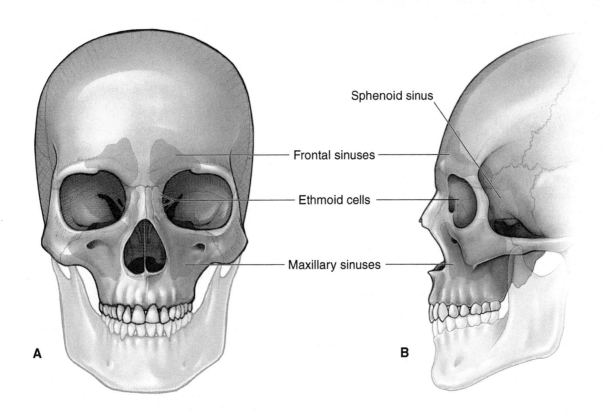

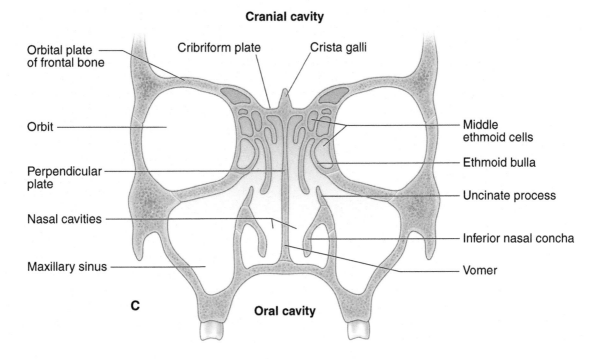

Figure 23-2: Anterior (**A**) and lateral (**B**) views of the paranasal sinuses. **C**. Coronal section of the skull revealing the cranial, orbital, and nasal cavities and their relationships to the paranasal sinuses.

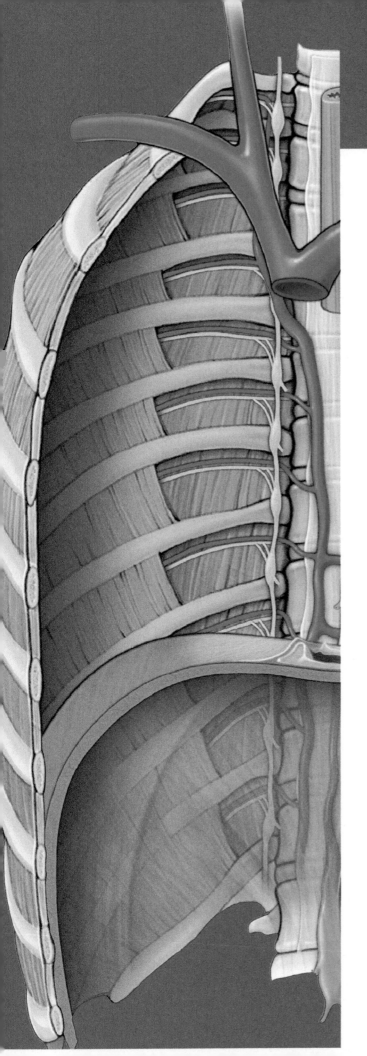

ORAL CAVITY

OVERVIEW OF THE ORAL CAVITY

BIG PICTURE

The cheeks and lips border the oral cavity, with a space (the vestibule) between the cheeks, lips, and teeth. The cheeks contain the buccinator muscle, which, along with the tongue, holds food between the teeth during mastication (chewing).

PALATE

BIG PICTURE

The palate forms both the roof of the oral cavity and the floor of the nasal cavity and consists of a hard and a soft palate. All the muscles that act upon the soft palate are innervated by the vagus nerve [cranial nerve (CN) X], with the exception of the tensor veli palatini, which is innervated by a small motor branch from CN V-3. The difference in innervation reflects the embryologic origins of the branchial arches.

HARD PALATE

The hard palate consists of the **palatine process** of the **maxillary bone** and the **horizontal plate** of the **palatine bone** (Figure 24-1A and B). The **incisive canal** is in the anterior midline and transmits the following branches (Figure 24-1B):

- **Nasopalatine and greater palatine nerves.** Branches of the maxillary nerve (CN V-2); provides general sensation to the palate.
- **Sphenopalatine and greater palatine arteries.** Branches of the maxillary artery originating from the infratemporal fossa.

SOFT PALATE

The **soft palate** forms the soft, posterior segment of the palate. The soft palate has a structure called the **uvula**, which is suspended from the midline (Figure 24-1A–C). The soft palate is continuous with the **palatoglossal** and **palatopharyngeal folds**. Functionally, the soft palate ensures that food moves inferiorly down into the esophagus when swallowing, rather than up into the nose. By moving posteriorly against the pharynx, which separates the oropharynx from the nasopharynx, the soft palate acts like a flap valve. The vascular supply is bilaterally derived from the **lesser palatine artery** (maxillary artery) and from smaller arteries, including the **ascending palatine artery** of the facial artery and the **palatine branch** of the ascending pharyngeal artery. The soft palate receives general sensory innervation via the **lesser palatine nerves** (CN V-2) (Figure 24-1B).

MUSCLES OF THE SOFT PALATE

The muscles of the soft palate are as follows (Figure 24-1B and C; Table 24-1):

- **Tensor veli palatini muscle.** Attaches laterally to the pterygoid plate of the sphenoid bone, hooks around the **hamulus**, and inserts in the soft palate. This muscle is innervated by **CN V-3**. As its name implies, contraction results in tensing the soft palate.
- **Levator veli palatini muscle.** Originates along the cartilaginous portion of the auditory tube and inserts into the superior aspect of the soft palate. Contraction of this muscle elevates the soft palate and is innervated by **CN X**.

▽ To test the function of CN X, the physician will ask the patient to open his mouth wide to determine if the palate deviates to one side or the other during a yawning motion. A **lesion of CN X** causes **paralysis of the ipsilateral levator veli palatini muscle**, resulting in the uvula being pulled superiorly to the opposite side of the lesion. ▼

- **Palatoglossus muscle.** Attaches between the soft palate and the tongue and is innervated by **CN X**. The palatoglossus muscle and the **palatopharyngeus muscle** surround the **palatine tonsil**, which aids the immune system in combating pathogens entering the oral cavity.

▽ Inflammation of the palatine tonsils (**tonsillitis**) is associated with difficulty swallowing and sore throat. Because the palatine tonsils are visible when inspecting the oral cavity, the tonsils of a patient who has tonsillitis will appear enlarged and red. In cases of chronic tonsillitis, the tonsils may be surgically removed (**tonsillectomy**) to ensure that the patient can swallow and breathe properly. ▼

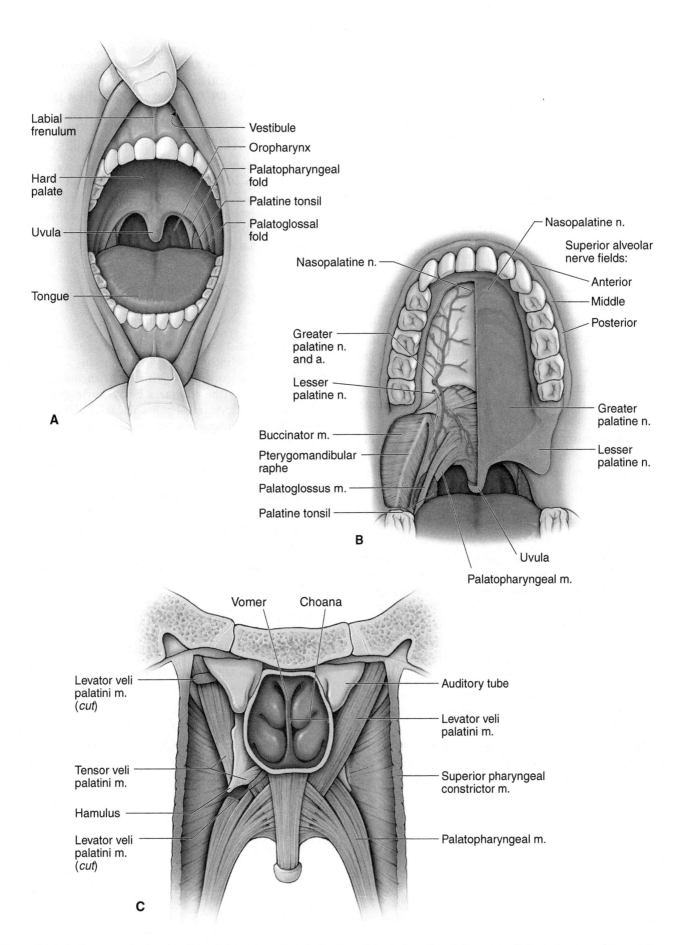

Figure 24-1: A. Open mouth showing the palatal arches. **B**. Anterior view of the innervation of the palate. **C**. Posterior view of the palate.

TONGUE

BIG PICTURE

The tongue consists of skeletal muscle, which has a surface covered with taste buds (special sensory) and general sensory nerve endings. The tongue is supported in the oral cavity by muscular connections to the hyoid bone, mandible, styloid process, palate, and pharynx. The V-shaped **sulcus terminalis** divides the tongue into anterior and posterior divisions, which differ developmentally, structurally, and by innervation. The **foramen cecum** is located at the apex of the "V" and indicates the site of origin of the embryonic **thyroglossal duct**.

MUSCLES OF THE TONGUE

Tongue muscles are bilaterally paired and **innervated by the hypoglossal nerve** (CN XII) (Figure 24-2A and C; Table 24-1):

- **Genioglossus muscle.** Attaches between the internal surface of the mandible and the tongue. Contraction causes the tongue to protrude out of the oral cavity ("sticking out your tongue").

- **Hyoglossus muscle.** Courses lateral to the genioglossus muscle, with attachments from the hyoid bone to the tongue. The **lingual artery**, en route to the tongue, courses between the hyoglossus and genioglossus muscles. In contrast, **CN XII** and **CN V-3** course from the infratemporal fossa to the tongue, along the external surface of the hyoglossus muscle.

- **Styloglossus muscle.** Originates on the styloid process of the temporal bone and courses between the superior and middle pharyngeal constrictors to insert on the lateral surface of the tongue.

INNERVATION OF THE TONGUE

The following nerves innervate the tongue (Figure 24-2B and C):

- **CN V-3** (lingual nerve). The anterior two-thirds of the tongue receives its **general sensory** innervation from the **lingual branch of CN V-3**. The **submandibular duct** ascends from the submandibular gland to open adjacent to the **lingual frenulum**. The lingual nerve courses inferior to the submandibular duct.

- **CN VII** (facial nerve). The anterior two-thirds of the tongue receives its **special sensory (taste)** innervation from the **chorda tympani nerve (CN VII)**.

- **CN IX** (glossopharyngeal nerve). The posterior third of the tongue receives both its **general sensory** and **special sensory (taste)** innervation from **CN IX**.

- **CN XII** (hypoglossal nerve). **All of the tongue muscles** receive their **somatic motor innervation** via **CN XII** (except the palatoglossus, which is innervated by CN X).

▽ To clinically test CN XII, the physician will ask the patient to stick out her tongue. If CN XII is functioning normally, the tongue should extend out evenly in the midline. If there is a lesion of **CN XII**, the tongue will deviate toward the same side of the face as the lesion. An easy way to remember this is that a lesion of **CN XII** will **cause the patient to "lick her wounds."** ▽

VASCULAR SUPPLY OF THE TONGUE

The **lingual artery** arises from the external carotid artery at the level of the tip of the greater horn of the hyoid bone in the carotid triangle. The **lingual artery** courses anteriorly between the **hyoglossus** and the **genioglossus muscles**, giving rise to the dorsal lingual and sublingual branches and terminating as the **deep lingual artery**.

SALIVARY GLANDS

BIG PICTURE

Saliva is essential for maintaining healthy oral tissue. It is composed primarily of water as well as mucins, enzymes, and immune components. Saliva initiates the chemical digestion of carbohydrates (the enzyme salivary amylase) and lubricates masticated food into a bolus. The three glands that produce saliva are the parotid, submandibular, and sublingual glands.

DUCTAL OPENINGS INTO THE ORAL CAVITY

The bilateral openings of ducts of the salivary glands into the oral cavity are as follows:

- **Parotid gland.** The duct opens opposite the second maxillary molar.

- **Submandibular gland.** The duct courses medial to the lingual nerve and opens into the area adjacent to the lingual frenulum (Figure 24-2B).

- **Sublingual gland.** The duct opens at the base of the tongue.

INNERVATION OF THE SALIVARY GLANDS

The **parasympathetic nervous system** is responsible for providing visceral motor innervation to the **salivary glands** via the following cranial nerves (Figure 24-2C):

- **Parotid gland.** CN IX, via the otic ganglion.

- **Submandibular gland.** CN VII, via the submandibular ganglion.

- **Sublingual gland.** CN VII, via the submandibular ganglion.

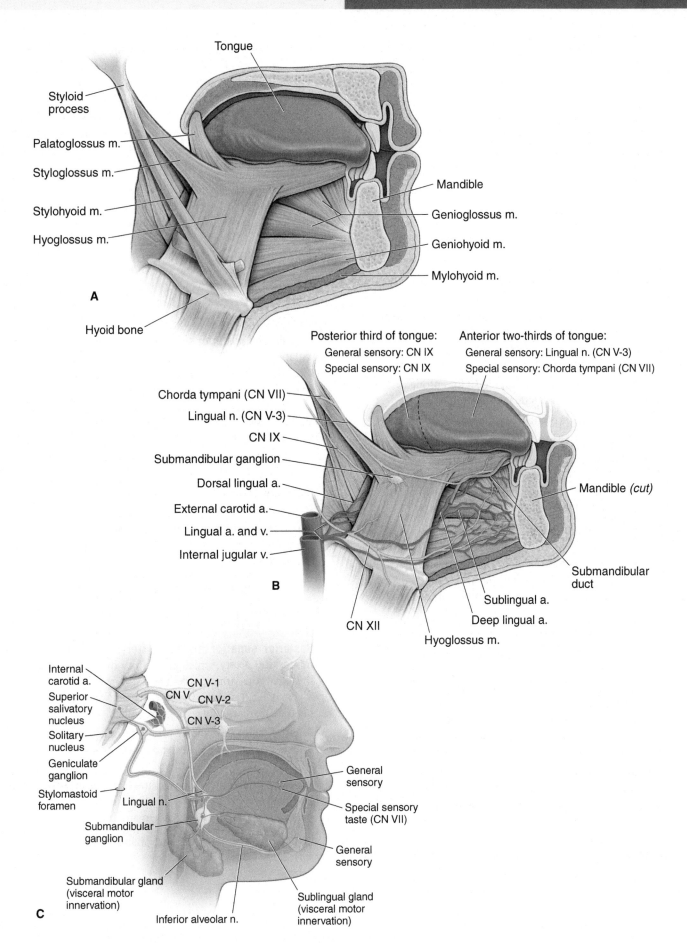

Figure 24-2: A. Tongue muscles. **B**. Neurovascular supply of the tongue. **C**. Innervation of the tongue.

TEETH AND GINGIVAE

BIG PICTURE

The teeth cut, grind, and mix food during mastication. In adults, there are 16 teeth in the maxilla and 16 in the mandible. Branches of CN V-2 and the maxillary artery and veins supply the maxillary teeth and gingivae. Branches of CN V-3, and the inferior alveolar artery and veins supply the mandibular teeth and gingivae.

TYPES OF TEETH

There are **20 teeth in a child** and **32 in an adult**. Adult teeth are typically numbered in a progressing clockwise fashion, with tooth number 1 (upper right maxillary molar) across to tooth number 16 (upper left maxillary molar). Tooth number 17 is the left third mandibular molar and continues to tooth number 32, the right third mandibular molar (Figure 24-3A). The teeth are divided into four quadrants with eight teeth located in the upper left, upper right, lower left, and lower right halves of the maxilla and mandible. **Each quadrant** consists of the following teeth:

- **Two Incisors.** Chisel-shaped teeth ideal for cutting or biting. There is a central and a lateral incisor.
- **One Canine.** A single pointed tooth used for tearing food.
- **Two Premolars.** Have two cusps, which are used for grinding.
- **Three Molars.** Have three cusps, which also are used for grinding. The third and most posterior molar tooth emerges last, usually in the late adolescent years, and is often referred to as the **wisdom tooth**.

▽ **Cavities (dental carries)** are holes in the teeth. Cavities form through the deposit of food products on teeth, known as **plaque**. **Bacteria** inhabit the plaque and metabolize carbohydrates into acids. Over time, the acids dissolve the outer protection of the tooth, the **enamel**, resulting in cavities. ▽

INNERVATION OF THE TEETH

The innervation of the teeth and gingivae are as follows (Figure 24-3B and C):

- **Maxillary teeth. CN V-2** enters the pterygopalatine fossa via the **foramen rotundum**. CN V-2 branches provide general sensory innervation to the maxillary teeth in a plexus of nerves formed by the **anterior, middle**, and **posterior superior alveolar nerves**.
  - **Maxillary gingivae.** The posterior, middle, and anterior superior alveolar nerves innervate the buccal surface of the gingivae, whereas the **greater palatine** and **nasopalatine nerves** innervate the lingual surface.
- **Mandibular teeth. CN V-3** enters the infratemporal fossa via the **foramen ovale**. A branch of CN V-3, the **inferior alveolar**

nerve, enters the **mandibular foramen** and provides general sensory innervation to the teeth on that side of the mandible. The inferior alveolar nerve exits the mandibular canal as the **mental nerve** by traversing the **mental foramen** and providing general sensory innervation to the bottom lip.
  - **Mandibular gingivae.** The buccal and mental nerves innervate the buccal surface of the gingivae, whereas the lingual nerve innervates the lingual surface.

▽ An **inferior alveolar nerve block** is administered to patients who require dental work in the mandible. Injection of local anesthetics (e.g., **lidocaine**) into the oral mucosa at the **lingula** of the mandible will anesthetize the inferior alveolar nerve and potentially the proximal lingual nerve, resulting in numbness of the tongue and oral mucosa. Care must be given when extracting the **third molar tooth** because the **lingual nerve** is closely associated with this nerve. ▽

▽ A plexus formed by the **anterior superior, middle superior, and posterior superior alveolar nerves** supplies the **maxillary teeth**. Unlike the mandibular teeth, where anesthesia can be administered in one location to effectively block all the teeth on that side, the maxillary teeth must be considered separately. The palate must also be considered; therefore, a **greater palatine** as well as a **nasopalatine nerve block** is often administered during dental procedures.

- **Maxillary molars.** Injection is given at the region of the second molar.
- **Maxillary premolars.** Injection is given at the region of the second premolar.
- **Maxillary canine and incisors.** Injection can be given above the roots of the anterior teeth or via an infraorbital nerve block. ▽

VASCULAR SUPPLY TO THE TEETH

The arteries supplying the teeth mirror their accompanying nerves.

- **Posterior superior alveolar artery.** Originates from the **infraorbital artery**, which arises from the maxillary artery in the pterygopalatine fossa. The infraorbital artery exits the fossa and enters the infraorbital canal, where branches supply the maxillary molars and premolars.
- **Anterior superior alveolar artery.** Originates from the **infraorbital artery**, where branches course through the maxilla to supply the incisor and canines.
- **Inferior alveolar artery.** Originates from the **maxillary artery** and courses with the inferior alveolar nerve through the mandibular canal to supply the mandibular teeth.

Veins from the maxillary and mandibular teeth generally mirror the arteries.

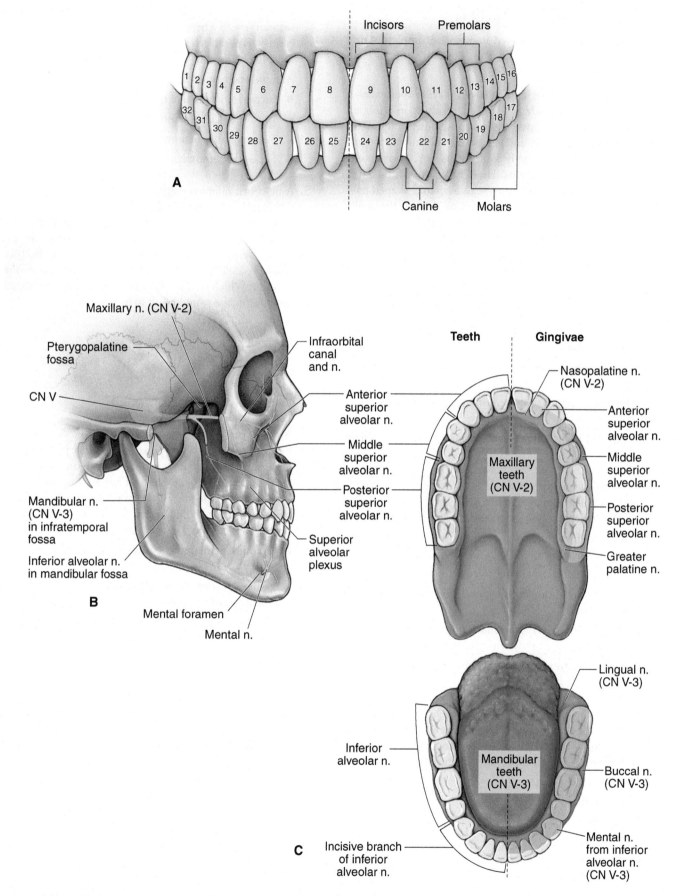

Figure 24-3: A. Teeth. Lateral (**B**) and anterior (**C**) views of the innervation of the teeth and gingiva.

STUDY QUESTIONS

Directions: Each of the numbered items or incomplete statements is followed by lettered options. Select the **one** lettered option that is **best** in each case.

1. During a rugby match, a 22-year-old player experiences a violent blow to the side of his head, resulting in a loss of consciousness. The player regains consciousness approximately 30 seconds later and sits on the sideline recuperating. A half-hour later, he becomes nauseous, disoriented, and falls over unconscious. He is taken to the hospital where a CT scan of the head reveals a discoid-shaped collection of blood confined by the suture lines of the skull, consistent with an epidural hematoma. Which vascular structure is most likely associated with this injury?

   A. Bridging cerebral vein hemorrhaging into the subarachnoid space

   B. Bridging cerebral vein hemorrhaging into the subdural space

   C. Middle meningeal artery hemorrhaging into the epidural space

   D. Middle meningeal artery hemorrhaging into the subdural space

   E. Superior sagittal sinus hemorrhaging into the epidural space

   F. Superior sagittal sinus hemorrhaging into the subdural space

2. A 19-year-old woman is taken to the emergency department after falling and lacerating her scalp. The scalp bleeds profusely when cut because the arteries most likely

   A. are held open due to the course of the arteries through the spongy bone of the skull

   B. the diameter of a pencil eraser

   C. bleed from both cut ends due to rich anastomoses of scalp vessels

   D. collapse and contract within the surrounding connective tissue

   E. course side to side over the vertex of the skull

3. A 31-year-old man is brought to the emergency department after being involved in a motor vehicle collision that resulted in head trauma. Radiographic imaging reveals a fracture to the bone deep to the man's moustache. Which of the following bones is most likely fractured in this patient?

   A. Frontal

   B. Mandible

   C. Maxilla

   D. Temporal

   E. Zygomatic

4. Radiographic imaging of the brain of an 84-year-old woman reveals a berry aneurysm in the anterior communicating cerebral artery. Rupture of this aneurysm would most likely result in what type of hemorrhage?

   A. Epidural hemorrhage

   B. Intraparenchymal hemorrhage

   C. Intraventricular hemorrhage

   D. Subarachnoid hemorrhage

   E. Subdural hemorrhage

5. A newborn infant is diagnosed with hydrocephalus, a condition where cerebrospinal fluid (CSF) builds up within the ventricular system of the brain. The skull bones have not yet fused because of the fontanelles; therefore, when the brain swells, the cranium swells as well. A cause of hydrocephalus would be a blockage of CSF in the ventricular system due to a tumor. Which of the following is the most likely location of a tumor that would result in hydrocephalus?

   A. Choroid plexus

   B. Cerebral aqueduct

   C. Lateral aperture

   D. Medial aperture

   E. Central canal of the spinal cord

6. An 81-year-old woman complains of slow onset of headache, dizziness, and nausea. Her history of the present illness reveals that 5 days ago she slipped on ice and hit her head on the pavement. She still has a bump on her head. Radiographic imaging of the brain reveals diffuse extracerebral bleeding. Which of the following is the most likely hematoma identified in this patient?

   A. Epidural hematoma because of a torn middle meningeal artery

   B. Subarachnoid hematoma because of a ruptured aneurism of the anterior communicating artery

   C. Subdural hematoma because of a torn bridging vein between the brain and superior sagittal sinus

7. A 56-year-old man develops a paraganglioma and is diagnosed with Vernet's syndrome. The tumor compresses structures that enter or exit the jugular foramen. The tumor would most likely compress which of the following?

A. Abducens n.	Facial n.	Vestibulocochlear n.	Internal carotid a.
B. Abducens n.	Facial n.	Vestibulocochlear n.	Sigmoid sinus
C. Abducens n.	Facial n.	Vestibulocochlear n.	Vertebral a.
D. Glossopharyngeal n.	Spinal accessory n.	Vagus n.	Internal carotid a.
E. Glossopharyngeal n.	Spinal accessory n.	Vagus n.	Sigmoid sinus
F. Glossopharyngeal n.	Spinal accessory n.	Vagus n.	Vertebral a.
G. Oculomotor n.	Trigeminal n.	Trochlear n.	Internal carotid a.
H. Oculomotor n.	Trigeminal n.	Trochlear n.	Sigmoid sinus
I. Oculomotor n.	Trigeminal n.	Trochlear n.	Vertebral a.

8. During a physical examination, the motor activity of extraocular muscles is tested along with the associated cranial nerves, which include CN III, CN VI, and

A. CN I

B. CN II

C. CN IV

D. CN V

E. CN VII

9. A 5-year-old boy is brought to the pediatrician with a complaint of severe pain, swelling, and redness around his right eye. He is diagnosed with periorbital cellulitis. The pediatrician tells the boy's parents that there is a possibility that the infection could spread to the boy's brain. The most probable route of spread to the brain would be through which of the following structures?

A. Cribriform plate into the meningeal space

B. Facial canal through the internal auditory meatus to the posterior cranial fossa

C. Frontal sinus into the sagittal sinus and into the subarachnoid space

D. Facial vein to the superior ophthalmic vein to the cavernous sinus

E. Orbital lymphatics to superficial cervical lymph nodes to the thoracic duct to the brain

10. When the corneal reflex in a patient is examined, sensory information is conducted from the cornea to the brain via the long ciliary nerve, a branch of CN V-1. The sensory input causes a motor response resulting in the closure of the patient's eyelids. Which motor nerve is being tested when the corneal reflex is being examined?

A. Abducens nerve (CN VI)

B. Facial nerve (CN VII)

C. Maxillary nerve (CN V-2)

D. Oculomotor nerve (CN III)

E. Trochlear nerve (CN IV)

11. A blue dye is placed into the right eye of a patient to assess the patency of the tear duct system. Assuming the lacrimal system is patent, at which structure would the physician see the eventual flow of the dye?

A. Inferior nasal meatus

B. Oral cavity

C. Pharynx

D. Sphenoethmoidal recess

E. Superior nasal meatus

12. A 32-year-old man sees his physician because of complaints of "double vision." On examination, findings in the right eye are consistent with a trochlear nerve injury. During the examination, the patient was most likely unable to accomplish which of the following movements?

A. Inward

B. Inward and downward

C. Inward and upward

D. Outward

E. Outward and downward

F. Outward and upward

13. A 42-year-old woman complains of steadily worsening pain and discomfort on the right side of the head. A CT scan shows a discrete tumor immediately lateral to the atlas and axis of the vertebral column that involves the superior cervical ganglion. The loss of autonomic innervation provided by this ganglion would most likely cause which of the following symptoms (assume the right side for each choice)?

	Eyelid	Pupil	Skin
A.	Normal	Miosis	Anhydrosis
B.	Normal	Mydriasis	Normal
C.	Normal	Normal	Oily
D.	Ptosis	Miosis	Anhydrosis
E.	Ptosis	Mydriasis	Normal
F.	Ptosis	Normal	Oily

14. A 42-year-old man sees his physician because of hearing loss and a sensation of the room spinning while he is standing. A lesion to which cranial nerve would most likely result in these symptoms?

A. CN IV

B. CN V

C. CN VI

D. CN VII

E. CN VIII

F. CN IX

15. Radiographic imaging reveals puss building up around the ear ossicles. Which of the following is the most likely location of the puss?

A. External ear

B. Middle ear

C. Internal ear

16. Tic douloureux is a neuropathic disorder characterized by sudden attacks of excruciating, lightening-like jabs of facial pain (paroxysm). Touching the face, brushing the teeth, shaving, or chewing often set off the paroxysms of sudden stabbing pain. The cause of the condition is unknown. Tic douloureux most likely results from deficits in which cranial nerve?

A. Oculomotor nerve

B. Facial nerve

C. Glossopharyngeal nerve

D. Trigeminal nerve

E. Vagus nerve

17. A 26-year-old woman presents with unilateral paralysis of facial muscles consistent with Bell's palsy. Which of the following cranial nerves is most likely affected that would result in this patient's condition?

A. Facial nerve

B. Glossopharyngeal nerve

C. Oculomotor nerve

D. Trigeminal nerve

E. Vestibulocochlear nerve

18. Branches of the maxillary artery gain entrance to the pterygopalatine fossa via which of the following structures?

A. Foramen rotundum

B. Foramen spinosum

C. Mandibular foramen

D. Pterygomaxillary fissure

E. Superior orbital fissure

19. After examination of a 60-year-old man, the dentist determines that the man has a cavity in a mandibular molar that needs to be filled. Which of the following nerves is the dentist most likely attempting to anesthetize to perform this procedure?

A. Chorda tympani

B. Hypoglossal

C. Inferior alveolar

D. Lingual

E. Superior alveolar

20. A herpes zoster virus infects the maxillary nerve of a 52-year-old woman. Blisters have formed on the lower eyelid and skin flanking the woman's nostrils. In addition to the areas observed on this patient, in what other locations would blisters most likely be seen?

A. Anterior portion of the tongue

B. Bridge of the nose

C. Chin

D. Forehead

E. Palatal mucosa

F. Upper eye lid

21. The pterygopalatine ganglion houses postganglionic neuronal cell bodies for visceral motor (parasympathetic) components of which of the following cranial nerves?

A. CN III

B. CN V

C. CN VII

D. CN IX

E. CN X

22. Irrigation of the maxillary sinus through its opening is a supportive measure to accelerate the resolution of a maxillary sinus infection. Which of the following nasal spaces is the approach to the opening of the maxillary sinus?

A. Choana

B. Inferior meatus

C. Middle meatus

D. Sphenoethmoidal recess

E. Superior meatus

23. A 7-year-old boy experiences acute speech difficulties. The findings on physical examination were unremarkable except that each time the boy protruded his tongue it deviated to the left. The results of the clinical and laboratory evaluations were consistent with the presence of infectious mononucleosis. Cranial nerve palsy is a rare complication of acute infectious mononucleosis in childhood. These findings are most likely the result of a deficit on which of the following cranial nerves?

A. Left CN VII

B. Left CN IX

C. Left CN XII

D. Right CN VII

E. Right CN IX

F. Right CN XII

24. The uvula of a 62-year-old patient deviates to the upper left when she is asked to say "Ahhh." This would most likely indicate a lesion on which of the following cranial nerves?

A. Left side of CN V-3

B. Left side of CN X

C. Left side of CN XII

D. Right side of CN V-3

E. Right side of CN X

F. Right side of CN XII

25. A 49-year-old woman presents with loss of sweet sensation on the right side of the anterior part of the tongue. Which additional findings may also be seen on the right side of this patient?

A. Adducted eye

B. Muscle of mastication weakness

C. Loss of corneal reflex

D. Reduced gag reflex

E. Tongue deviation during protrusion

ANSWERS

1—C: The middle meningeal artery courses along the internal surface of the lateral skull along the pterion. This artery courses in the dura mater, and when it ruptures, blood pools in the epidural space. Bridging cerebral veins hemorrhage into the subdural space. The superior sagittal sinus is a vein and, as such, low vascular pressure will require a much longer time to present symptoms. A superior sagittal sinus hemorrhage rarely will occur; however, more likely a superior sagittal sinus thrombosis will occur, which is not what this patient experienced.

2—C: The scalp consists of five layers. The vessels primarily course within the second, subcutaneous connective tissue layer. The rich anastomoses of internal and external carotid artery branches result in blood hemorrhaging from both cut ends. The dense connective tissue surrounding the cut vessels keeps the cut vessels patent. As a result, the scalp bleeds profusely.

3—C: The maxilla is the bone of the upper jaw and, therefore, would be associated with the location of a moustache. The frontal bone is in the forehead; the mandible is the lower jaw; the temporal bone is along the side of the head; and the zygomatic bone is the cheekbone.

4—D: The anterior communicating cerebral artery connects the paired anterior cerebral arteries within the subarachnoid space. Therefore, if the berry aneurysm was to rupture, it would cause an extracerebral hemorrhage in the subarachnoid space. Intraparenchymal and intraventricular hemorrhages are intracerebral (occur within the brain tissue). An epidural hemorrhage occurs between the skull and dura as a result of a ruptured middle meningeal artery. A subdural hemorrhage occurs between the dura and arachnoid mater, usually as a result of a torn bridging vein.

5—B: The cerebral aqueduct serves as the only drainage of CSF from the lateral and third ventricles. The choroid plexus continually filters plasma to create CSF, and therefore, a block in the aqueduct would result in CSF buildup in the lateral and third ventricles and result in hydrocephalus. A blockage in choroid plexus would result in a decrease in CSF production. A blockage in one of the apertures would not result in hydrocephalus because CSF could exit through another aperture. The central canal of the spinal cord would not result in hydrocephalus.

6—C: The patient injured her head 5 days ago. A slow bleed indicates a venous hemorrhage, which often occurs when small bridging veins are torn or sheared as they travel from the brain through the arachnoid mater and meningeal dura to dump into a dural sinus. The diffuse bleeding indicates that the blood is in the subdural space with no limitations to spread around the brain. A ruptured aneurysm of any artery would present in hours, not days, due to the increased blood pressure.

7—E: This patient has a paraganglioma, a jugular glomus tumor. The jugular foramen transmits CNN IX, X, and XI (glossopharyngeal, vagus, and spinal accessory nerves, respectively). In addition, the sigmoid sinus traverses the jugular foramen, where it continues as the internal jugular vein. The internal carotid artery enters the base of the skull through the carotid canal; the vertebral artery enters via the foramen magnum and, therefore, would not be affected by a jugular glomus tumor.

8—C: The seven extraocular muscles are the superior rectus, medial rectus, inferior rectus, and inferior oblique (innervated by CN III, the oculomotor nerve), the lateral rectus (innervated by CN VI, the abducens nerve), and the superior oblique (innervated by CN IV, the trochlear nerve). CN I (olfactory nerve) is for smell; CN II (optical nerve) is for sight; CN V (trigeminal nerve) is for general sensation; and CN VII (facial nerve) closes the eye and innervates the lacrimal gland.

9—D: The facial vein drains blood from the orbital region down to the internal jugular vein. However, blood from the orbit also spreads from the orbit via the superior ophthalmic vein through the superior orbital fissure to the cavernous sinus within the skull.

10—B: The orbicularis oculi muscle, the muscle that closes and blinks the eye, is innervated by CN VII, the facial nerve.

11—A: The nasolacrimal duct drains into the nasal cavity, into the space inferior to the inferior nasal concha called the inferior nasal meatus.

12—B: This patient is complaining of double vision, or diplopia. The superior oblique muscle is innervated by the trochlear nerve (CN IV). To clinically test the superior oblique muscle, the patient adducts his eye and then looks inferiorly. Therefore, Choice B (inward and downward) is the correct answer. Choice E (outward and downward) is the position for the anatomic action of the superior oblique muscle; however, outward and downward is the location for clinically testing the inferior rectus muscle.

13—D: The superior cervical ganglion is responsible for distributing postganglionic sympathetic neurons to the head, including sweat glands in the skin, superior tarsal muscle in the upper eyelid, and the pupillary dilator muscle. If sympathetics are absent, the eyelid droops (ptosis) due to losing the superior tarsal muscle. The pupil constricts due to losing tone in the pupillary dilator muscle and skin is red and dry from lack of innervation of sweat and sebaceous glands.

14—E: The vestibulocochlear nerve (CN VIII) is responsible for hearing and equilibrium. CN IV (trochlear nerve) and CN VI (abducens nerve) both innervate muscles in the orbit. CN V (trigeminal nerve) is general sensory to the head. CN VII (facial nerve) provides motor innervation to the facial muscles and to the stapedius muscle in the ear. CN IX (glossopharyngeal nerve) innervates the posterior third of the tongue and oropharynx.

15—B: The middle ear is the location of the three ear ossicles.

16—D: Tic douloureux is a condition that is associated with general sensation to the face, which is associated with CN V (trigeminal nerve). The facial nerve (CN VII) is only associated with branchial motor innervation to muscles of facial expression.

17—A: The facial nerve (CN VII) innervates muscles of facial expression. Therefore, a lesion of CN VII would result in unilateral facial paralysis. The trigeminal nerve (CN V) is responsible for conducting sensory information from the skin of the face, but does not provide motor innervation.

18—D: The maxillary artery branches off the external carotid artery and courses through the infratemporal fossa. The infratemporal fossa communicates with the pterygopalatine fossa via the pterygomaxillary fissure.

19—C: The inferior alveolar nerve provides sensory innervation to the mandibular teeth on the ipsilateral side. The chorda tympani provides taste sensation to the anterior part of the tongue via the lingual nerve. The superior alveolar provides sensory innervation for the maxillary teeth. The hypoglossal nerve provides motor innervation to the tongue muscles.

20—E: The mucosa lining the palate is innervated by the greater and lesser palatine nerves, which are branches of the maxillary nerve (CN V-2). Therefore, this patient with herpes zoster virus will most likely have blisters on the palatal mucosa.

21—C: The pterygopalatine ganglion houses the postganglionic parasympathetic neuronal cell bodies for the facial nerve, CN VII. As CN VII enters the internal acoustic meatus, it gives rise to the greater petrosal nerve, which courses into the pterygopalatine ganglion and synapses. The postganglionic parasympathetics from CN VII then course to the lacrimal gland, nasal glands, and palatal glands.

22—C: The hiatus semilunaris is located in the middle meatus, inferior to the middle nasal concha, and forms a communication with the maxillary sinus.

23—C: In a patient with a lesion on the left side of CN XII (hypoglossal nerve), the left genioglossus muscle will not cause the tongue to protrude, and therefore, the right genioglossus muscle will cause the tongue to stick out of the mouth and deviate to the side of the lesion ("lick your wounds").

24—E: The levator veli palatini muscles bilaterally elevate the soft palate, including the uvula. If the right levator veli palatini muscle is not innervated as a result of a lesion on CN X (vagus nerve), then the left levator veli palatini muscle contracts and pulls the uvula toward the left.

25—C: CN VII (facial nerve) provides special sensory innervation for taste in the anterior part of the tongue. In addition, CN VII also innervates the muscles of facial expression, including the orbicularis oculi muscle, which is the efferent limb of the corneal reflex.

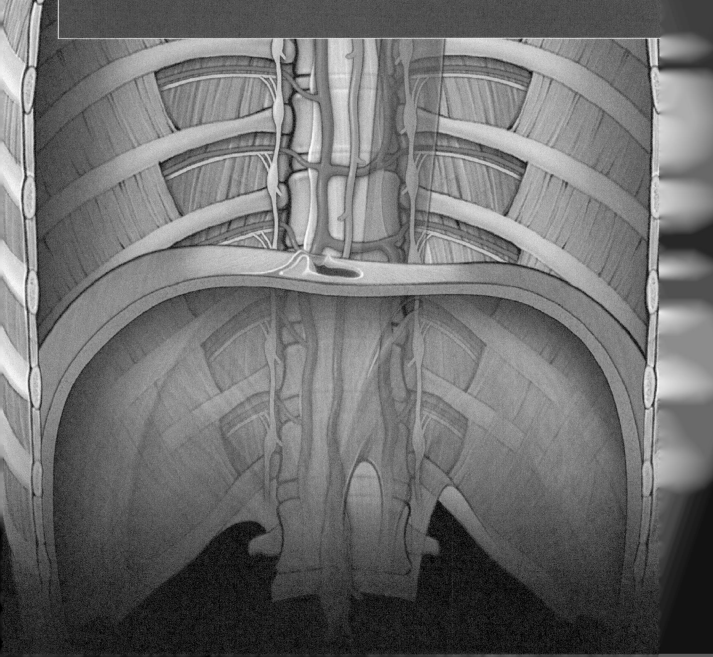

SECTION 5

NECK

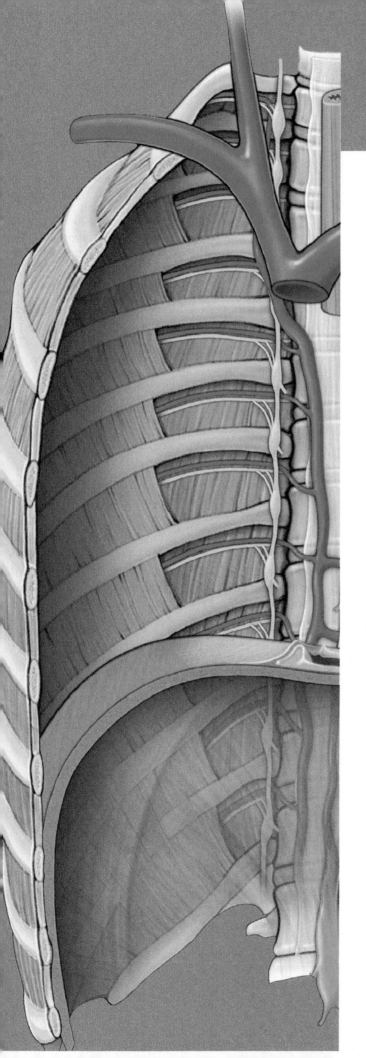

OVERVIEW OF THE NECK

FASCIA OF THE NECK

BIG PICTURE

The cervical fascia consists of concentric layers of fascia that compartmentalize structures in the neck (Figure 25-1). These fascial layers are defined as the superficial fascia and the deep fascia, with sublayers within the deep fascia. The fascia of the neck can determine the direction in which infection in the neck may spread.

SUPERFICIAL CERVICAL FASCIA

The superficial cervical fascia is the subcutaneous layer of the skin in the neck. This thin layer contains the muscles of facial expression, including the platysma muscle in the neck. The cutaneous nerves, superficial vessels, and superficial lymph nodes course within the superior cervical fascia.

DEEP CERVICAL FASCIA

The deep cervical fascia is deep to the superficial fascia. The deep cervical fascia is condensed in various regions to form the following sublayers: the investing layer of the deep cervical fascia, the pretracheal fascia, the prevertebral fascia, and the carotid sheath. The function of the deep fascia is to provide containment of muscles and viscera in compartments, to enable structures to slide over each other, and to serve as a conduit for neurovascular bundles.

INVESTING FASCIA   The investing fascia attaches as follows:

Posteriorly to the nuchal ligament, completely encircling the neck and splitting to enclose the sternocleidomastoid and trapezius muscles.

Superiorly to the hyoid bone and then splitting to enclose the submandibular gland.

Along the mandible and splitting to enclose the parotid gland.

Superiorly to the mastoid process, occipital bone, and zygomatic arch.

Inferiorly along the acromion, scapular spine, clavicle, and manubrium.

PRETRACHEAL FASCIA   The pretracheal fascia forms a tubular sheath in the **anterior part of the neck**. The pretracheal fascia extends superiorly from the hyoid bone and inferiorly to the

thorax, where it blends with the fibrous pericardium. The **pretracheal fascia:**

Encloses the infrahyoid muscles.

Encloses separately the thyroid gland, trachea, and esophagus.

Blends laterally with the carotid sheath.

Is contiguous inferiorly with the buccopharyngeal fascia of the pharynx.

PREVERTEBRAL FASCIA   The prevertebral fascia forms a tubular sheath around the vertebral column and the prevertebral muscles, which are attached to the vertebral column. The **prevertebral muscles** are:

Anteriorly, the longus colli and capitis muscles.

Laterally, the anterior, middle, and posterior scalene muscles.

Posteriorly, the deep cervical muscles.

The **prevertebral fascia:**

Attaches superiorly from the base of the skull and inferiorly to the endothoracic fascia in the thorax.

Extends laterally as the axillary sheath, which surrounds the axillary vessels and brachial plexus of nerves to the upper limb.

Contains, within its connective tissue fibers, the cervical sympathetic trunk and ganglia.

CAROTID SHEATH   The carotid sheath also is a tubular fascial investment that extends superiorly between the cranial base and inferiorly to the root of the neck.

The carotid sheath blends with the investing, pretracheal, and prevertebral layers of the deep cervical fascia.

The carotid sheath contains the **common and internal carotid arteries, internal jugular vein, and vagus nerve [cranial nerve (CN) X]**. In addition, the carotid sheath contains deep cervical lymph nodes, sympathetic fibers, and the carotid sinus nerve.

RETROPHARYNGEAL SPACE   The retropharyngeal space is a potential space consisting of loose connective tissue between the prevertebral and the buccopharyngeal fascia.

▼ The retropharyngeal space serves as a potential conduit for the **spread of infection** from the pharyngeal region to the mediastinum. ▼

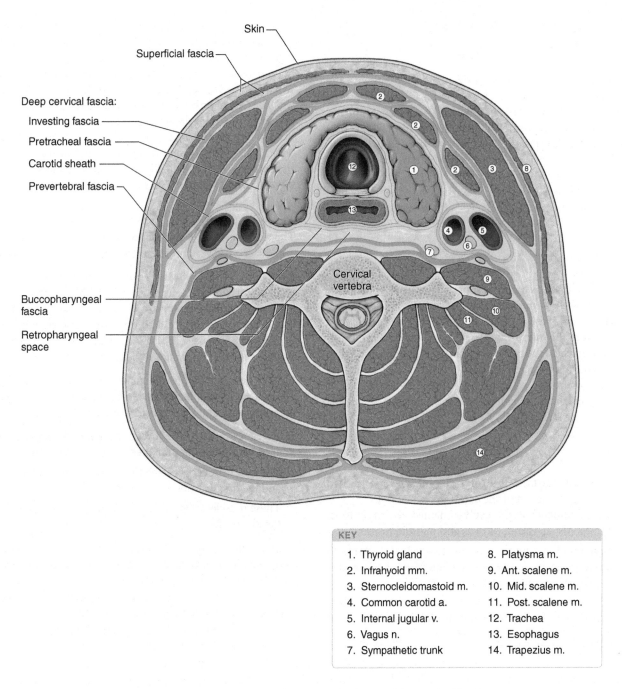

Skin

Superficial fascia

Deep cervical fascia:

Investing fascia

Pretracheal fascia

Carotid sheath

Prevertebral fascia

Buccopharyngeal fascia

Retropharyngeal space

Cervical vertebra

KEY

1. Thyroid gland
2. Infrahyoid mm.
3. Sternocleidomastoid m.
4. Common carotid a.
5. Internal jugular v.
6. Vagus n.
7. Sympathetic trunk
8. Platysma m.
9. Ant. scalene m.
10. Mid. scalene m.
11. Post. scalene m.
12. Trachea
13. Esophagus
14. Trapezius m.

Figure 25-1: Cross-section of the neck through the thyroid gland, showing the layers of the cervical fascia.

MUSCLES OF THE NECK

BIG PICTURE

The muscles of the neck are organized and grouped with the cervical fascia (Table 25-1). The platysma muscle is located within the superficial fascia, and the sternocleidomastoid and trapezius muscles are located within the investing fascia (part of the deep cervical fascia). Vertebral muscles (prevertebral, scalene, and deep cervical) are located within the prevertebral fascia. The suprahyoid muscles are deep to the investing fascia, whereas the infrahyoid muscles are within the pretracheal fascia. The vertebral muscles are within the prevertebral fascia.

PLATYSMA MUSCLE

The platysma muscle is the most superficial muscle of the neck. Unlike most skeletal muscles, the platysma is located in the superficial fascia (Figure 25-2A and B). The muscle extends superiorly from the inferior border of the mandible and inferiorly to the clavicle to the fascia of the anterior shoulder and thorax. The platysma muscle is a muscle of facial expression and therefore is innervated by the facial nerve (cervical branch of CN VII). Upon contraction, the platysma depresses the mandible and wrinkles the skin of neck.

STERNOCLEIDOMASTOID AND TRAPEZIUS MUSCLES

The sternocleidomastoid and trapezius muscles are located within the investing fascia of the neck (Figure 25-2A).

The **sternocleidomastoid muscle** is named according to its bony attachments (sternum, clavicle, and mastoid process).

- The sternocleidomastoid muscle creates the borders for both the anterior and the posterior triangles of the neck, and is innervated by the spinal accessory nerve (CN XI).
- This muscle flexes the neck, pulls the chin upward, and assists in elevating the rib cage during inspiration.

The **trapezius muscle** creates the anterior border of the posterior cervical triangle.

- The trapezius muscle attaches to the occipital bone, nuchal ligament, spinous processes of C7–T12, scapular spine, acromion, and the lateral part of the clavicle.
- The trapezius muscle is innervated by CN XI.

▽ During a physical examination, the functions of the sternocleidomastoid and trapezius muscles are evaluated together because they share the same innervation. During the examination, the physician looks for signs of muscle atrophy or weakness. To test the sternocleidomastoid muscle, the physician will place her hand on the patient's chin and instruct him to rotate his head to the opposite side against resistance. If acting normally, the patient's muscle can be seen and palpated. The physician next will instruct the patient to shrug his shoulders against resistance. Patients with **damage to CN XI** will have **diminished shoulder strength while shrugging on the injured side.** ▼

VERTEBRAL MUSCLES

The vertebral muscles are located between the prevertebral fascia and the cervical vertebrae, and are organized as described below.

PREVERTEBRAL MUSCLES The longus colli and the longus capitis muscles are anterior muscles and course superiorly between the anterior aspects of the base of the skull and inferiorly between the cervical and upper thoracic vertebrae (Figure 25-2B). The longus colli and longus capitis muscles help stabilize the cervical vertebrae and flex the neck.

▽ The longus colli and longus capitis muscles may be damaged in **whiplash injuries** because they course vertically and anteriorly along the cervical vertebrae. ▼

SCALENE MUSCLES The three scalene muscles are the **anterior, middle, and posterior scalenes** (Figure 25-2A and B).

- The scalene muscles attach to the cervical transverse processes and the first two ribs.
- These muscles elevate the ribs during breathing and laterally flex the neck.

The scalene muscles are homologous to the muscles of the body wall in that neurovascular structures course between the middle layer (middle scalene) and the deep layer (anterior scalene). As a result, both the cervical and brachial plexuses exit the vertebral column between the anterior and middle scalenes. In addition, the subclavian vein courses horizontally across the anterior scalene, whereas the phrenic nerve descends vertically on the anterior scalene.

DEEP CERVICAL MUSCLES The deep cervical muscles are located posteriorly on the neck and are part of the erector spinae and transversospinalis muscles. As such, they are innervated by the dorsal primary rami.

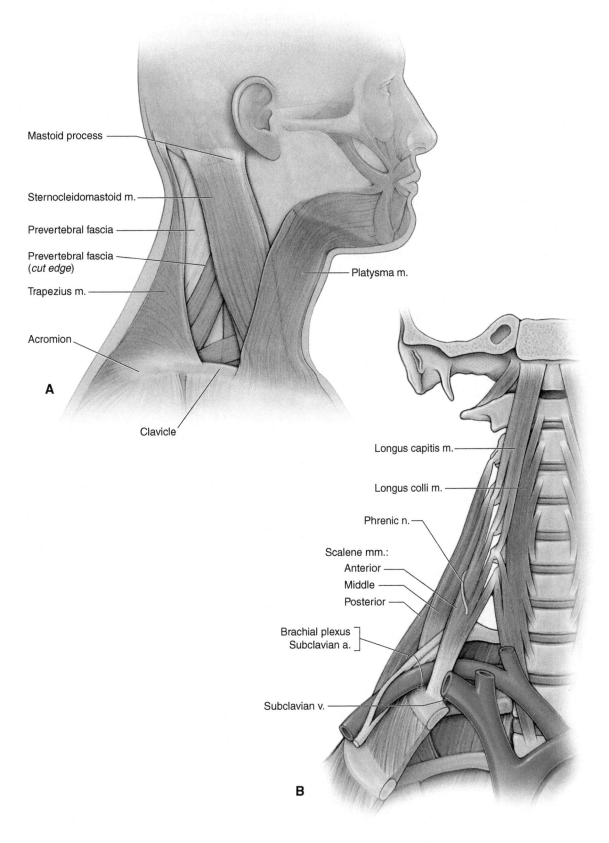

Figure 25-2: A. Muscles of the neck. **B**. Anterior view of the scalene and prevertebral muscles.

SUPRAHYOID MUSCLES

The suprahyoid muscles (i.e., digastric, stylohyoid, mylohyoid, and geniohyoid muscles) are located deep to the investing fascia of the deep cervical fascia (Figure 25-3A; Table 25-1). These muscles raise the hyoid bone during swallowing because the mandible is stabilized.

Digastric muscle. Has two bellies connected by a central tendon, which is attached to the hyoid bone. The posterior belly arises from the mastoid process, whereas the anterior belly arises from the mandible. Because of the two bellies, the digastric muscle can raise the hyoid bone or open the mouth. The digastric muscle originates embryologically from both the first and second pharyngeal arches; as such, it has a dual innervation [the anterior belly from the mandibular branch of the trigeminal nerve (CN V-3) and the posterior belly from CN VII].

Stylohyoid muscle. Attaches from the styloid process, bifurcates around the posterior belly of the digastric muscle, and inserts on the hyoid bone. The stylohyoid muscle elevates the hyoid bone and is innervated by CN VII.

Mylohyoid muscle. Located superiorly to the anterior belly of the digastric muscle and forms the floor of the mouth. The mylohyoid muscle elevates the floor of the mouth and is innervated by CN V-3.

Geniohyoid muscle. Located superiorly to the mylohyoid muscle. The geniohyoid muscle elevates the hyoid bone and is innervated by the cervical plexus (C1) (not shown in the illustration).

INFRAHYOID MUSCLES

The infrahyoid muscles include four pairs of muscles that are located within the muscular layer of the pretracheal fascia, inferior to the hyoid bone (hence the name) (Figure 25-3B; Table 25-1). Each muscle is innervated by the ansa cervicalis from the cervical plexus (ventral rami C1–C3). Collectively, these muscles function to depress the hyoid bone and larynx during swallowing and speaking. They receive their names according to their attachments.

Sternothyroid muscle. Sternum and thyroid cartilage.

Sternohyoid muscle. Sternum and hyoid bone.

Thyrohyoid muscle. Thyroid cartilage and hyoid bone.

Omohyoid muscle. Has two bellies. The inferior belly attaches to the superior border of the scapula ("omo" for shoulder), and the superior belly attaches to the hyoid bone.

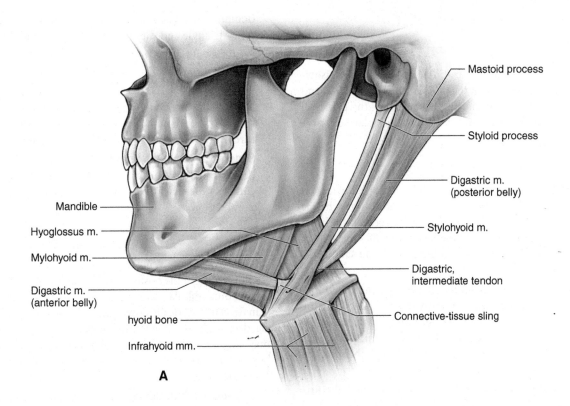

Mastoid process

Styloid process

Digastric m.
(posterior belly)

Mandible

Hyoglossus m.

Mylohyoid m.

Stylohyoid m.

Digastric m.
(anterior belly)

Digastric,
intermediate tendon

hyoid bone

Infrahyoid mm.

Connective-tissue sling

A

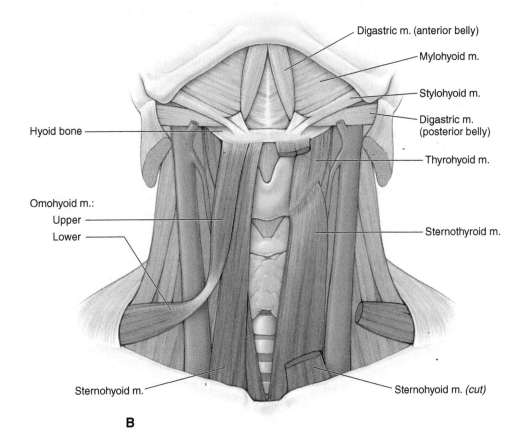

Digastric m. (anterior belly)

Mylohyoid m.

Stylohyoid m.

Digastric m.
(posterior belly)

Hyoid bone

Thyrohyoid m.

Omohyoid m.:

Upper

Lower

Sternothyroid m.

Sternohyoid m.

Sternohyoid m. *(cut)*

B

Figure 25-3: A. Lateral view of the floor of the mouth, highlighting the suprahyoid muscles (the geniohyoid muscle is not shown). **B**. Anterior view of a step dissection, highlighting the infrahyoid muscles.

VESSELS OF THE NECK

BIG PICTURE

The subclavian and common carotid arteries and their associated branches provide most of the blood supply to the head and neck. The external and anterior jugular veins are the principal venous return for the neck, and the internal jugular vein provides venous return for the head.

SUBCLAVIAN ARTERY

The subclavian arteries branch from the brachiocephalic artery on the right side and directly from the aortic arch on the left side (Figure 25-4A). The subclavian arteries course between the anterior and middle scalene muscles, where each becomes the axillary artery at the lateral edge of the first rib. **Branches of the subclavian artery** are as follows:

Vertebral artery. Arises from the first part of the subclavian artery and ascends between the anterior scalene and the longus coli muscles, through the **transverse foramina** of C6 to C1. At the superior border of C1, the vertebral artery turns medially and crosses the posterior arch of C1, through the **foramen magnum** en route to the brain.

Thyrocervical trunk. A short trunk that arises from the first part of the subclavian artery. Branches of the thyrocervical trunk are the **suprascapular, transverse cervical**, and **inferior thyroid arteries**.

COMMON CAROTID ARTERY

The common carotid artery branches from the brachiocephalic artery on the right side and directly from the aortic arch on the left side. The common carotid artery ascends within the carotid sheath, along with the internal jugular vein and the vagus nerve (Figure 25-4A). The common carotid artery bifurcates at the upper border of the thyroid cartilage into an internal and an external carotid artery.

INTERNAL CAROTID ARTERY The internal carotid artery gives off no branches in the neck, but within the skull it provides vascular supply to the anterior and middle regions of the brain, the orbit and the scalp.

EXTERNAL CAROTID ARTERY The external carotid artery supplies the neck and face through the following **branches**:

Superior thyroid artery. Arises at the level of the hyoid bone and supplies the larynx and the thyroid gland.

Lingual artery. Courses deep to the hyoglossus muscle, becoming the principal blood supply to the tongue.

Facial artery. Ascends deep to the posterior belly of the digastric and stylohyoid muscles and the submandibular gland, where the facial artery hooks around the mandible along the anterior border of the masseter muscle. The facial artery contributes to the blood supply to the face.

Ascending pharyngeal artery. Arises from the posterior surface of the external carotid artery and ascends to provide vascular supply to the pharynx and the palatine tonsils.

Occipital artery. Courses posteriorly to the apex of the posterior triangle and supplies the occipital region of the scalp.

Posterior auricular artery. Arises from the posterior surface of the external carotid artery and supplies the scalp posterior to the ear.

Maxillary artery. Arises from the external carotid artery, posterior to the mandibular neck and supplies the deep structures of the face such as the infratemporal fossa.

Superficial temporal artery. Arises as a terminal branch of the external carotid artery within the parotid gland and courses superficial to the zygomatic arch supplying the temporal region.

EXTERNAL JUGULAR VEIN

The external jugular vein is forward at the angle of the mandible via the joining of the posterior auricular and posterior branch of the retromandibular veins (Figure 25-4B). The external jugular vein descends vertically down the neck within the superficial fascia, deep to the platysma muscle. After crossing the sternocleidomastoid muscle, the external jugular vein pierces the deep investing fascia posterior to the clavicular head and enters the subclavian vein.

The **posterior auricular vein** drains the scalp behind and above the ear.

The **retromandibular vein** is formed by the superficial temporal and maxillary veins within the parotid gland. The retromandibular vein divides into anterior and posterior divisions. The anterior division joins the facial vein to form the common facial vein. The posterior division contributes to the external jugular vein.

INTERNAL JUGULAR VEIN

The internal jugular vein originates at the jugular foramen by the union of the sigmoid and inferior petrosal sinuses and serves as the principal drainage of the skull, brain, superficial face, and parts of the neck. After exiting the skull via the jugular foramen, along with the glossopharyngeal, vagus, and accessory nerves (CNN IX, X, and XI, respectively), the internal jugular vein traverses the neck within the carotid sheath. The internal jugular vein joins with the subclavian vein to form the brachiocephalic vein. Tributaries include the facial, lingual, pharyngeal, and occipital veins, and the superior and middle thyroid veins. The right internal jugular vein lacks valves where as the left has one valve.

▽ Because the right internal jugular vein lacks valves, two pulsations, known as the **jugular venous pulse,** are observed due to right atrial contraction and closure of the tricuspid value. The jugular venous pulse assists the physician in assessing the cardiac health of the patient. For example, an elevated jugular venous pulse may suggest right-sided congestive heart failure or stenosis of the tricuspid valve. ▼

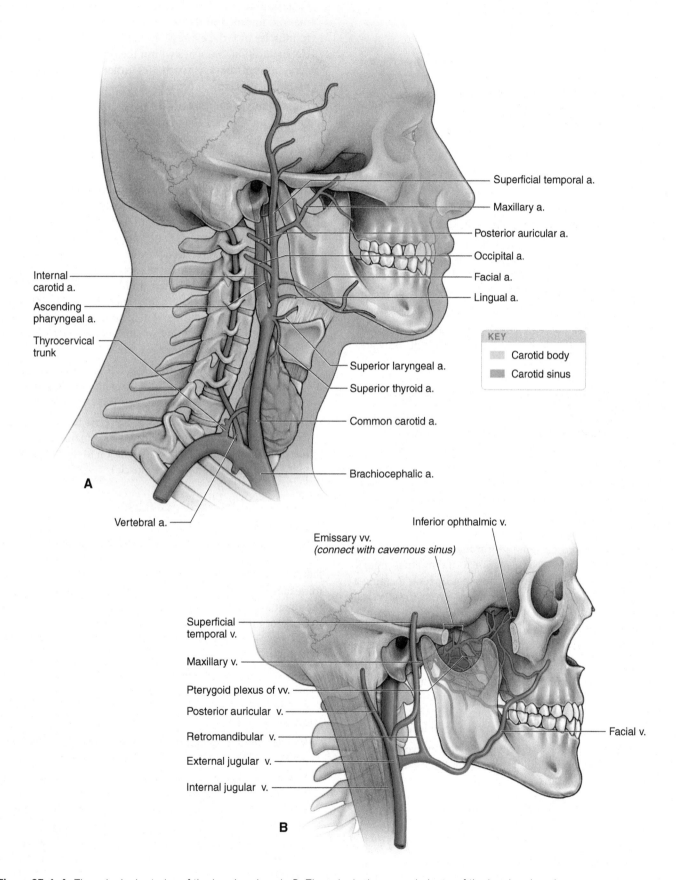

Superficial temporal a.

Maxillary a.

Posterior auricular a.

Occipital a.

Facial a.

Lingual a.

Internal carotid a.

Ascending pharyngeal a.

Thyrocervical trunk

Superior laryngeal a.

Superior thyroid a.

Common carotid a.

Brachiocephalic a.

Vertebral a.

KEY

Carotid body

Carotid sinus

A

Inferior ophthalmic v.

Emissary vv. (connect with cavernous sinus)

Superficial temporal v.

Maxillary v.

Pterygoid plexus of vv.

Posterior auricular v.

Retromandibular v.

External jugular v.

Internal jugular v.

Facial v.

B

Figure 25-4: A. The principal arteries of the head and neck. **B**. The principal venous drainage of the head and neck.

INNERVATION OF THE NECK

BIG PICTURE

The cervical plexus of nerves is responsible for much of the sensory and motor innervation of the neck. CNN VII, IX, X, XI, and XII also play an important role. In addition, sympathetic innervation of the neck and head is via the cervical sympathetic trunk.

CERVICAL PLEXUS

The cervical plexus of nerves arises from the ventral rami of cervical nerves C1 to C4 and exits the vertebral column between the anterior and posterior scalene muscles. **Branches of the cervical plexus** are as follows (Figure 25-5A):

Cutaneous (sensory) branches. Pierce the prevertebral fascia at the central region of the posterior border of the sternocleidomastoid muscle serving various regions of the skin of the neck. The cutaneous branches of the cervical plexus are as follows:

- **Lesser occipital nerve (C2).** Innervates the skin over the lower, lateral region of the scalp.
- **Great auricular nerve (C2–C3).** Innervates the skin over the parotid gland and angle of the jaw (this is the only area of the face not supplied by CN V); also contributes some cutaneous sensation to the external ear.
- **Transverse cervical nerve (C2–C3).** Innervates the skin over the anterior part of the neck.
- **Supraclavicular nerve (C3–C4).** Innervates the skin over the lower portion of the neck, upper part of the chest, and the shoulder.

Motor nerves. The deep branches of the cervical plexus innervate muscles. The fibers from C1 travel briefly with CN XII and innervate the thyrohyoid and geniohyoid muscles. Fibers from C1 form the superior root of the ansa cervicalis. Fibers from C2 and C3 join to form the inferior root of the ansa cervicalis, where it lies anterior to the internal jugular vein and passes upward to join the superior root, forming a loop (ansa). Most of the motor nerves from the cervical plexus branch from the ansa cervicalis, supplying the infrahyoid muscles (sternothyroid, sternohyoid, and omohyoid). The cervical plexus also gives rise to the phrenic nerve.

- **Phrenic nerve (C3–C5).** Courses vertically along the anterior scalene muscle between the subclavian artery and the subclavian vein en route to innervate the diaphragm. ("C3–C5, keep the diaphragm alive" is a pneumonic that is used to remember the spinal nerve levels of the phrenic nerve.)
- The cervical plexus also provides motor nerves to other cervical muscles, including the scalenes, the longus colli, and the longus capitis.

CRANIAL NERVES IN THE NECK

The following cranial nerves travel through the neck (Figure 25-5B and C).

Trigeminal nerve (CN V). CN V courses through the submandibular triangle to innervate the **mylohyoid** and the **anterior digastric muscles.**

Facial nerve (CN VII). CN VII exits the skull, via the stylomastoid foramen, and provides branchial motor innervation to muscles of facial expression including the platysma muscle. In addition, branches within the oral cavity provide visceral motor (parasympathetic) innervation to the submandibular gland.

Glossopharyngeal nerve (CN IX). CN IX exits the skull with CNN X and XI, via the jugular foramen, and provides visceral sensory innervation to the **carotid sinus** (baroreceptor) and **carotid body** (chemoreceptor) monitoring blood pressure.

Vagus nerve (CN X). CN X exits the skull with CN IX and CN XI, via the jugular foramen, and descends within the carotid sheath, deep to the carotid artery and internal jugular vein. The vagus nerve contains both motor and sensory neurons supplying the gut tube (as far as the transverse colon), which includes the pharynx and larynx for swallowing and phonation, as well as sensory innervation from this same region. Important branches of CN X in the neck include the **superior** and **recurrent laryngeal nerves**. In addition, CN X provides visceral sensory innervation from the **carotid body** (chemoreceptor), monitoring blood pressure.

Spinal accessory nerve (CN XI). CN XI exits the jugular foramen with CN IX and CN X, providing motor innervation to the **sternocleidomastoid** and **trapezius muscles**.

Hypoglossal nerve (CN XII). CN XII exits the skull via the hypoglossal canal and courses into the submandibular triangle, providing somatic motor innervation to the **tongue muscles** (except the palatoglossus muscle).

SYMPATHETIC NERVES OF THE NECK

The sympathetic trunk (chain) ascends from the thorax into the cervical region, within the prevertebral fascia along the longus colli and longus capitis muscles (Figure 25-5C). The **sympathetic trunk in the cervical region** receives only gray rami communicantes (no white rami). The sympathetic trunk innervates the sweat and sebaceous glands, blood vessels, and the errector pili, dilator pupillae, and superior tarsal muscles.

The sympathetic trunk gives rise to the following **three cervical ganglia**:

Inferior cervical ganglion. Fuses with the first thoracic paravertebral ganglion to become the cervicothoracic (stellate) ganglion at the level of rib 1. The inferior cervical ganglion gives rise to the inferior cervical cardiac nerves.

Middle cervical ganglion. Lies at the C6 vertebral level and gives rise to the middle cervical cardiac nerves.

Superior cervical ganglion. Lies anterior to the C1–C2 transverse processes, between the internal carotid artery and the longus capitis muscle. The superior cervical ganglion contains cell bodies of postganglionic sympathetic fibers en route to the head and gives rise to the internal and external carotid plexuses and the superior cervical cardiac nerves.

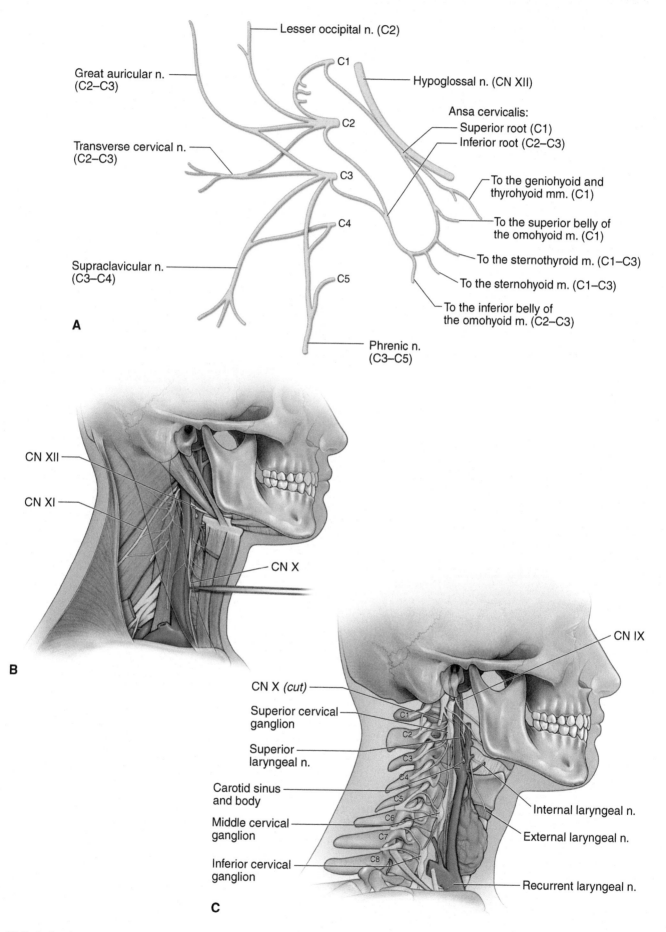

Figure 25-5: A. Cervical plexus. **B** and **C**. Cranial nerves and autonomics of the neck.

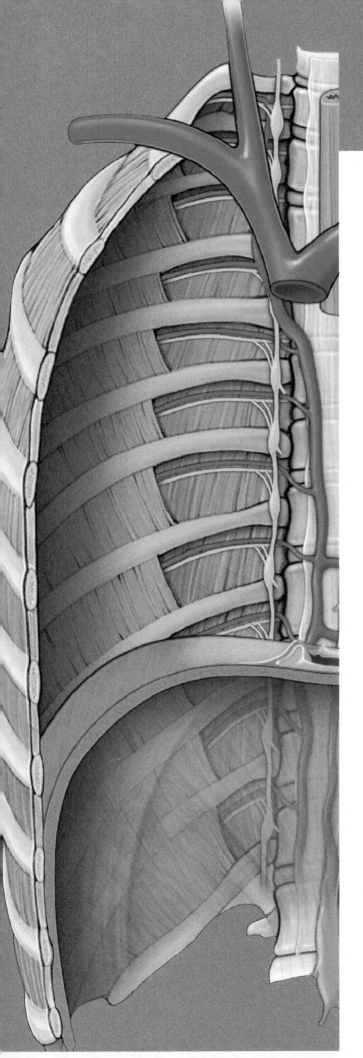

CHAPTER 26

TRIANGLES AND ROOT OF THE NECK

POSTERIOR TRIANGLE OF THE NECK

BIG PICTURE

The neck region is divided into triangles to compartmentalize the contents. The sternocleidomastoid muscle divides the neck region into posterior and anterior triangles. The posterior triangle of the neck is located on the lateral aspect of the neck.

BOUNDARIES OF THE POSTERIOR TRIANGLE

The boundaries of the posterior triangle are as follows (Figure 26-1A):

Anteriorly. The sternocleidomastoid muscle.

Posteriorly. The trapezius muscle.

Inferiorly. The clavicle.

ROOF OF THE POSTERIOR TRIANGLE

The roof of the posterior triangle of the neck consists of the investing layer of the deep cervical fascia that surrounds the neck, enveloping the sternocleidomastoid and trapezius muscles (Figure 26-1B). The sensory branches of the cervical plexus and the external jugular vein pierce the investing layer of cervical fascia.

SENSORY BRANCHES OF THE CERVICAL PLEXUS The sensory branches of the cervical plexus course between the anterior and middle scalene muscles piercing the investing fascia at the posterior border of the sternocleidomastoid muscle en route to their respective cutaneous fields (Figure 26-1B).

Lesser occipital nerve (C2 contribution). Ascends along the posterior border of the sternocleidomastoid muscle en route to the skin of the neck and scalp, behind the ear.

Great auricular nerve (C2–C3 contributions). Ascends over the sternocleidomastoid muscle en route to the skin of the parotid region and ear.

Transverse cervical nerve (C2–C3 contributions). Courses horizontally over the sternocleidomastoid muscle en route to the skin on the anterior and lateral sides of the neck.

Supraclavicular nerve (C3–C4 contributions). Descends obliquely across the posterior triangle of the neck to the skin over the clavicle and superior region of the thorax.

EXTERNAL JUGULAR VEIN The **external jugular vein** descends vertically across the superficial surface of the sternocleidomastoid muscle, between the platysma muscle and the investing fascia. The external jugular vein pierces the investing fascia where it joins with the subclavian vein.

FLOOR OF THE POSTERIOR TRIANGLE

The prevertebral fascia forms the floor of the posterior triangle of the neck (Figure 26-1C and D).

The following structures are **superficial to the prevertebral fascia:**

Inferior belly of the omohyoid muscle. Courses from the hyoid bone en route to the scapula within the **pretracheal fascia**.

Transverse cervical artery. A branch from the thyrocervical trunk that courses along the floor of the posterior triangle en route to the deep surface of the trapezius muscle. The artery bifurcates into a superficial branch, which courses superficial to the rhomboid muscles, and a deep branch, which courses deep to the rhomboid muscles.

Suprascapular artery. The most inferior branch from the thyrocervical trunk, the suprascapular artery courses across the anterior scalene muscle and the phrenic nerve, where it crosses over the superior transverse scapular ligament to enter the supraspinous fossa.

Spinal accessory nerve [cranial nerve (CN) XI]. Exits the jugular foramen and obliquely descends along the prevertebral fascia en route to the sternocleidomastoid and trapezius muscles.

▽ The spinal accessory nerve is one of the more vulnerable structures during any surgical procedure involving the posterior triangle of the neck. During surgery, careful isolation and protection of this nerve is essential to avoid **spinal accessory nerve damage**.

The following structures are **deep to the prevertebral fascia:**

Cervical muscles, from superior to inferior:

- Splenius capitis
- Levator scapulae
- Posterior scalene
- Middle scalene
- Anterior scalene (notice that the phrenic nerve descends vertically along the anterior surface of the anterior scalene muscle en route to the thoracic cavity). ▽

Brachial plexus and subclavian artery. The brachial plexus and the subclavian artery course between the anterior and middle scalene muscles, and as the nerve plexus and artery emerge from those muscles, they carry an extension of the prevertebral fascia along to form the axillary sheath.

▽ **Thoracic outlet syndrome** consists of a group of disorders related to the exit of the brachial plexus and the subclavian artery between the neck and the upper limb. Most disorders are caused by compression of the brachial plexus or subclavian artery. Compression may be caused by the clavicle, movement of the shoulder joint, or an enlargement of muscles surrounding the plexus or vessel. ▽

▽ The region between the anterior and middle scalene muscles where the brachial plexus exits is referred to as the **interscalene triangle**. This region serves as a chosen site to administer nerve blocks for shoulder surgeries because the sensory branches to the shoulder and upper limb exit within the brachial plexus at this point. ▽

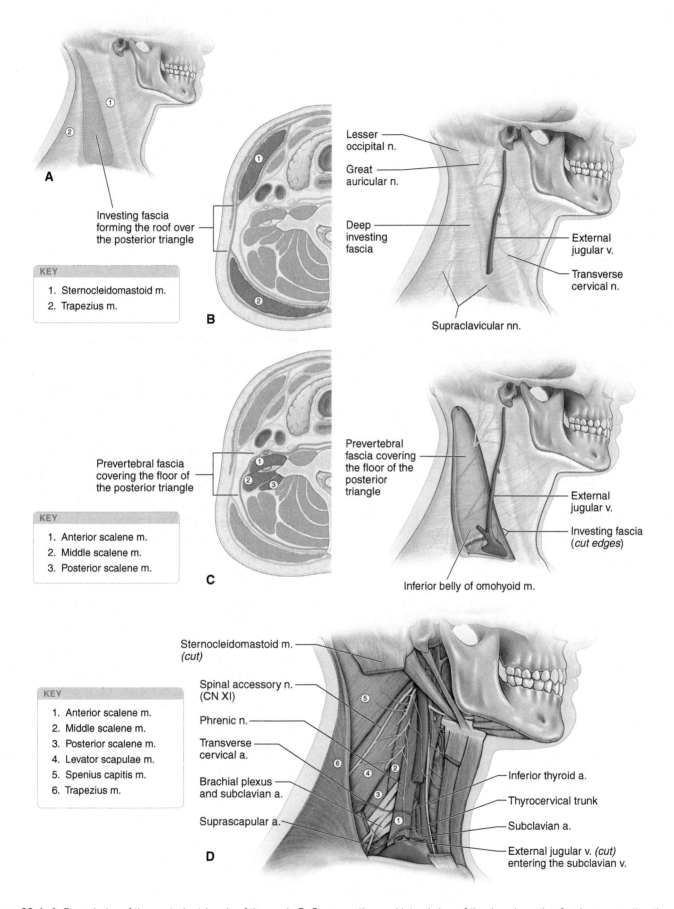

Figure 26-1: A. Boundaries of the posterior triangle of the neck. **B**. Cross-section and lateral view of the deep investing fascia surrounding the sternocleidomastoid and trapezius muscles. **C**. Cross-section and lateral view of the prevertebral fascia covering the prevertebral muscles. **D**. Prevertebral fascia removed from the posterior triangle of the neck.

ANTERIOR TRIANGLE OF THE NECK

BIG PICTURE

The anterior triangle of the neck contains structures that enter the cranial vault and the thorax.

BOUNDARIES AND DIVISIONS OF THE ANTERIOR TRIANGLE

The borders, or boundaries, of the anterior triangle are as follows (Figure 26-2A):

Anteriorly. The midline down the neck.

Posteriorly. The sternocleidomastoid muscle.

Superiorly. The lower border of the mandible.

The anterior triangle is subdivided by three structures—the omohyoid muscle, the hyoid bone, and the digastric muscle—into submandibular, carotid, submental, and muscular triangles (Figure 26-2B). The submental and muscular triangles have little clinical value and, therefore, will not be discussed in this text.

SUBMANDIBULAR TRIANGLE The submandibular triangle is bounded by the mandible, digastric, stylohyoid, and mylohyoid muscles (Figure 26-2C). The triangle consists of the following structures:

Submandibular salivary gland. Fills most of the space of the triangle and is one of three paired salivary glands. The facial vein courses superficially to the submandibular salivary gland, whereas the facial artery courses deep to the gland. The submandibular gland is innervated by parasympathetic fibers from the facial nerve (CN VII).

Hypoglossal nerve (CN XII). Provides motor innervation to the tongue muscles. The nerve courses deep to the posterior digastric and stylohyoid muscles and deep to the hyoglossus muscle.

CAROTID TRIANGLE The superior belly of the omohyoid and the posterior belly of the digastric and sternocleidomastoid muscles form the boundaries of the carotid triangle (Figure 26-2B). The triangle contains the superior portion of the common carotid artery (hence, the triangle's name), which bifurcates at the thyroid cartilage into the external and internal carotid arteries.

The **internal carotid artery** gives off no branches in the neck as it ascends to the carotid canal in the base of the skull. However, the following structures are related to the internal carotid artery in the neck:

Carotid sinus. A swelling in the origin of the internal carotid artery containing baroreceptors that monitor blood pressure. The carotid sinus is innervated by visceral sensory neurons from the glossopharyngeal nerve (CN IX).

Carotid body. A chemoreceptor at the bifurcation of the common carotid artery that monitors the partial pressure of oxygen. The carotid body is innervated by visceral sensory neurons from CN IX and the vagus nerve (CN X).

The **external carotid artery** sends off the following branches (Figure 26-2C):

Superior thyroid artery. Courses inferiorly to the thyroid gland.

Lingual artery. Courses anteriorly, deep to the hyoglossus muscle to reach the tongue.

Facial artery. Courses deep to the submandibular gland, within the submandibular triangle, to exit anterior to the masseter muscle en route to the face.

Occipital artery. Courses deep to the posterior belly of the digastric muscle en route to the occipital region.

Ascending pharyngeal artery. Arises posteriorly from the external carotid artery and ascends vertically between the internal carotid artery and side of the pharynx.

Veins within the carotid triangle include the following:

Internal jugular vein, which courses lateral to the common carotid artery.

Tributaries of the internal jugular vein are the superior thyroid, lingual, common facial, ascending pharyngeal, and the occipital veins.

Nerves within this space include the following:

Glossopharyngeal nerve (CN IX). Before entering the base of the tongue, CN IX sends off visceral sensory fibers to the carotid sinus and the carotid body.

Hypoglossal nerve (CN XII). CN XII crosses both the internal and external carotid arteries en route to the tongue muscles through the submandibular triangle.

Ansa cervicalis. This motor loop from the cervical plexus is found within the fascia of the carotid sheath and innervates the infrahyoid muscles. The superior limb originates from the ventral rams of C1, whereas the inferior limb originates from ventral rami of C2 and C3.

Vagus nerve (CN X). CN X exits the jugular foramen and courses inside the carotid sheath, deep between the internal jugular and common carotid vessels. CN X gives off a superior laryngeal nerve, which divides into internal and external branches. The internal branch courses through the thyrohyoid membrane to provide sensory innervation to the mucosa of the larynx superior to the vocal folds. The external branch provides motor innervation to the cricothyroid muscle of the larynx.

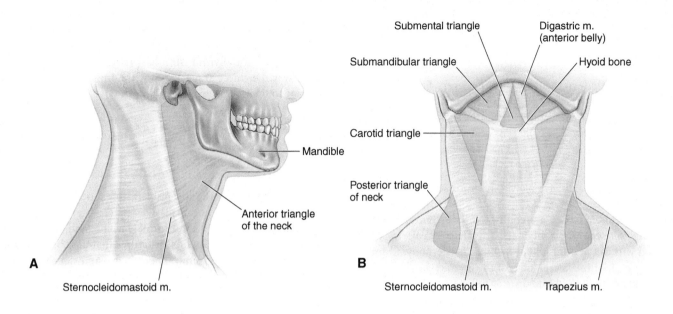

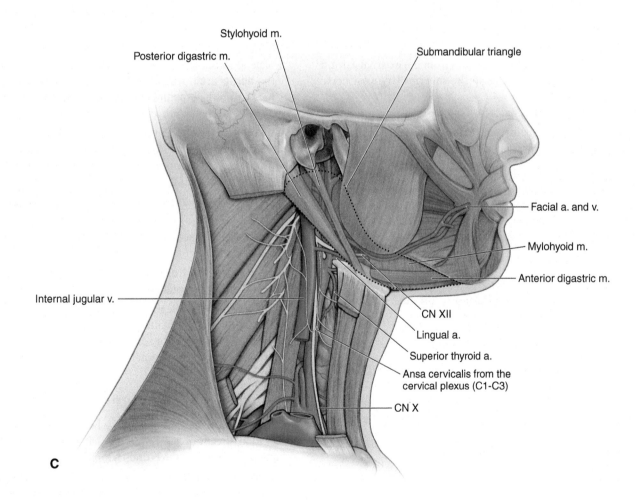

Figure 26-2: A. Boundaries of the anterior triangle of the neck. **B**. The anterior triangle of the neck further divided into the submandibular, carotid, submental, and muscular triangles. **C**. Contents of the submandibular and carotid triangles.

VISCERAL TRIANGLE OF THE NECK

BIG PICTURE

The visceral triangle of the neck has three layers. From anterior to posterior, the layers are an endocrine layer (the thyroid and parathyroid glands), a respiratory layer (the trachea and larynx), and an alimentary layer (the pharynx and esophagus).

THYROID GLAND

The thyroid gland regulates metabolic processes and decreases blood calcium concentration (Figure 26-3A–C). The gland has two **lateral lobes**, which are connected by a central **isthmus** overlying the second through the fourth tracheal rings, all enclosed within the **pretracheal fascia**. The sternohyoid muscles cover each of the lateral lobes of the thyroid gland.

Vessels of the thyroid gland are as follows:

Inferior thyroid artery. Originates from the thyrocervical trunk and is closely related to the recurrent laryngeal nerve (CN X) as the artery enters the thyroid gland.

Inferior thyroid veins. Varied number of vessels that drain into the brachiocephalic veins.

Superior thyroid artery. Originates from the external carotid artery and is closely related to the superior laryngeal nerve (CN X).

Middle and superior thyroid veins. Generally, middle and superior thyroid veins course anterior to the common carotid artery and drain into the internal jugular vein.

▽ A **goiter** is a pathologic enlargement of the thyroid gland. Consequently, a goiter presents as a swelling in the anterior part of the neck, inferior to the thyroid cartilage. A goiter is usually caused by iodine deficiency. Iodine is necessary for the synthesis of thyroid hormones; when there is a deficiency of iodine, the gland is unable to produce thyroid hormones. When the levels of thyroid hormones decrease, the pituitary gland secretes more thyroid-stimulating hormone, which stimulates the thyroid gland to produce thyroid hormone, causing the gland to enlarge. ▽

PARATHYROID GLANDS

Parathyroid glands work antagonistically with the thyroid by increasing the blood calcium concentration (Figure 26-3B). Usually four parathyroid glands (two located on each side of the thyroid gland) are present on the deep surface of the thyroid gland.

TRACHEA AND ESOPHAGUS

The trachea extends inferiorly from the cricoid cartilage in the midline (Figure 26-3A–C). At the level of the jugular notch of the manubrium the trachea is halfway between the sternum and the vertebral column. Sympathetic nerves from the T1 to T4 spinal nerve levels cause airway smooth muscle relaxation and thus dilation of the airways, whereas parasympathetic innervation from the recurrent laryngeal nerves (CN X) causes airway smooth muscle constriction and thus narrowing of the airways.

The esophagus is posterior to the trachea.

▽ For long-term access to the trachea, a **tracheostomy** is performed. A tracheostomy is a surgical incision in the trachea below the thyroid isthmus, providing an opening into the airway. During a tracheostomy, the inferior thyroid veins anterior to the trachea must be avoided. ▽

ROOT OF THE NECK

BIG PICTURE

The root of the neck is the area immediately superior to the superior thoracic aperture and axillary inlets (Figure 26-3A and B) and is bounded by the manubrium, clavicles, and T1 vertebra. The root of the neck contains structures that pass between the neck, thorax, and upper limb. There is also an extension of the thoracic cavity projecting into the root of the neck.

VASCULAR SUPPLY

The **vascular supply of the root of the neck** includes the following vessels:

Subclavian artery. Course between the anterior and middle scalene muscles. The left subclavian artery originates from the **aortic arch**, whereas the right subclavian artery originates from the **brachiocephalic trunk. Branches of the subclavian artery** include the following:

- **Vertebral artery.** Courses superiorly between the anterior scalene and the longus colli muscles before entering the transverse foramina of the cervical vertebrae (C6-C1) en route to the brain.

- **Thyrocervical trunk.** Gives rise to the suprascapular, transverse cervical, and inferior thyroid arteries.

- **Internal thoracic artery.** Descends deep to the thoracic cage and gives off anterior intercostal branches.

- **Costocervical trunk.** Gives rise to the deep cervical and supreme intercostal arteries.

Subclavian vein. Begins at the lateral edge of rib 1 as the continuation of the axillary vein. The subclavian vein courses anterior to the anterior scalene muscle, where the subclavian vein is joined by the internal jugular vein to form the brachiocephalic vein.

LYMPHATICS

All lymphatic vessels throughout the body return their lymph to the blood stream by either the thoracic or the right lymphatic ducts. Both lymphatic ducts drain lymph into the subclavian veins.

Thoracic duct. This duct is a major lymphatic vessel that begins in the abdomen and passes superiorly through the thorax entering the root of the neck, on the left side. Arching laterally, the duct passes deep to the carotid sheath and courses inferiorly to terminate in the junction between the left internal jugular and the subclavian veins (Figure 26-3A).

Right lymphatic duct. Lymphatic vessels from the right side of the thorax, upper limb, neck, and head connect together to form the right thoracic duct, which drains into the junction between the right internal jugular and subclavian veins.

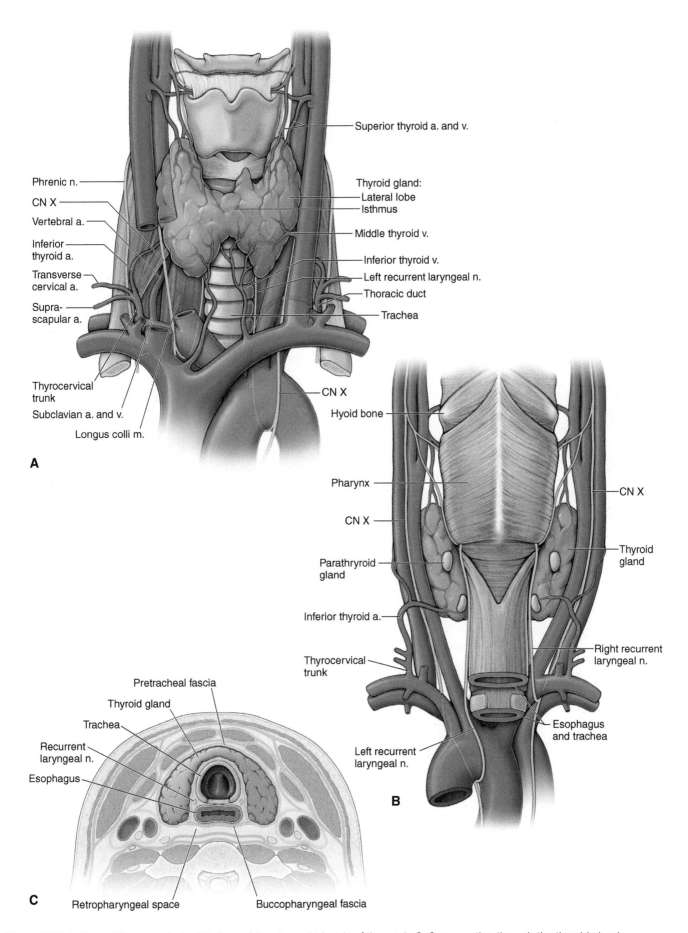

Superior thyroid a. and v.

Phrenic n.

CN X

Vertebral a.

Inferior thyroid a.

Transverse cervical a.

Supra-scapular a.

Thyroid gland:
Lateral lobe
Isthmus

Middle thyroid v.

Inferior thyroid v.

Left recurrent laryngeal n.

Thoracic duct

Trachea

Thyrocervical trunk

Subclavian a. and v.

Longus colli m.

CN X

A

Hyoid bone

Pharynx

CN X

CN X

Parathyroid gland

Thyroid gland

Inferior thyroid a.

Thyrocervical trunk

Right recurrent laryngeal n.

Esophagus and trachea

Left recurrent laryngeal n.

B

Pretracheal fascia

Thyroid gland

Trachea

Recurrent laryngeal n.

Esophagus

C

Retropharyngeal space

Buccopharyngeal fascia

Figure 26-3: Anterior (**A**) and posterior (**B**) views of the visceral triangle of the neck. **C**. Cross-section through the thyroid gland.

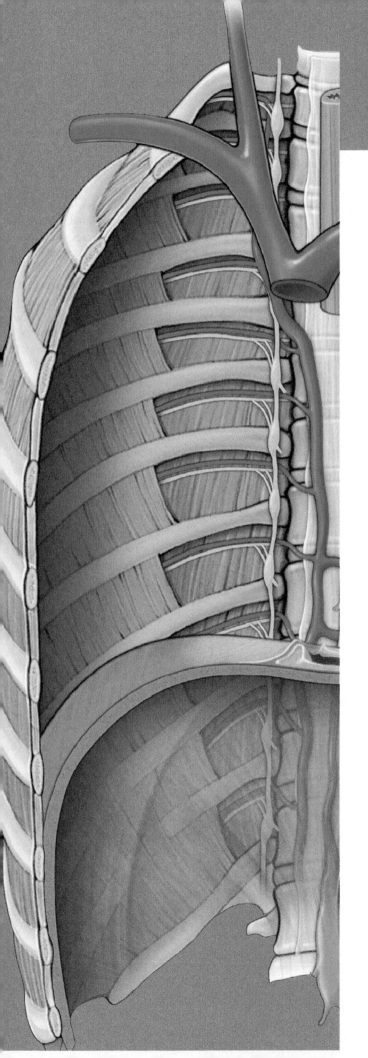

CHAPTER 27

PHARYNX

OVERVIEW OF THE PHARYNX

BIG PICTURE

The pharynx is a funnel-shaped, fibromuscular tube that extends from the base of the skull to the cricoid cartilage, where the pharynx continues as the esophagus. The pharynx is commonly called the throat and serves as a common pathway for food and air.

SUBDIVISIONS OF THE PHARYNX

The pharynx is classically divided into three compartments, based on location: the nasopharynx, the oropharynx, and the laryngopharynx (Figure 27-1A–C).

NASOPHARYNX The nasopharynx is posterior to the nasal cavity, inferior to the sphenoid bone, and superior to the soft palate. During swallowing, the soft palate elevates and the pharyngeal wall contracts anteriorly to form a seal, preventing food from refluxing into the nasopharynx and nose. When we laugh, this sealing action can fail, and fluids that are being swallowed while we laugh can end up in the nasal cavity.

The nasopharynx is continuous with the nasal cavity through the arched openings called **choanae**. The **auditory tubes**, also known as the pharyngotympanic or eustachian tubes, connect the middle ear to the pharynx and open into the lateral walls of the nasopharynx. The auditory tubes allow middle ear pressure to equalize with atmospheric pressure.

The mucosa that lines the superior part of the posterior wall contains a ring of lymphatic tissue called the **pharyngeal tonsil (adenoids)**, which traps and destroys pathogens that enter the nasopharynx in air (Figure 27-1B).

▽ When the **pharyngeal tonsil (adenoids)** is infected and swollen, it can completely block airflow through the nasal cavity so that breathing through the nose requires an uncomfortable amount of effort. As a result, inhalation occurs through an open mouth. Surgical removal of the adenoids (adenoidectomy) may be necessary if infections, earaches, or breathing problems become chronic. ▽

Much of the mucosa of the nasopharynx posterior to the auditory tubes is supplied by the pharyngeal branch of **cranial nerve (CN) V-2** (maxillary nerve), which traverses the palatovaginal canal with the pharyngeal branch of the maxillary artery.

OROPHARYNX The oropharynx is the region of the pharynx located between the soft palate and the epiglottis and communicates with the oral cavity. The **palatoglossal arches** mark the boundary between the oral cavity anteriorly and the oropharynx posteriorly. The posterior third of the tongue forms a partial anterior wall of the oropharynx. The mucosa of the oropharynx and the posterior third of the tongue are innervated by the glossopharyngeal nerve **(CN IX)**.

LARYNGOPHARYNX The laryngopharynx extends between the epiglottis and the cricoid cartilage, with the larynx forming the anterior wall. The laryngopharynx serves as a common passageway for food and air. The laryngopharynx communicates posteriorly with the esophagus, where food and fluids to the stomach pass (Figure 27-1B and C). In addition, the laryngopharynx communicates anteriorly with the larynx, where air is conducted in and out of the lungs during breathing. During swallowing, food has the "right of way" and air passage stops temporarily. Innervation of the mucosa of the laryngopharynx is provided by the vagus nerve **(CN X)**.

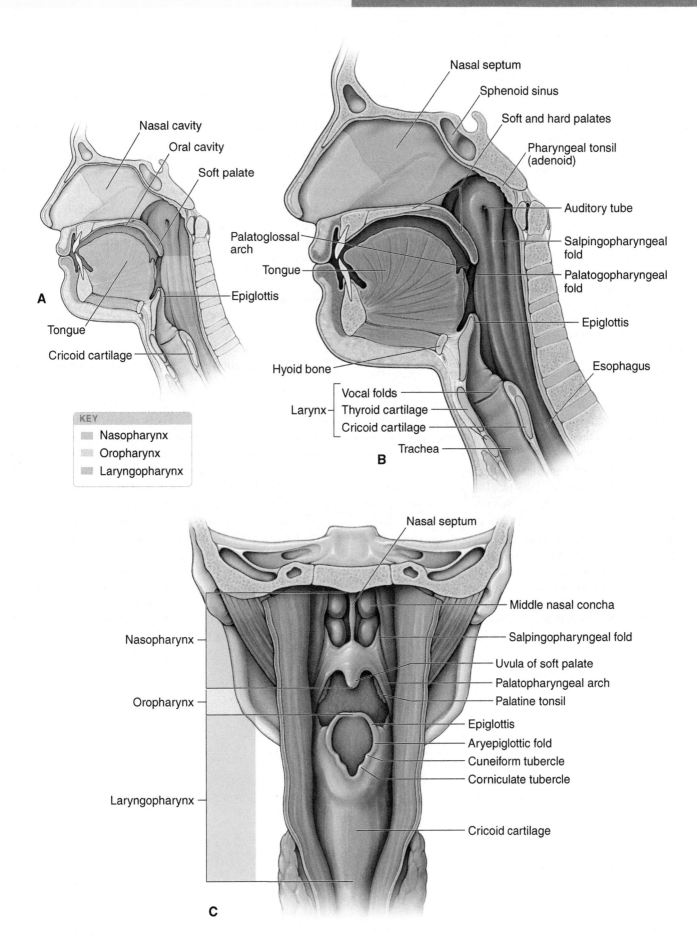

Figure 27-1: A. Regions of the pharynx. **B**. Sagittal section of the head. **C**. Posterior view of the pharynx (midsagittal incision through the pharyngeal constrictor muscles).

FUNCTIONS OF THE PHARYNX

BIG PICTURE

The pharynx is composed of skeletal muscle that is lined internally by a mucous membrane. The muscles aid in swallowing and speaking.

PHARYNGEAL CONSTRICTORS

The constrictor muscles form the lateral and posterior walls of the pharynx and are attached posteriorly to the **median pharyngeal raphe**. The median pharyngeal raphe extends downward from the pharyngeal tubercle, on the base of the occipital bone anterior to the foramen magnum, and blends inferiorly with the posterior wall of the laryngopharynx and esophagus. The **pharyngobasilar fascia** separates the mucosa and the muscle layer, and blends with the periosteum of the base of the skull.

The constrictors are arranged as the following **three overlapping muscles** (Figure 27-2A and B):

Superior pharyngeal constrictor. Attaches to the medial pterygoid plate, the pterygomandibular raphe, and the lingula of the mandible. The levator veli palatini muscle, the auditory (eustachian) tube, and the ascending palatine artery course between the floor of the sphenoid bone and the superior pharyngeal constrictor.

Middle pharyngeal constrictor. Attaches to the side of the body and the lesser horn of the hyoid bone. The **stylopharyngeus muscle**, **CN IX**, and the **stylohyoid ligament** course between the superior and the middle pharyngeal constrictor.

Inferior pharyngeal constrictor. Attaches to the lateral surface of the thyroid and cricoid cartilages. The lowest fibers of the inferior pharyngeal constrictor are thought to constitute a **cricopharyngeus muscle**, which must relax if food is to enter the esophagus. The **internal laryngeal nerve** and the **superior laryngeal artery and vein** course between the middle and inferior pharyngeal constrictors.

The pharyngeal constrictor muscles narrow the pharynx when swallowing and are activated in a sequence, from top to bottom, to propel food toward the esophagus.

ACCESSORY PHARYNGEAL MUSCLES

There are three small accessory pharyngeal muscles that elevate the larynx and pharynx when swallowing is initiated. These muscles have separate origins and insert onto the posterior part of the pharynx (Figure 27-2A and B). The names of these muscles identify their origins and insertions.

Stylopharyngeus muscle. Attaches to the styloid process of the temporal bone and the pharyngeal wall, between the superior and middle pharyngeal constrictors.

Palatopharyngeus muscle. Attaches to the soft and hard palates and the pharyngeal wall.

Salpingopharyngeus muscle. Attaches to the auditory tube and the pharyngeal wall.

SWALLOWING

The **stages of swallowing** (deglutition) are as follows (Figure 27-2C):

1. The tongue pushes the bolus of food back toward the oropharynx.

2. The palatoglossus and palatopharyngeus muscles contract to squeeze the bolus backward into the oropharynx. The tensor veli palatini and levator veli palatini muscles elevate and tense the soft palate to close the entrance into the nasopharynx.

3. The palatopharyngeus, stylopharyngeus, and salpingopharyngeus muscles elevate the walls of the pharynx in preparation to receive the food. The suprahyoid muscles elevate the hyoid bone and the larynx to close the opening into the larynx, thus preventing the food from entering the respiratory passageways.

4. The sequential contraction of the superior, middle, and inferior pharyngeal constrictor muscles moves the food through the oropharynx and the laryngopharynx into the esophagus, where it is propelled via peristalsis.

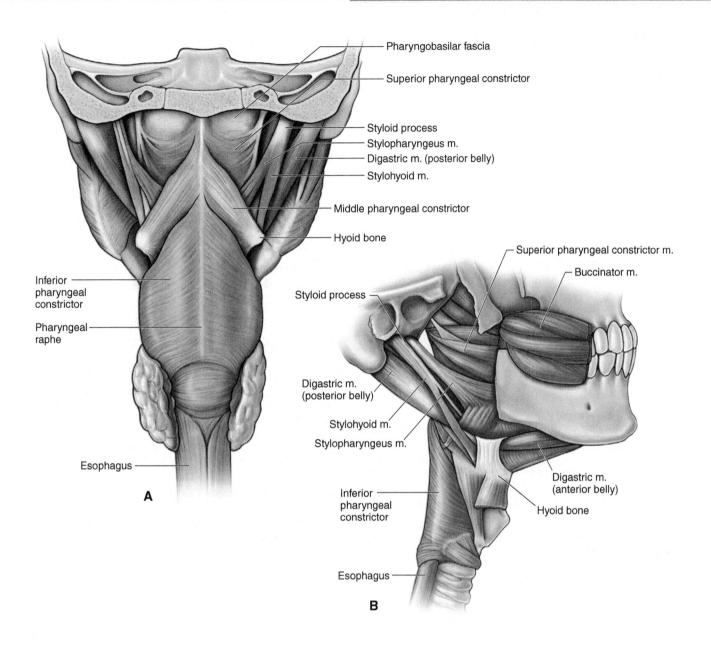

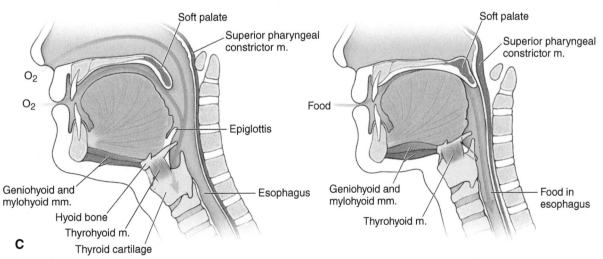

Figure 27-2: Posterior (**A**) and lateral (**B**) views of the muscles of the pharynx. **C**. Swallowing mechanism.

NEUROVASCULAR SUPPLY OF THE PHARYNX

BIG PICTURE

The mucosa of the pharynx receives its blood supply from branches of the external carotid artery and sensory innervation from CN V-2, CN IX, and CN X. The motor supply to the pharynx is primarily from CN X, with the exception of the stylopharyngeus muscle, which is CN IX.

VASCULAR SUPPLY AND LYMPHATICS OF THE PHARYNX

The arterial supply of the pharynx includes branches from the following arteries (Figure 27-3A):

Pharyngeal artery. The pharyngeal artery is a branch from the maxillary artery. The pharyngeal artery courses through the palatovaginal canal en route to the nasopharynx.

Facial artery. Two branches course from the facial artery en route to the pharynx. The ascending palatine artery ascends along the pharynx and passes superior to the superior pharyngeal constrictor. The tonsillar artery branches from the ascending palatine artery, penetrates the superior pharyngeal constrictor muscle, and supplies the palatine tonsil.

Ascending pharyngeal artery. The ascending pharyngeal artery branches directly off the external carotid artery and courses with the ascending palatine artery.

In addition, branches from the external carotid artery also contribute to the vascular supply of the pharynx.

The venous drainage of the pharynx includes tributaries of the internal and external jugular veins.

The lymph drainage is through the deep cervical and retropharyngeal lymph nodes.

PHARYNGEAL PLEXUS OF NERVES

Pharyngeal nerves from CN IX and CN X and a small contribution from CN V-2 form the pharyngeal plexus (Figure 27-3B).

The plexus lies along the middle pharyngeal constrictor muscle and is responsible for sensory and motor innervation.

SENSORY INNERVATION The three regions of the pharynx each receive a unique cranial nerve supply.

Nasopharynx. Sensory neurons travel from the mucosa of the nasopharynx, through the palatovaginal canal, and on through the pterygopalatine ganglion to CN V-2. CN V-2 travels through the foramen rotundum and back to the brainstem. Therefore, CN V-2 provides the pharyngeal branch of the maxillary nerve.

Oropharynx. Sensory neurons from CN IX enter the pharyngeal plexus and provide the sensory innervation for the oropharynx.

Laryngopharynx. Sensory neurons from CN X enter the pharyngeal plexus and provide the sensory innervation for the laryngopharynx.

MOTOR INNERVATION CN X originates in the brainstem and exits the jugular foramen to the posterior region of the middle pharyngeal constrictor muscle. Neurons originating from the nucleus ambiguous in CN X supply all pharyngeal muscles, except the stylopharyngeus. CN IX originates in the brainstem and exits the jugular foramen en route to the stylopharyngeus muscle.

▼ The **gag reflex** tests both the sensory and motor components of CN IX and CN X, respectively. This reflex contraction of the back of the throat is evoked by touching the soft palate. The sensory limb of the gag reflex is supplied by CN IX to the brainstem. The motor limb is supplied by CN X, which results in contraction of the pharyngeal muscles. Absence of the gag reflex can be symptomatic of CN IX or CN X injury. ▼

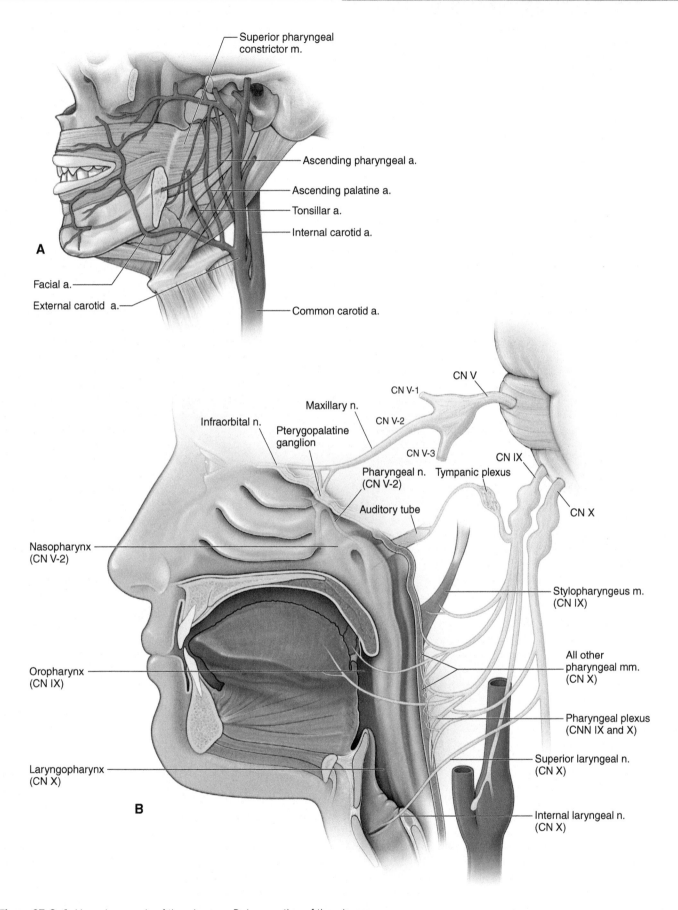

Figure 27-3: A. Vascular supply of the pharynx. **B**. Innervation of the pharynx.

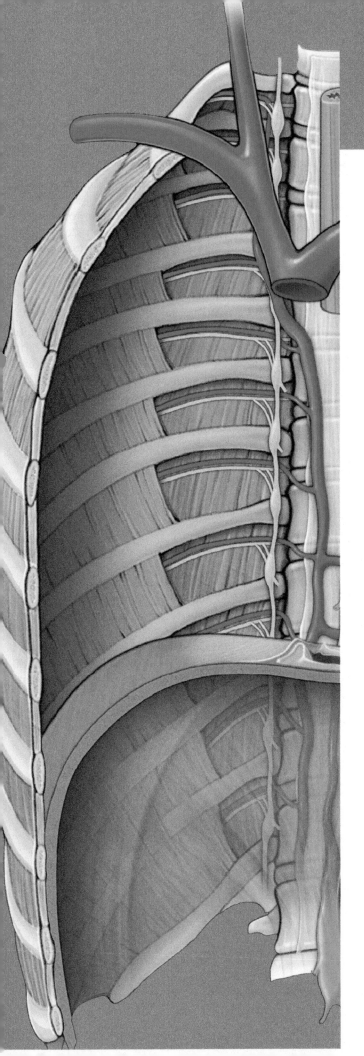

CHAPTER 28

LARYNX

LARYNGEAL FRAMEWORK

BIG PICTURE

The larynx forms the air passageway from the hyoid bone to the trachea. The larynx is continuous with the laryngopharynx superiorly and with the trachea inferiorly. The larynx provides a patent (open) airway and acts as a switching mechanism to route air and food into the proper channels. The larynx is commonly known as the voice box and provides the cartilaginous framework for vocal fold and muscle attachments, which vibrate to produce sound.

HYOID BONE

The hyoid bone consists of a body, two greater horns, and two lesser horns, and is the only bone that does not articulate with another bone. The hyoid bone is U-shaped and is suspended from the tips of the styloid processes of the temporal bones by the **stylohyoid ligaments** (Figure 28-1A and C). The hyoid bone is connected to the thyroid cartilage and supported by the suprahyoid and infrahyoid muscles and by the middle pharyngeal constrictor muscle. In addition, the hyoid bone supports the root of the tongue.

LARYNGEAL CARTILAGES

The framework of the larynx is an intricate arrangement of nine cartilages connected by membranes and ligaments (Figure 28-1A–C).

Thyroid cartilage. The thyroid cartilage forms a median elevation, called the **laryngeal prominence** ("Adam's apple"), and lies inferior to the hyoid bone. The thyroid cartilage typically is larger in males than in females because the male sex hormones stimulate its growth during puberty. The superior horn of the laryngeal cartilage is attached to the tip of the greater horn of the hyoid bone, whereas its inferior horn articulates with the cricoid cartilage, forming the cricothyroid joint.

- The **thyrohyoid membrane**, as its name implies, stretches between the thyroid cartilage and the hyoid bone. The superior laryngeal vessels and the internal laryngeal nerve pierce the membrane en route to providing vascular supply and sensory information, respectively, to the mucosa superior to the vocal folds.

Cricoid cartilage. The cricoid cartilage is shaped like a signet ring, with the broad part of the ring facing posteriorly. The lower border marks the inferior limits of the larynx and pharynx. The function of the cricoid cartilage is to provide attachments for laryngeal muscles, cartilages, and ligaments involved in opening and closing of the airway to produce sound.

Epiglottis. The epiglottis is elastic cartilage, shaped like a spoon, that is posterior to the root of the tongue. The lower end of the epiglottis is attached to the deep surface of the thyroid cartilage. When only air is flowing into the larynx, the inlet to the larynx is open wide, with the free edge of the epiglottis projecting superiorly and anteriorly. During swallowing, the larynx is pulled superiorly and the epiglottis tips posteriorly to cover the laryngeal inlet. As a result, the epiglottis acts as a deflector to keep food out of the larynx (and trachea) during swallowing.

Arytenoid cartilages. The arytenoid cartilages are shaped like a pyramid. Their base articulates with the cricoid cartilage. Each arytenoid cartilage has a vocal process, which gives attachment to the vocal ligaments and vocalis muscle, and a muscular process, which gives attachment to the thyroarytenoid and the lateral and posterior cricoarytenoid muscles.

Cuneiform and corniculate cartilages. These tiny cartilages lie on the apices of the arytenoid cartilages and are enclosed within the aryepiglottic folds.

VOCAL LIGAMENTS AND THEIR MOVEMENTS

From the vocal process of each arytenoid cartilage, a fibrous band extends anteriorly to attach to the deep surface of the thyroid cartilage (Figure 28-1C). These two fibrous ligaments are composed largely of elastic fibers and form the core of the mucosal folds called the **vocal folds, or true vocal cords.** Consequently, the vocal folds vibrate, producing sound as air rushes up from the lungs through the rima glottis, the opening between the vocal cords. Superior to the true vocal folds are a similar pair of mucosal folds, called the **vestibular (false) vocal folds,** which play no part in the production of sound because the false vocal folds are not opposable.

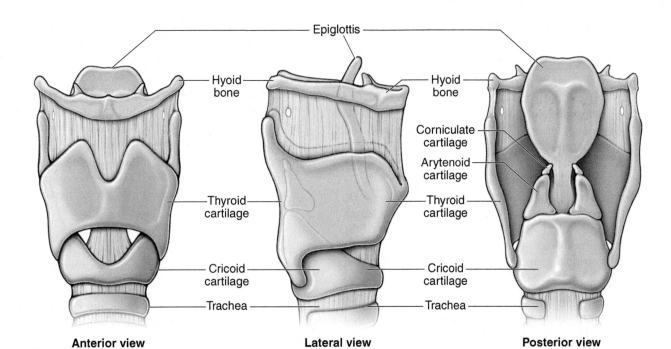

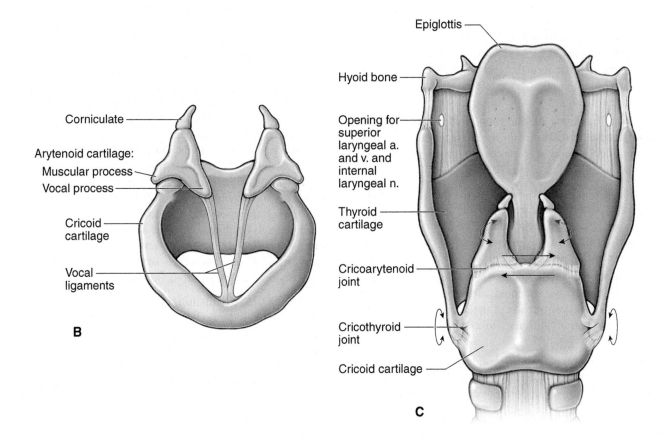

Figure 28-1: A. The cartilaginous skeleton **B**. Vocal ligament anatomy. **C**. Posterior view of the movements of the laryngeal cartilage joints.

FUNCTION OF THE LARYNX

BIG PICTURE

Laryngeal muscles are innervated by the vagus nerve [cranial nerve (CN) X] and move the laryngeal skeleton. In turn, this movement changes the width and the tension on the vocal folds so that air passing between the vocal folds causes them to vibrate, producing sound.

LARYNGEAL MUSCLES

The intrinsic laryngeal muscles move the laryngeal framework, altering the size and shape of the **rima glottidis** and the length and tension of the vocal folds (Figure 28-2). The actions of the laryngeal muscles are best understood when considered in the following functional groups: adductors and abductors and tensors and relaxers.

ADDUCTORS AND ABDUCTORS
Rotation of the cricoarytenoid cartilages results in medial or lateral displacement of the vocal folds, thereby decreasing or increasing, respectively, the aperture of the rima glottis.

- **Adduction (closing) of the vocal folds.** The **lateral cricoarytenoid muscles** pull the muscular processes anteriorly, rotating the arytenoids so their vocal processes swing medially. When this action is combined with that of the **transverse arytenoid muscles**, which pull the arytenoid cartilages together, the gap between the vocal folds is decreased. Air pushed through the rima glottidis causes vibration of the vocal ligaments.
- **Abduction (opening) of the vocal folds.** The **posterior cricoarytenoid muscles** pull the muscular processes posteriorly, rotating the vocal processes laterally and thus widening the rima glottidis.

GLIDING MOVEMENT OF THE ARYTENOID CARTILAGES
Horizontal gliding action of the arytenoid cartilages permits the bases of these cartilages to move side to side. Medial gliding and medial rotation of the arytenoid cartilages occur simultaneously, as do the two lateral movements.

- **Adduction of the vocal folds** also is aided by horizontal medial gliding of the arytenoid cartilages. This action is caused by bilateral contraction of the lateral cricoarytenoid, the transverse and oblique arytenoids, and the aryepiglottic muscles.
- **Abduction of the vocal folds** also is aided by horizontal lateral gliding of the arytenoid cartilages. This action is caused by bilateral contraction of the posterior cricoarytenoid muscles.

LENGTH AND TENSION OF THE VOCAL FOLDS
The **cricothyroid joint** is a synovial articulation between the side of the cricoid cartilage and the inferior horn of the thyroid cartilage. This joint enables the thyroid cartilage to tilt back and forth upon the cricoid cartilage, altering the length and tension of the vocal folds.

- **Tensing of the vocal folds.** The principal tensors are the **cricothyroid muscles**, which tilt the thyroid cartilage anteriorly and inferiorly, increasing the distance between the thyroid cartilage and the arytenoid cartilage. This movement elongates and tightens the vocal ligaments, raising the pitch of the voice.
- **Relaxing the vocal folds.** The principal relaxers are the **thyroarytenoid muscles**, which pull the arytenoid cartilages anteriorly toward the thyroid angle (prominence), thereby relaxing the vocal ligaments.
  - **Vocalis muscles.** Produce minute adjustments of the vocal ligaments, selectively tensing and relaxing parts of the vocal folds during animated speech and singing.

VOICE PRODUCTION

To understand how we speak and sing, we first need to understand how sound is produced, and then we need to understand how that sound is articulated.

PHONATION
Phonation, or the production of sound, involves the intermittent release of expired air coordinated with opening and closing the rima glottis. The length of the true vocal folds and the size of the rima glottis are altered by the action of the intrinsic laryngeal muscles, most of which move the arytenoid cartilages. As the length and tension of the vocal folds change, the pitch of the sound is altered. Generally, the more tense the vocal folds, the faster they vibrate and thus the higher the pitch. The rima glottis is wide when deep tones are produced and narrows to a slit when high-pitched sounds are produced. As a young boy's larynx enlarges during puberty, the vocal folds become both longer and thicker, causing them to vibrate more slowly and thus the voice becomes deeper.

Loudness of the voice depends on the force with which air rushes across the vocal folds. The greater the force of air across the vocal folds, the stronger the vibration, which results in louder sounds. The vocal folds do not move at all when we whisper, but they vibrate vigorously when we yell.

Although the vocal folds produce sounds, the quality of the voice depends on the coordinated activity of many other structures. For example, the pharynx acts as a resonating chamber to amplify and enhance the quality of the sound, as do the oral and nasal cavities and the paranasal sinuses.

ARTICULATION
Articulation, or the production of intelligible sounds, involves the actions of the pharyngeal muscles (vagus nerve, CN X), the tongue (hypoglossal nerve, CN XII), the muscles of facial expression (facial nerve, CN VII), mandibular movements (mandibular branch of the trigeminal nerve, CN V-3), and the soft palate (CN X). Each of these structures modifies the crude sounds that are produced by the larynx and convert them into recognizable consonants and vowels.

▽ Inflammation of the vocal folds, or **laryngitis**, results in hoarseness or inability to speak above a whisper. Overuse of the voice, very dry air, bacterial infections, or inhalation of irritating chemicals can cause laryngitis. Whatever the cause, irritation of the laryngeal tissues causes swelling and prevents the vocal folds from moving freely. ▽

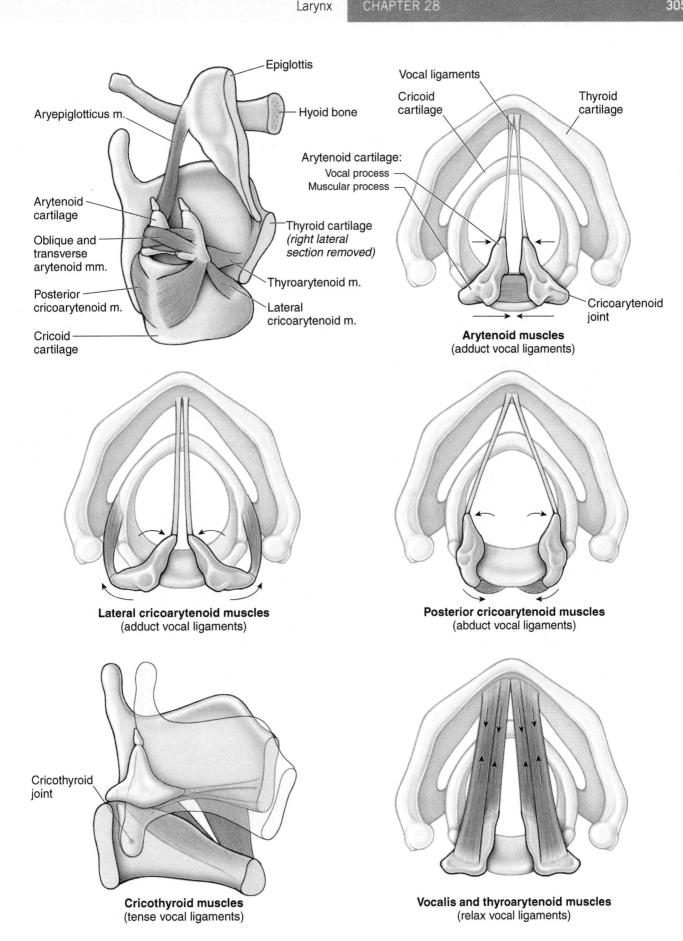

Figure 28-2: Muscles and actions of laryngeal muscles.

VASCULAR SUPPLY AND INNERVATION OF THE LARYNX

BIG PICTURE

Blood supply to the larynx is derived principally from the superior and inferior laryngeal arteries. Innervation is both sensory and motor (CN X).

LARYNGEAL VESSELS

The blood supply to the larynx is derived from the superior and inferior laryngeal arteries (Figure 28-3A and B). Rich anastomoses exist between the corresponding contralateral and ipsilateral arteries.

Superior thyroid artery. Branching from the external carotid artery, the superior thyroid artery descends en route to the thyroid gland. In its course, the superior thyroid artery gives rise to the superior laryngeal arteries, which penetrate the thyrohyoid membrane to gain entrance to the interior of the larynx. The artery supplies most of the tissues of the larynx above the vocal folds, including almost all of the laryngeal muscles.

Inferior laryngeal artery. Branching from the thyrocervical trunk, the inferior laryngeal artery supplies the region inferior to the vocal folds.

Venous return from the larynx occurs via the **superior and inferior laryngeal veins**, which are tributaries of the superior and inferior thyroid veins, respectively. The superior thyroid vein drains into the internal jugular vein, whereas the inferior thyroid vein drains into the brachiocephalic vein.

LARYNGEAL INNERVATION

Innervation of the larynx is from CN X via the superior and recurrent laryngeal nerves (Figure 28-3A and B).

SUPERIOR LARYNGEAL NERVE

The superior laryngeal nerve arises from the inferior vagal ganglion and divides into a smaller external laryngeal branch and a larger internal laryngeal branch.

External laryngeal nerve. The external laryngeal nerve courses laterally to the external surface of the larynx to provide motor innervation to the cricothyroid muscle.

Internal laryngeal nerve. The internal laryngeal nerve, along with the superior thyroid artery and vein, pierces the thyrohyoid membrane and provides general sensory innervation to the mucosa superior to the vocal folds.

▽ The **cough reflex** mediates coughing in response to irritation of the laryngeal mucosa above the vocal folds. The internal laryngeal nerve provides the sensory limb of the cough reflex above the vocal folds. ▽

RECURRENT LARYNGEAL NERVE

All intrinsic laryngeal muscles are supplied by CN X. The recurrent laryngeal nerve of CN X innervates all of the intrinsic muscles, except for the cricothyroid muscle, which is supplied by the external laryngeal nerve of CN X. The recurrent laryngeal nerve also provides visceral sensory innervation from the mucosa inferior to the vocal folds. The superior portion of the recurrent laryngeal nerve has a close but variable relationship to the inferior thyroid artery and vein.

▽ The recurrent laryngeal nerve innervates muscles involved in moving the vocal folds. As a result, paralysis of the vocal folds results when there is a **lesion of the recurrent laryngeal nerve**. The voice is weak because the paralyzed vocal fold on the side of the lesion cannot meet the contralateral vocal fold. When bilateral paralysis of the vocal folds occurs, the voice is almost absent.

Hoarseness is the most common symptom of disorders of the larynx, including inflammation or carcinoma of the larynx.

Paralysis of the superior laryngeal nerve causes anesthesia of the laryngeal mucosa superior to the vocal folds. As a result, the protective mechanism designed to keep food out of the larynx (the sensory limb of the cough reflex) is inactive. ▽

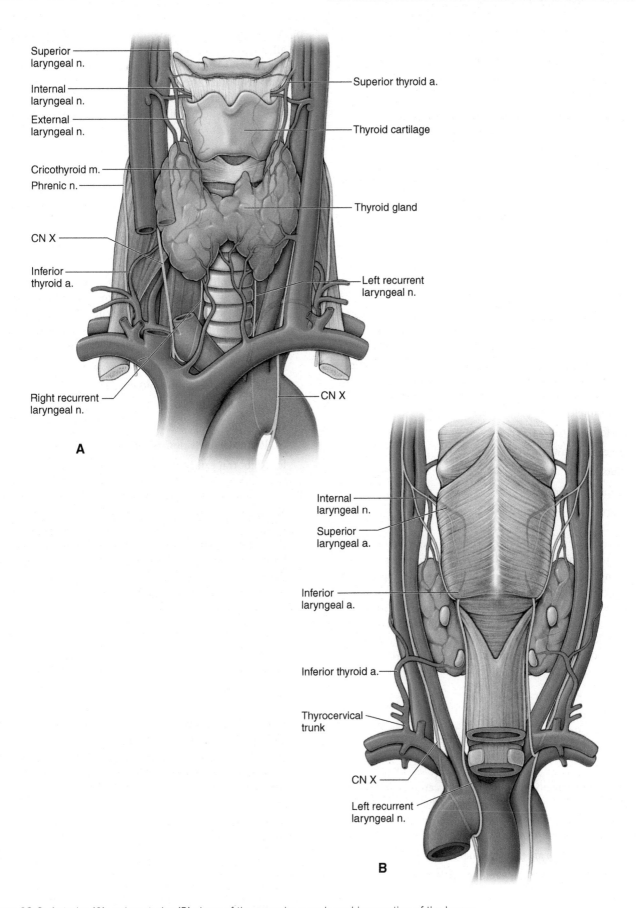

Figure 28-3: Anterior (**A**) and posterior (**B**) views of the vascular supply and innervation of the larynx.

STUDY QUESTIONS

Directions: Each of the numbered items or incomplete statements is followed by lettered options. Select the **one** lettered option that is **best** in each case.

1. A surgeon is performing an endarterectomy on a 64-year-old man who has stenosis of the carotid artery. In approaching the internal carotid artery, the surgeon severs a nerve embedded within the fascia of the carotid sheath. As a result of this complication, which muscle would most likely be paralyzed?

   A. Digastric muscle (anterior belly)
   B. Digastric muscle (posterior belly)
   C. Masseter muscle
   D. Mylohyoid muscle
   E. Sternohyoid muscle
   F. Stylohyoid muscle

2. A 62-year-old woman sees her physician with the complaint of severe pain on the left side of her face. Physical examination shows pallor of the left ear, preauricular region, and tongue. Angiography would most likely confirm complete occlusion of which artery?

   A. External carotid
   B. Internal carotid
   C. Subclavian
   D. Superior thyroid
   E. Vertebral

3. A 46-year-old woman undergoes surgery to fuse (stabilize) the C4–C5 vertebrae due to a herniated disc that is compressing spinal nerves in the neck. The surgeon approaches the cervical vertebrae laterally between the sternocleidomastoid and trapezius muscles. After dissection through the skin and superficial fascia, which of the fascial layers of the neck, from superficial to deep, is the surgeon most likely to dissect through to reach the cervical vertebrae?

   A. Carotid to deep investing
   B. Investing to pretracheal
   C. Investing to prevertebral
   D. Prevertebral to carotid to pretracheal
   E. Pretracheal to prevertebral

4. During a subclavian venipuncture, the cannula is inserted into the patient's neck just superior to the clavicle. Which of the following is the relative position of the subclavian vein in the root of the neck?

   A. Anterior to the anterior scalene muscle
   B. Anterior to the longus capitis muscle
   C. Medial to the trapezius muscle
   D. Medial to the sternocleidomastoid muscle
   E. Posterior to the anterior scalene muscle
   F. Posterior to the subclavian artery

5. A 22-year-old student stands up after studying at her desk all morning. As soon as she stands, she becomes light headed and has to hold the back of her chair because she feels as though she may faint. Within a few seconds, she feels fine and walks to the water fountain. What anatomic structure was responsible for measuring her drop in blood pressure?

   A. Carotid body
   B. Carotid sinus
   C. Cervical plexus
   D. Choroid plexus
   E. Ciliary ganglion
   F. Submandibular ganglion

6. During a physical examination, the physician stands behind the patient to feel the thyroid gland. To palpate the left thyroid lobe, the patient is instructed to look to the left. The action of turning the head to the left is accomplished by contraction of which of the following muscles?

   A. Left anterior scalene muscle
   B. Left sternocleidomastoid muscle
   C. Left trapezius muscle
   D. Right anterior scalene muscle
   E. Right sternocleidomastoid muscle
   F. Right trapezius muscle

7. A surgeon dissects through subcutaneous fat in the neck and identifies lobulated, slightly paler glandular tissue that will be surgically removed. A vein coursing superficial to the gland and an artery coursing deep to the gland are isolated. The hypoglossal nerve is retracted to avoid risk of damage during the procedure. This surgery is most likely occurring in which of the following cervical triangles?

   A. Carotid
   B. Muscular
   C. Posterior
   D. Submandibular
   E. Submental

8. While eating popcorn, a child inhaled a kernel into her laryngeal cavity. The popcorn kernel touched the top of her vocal folds, initiating her cough reflex. Which sensory nerve is responsible for relaying the message to the brain that a popcorn kernel has touched the top of the vocal folds?

   A. External laryngeal nerve
   B. Facial nerve
   C. Glossopharyngeal nerve
   D. Internal laryngeal nerve
   E. Recurrent laryngeal nerve

9. For general surgical procedures, anesthetics and muscle relaxants are used routinely. However, anesthetics and muscle relaxants may decrease nerve stimulation to skeletal muscles, including the intrinsic muscles of the larynx, which results in closure of the vocal folds. Therefore, tracheal intubation is necessary. Which of the following intrinsic muscles of the larynx may be unable to maintain an open glottis because of the anesthetics?

    A. Cricothyroid muscles

    B. Lateral cricoarytenoid muscles

    C. Posterior cricoarytenoid muscles

    D. Thyroarytenoid muscles

    E. Transverse arytenoid muscles

10. A 37-year-old woman complains of cough and hoarseness of several weeks' duration. Upon further examination, the physician notes that the patient has partial paralysis of her vocal cords. Radiographic studies confirm an aneurysm of the aortic arch. Which of the following would account for the relationship between symptoms of cough and hoarseness and this finding?

    A. Direct contact of the aneurysm with the trachea in the superior mediastinum

    B. Injury to that part of the sympathetic chain that provides sensory innervation to the larynx

    C. Irritation of the left phrenic nerve as it crosses the arch of the aorta on its way to the diaphragm

    D. Pressure of the aneurysm on the esophagus in the posterior mediastinum

    E. Pressure on the left recurrent laryngeal nerve, which wraps around the aortic arch

11. A 55-year-old man who has been diagnosed with colon cancer is noted to have a probable metastatic mass in the neck at the thoracic duct. In which region is the metastasis most likely to be located?

    A. Left subclavicular region

    B. Left supraclavicular region

    C. Right subclavicular region

    D. Right supraclavicular region

12. The phrenic nerve in the cervical region courses along the anterior surface of which of the following muscles?

    A. Anterior scalene muscle

    B. Middle scalene muscle

    C. Posterior scalene muscle

    D. Sternocleidomastoid muscle

    E. Trapezius muscle

13. The parietal peritoneum covering the inferior surface of the diaphragm transmits its sensory information via the phrenic nerve. In the case of peritonitis in the parietal peritoneum on the inferior surface of the diaphragm, pain may be referred through which of the following nerves?

    A. Greater occipital

    B. Lesser occipital

    C. Superior division of the ansa cervicalis

    D. Supraclavicular

ANSWERS

1—E: The ansa cervicalis of the cervical plexus is embedded within the fascia of the carotid sheath and, as such, infrahyoid muscles would be affected if a nerve is severed within this fascia. The only infrahyoid muscle in the list of choices is the sternohyoid muscle. The anterior belly of the digastric muscle and mylohyoid muscle are innervated by the mandibular nerve (CN V-3); the stylohyoid and posterior belly of the digastric muscles are innervated by the facial nerve (CN VII). The masseter is a muscle of mastication and is innervated by CN V-3.

2—A: The external carotid artery supplies the ipsilateral face, ear region, and tongue. The internal carotid artery does not supply any of the described areas in the question, but instead enters the base of the skull at the carotid canal to supply the brain. The superior thyroid artery is a branch of the external carotid artery, but it does not supply the face, ear, or tongue. The subclavian and vertebral arteries do not directly distribute to the affected areas.

3—C: Deep to the skin and superficial fascia, the next layer of tissues the surgeon will reach is the investing fascia between the trapezius and the sternocleidomastoid muscles and then the prevertebral fascia around the cervical muscles. The cervical vertebrae are located deep to the prevertebral fascia.

4—A: The subclavian artery and brachial plexus course between the anterior and middle scalene muscles. However, the subclavian vein courses anterior to the anterior scalene muscle.

5—B: The carotid sinus is a baroreceptor that measures blood pressure to the brain. The carotid body is a chemoreceptor that measures blood oxygen concentration. The choroid plexus has nothing to do with blood pressure; it produces cerebrospinal fluid in the ventricular system. The ciliary and the submandibular ganglia are parasympathetic ganglia in the orbit and oral cavity, respectively. The circle of Willis is the anastomosis between the paired internal carotid and vertebral arteries around the pituitary gland.

6—E: Unilateral contraction of the sternocleidomastoid muscle results in the head rotating to the contralateral side. Therefore, contraction of the right sternocleidomastoid muscle results in the head rotating to the left. Unilateral contraction of the trapezius muscle results in minor rotation, but it is not the primary muscle responsible for this movement. The anterior scalene muscle attaches between rib 1 and the cervical vertebrae and, therefore, will not directly move the skull.

7—D: The question outlines the course of the facial vessels in relation to the submandibular salivary gland (the vein is superficial and the artery is deep). The hypoglossal nerve (CN XII) courses within the submandibular triangle. Therefore, the relation of the facial vessels to the submandibular gland and identification of CN XII indicate the location of the surgery within the submandibular triangle.

8—D: The internal laryngeal nerve is the branch of the vagus nerve (CN X) that provides general sensory innervation to the laryngopharynx and mucosal lining superior to the vocal folds. As such, when a stimulus touches the superior element of the vocal folds, the sensory stimulation is conducted along the internal laryngeal nerve to the brainstem, initiating the cough reflex. The external laryngeal nerve innervates the cricothyroideus muscle. The facial nerve (CN VII) innervates muscles of facial expression. The glossopharyngeal nerve (CN IX) provides sensory innervation for the oropharynx and the posterior third of the tongue. The recurrent laryngeal nerve provides visceral sensory innervation for the mucosal lining inferior to the vocal folds.

9—C: The posterior cricoarytenoid is the only muscle in the list of choices, which, when stimulated to contract, will open the vocal folds and therefore open the glottis. The other muscles (i.e., cricothyroid, lateral cricoarytenoid, thyroarytenoid, and the transverse arytenoids) will either tense or close the vocal folds.

10—E: The left recurrent laryngeal nerve courses back up the neck between the trachea and esophagus and will provide motor innervation to all laryngeal muscles (except the cricothyroideus). Therefore, pressure from the aneurysm may inhibit conduction of motor impulses and, therefore, result in paralysis of the laryngeal muscles.

11—B: The thoracic duct collects lymph from all regions of the body (including the colon), except for the right side of the head, neck, and right arm. Therefore, if there were a mass, it could manifest in the root of the neck where the thoracic duct enters the junction of the left internal jugular and subclavian veins in the supraclavicular region.

12—A: The phrenic nerve courses along the anterior surface of the anterior scalene muscle en route to the thoracic cavity.

13—D: The phrenic nerve consists of contributions from spinal nerve levels C3 to C5. Therefore, when sensory information comes from the parietal peritoneum on the inferior diaphragmatic surface, it may refer through spinal nerves at the same levels. Therefore, the supraclavicular nerve shares levels with the C3 and C4 levels. The greater and lesser occipital nerves both originate at the C2 level, and the superior division of the ansa cervicalis originates from the C1 level.

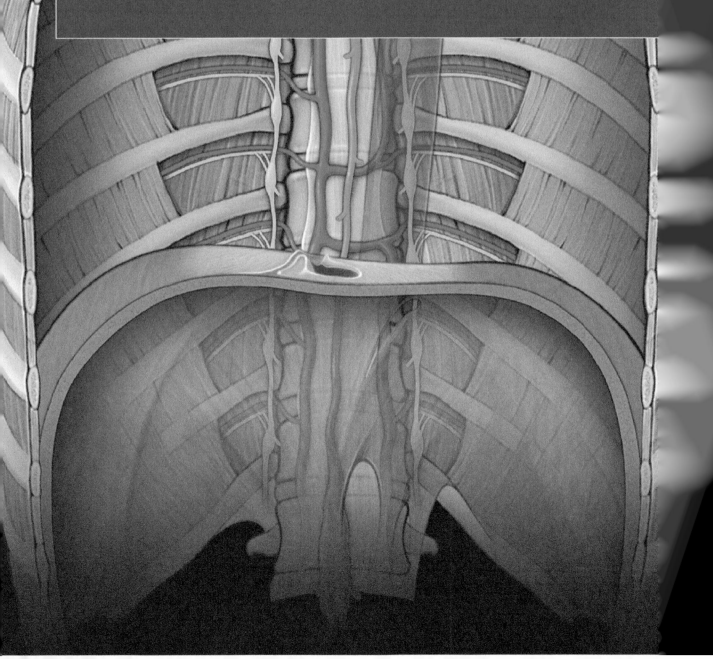

SECTION 6

UPPER LIMB

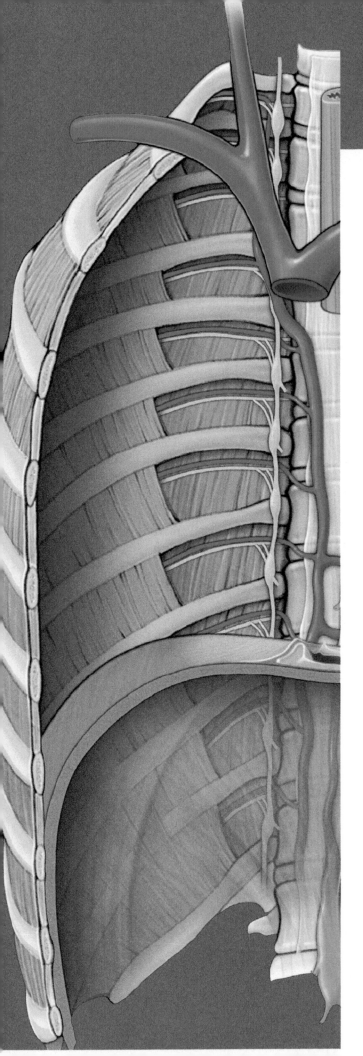

CHAPTER 29

OVERVIEW OF THE UPPER LIMB

BONES OF THE SHOULDER AND ARM

BIG PICTURE

The bones of the skeleton provide a framework to which soft tissues (e.g., muscles) can attach. The bony structure of the shoulder and arm, from proximal to distal, consists of the clavicle, scapula, and humerus (Figure 29-1A). Synovial joints and ligaments connect bone to bone.

CLAVICLE

The **clavicle**, or **collarbone**, is the only bony attachment between the upper limb and the axial skeleton (Figure 29-1B and C). It is superficial along its entire length and shaped like an "**S**." The clavicle provides an attachment for muscles that connect the clavicle to the trunk and the upper limb. The following **landmarks** are found on the clavicle:

- **Acromial end.** Articulates laterally with the acromion of the scapula and forms the **acromioclavicular joint**.
- **Sternal end.** Articulates medially with the manubrium and forms the **sternoclavicular joint**.
- **Conoid tubercle.** Located on the inferior surface of the lateral clavicle and serves as an attachment for the **coracoclavicular ligament**.

SCAPULA

The **scapula**, or **shoulder blade**, is a large, flat triangular bone with **three angles** (lateral, superior, and inferior), **three borders** (superior, lateral, and medial), **two surfaces** (costal and posterior), and **three processes** (acromion, spine, and coracoid) (Figure 29-1D and E). The following **landmarks** are found on the scapula:

- **Subscapular fossa.** Located anteriorly and characterized by a shallow, concave fossa. Because the subscapular fossa glides upon the ribs, it is also known as the **costal surface**.
- **Acromion.** A relatively large projection of the anterolateral surface of the spine; the acromion arches over the glenohumeral joint and articulates with the clavicle.
- **Spine.** Very prominent and palpable; the spine subdivides the posterior surface of the scapula into a small supraspinous fossa and a larger infraspinous fossa.
- **Supraspinous fossa.** Located on the posterior surface of the scapula and superior to the spine of the scapula.
- **Infraspinous fossa.** Located on the posterior surface of the scapula and inferior to the spine of the scapula.
- **Suprascapular notch.** A small notch medial to the root of the coracoid process where the suprascapular nerve, artery, and vein course.
- **Glenoid cavity (fossa).** A shallow cavity that articulates with the head of the humerus to form the **glenohumeral joint**.

- **Supraglenoid tubercle.** Located superior to the glenoid cavity and serves as the attachment for the long head of the biceps brachii muscle.
- **Infraglenoid tubercle.** Located inferior to the glenoid cavity and serves as the attachment for the long head of the triceps brachii muscle.
- **Coracoid process.** A prominent and palpable hook-like structure inferior to the clavicle. The coracoid process serves as an attachment for the pectoralis minor, coracobrachialis, and short head of the biceps brachii muscles.

HUMERUS

The humerus is the longest bone of the arm and is characterized by many distinct features that help to allow the upper extremity to move through a significant range of motion. The following **landmarks** are found on the humerus (Figure 29-1F and G):

- **Head.** A ball-shaped structure that articulates with the glenoid cavity.
- **Anatomical neck.** Formed by a narrow constriction immediately distal to the head of the humerus.
- **Surgical neck.** Lies distal to the anatomical neck and tubercles of the humerus. The axillary nerve and the posterior humeral circumflex artery course into the posterior compartment of the arm, deep to the surgical neck.
- **Greater and lesser tubercles.** Enlarged areas for muscle attachments.
- **Intertubercular (bicipital) groove.** A deep sulcus between the greater and lesser tubercles, where the long head of the biceps brachii tendon courses en route to the supraglenoid tubercle.
- **Radial (spiral) groove.** A distinct groove on the posterior surface of the humerus, where the radial nerve and the deep brachial artery course.
- **Deltoid tuberosity.** A large V-shaped protrusion on the lateral surface of the humerus, midway along its length where the deltoid muscle attaches.
- **Lateral epicondyle.** Located on the distal lateral end of the humerus and provides an attachment surface for the posterior forearm muscles (extensors).
- **Medial epicondyle.** Located on the distal medial end of the humerus and provides an attachment surface for the anterior forearm muscles (flexors).
- **Trochlea.** Characterized by a pulley shape; it helps to guide the hinge joint. The trochlea of the humerus articulates with the trochlear notch of the ulna.
- **Capitulum.** Characterized by its oval, convex shape for articulation with the radial head.
- **Coronoid fossa.** Located on the distal anterior surface of the humerus, where the coronoid process of the ulna articulates.
- **Olecranon fossa.** Located on the distal posterior surface of the humerus, where the olecranon process of the ulna articulates.

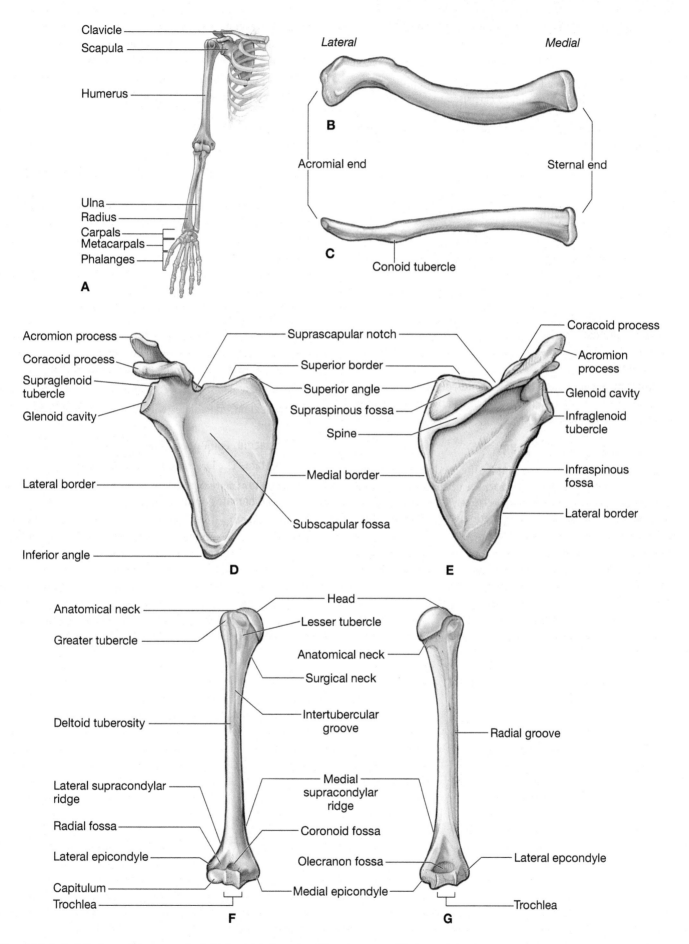

Figure 29-1: A. Osteology of the upper limb (right side). Superior (**B**) and anterior (**C**) views of the clavicle. Anterior (**D**) and posterior (**E**) views of the scapula. Anterior (**F**) and posterior (**G**) views of the humerus.

BONES OF THE FOREARM AND HAND

BIG PICTURE

The bony structure of the forearm and hand, from proximal to distal, consists of the radius, ulna, 8 carpals, 5 metacarpals, and 14 phalanges (Figure 29-2A). The radius and ulna are bound together by a tough fibrous sheath known as the **interosseous membrane**.

RADIUS

In the anatomic position, the **radius** is the lateral bone of the forearm. It articulates with the capitulum of the humerus and with the ulna. The radius is primarily a bone of movement in the forearm during rotation (supination and pronation) relative to the fixed ulna. The following **landmarks** are found on the radius (Figure 29-2B):

- **Head.** A disc-shaped structure that enables the synovial pivot joint in the forearm.
- **Radial tuberosity.** A swelling inferior to the radial neck on the medial surface that serves as an attachment for the biceps brachii muscle.
- **Radial styloid process.** Prominent and palpable process on the distal and lateral end that serves as an attachment for the brachioradialis muscle.

ULNA

In the anatomic position, the **ulna** is the medial bone of the forearm. It articulates with the trochlea of the humerus and with the radius. The ulna remains relatively fixed during forearm rotation of pronation and supination. The following **landmarks** are found on the ulna (Figure 29-2B):

- **Olecranon process.** A large posterior projection that contributes to the trochlear notch and articulates with the humerus (trochlea and the olecranon fossa).
- **Coronoid process.** A small anterior projection that contributes to the trochlear notch and articulates with the trochlea and the coronoid fossa of the humerus.

- **Trochlear notch.** A large notch on the proximal end of the ulna that is formed by the olecranon and coronoid processes. It articulates with the trochlea on the humerus.
- **Ulnar head.** A distal rounded surface at the end of the ulna.
- **Ulnar styloid process.** A palpable distal projection from the dorsal medial ulna.

HAND

The hand is subdivided into the carpus (wrist), metacarpus, and digits (Figure 29-2C).

- **Carpus.** Formed by eight small carpal bones arranged as a proximal row and a distal row, with each row consisting of four bones.
  - **Proximal row**
    - **Pisiform.** A prominent pea-shaped sesamoid bone lying in the tendon of the flexor carpi ulnaris muscle.
    - **Triquetrum.** Three-sided bone.
    - **Lunate.** Characterized by its crescent shape.
    - **Scaphoid.** Most commonly fractured carpal bone located in the floor of the **anatomical snuffbox**.
  - **Distal row**
    - **Trapezium.** Articulates with the metacarpal bone of the thumb.
    - **Trapezoid.** Four-sided bone.
    - **Capitate.** The largest of the carpal bones.
    - **Hamate.** Characterized by a prominent hook (the **hook of the hamate**) on its palmar surface.
- **Metacarpals.** Each of the five metacarpal bones is related to one digit. The first metacarpal is related to the thumb (digit 1), and metacarpals 2 through 5 are related to the index, middle, ring, and little finger, respectively.
- **Phalanges.** The phalanges are the bones of the five digits (numbered 1–5, beginning at the thumb). Digits 2 through 5 consist of a proximal, a middle, and a distal phalanx. Digit 1 (thumb) contains only a proximal and a distal phalanx.

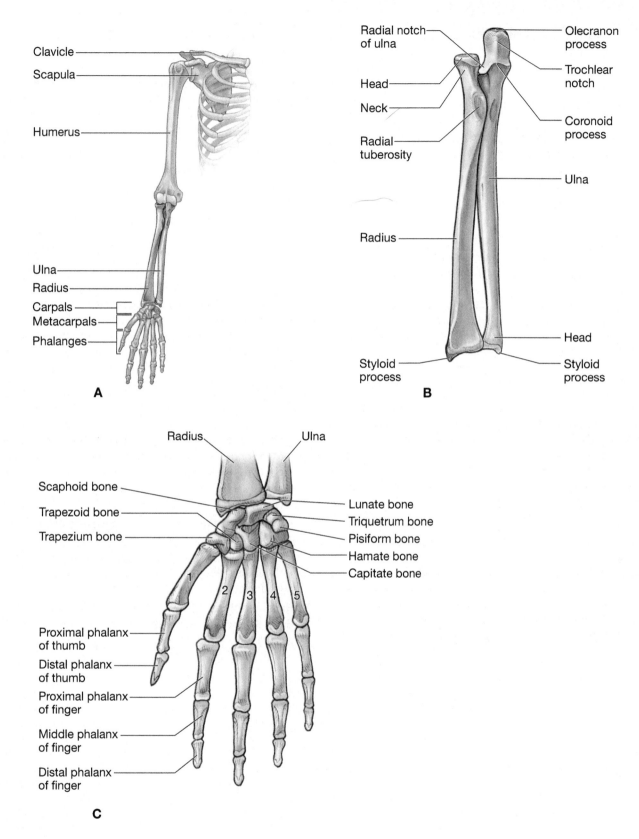

Figure 29-2: A. Osteology of the upper limb. **B**. Radius and ulna. **C**. Hand.

FASCIAL PLANES AND MUSCLES

BIG PICTURE

Two fascial layers, defined as the superficial and the deep fascia, lie between the skin and the bone of the upper limb. The deep fascia divides the upper limb into anterior and posterior compartments. Muscles are organized into these compartments and have common attachments, innervations, and actions.

FASCIA OF THE UPPER LIMB

The upper limb consists of superficial and deep fascia (Figure 29-3A).

- **Superficial fascia.** Referred to as the subcutaneous or hypodermis layer; the superficial fascia is located deep to the skin and primarily contains fat, superficial veins, lymphatics, and cutaneous nerves.
- **Deep fascia.** Lies deep to the superficial fascia; the deep fascia primarily contains muscles, nerves, vessels, and lymphatics. The deep fascia of the upper limb is a continuation of the deep fascia covering the deltoid and pectoralis major muscles. It extends distally and gives off intermuscular septae, which extend to the bones, dividing the arm and forearm into anterior and posterior compartments. Each compartment contains muscles that perform similar movements and have a common innervation.

MUSCLES OF THE UPPER LIMB

The muscles of the upper limb can be organized into the following groups (Figure 29-3B):

- **Scapular muscles.** The muscles of the shoulder are primarily responsible for stability and movement of the scapulothoracic and glenohumeral joints. Muscular stability of the scapula is important because of the lack of bony stability. These muscles consist of the trapezius, deltoid, rhomboid major, rhomboid minor, serratus anterior, levator scapulae, pectoralis minor, and subclavius.
- **Rotator cuff muscles.** These muscles are considered a cuff because the inserting tendons blend with the **glenohumeral joint capsule** and provide stability and movement to the joint. These muscles consist of the supraspinatus, infraspinatus, teres minor, and subscapularis.
- **Intertubercular groove muscles.** The muscles of the intertubercular sulcus attach proximally to the scapula (pectoralis major and teres major) or the thorax (latissimus dorsi) and cross the anterior glenohumeral joint to attach to the humerus, medial to the intertubercular sulcus.
- **Arm muscles.** The deep fascia divides the arm into anterior and posterior compartments, with common actions and innervation.
  - **Muscles of the anterior compartment of the arm.** Include the coracobrachialis, biceps brachii, and brachialis muscles. The muscles of the anterior compartment of the arm share common actions (flexion of the glenohumeral joint and/or elbow) and innervation (musculocutaneous nerve).
  - **Muscles of the posterior compartment of the arm.** Consist of the triceps brachii muscle. The triceps brachii is a three-headed muscle that extends from the glenohumeral joint to the elbow and receives motor innervation via the radial nerve.
- **Forearm muscles.** The deep fascia divides the forearm into anterior and posterior compartments with common attachments, actions, and innervation.
  - **Muscles of the anterior compartment of the forearm.** Many of these muscles share a common origin (medial epicondyle of the humerus), common actions (flexion of elbow, wrist, and digits), and common innervation (median and ulnar nerves).
  - **Muscles of the posterior compartment of the forearm.** Many of these muscles share a common origin (lateral epicondyle of the humerus), common actions (extension of the elbow, wrist, and digits), and common innervation (radial nerve).
- **Hand muscles.** The intrinsic muscles of the hand consist of those that act on the thumb (**thenar muscles**), the little finger (**hypothenar muscles**), and lumbricals, dorsal interossei, and palmar interossei muscles.

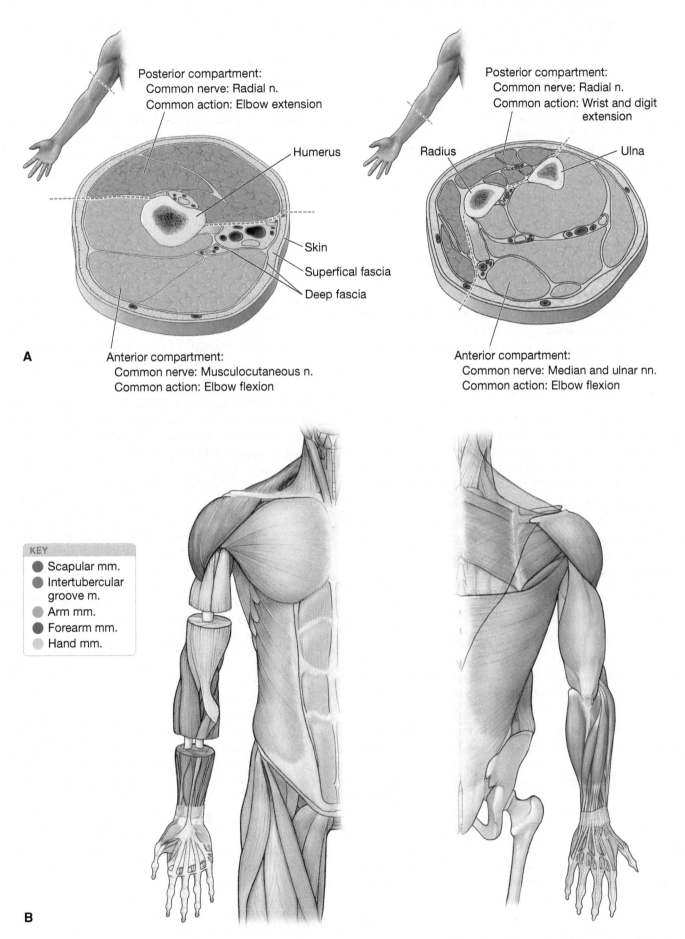

Figure 29-3: A. Cross-section of the arm and forearm showing the anterior compartments (flexors) and the posterior compartments (extensors). **B**. Upper limb divided into compartments.

INNERVATION OF THE UPPER LIMB BY THE BRACHIAL PLEXUS

BIG PICTURE

The upper limb is innervated by anterior rami originating from spinal nerve levels C5–T1. These rami form a network of nerves referred to as the **brachial plexus** (Figure 29-4). The brachial plexus extends from the neck and courses distally through the axilla, providing motor and sensory innervation to the upper limb. As the brachial plexus courses distally, it forms roots, trunks, divisions, cords, and terminal branches.

ROOTS OF THE BRACHIAL PLEXUS

The five **roots** are the ventral rami originating from spinal nerve levels **C5–T1**. The roots course between the **anterior and middle scalene muscles**, along with the **subclavian artery**. The following two nerves originate from the roots:

- **Dorsal scapular nerve (C5)**. Innervates the rhomboid and levator scapular muscles.
- **Long thoracic nerve (C5–C7)**. Innervates the serratus anterior muscle.

TRUNKS OF THE BRACHIAL PLEXUS

Once the roots exit between the anterior and middle scalene muscles, they unite to form three trunks.

- **Superior trunk.** Formed from the union of the C5 and C6 roots at the lateral border of the middle scalene muscle. The superior trunk gives rise to the following two nerves:
  - **Nerve to subclavius (C5).** Provides innervation to the subclavius muscle.
  - **Suprascapular nerve (C5–C6).** Provides innervation to the supraspinatus and infraspinatus muscles.
- **Middle trunk.** A continuation of the C7 root.
- **Inferior trunk.** Located posterior to the anterior scalene and formed from the union of C8 and T1 roots.

DIVISIONS OF THE BRACHIAL PLEXUS

The trunks divide into three anterior and three posterior divisions, which are associated with the ventral and dorsal musculature, respectively. The **axillary artery** separates the anterior and posterior divisions of the brachial plexus deep to the clavicle.

- **Anterior divisions.** Give rise to the nerves that will eventually innervate the flexors of the arm and forearm in the anterior compartments.
- **Posterior divisions.** Give rise to the nerves that will eventually innervate the extensors of the arm and forearm in the posterior compartments.

CORDS OF THE BRACHIAL PLEXUS

The anterior and posterior divisions form three cords, named according to their anatomic position relative to the axillary artery.

- **Lateral cord.** Gives rise to the **lateral pectoral nerve (C5–C7)**, which innervates the pectoralis minor muscle.
- **Medial cord.** Gives rise to the following nerves:
  - **Medial pectoral nerve (C8–T1).** Innervates the pectoralis major and pectoralis minor muscles.
  - **Medial cutaneous nerve of the arm (C8–T1).** Provides cutaneous innervation to the medial surface of the arm.
  - **Medial cutaneous nerve of the forearm (C8–T1).** Provides cutaneous innervation to the medial surface of the forearm.
- **Posterior cord.** Gives rise to the following nerves:
  - **Upper subscapular nerve (C5–C6).** Innervates the subscapularis muscle.
  - **Thoracodorsal** (middle subscapular) **nerve (C6–C8).** Innervates the latissimus dorsi muscle.
  - **Lower subscapular nerve (C5–C6).** Innervates the subscapularis and teres major muscles.

TERMINAL BRANCHES OF THE BRACHIAL PLEXUS

The brachial plexus terminates in the following branches, discussed briefly below and in greater detail in Chapters 30 to 33.

- **Musculocutaneous nerve (C5–C7).** Provides most of the motor innervation to the anterior compartment of the arm and the sensory innervation to the lateral forearm.
- **Median nerve (C6–T1).** Provides most of the innervation to the anterior forearm, excluding one and one half muscles, the flexor carpi ulnaris and the ulnar half of the flexor digitorum profundus muscle, which are innervated by the ulnar nerve. The median nerve continues into the hand to innervate the thenar eminence and lumbricals 1 and 2. It provides cutaneous innervation to the medial palmar side of the hand and the palmar surface of digits 1 through 3 and half of digit 4.
- **Ulnar nerve (C7–T1).** Provides motor innervation to one and one half muscles of the forearm (the flexor carpi ulnaris and the ulnar half of the flexor digitorum profundus muscle) and all of the hand musculature except the thenar eminence and lumbricals 1 and 2. The ulnar nerve provides cutaneous innervation to the medial half of digit 4, to digit 5, and to the medial palmar surface of the hand.
- **Radial nerve (C5–T1).** Provides motor innervation to the posterior compartment of the arm and forearm. The radial nerve also provides cutaneous innervation to the posterior and inferior lateral portion of the arm, posterior forearm, and lateral dorsum of the hand and the dorsum of digits 1 through 3 and half of digit 4.
- **Axial nerve (C5–C6).** Provides motor innervation to the deltoid and the teres minor muscles. The axial nerve also provides cutaneous innervation to the superior portion of the lateral arm.

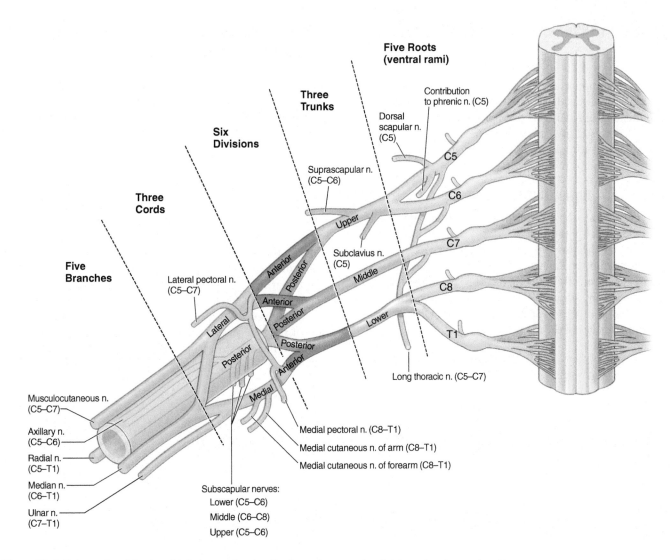

Figure 29-4: Schematic of the brachial plexus showing the branches, cords, divisions, trunks, and roots.

SENSORY INNERVATION OF THE UPPER LIMB

BIG PICTURE

Sensory innervation to the upper limb is provided by branches of the brachial plexus that provide **cutaneous** and **dermatomal distributions**. The cutaneous distribution consists of multiple nerve root levels carried by a single nerve, whereas the dermatomal distribution consists of innervation from a single nerve root carried by multiple nerves.

CUTANEOUS DISTRIBUTION

Peripheral nerves supply **cutaneous innervation** to an area of skin on the surface of the body (Figure 29-5A). For example, the lateral part of the forearm and hand receives its cutaneous innervation from the C6 nerve root (Figure 29-5B). However, the sensory neurons from C6 are distributed to the lateral part of the forearm and hand via three cutaneous nerves: the lateral cutaneous nerve of the forearm, the superficial radial nerve, and the digital branches of the median nerve.

DERMATOMAL DISTRIBUTION

The **dermatomal innervation** is from multiple peripheral nerves that carry the same spinal levels (Figure 29-5C). For example, the spinal root of C6 provides cutaneous innervation to the lateral part of the forearm and hand. In other words, all sensory neurons originating in this region will terminate at the C6 spinal nerve level (Figure 29-5D).

▽   The two sensory distributions (i.e., cutaneous and dermatomal) provide different sensory patterns to the upper limb although the same nerve root supplies both distributions. This is especially evident when examining a patient with nerve root injuries. A **lesion to a cutaneous nerve** will present differently than will a **lesion to a spinal root** (dermatomal distribution). If the lateral cutaneous nerve of the forearm is damaged, a loss of sensation will occur on only part of the lateral side of the forearm, whereas sensation to the side of the hand will remain intact. Therefore, the result is partial loss of the C6 dermatome. In contrast, if the C6 nerve root is severed, the result is loss of all of the sensory innervation of the skin served by the C6 dermatome in the forearm and hand. ▼

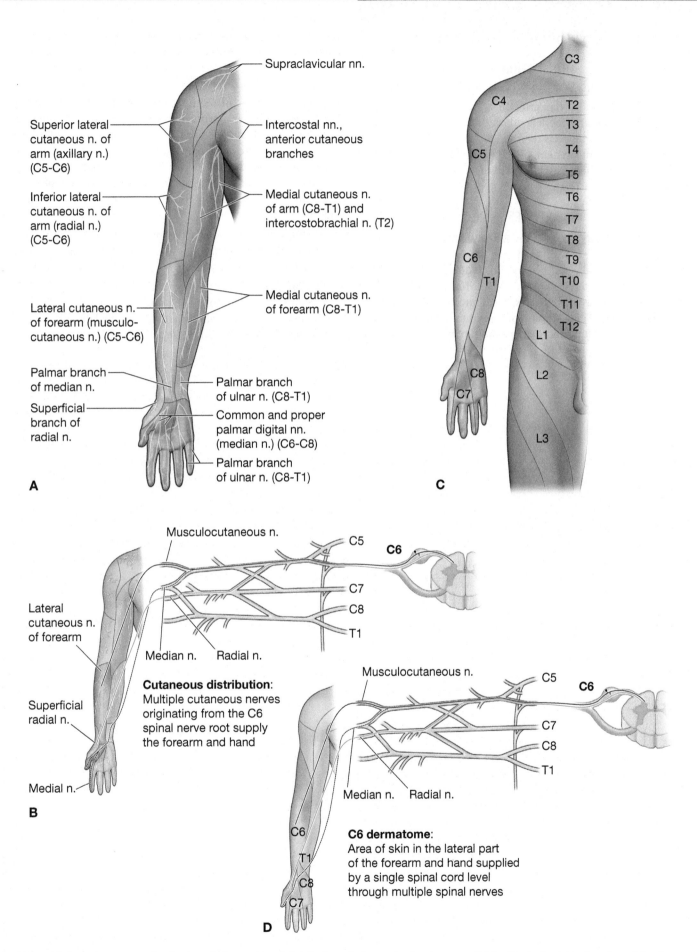

Figure 29-5: A. Cutaneous distribution. **B**. Dermatomal distribution. **C**. Example of cutaneous distribution to the lateral portion of the forearm and hand. **D**. Example of C6 dermatome.

VASCULARIZATION OF THE UPPER LIMB

BIG PICTURE

Blood supply to both upper limbs is provided by the subclavian arteries. The right subclavian artery is a branch from the brachiocephalic artery, and the left subclavian artery is a branch from the aortic arch. Blood is returned to the heart via a superficial and a deep venous system. Given that the deep venous system follows the arteries, most deep veins have the same name as their accompanying arteries.

ARTERIES

The **subclavian artery** becomes the **axillary artery** as it crosses over the lateral border of the first rib (Figure 29-6A). The axillary artery continues distally and becomes the **brachial artery** at the inferior border of the teres major muscle. The brachial artery continues distally, passing over the elbow, and bifurcates into the **ulnar** and **radial arteries**. These arteries continue into the hand, where they form the **superficial** and **deep palmar arches**. Throughout the upper limb, smaller vessels branch from the larger vessels to supply structures such as muscle, bone, and joints.

VEINS

Generally, the veins of the upper limb drain into veins of the back, neck, axilla, and arm, and eventually reach the superior vena cava. The **deep veins** follow the arteries and usually consist of two or more veins that wrap around the accompanying artery (**vena comitantes**) (Figure 29-6B). The **superficial veins** originate in the hand and primarily consist of the **basilic** and **cephalic veins**, joined in the elbow region by the **median cubital vein** (Figure 29-6C).

- **Basilic vein.** Travels along the medial side of the arm until it reaches the inferior border of the teres major muscle, where it becomes the axillary vein.

- **Cephalic vein.** Runs more laterally along the arm and remains superficial until it joins the axillary vein in the shoulder. After the axillary vein crosses the lateral border of the first rib, it becomes the subclavian vein.

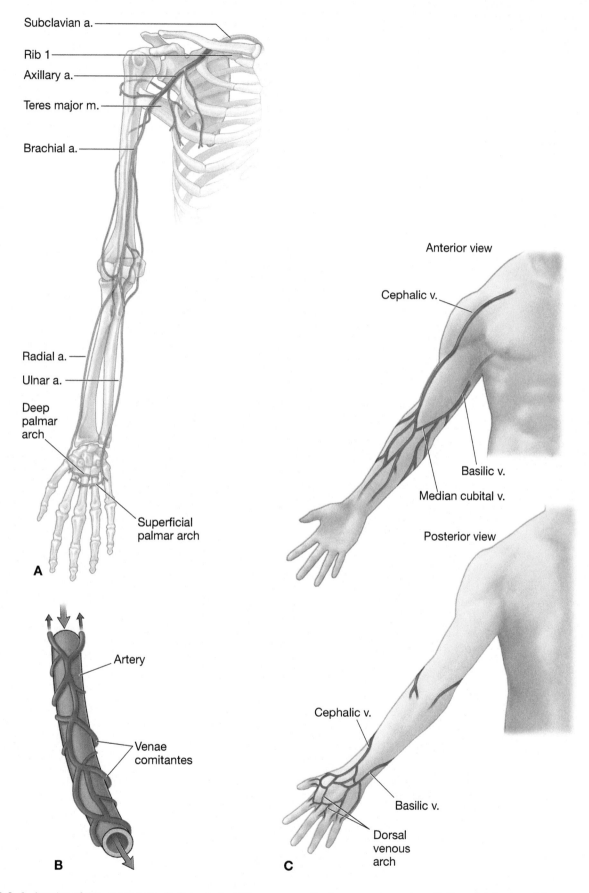

Figure 29-6: A. Arteries of the upper limb. **B**. Schematic of the vena comitantes around an artery. **C**. Superficial veins of the upper limb.

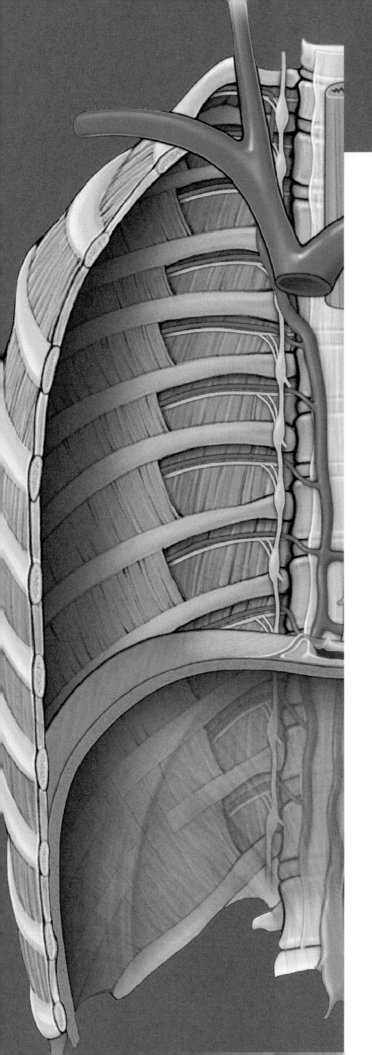

SHOULDER AND AXILLA

SHOULDER COMPLEX

BIG PICTURE

The combined joints connecting the **scapula** (scapulothoracic joint), **clavicle** (sternoclavicular and acromioclavicular joints), and **humerus** (glenohumeral joint) form the shoulder complex and anchor the upper limb to the trunk. The only boney stability of the upper limb to the trunk is through the connection between the clavicle and the sternum. The remaining stability of the shoulder complex depends on muscles, and as a result, the shoulder complex has a wide range of motion.

SCAPULAR SUPPORT

To best understand the actions of the scapula, it is important to understand the scapulothoracic, acromioclavicular, and sternoclavicular joints (Figure 30-1A).

- **Scapulothoracic joint.** Formed by the articulation of the scapula with the thoracic wall through the scapular muscles, including the trapezius and the serratus anterior muscles. The scapulothoracic joint is not considered a true anatomic joint; as such, it is frequently referred to as a "pseudo joint" because it does not contain the typical joint characteristics (e.g., synovial fluid and cartilage).
- **Acromioclavicular joint.** A synovial joint formed by the articulations of the scapula (acromion) and the clavicle.
- **Sternoclavicular joint.** A synovial joint formed by the articulations between the clavicle and sternum.

The scapulothoracic, sternoclavicular, and acromioclavicular joints are interdependent. For example, if motion occurs at one joint (e.g., the scapula elevates), the movement will directly affect the other two joints. Therefore, the motions produced frequently involve more than a single joint. Although the scapular movements include the scapulothoracic, acromioclavicular, and sternoclavicular joints, we will refer only to the scapula in the following text.

MOVEMENTS OF THE SCAPULA

The following terms describe the movements of the scapula (Figure 30-1B):

- **Protraction.** Anterior movement of the scapula on the thoracic wall (e.g., reaching in front of the body).
- **Retraction.** Posterior movement of the scapula on the thoracic wall (e.g., squeezing the shoulder blades together).
- **Elevation.** Raising the entire scapula in a superior direction without rotation.
- **Depression.** Lowering the entire scapula in an inferior direction without rotation.
- **Upward rotation.** Named according to the upward rotation and direction that the glenoid fossa faces.
- **Downward rotation.** Named according to the downward rotation and direction that the glenoid fossa faces.

ACTIONS OF THE GLENOHUMERAL JOINT

The glenohumeral joint is a synovial, ball-and-socket joint. The "ball" is the head of the humerus, and the "socket" is the glenoid fossa of the scapula. The glenohumeral joint is considered to be the most mobile joint in the body and produces the following actions (Figure 30-1C and D):

- **Flexion.** Movement anterior in the sagittal plane.
- **Extension.** Movement posterior in the sagittal plane.
- **Abduction.** Movement away from the body in the frontal plane.
- **Adduction.** Movement toward the body in the frontal plane.
- **Medial rotation.** Movement toward the body in the transverse or axial plane.
- **Lateral rotation.** Movement away from the body in the transverse or axial plane.
- **Circumduction.** A combination of glenohumeral joint motions that produce a circular motion.

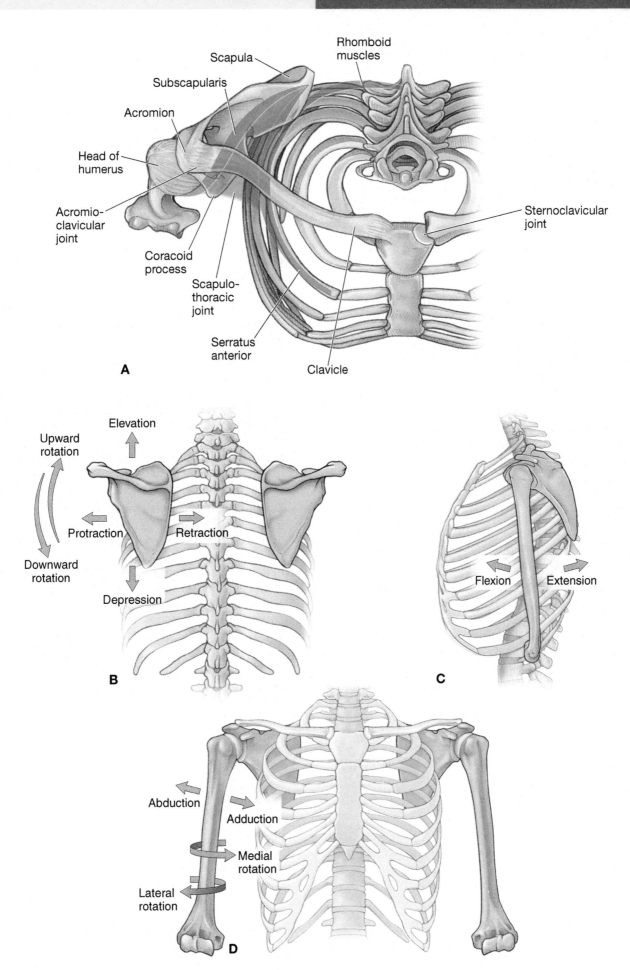

Figure 30-1: A. Superior view of the scapulothoracic joint. **B.** Scapular actions. **C, D.** Glenohumeral joint actions.

MUSCLES OF THE SHOULDER COMPLEX

BIG PICTURE

The musculature of the scapulothoracic joint is responsible primarily for the stability of the scapula to provide a stable base for the muscles acting on the glenohumeral joint. In other words, the upper limb must have proximal stability to have distal mobility. In addition, the scapulothoracic joint works in conjunction with the glenohumeral joint to produce movements of the shoulder. For example, the available range of motion for shoulder abduction is 180 degrees. This motion is produced by approximately 120 degrees from abduction at the glenohumeral joint and by approximately 60 degrees of upward rotation from the scapulothoracic joint.

MUSCLES OF THE SCAPULA

The following are muscles of the scapula (Figure 30-2A–C and Table 30–1):

Trapezius muscle. Attaches to the occipital bone, nuchal ligament, spinous processes of C7–T12, spine of the scapula, acromion, and the clavicle. The trapezius muscle is a triangular shape and has the following **muscle fiber orientations:**

- **Superior fibers.** Course obliquely from the occipital bone and upper nuchal ligament to the scapula, producing scapular elevation and upward rotation.

- **Middle fibers.** Course horizontally from the lower nuchal ligament and thoracic vertebrae to the scapula, producing scapular retraction.

- **Inferior fibers.** Course superiorly from the lower thoracic vertebrae to the scapula, producing scapular depression and upward rotation.

The multiple fiber orientations of the trapezius muscle stabilize the scapula to the posterior thoracic wall during upper limb movement, which is innervated by the **spinal accessory nerve [cranial nerve (CN) XI]**. The vascular supply is provided by the **superficial branch of the transverse cervical artery**.

Levator scapulae muscle. Located deep to the trapezius muscle and superior to the rhomboid muscles. The levator scapula muscle attaches to the cervical vertebrae (C1–C4) and the superior angle of the scapula, producing elevation and downward rotation of the scapula. The nerve supply is provided by branches of the ventral rami from spinal nerves C3 and C4 and occasionally by C5 via the **dorsal scapular nerve**. The vascular supply is provided by the **deep branch of the transverse cervical artery**.

Rhomboid major and minor muscles. The rhomboid minor is superior to the rhomboid major, with both positioned deep to the trapezius muscle. The rhomboid minor muscle attaches to the spinous processes of C7–T1. The rhomboid major muscle attaches to the spinous processes of T2–T5. Both muscles attach to the medial border of the scapula, resulting in scapular retraction. They are innervated by the **dorsal scapular nerve** (ventral ramus of C5) and the vascular supply from the deep branch of the transverse cervical artery. In some instances, the **dorsal scapular artery** will replace the deep branch of the transverse cervical artery.

Pectoralis minor muscle. Attaches anteriorly on the thoracic skeleton to ribs 3 to 5 and superiorly to the coracoid process of the scapula. The pectoralis minor muscle protracts, depresses, and stabilizes the scapula against the thoracic wall. The **axillary vessels and the brachial plexus** travel posteriorly to the pectoralis minor muscle. The deltoid and pectoral branches of the thoracoacromial trunk and the superior and lateral thoracic arteries provide the vascular supply to this muscle. The **medial pectoral nerve** (C8–T1) provides innervation to this muscle.

Serratus anterior muscle. Attaches to ribs 1 to 8 along the midaxillary line and courses posteriorly to the medial margin of the scapula. The serratus anterior muscle primarily protracts and rotates the scapula and stabilizes the medial border of the scapula against the thoracic wall. The **lateral thoracic arteries** provide the vascular supply to the serratus anterior muscle, and the **long thoracic nerve** provides innervation (ventral rami of C5–C7).

Subclavius muscle. Attaches to the first rib and clavicle. The subclavius muscle depresses the clavicle and provides stability to the sternoclavicular joint. The **nerve to the subclavius muscle** (C5–C6) innervates this muscle.

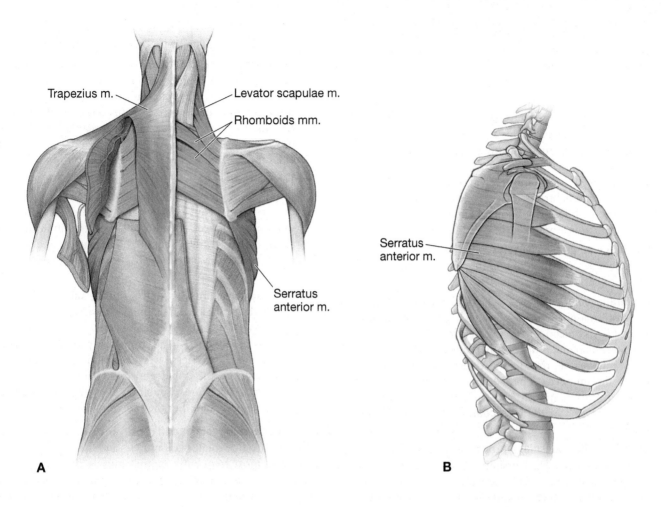

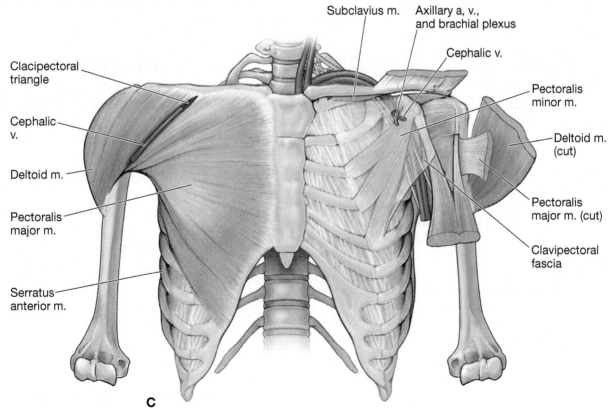

Figure 30-2: A. Back muscles in step dissection. **B**. Lateral view of the thorax. **C**. Anterior view of thoracic muscles.

MUSCLES OF THE GLENOHUMERAL JOINT

The following muscles and muscle groups comprise the muscles of the glenohumeral joint (Figure 30-3A–C and Table 30–2)

Deltoid muscle. The deltoid muscle attaches to the spine of the scapula, acromion, clavicle, and the deltoid tuberosity of the humerus. It has a triangular shape, with the muscle fibers coursing anteriorly to the glenohumeral joint producing flexion, laterally producing abduction, and posteriorly producing extension. The deltoid muscle is innervated by the **axillary nerve** (C5–C6) and receives its blood supply from the thoracoacromial trunk (deltoid and acromial branches), the anterior and posterior humeral circumflex arteries, and the subscapular artery.

Intertubercular groove muscles. This group of muscles is named because of their common insertion into the intertubercular sulcus of the humerus.

• **Pectoralis major muscle.** Attaches to the sternum, clavicle, and costal margins and laterally attaches over the long head of the biceps brachii tendon to insert into the lateral lip of the intertubercular groove of the humerus. The pectoralis major muscle is a prime flexor, adductor, and medial rotator of the humerus. The **medial and lateral pectoral nerves** (ventral rami of C5–T1) provide innervation. The pectoral branch of the **thoracoacromial trunk** provides most of the blood supply to the pectoralis major muscle.

• **Latissimus dorsi muscle.** A broad, flat muscle of the lower region of the back. The latissimus dorsi muscle attaches to the spinous processes of T7 inferiorly to the sacrum via the **thoracolumbar fascia**, and inserts laterally into the intertubercular groove of the humerus. The latissimus dorsi muscle acts on the humerus (arm), causing powerful adduction, extension, and medial rotation of the arm. It is innervated by the **thoracodorsal nerve** (ventral rami of C6–C8) and receives its blood supply from the **thoracodorsal artery** (branch off the axillary artery).

• **Teres major muscle.** Attaches to the inferior angle of the scapula and the medial lip of the intertubercular sulcus. The teres major muscle medially rotates the humerus and is innervated by the **lower subscapular nerve** (C5–C6).

Rotator cuff muscles. The rotator cuff muscles consist of **four muscles** (supraspinatus, infraspinatus, teres minor, and subscapularis) that form a musculotendinous cuff around the glenohumeral joint. The cuff provides muscular support primarily to the anterior, posterior, and superior aspects of the joint (the first letter of each muscle forms an acronym known as **SITS**).

• **Supraspinatus muscle.** Attaches to the supraspinous fossa and courses under the acromion to attach to the greater tubercle of the humerus. Humeral abduction is the primary action of the supraspinatus muscle. The suprascapular nerve (C5–C6) and the suprascapular artery provide innervation and blood supply to the supraspinatus muscle.

• **Infraspinatus muscle.** Attaches to the infraspinous fossa of the scapula and courses posteriorly to the glenohumeral joint to attach to the greater tubercle of the humerus. The primary action of the infraspinatus muscle is lateral rotation of the humerus. The suprascapular nerve (C5–C6) provides innervation, and the suprascapular artery provides the vascular supply to the infraspinatus muscle.

• **Teres minor muscle.** Attaches to the lateral border of the scapula and greater tubercle of the humerus. The teres minor muscle courses posteriorly to the glenohumeral joint and produces lateral rotation of the humerus. The axillary nerve (C5–C6) supplies innervation, and the circumflex scapular artery supplies blood to the teres minor muscle.

• **Subscapularis muscle.** Attaches to the subscapular fossa on the deep side of the scapula. The subscapularis muscle crosses the anterior glenohumeral joint to attach to the lesser tubercle of the humerus, producing medial rotation of the humerus with contraction. The muscle is innervated by the upper and lower subscapular nerves (C5–C6). The suprascapular, axillary, and subscapular arteries supply blood to the subscapularis muscle.

▽ A **tear of the rotator cuff** usually involves a tear of one or more of the rotator cuff muscles and their associated tendons. The most frequently injured muscle or tendon is the supraspinatus muscle. Most injuries occur as the result of overuse (usually repetitive overhead activities). However, traumatic injuries also occur as the result of a shoulder dislocation, lifting injury, or a fall. ▽

AXILLARY BORDERS

The axilla is a passageway that connects the structures that run from the trunk to the arm. The **borders of the axillary region** are as follows:

Anterior border. Pectoralis major and minor muscles and the clavicopectoral fascia.

Posterior border. Subscapularis, teres major, and latissimus dorsi muscles.

Lateral border. Intertubercular groove of the humerus and the coracobrachialis muscle.

Medial border. Serratus anterior (ribs 1–4).

Apex. Rib 1, the clavicle, and the proximal edge of the subscapularis muscle.

Base. Skin, subcutaneous tissue, and axillary fascia that spans from the arm to the thorax.

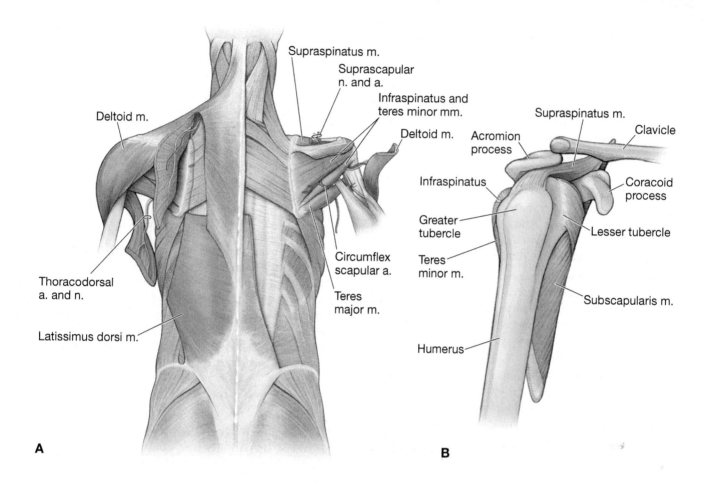

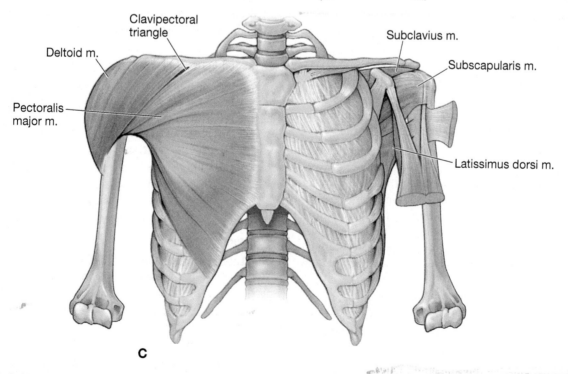

Figure 30-3: A. Muscles of the glenohumeral joint (posterior view). **B**. Lateral view of the rotator cuff muscles supporting the glenohumeral joint. **C**. Muscles of the glenohumeral joint (anterior view).

BRACHIAL PLEXUS OF THE SHOULDER

BIG PICTURE

The upper limb is innervated by the ventral rami from nerve roots C5–T1, which form a network of nerves referred to as the brachial plexus. The brachial plexus is divided into five regions consisting of roots, trucks, divisions, cords, and terminal branches.

ROOTS

The roots of the brachial plexus are a continuation of the ventral rami of C5–T1 and pass between the anterior and middle scalene muscles with the subclavian artery (Figure 30-4A). The following **nerves branch off the roots of the brachial plexus:**

Dorsal scapular nerve (C5). Branches off the C5 nerve root. The dorsal scapular nerve pierces the middle scalene muscle and descends deep to the levator scapulae and rhomboid major and minor muscles, along with the deep branch of the transverse cervical artery. The nerve supplies both rhomboid muscles and occasionally supplies the levator scapulae.

Long thoracic nerve (C5–C7). Branches off the C5–C7 roots, descends posteriorly to the roots of the plexus and the axillary artery and along the lateral surface of the serratus anterior muscle, with the lateral thoracic artery, while supplying the muscle. The long thoracic nerve is one of the few nerves found superficial to the serratus anterior muscle.

▽ Injury to the long thoracic nerve results in paralysis of the serratus anterior muscle. This presents with the medial border of the scapula sticking straight out of the back (winged scapula) ▽

TRUNKS

The trunks of the brachial plexus (**upper, middle, and lower**) emerge laterally between the anterior and middle scalene muscles and descend toward the clavicle (Figure 30-4A). The following **nerves branch off the trunks of the brachial plexus:**

Suprascapular nerve (C5, C6). This branch from the upper trunk passes through the posterior triangle of the neck and the suprascapular foramen, inferior to the transverse scapular ligament (the suprascapular artery and vein pass *superior* to the transverse scapular ligament), entering the supraspinous fossa to supply the supraspinatus muscle. The suprascapular nerve then continues inferiorly through the greater scapular notch to the infraspinous fossa to supply the infraspinatus muscle.

Nerve to the subclavius (C5, C6). Branches off the upper trunk and descends anteriorly to the brachial plexus and subclavian artery and posteriorly to the clavicle to the subclavius muscle.

▽ **Erb's palsy (brachial plexus birth injury)** is caused by a stretch injury of the brachial plexus that occurs during a difficult birth, specifically to nerve roots C5 and C6 and sometimes C7. The severity of the injury varies from a stretch to complete avulsion from the spinal cord. Depending on the severity of the injury, some infants experience complete recovery, while others may not recover. The suprascapular, musculocutaneous, axillary, upper and lower subscapular, long thoracic, and dorsal scapular nerves may be involved, resulting in notable loss or limited function of the supraspinatus, infraspinatus, biceps brachii, brachialis, deltoid, subscapularis, rhomboids, and serratus anterior musculature. In addition, the affected limb will be internally rotated as a result of the lack of muscle support. ▽

▽ Klumple's palsy results from the inferior trunk of the brachial plexus. The intrinsic muscles of the hand are affected and a "claw hand" may result. ▽

DIVISIONS

The divisions of the brachial plexus (**anterior and posterior**) consist of three anterior divisions that travel anteriorly to the axillary artery and give rise to the nerves that innervate the anterior compartments of the limb (flexor muscles). In addition to the three anterior divisions, three posterior divisions travel posteriorly to the axillary artery. The posterior divisions give rise to the nerves that innervate the posterior compartments of the limb (extensor muscles) (Figure 30-4A and B).

CORDS

The cords of the brachial plexus (**medial, lateral, and posterior**) pass posteriorly to the pectoralis minor muscles. The medial and lateral cords arise from the anterior divisions and are named for their relation to the axillary artery; the posterior cord arises from the posterior division and runs posteriorly to the axillary artery. The cords give rise to the terminal branches of the plexus. The following nerves are **branches from the medial, lateral, and posterior cords** of the **brachial plexus** (Figure 30-4A–C):

Medial cord

- **Medial pectoral nerve (C8, T1).** Branches off the medial cord, receiving a contribution from the lateral pectoral nerve. The medial pectoral nerve innervates the pectoralis minor muscle as it pierces it and continues on to innervate the pectoralis major muscle.

- **Medial cutaneous nerve of the arm (C8–T1).** Provides cutaneous innervation to the medial side of the arm.

- **Medial cutaneous nerve of the forearm (C8–T1).** Provides cutaneous innervation to the medial side of the forearm.

Lateral cord

- **Lateral pectoral nerve (C5–C7).** Branches off the lateral cord, sending a branch to the medial pectoral nerve anteriorly to the axillary artery and passing proximally to the pectoralis minor muscle to reach the pectoralis major muscle.

Posterior cord

- **Upper subscapular nerve (C5, C6).** Branches off the posterior cord and enters the anterior surface of the subscapularis muscle.

- **Thoracodorsal nerve (middle subscapular) (C6–C8).** Branches off the posterior cord and descends to the posterolateral thorax to supply the latissimus dorsi muscle.

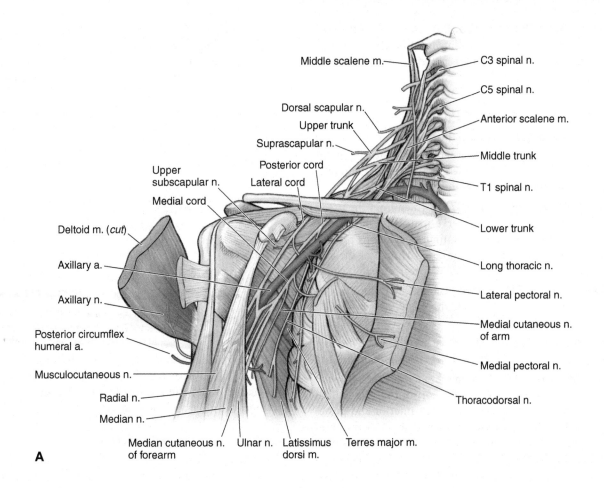

A

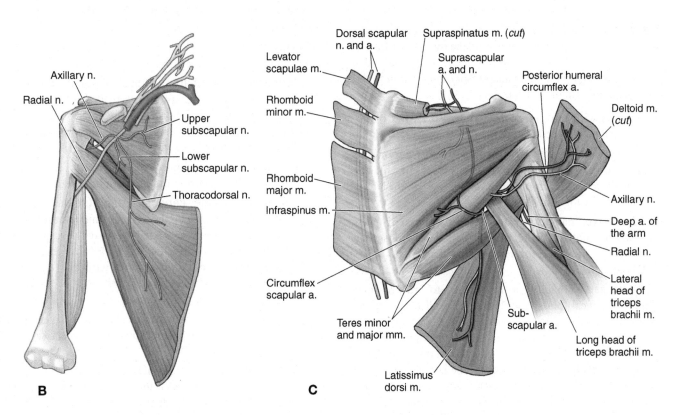

B

C

Figure 30-4: A. Brachial plexus and topography of the axillary artery. **B**. Posterior division of the brachial plexus. **C**. Posterior view of the shoulder joint.

- **Lower subscapular nerve (C5, C6).** Branches off the posterior cord and bifurcates, sending one branch to the anterior surface of the subscapularis muscle and one branch to the anterior surface of the teres major muscle.

TERMINAL BRANCHES

The terminal branches are the most distal region of the brachial plexus (Figure 30-4A–C). They travel distally to provide motor and sensory innervation to both the **anterior (flexor) and posterior (extensor) compartments of the arm and forearm**.

- **Musculocutaneous nerve (C5–C7).** Formed from the lateral cord. The musculocutaneous nerve frequently pierces the coracobrachialis muscle, as it continues distally to provide motor innervation to the anterior compartment of the arm and sensory innervation to the lateral forearm.
- **Median nerve (C6–T1).** Formed from the lateral and medial cords. There are no branches off the median nerve until it reaches the anterior forearm.
- **Ulnar nerve (C7–T1).** Formed from the medial cord. The ulnar nerve gives off no branches until it reaches the anterior forearm.
- **Axillary nerve (C5, C6).** Terminal branch off the posterior cord. The axillary nerve wraps posteriorly through the quadrangular space and then bifurcates into anterior and posterior branches. The posterior branch innervates the teres minor muscle. The anterior branch wraps anteriorly around the humerus, deep to the deltoid, while sending fibers into the muscle. Both branches contribute to the superior lateral cutaneous nerve of the arm, supplying the shoulder joint tissues and most of the skin surrounding the deltoid muscle.
- **Radial nerve (C5–T1).** Formed from the posterior cord. The radial nerve travels posteriorly around the humerus, through the radial groove, to provide motor innervation to the posterior compartment of the arm and sensory innervation to the inferior lateral skin of the arm.

▽ Patients with injuries of the terminal branches of the brachial plexus present with the following symptoms:

- **Musculocutaneous nerve.** Weakness in elbow flexion and supination as a result of paralysis of the biceps brachii and brachialis muscles. Sensory loss may be noted on the lateral surface of the forearm.
- **Radial nerve.** Inability to extend the wrist (**wrist drop**) as a result of paralysis of forearm extensors. Sensory loss may be noted on the dorsal surface of the hand and digits 1, 2, 3 and radial half of digit 4.
- **Median nerve.** Inability to flex the 1st to 3rd proximal interphalangeal joints, and 2nd and 3rd distal interphalangeal joints due to paralysis of the flexor digitorum superficialis and medial slips of the flexor digitorum profundus muscles. Flexion of the 4th and 5th distal interphalangeal joints is unaffected because the lateral slips of the flexor digitorum profundus muscle is supplied by the ulnar nerve. Flexion of the 2nd and 3rd metacarpophalangeal joints is also lost due to paralysis of the 1st and 2nd lumbrical muscles. Therefore, when a patient with a median nerve injury attempts to make a fist the 2nd and 3rd fingers

remain partially extended ("sign of benediction"). Strength in the thumb is also affected due to innervation of the flexor pollicus longus and thenar muscles ("carpal tunnel syndrome"). Sensory loss may be noted on the palmar surface of hand, digit 1, 2, 3 and radial half of digit 4.

- **Ulnar nerve.** Weakness with finger abduction due to paralysis of interinsic hand muscles such as the dorsal interossei muscles. Sensory loss may be noted on the palmar and dorsal surface of the hand, digit 5 and the ulnar half of digit 4. ▼

VASCULARIZATION OF THE SHOULDER AND AXILLA

BIG PICTURE

Blood supply to both upper limbs is provided by the subclavian arteries. The right subclavian artery is a branch from the brachiocephalic artery, and the left subclavian artery is a branch from the aortic arch. The subclavian artery becomes the axillary artery as it crosses over the lateral border of the first rib. The axillary artery continues distally and becomes the brachial artery at the inferior border of the teres major muscle. The brachial artery continues distally, passing over the elbow, and becomes the ulnar and radial arteries.

SUBCLAVIAN ARTERY

The subclavian artery branches directly from the aortic arch on the left and from the brachiocephalic artery on the right. The subclavian artery passes between the anterior and middle scalene muscles and then posteriorly to the clavicle (the subclavian vein travels anteriorly to the anterior scalene) (Figure 30-5A–C).

- **Thyrocervical trunk.** Branches from the subclavian artery, medially to the anterior scalene.
  - **Transverse cervical artery.** Branches from the thyrocervical trunk and crosses anteriorly to the anterior scalene muscle. At the lateral border of the anterior scalene, the transverse cervical artery branches into the deep and superficial transverse cervical arteries.
    - **Superficial transverse cervical artery.** Travels deep between the trapezius and the rhomboid muscles.
    - **Deep transverse cervical artery (dorsal scapular artery).** Travels deep to the trapezius and levator scapulae muscles and along the medial border of the scapula, deep to the rhomboid muscles. The deep transverse cervical artery forms collateral circuits with the circumflex scapular and suprascapular arteries.
      - **Dorsal scapular artery.** Occasionally, the dorsal scapular artery will branch directly off the subclavian artery. When this occurs, the deep transverse cervical artery may be absent.
  - **Suprascapular artery.** Branches from the thyrocervical trunk, and courses superior to the transverse scapular ligament. The suprascapular artery courses from the supraspinous fossa through the greater scapular notch to the infraspinous fossa and forms collateral circuits with the circumflex scapular and dorsal scapular arteries.

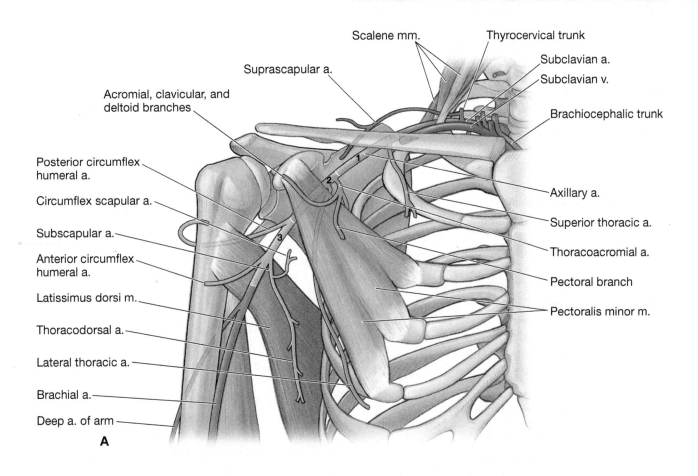

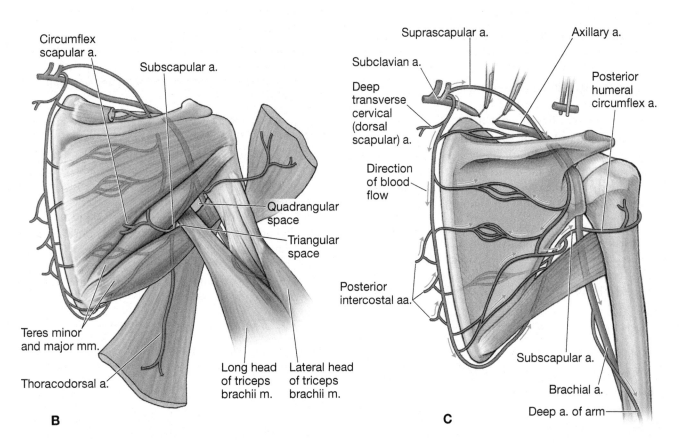

Figure 30-5: A. Branches of the subclavian and axillary arteries. **B**. Posterior view of the shoulder arteries. **C**. Anastomoses of the shoulder arteries.

AXILLARY ARTERY

The axillary artery is a continuation of the subclavian artery and is divided into three parts (Figure 30-5A–C):

First part (one branch). Lateral border of the first rib to the medial border of the pectoralis minor muscle.

- **Superior thoracic artery.** Supplies the upper part of the anterior and medial axillary walls.

Second part (two branches). Runs posteriorly to the pectoralis minor muscle.

- **Thoracoacromial artery (trunk).** Wraps around the proximal border of the pectoralis minor muscle and then branches into the **p**ectoral, **a**cromial, **c**lavicular, and **d**eltoid branches (*you can remember this by thinking of the acronym **PACD***). Each arterial branch of the thoracoacromial artery is named for the region it supplies.

- **Lateral thoracic artery.** Courses with the long thoracic nerve along the lateral surface of the thorax, supplying the serratus anterior muscle and the surrounding tissue.

Third part (three branches). Lateral border of the pectoralis minor muscle to the inferior border of the teres major muscle.

- **Subscapular artery.** Courses along the anterior surface of the subscapularis muscle and branches into the circumflex scapular artery and the thoracodorsal artery.

  - **Circumflex scapular artery.** Courses through the triangular space to the posterior side of the scapula, forming an anastomosis with the suprascapular and dorsal scapular (deep transverse cervical) arteries.

  - **Thoracodorsal artery.** Courses with the thoracodorsal nerve along the posterolateral thorax, supplying the latissimus dorsi muscle.

- **Anterior circumflex humeral artery.** Courses anteriorly around the surgical neck of the humerus and forms an anastomosis with the posterior circumflex humeral artery.

- **Posterior circumflex humeral artery.** Passes through the quadrangular space with the axillary nerve, wraps around the surgical neck of the humerus, and forms anastomoses with the anterior circumflex humeral. Supplies the surrounding muscles and the glenohumeral joint.

▽ If there is a blood clot blocking the subclavian artery or the artery is surgically clamped or a segment is removed, blood can bypass the blockage and reach the arm because of the rich **shoulder anastomosis** with the dorsal scapular, supraclavicular, and posterior humeral circumflex arteries (Figure 30-5C). ▼

ANATOMIC SPACES

The following spaces are helpful in locating the neurovascular structures in the posterior scapular region (Figure 30-5B):

Quadrangular space

- **Borders.** Teres major and teres minor muscles, long head of the triceps brachii muscle, and the humerus.

- **Contents. Axillary nerve** and the **posterior circumflex humeral artery.**

Triangular space

- **Borders.** Teres major and teres minor muscles and the long head of the triceps brachii muscle.

- **Contents.** Circumflex scapular artery.

LYMPHATICS OF THE SHOULDER AND AXILLA

BIG PICTURE

Lymphatic vessels and nodes in the shoulder and axillary region drain excess interstitial fluid as well as having an immunologic function (Figure 30-6). Lymphatics from the right upper limb drain into the right subclavian vein via the **right lymphatic duct.** Lymphatics from the left upper limb drain into the left subclavian vein via the **thoracic duct.**

LYMPH NODES

The lymphatics of the shoulder are organized into the following **axillary lymph nodes** (Figure 30-6A):

Humeral (lateral) nodes. Located posteriorly to the axillary vein and receive drainage from the arm.

Pectoral (anterior) nodes. Located along the distal border of the pectoral minor muscles and receive drainage from the abdominal wall, thoracic wall, and the mammary gland.

Subscapular (posterior) nodes. Located along the posterior axillary wall and receive drainage from the posterior axillary wall, neck, shoulder, and upper back.

Central nodes. Embedded in axillary fat and receive drainage from the humeral, pectoral, and subscapular nodes.

Apical nodes. Surrounds the axillary vein near the pectoralis minor muscle. The apical nodes drain all other axillary nodes and lymphatic vessels from the mammary gland into the subclavian veins.

▽ **Lymphedema** results from the accumulation of lymphatic fluid in tissues, which is caused by restricted lymphatic flow. Lymphedema is typically classified as primary or secondary. **Lymphedema** often results when lymph nodes or lymphatic vessels are damaged or surgically removed. A patient with breast cancer who has undergone axillary lymph node dissection or radiation therapy, or both, is at risk of developing lymphedema because of the removal or damage of the lymph nodes and small lymphatic vessels. Symptoms include persistent accumulation of a protein-rich fluid in the interstitial tissues and swelling of the upper limb on the affected side. ▼

VEINS OF THE SHOULDER AND AXILLA

The veins of the scapular region generally follow the arteries and have similar names (e.g., the axillary vein and the axillary artery) (Figure 30-6B and C). The veins of the scapular region generally drain into veins of the back, neck, axilla, and arm and eventually reach the superior vena cava. The upper limb contains a deep and a superficial venous system.

Deep venous system. Follows the arteries and usually consists of two or more veins.

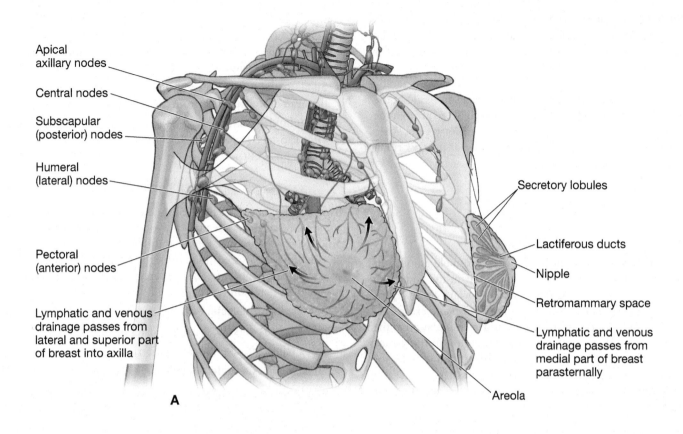

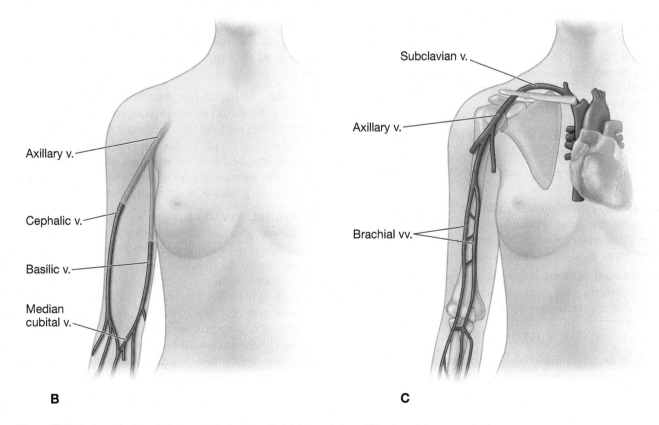

Figure 30-6: A. Lymphatics of the upper limb. Superficial (**B**) and deep (**C**) veins of the upper limb.

Superficial venous system. Consists of the basilic and cephalic veins. Both veins originate in the hand. The basilic vein travels along the medial side of the arm until it reaches the inferior border of the teres major muscle and becomes the axillary vein. The cephalic vein travels more laterally along the arm and remains superficial as it travels along the anterior border of the deltoid muscle and deep through the clavipectoral triangle to join with the axillary vein. After the axillary vein crosses the lateral border of the first rib, it becomes the subclavian vein.

GLENOHUMERAL JOINT

BIG PICTURE

The **glenohumeral joint** allows for a considerable amount of range of motion; more than any other joint in the body. The glenohumeral joint is a ball-and-socket synovial joint that produces a great deal of freedom, including flexion and extension, abduction and adduction, and medial and lateral rotation of the humerus. Because of the large range of motion in the glenohumeral joint it must depend upon ligaments and muscles for structural support. To minimize friction, bursae (synovial sacs) are positioned between the rotator cuff muscles and the joint capsule.

STRUCTURE OF THE GLENOHUMERAL JOINT

The articulating surface of the **glenoid cavity** is approximately one-third the size of the articulating surface of the **humeral head** (Figure 30-7A). This disproportionate articulation results in increased range of motion. However, it also results in a joint that is not as stable as other ball-and-socket joints (e.g., hip). To compensate for this lack of stability, a cartilaginous cuff called the **glenoid labrum** enhances and deepens the articulating surface of the glenoid fossa. The capsule of the **glenohumeral joint** is loose, which enables a major degree of motion.

LIGAMENTOUS SUPPORT TO THE GLENOHUMERAL JOINT

The following ligaments provide **support to the glenohumeral joint** (Figure 30-7B):

- **Joint capsule (glenohumeral ligaments).** Reinforce the superior, anterior, posterior, and inferior regions of the joint capsule. The inferior region of the joint capsule possesses an **axillary fold** to allow slack when the joint is fully abducted.
- **Coracoacromial ligament.** Supports the superior aspect of the glenohumeral joint to prevent superior dislocation by forming an arch over the superior aspect of the humeral head.

BURSAE

Bursae help to decrease the friction between two moving structures such as tendon and bone (Figure 30-7C). In the shoulder complex, and especially the glenohumeral joint, the bursae are important because of the complexity of the muscular stability and the high degree of mobility. The **two most important bursae in the shoulder complex** are as follows:

- **Subacromial bursa.** Separates the supraspinatus tendon and the head of the humerus from the acromion.
- **Subdeltoid bursa.** Separates the deltoid muscle from the joint capsule. The subdeltoid bursa is continuous with the subacromial bursa.

MUSCULAR SUPPORT

Any muscle that crosses the glenohumeral joint and produces a compressive force between the head of the humerus and the glenoid cavity will produce muscle stability (Figure 30-7D). Muscle stability is best exemplified by the **rotator cuff muscles**, which provide support to all sides, except the inferior aspect of the glenohumeral joint.

- **Supraspinatus muscle.** Provides superior support.
- **Infraspinatus muscle.** Provides posterior support.
- **Teres minor muscle.** Provides posterior support.
- **Subscapularis muscle.** Provides anterior support.

In addition to the support of the rotator cuff musculature, the long head of the biceps brachii and deltoid muscles assist in the support of the glenohumeral joint:

- **Long head of the biceps brachii muscle.** Provides superior and anterior support.
- **Deltoid muscle.** Provides superior support.

A combination of ligamentous and dynamic muscle support of multiple joints is critical for the stability of the shoulder complex because of the laxity of the capsule and the high degree of mobility.

▽ A **superior lateral anteroposterior (SLAP) tear** is an injury that usually results from an activity such as throwing an object over the head (e.g., pitching a baseball). The result is a tear of the labrum of the superior glenoid muscle. A SLAP tear is thought to be due to the long head of the biceps tendon pulling on the superior labrum when the humerus decelerates during a throwing motion, resulting in a tear. ▽

▽ **Shoulder separation** is not an injury to the glenohumeral joint itself, but rather is an injury to the acromioclavicular joint. The injury is usually caused by falling directly on the shoulder, resulting in damage or tearing of the ligaments that support the acromioclavicular joint. The clavicle may be out of alignment with the acromion of the scapula, resulting in a bump (i.e., the clavicle becomes more superior to the acromion). The severity of this injury is determined by the amount of ligamentous damage; usually if there is more ligamentous damage, the deformity is more noticeable. ▽

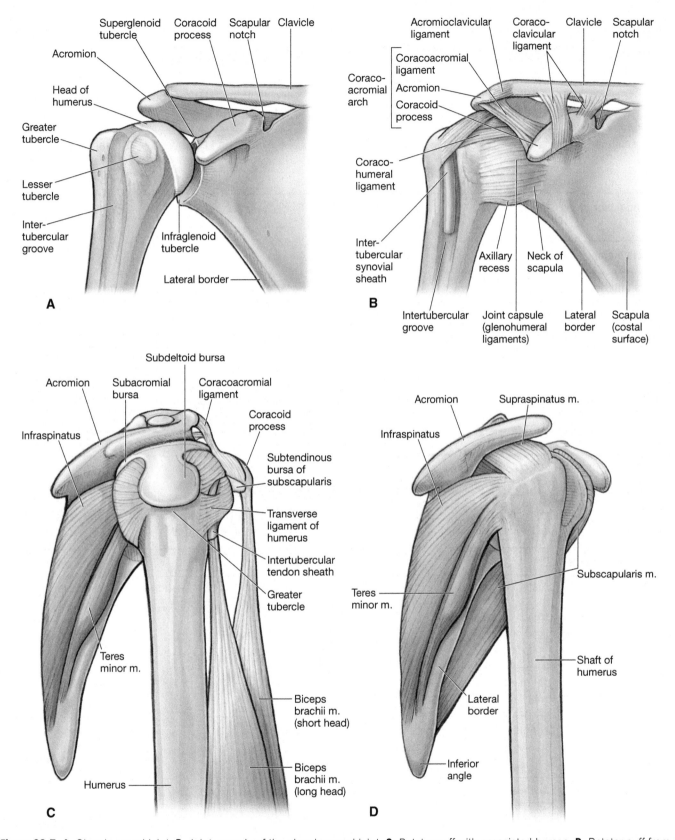

Figure 30-7: A. Glenohumeral joint. **B**. Joint capsule of the glenohumeral joint. **C**. Rotator cuff with associated bursae. **D**. Rotator cuff from a lateral degree.

TABLE 30-1. Muscles of the Scapula

Muscle	Proximal Attachment	Distal Attachment	Action	Innervation
Trapezius	Occipital bone, nuchal ligament, C7–T12 vertebrae	Spine, acromion, and lateral clavicle	Elevation, retraction, rotation, and depression of scapula	Spinal accessory n. and ventral rami of C3 and C4
Levator scapulae	Transverse processes of C1–C4	Superior angle of scapula	Elevation and downward rotation of scapula	Dorsal scapular n. (C5) and ventral rami of C3 and C4
Rhomboid minor	C7–T1 vertebrae	Medial margin of scapula	Retraction of scapula	Dorsal scapular n. (C5)
Rhomboid major	T2–T5 vertebrae			
Serratus anterior	Ribs 1–8		Protraction and rotation of the scapula	Long thoracic n. (C5–C7)
Pectoralis minor	Ribs 3–5	Coracoid process of scapula	Protraction, depression, and stabilization of scapula	Medial pectoral n. (C8–T1)
Subclavius	Rib 1	Clavicle	Depression and stabilization of clavicle	Nerve to the subclavius (C5–C6)

TABLE 30-2. Muscles of the Glenohumeral Joint

Muscle	Proximal Attachment	Distal Attachment	Action		Innervation
Deltoid	Spine, acromion, and lateral clavicle	Deltoid tuberosity of humerus	Flexion, extension, and abduction of the humerus		Axillary n. (C5–C6)
Rotator cuff muscles					
Supraspinatus	Supraspinous fossa	Greater tubercle of humerus	Abduction of humerus	Stabilization of shoulder joint	Suprascapular n. (C5–C6)
Infraspinatus	Infraspinous fossa		Lateral rotation of humerus		Suprascapular n. (C5–C6)
Teres minor	Lateral margin of scapula				Axillary n. (C5–C6)
Subscapularis	Subscapular fossa	Lesser tubercle of humerus	Medial rotation of humerus		Upper and lower subscapular nn. (C5–C6)
Intertubercular groove muscles					
Teres major	Inferior angle of scapula	Intertubercular groove of humerus	Adduction, extension, and medial rotation of humerus		Lower subscapular n. (C5–C6)
Pectoralis major	Clavicle, sternum, and costal cartilage		Adduction, medial rotation, extension, and flexion of humerus		Medial and lateral pectoral nn. (C5–T1)
Latissimus dorsi	T7–T12, sacrum, and thoracolumbar fascia		Adduction, extension, and medial rotation of humerus		Thoracodorsal n. (C6–C8)

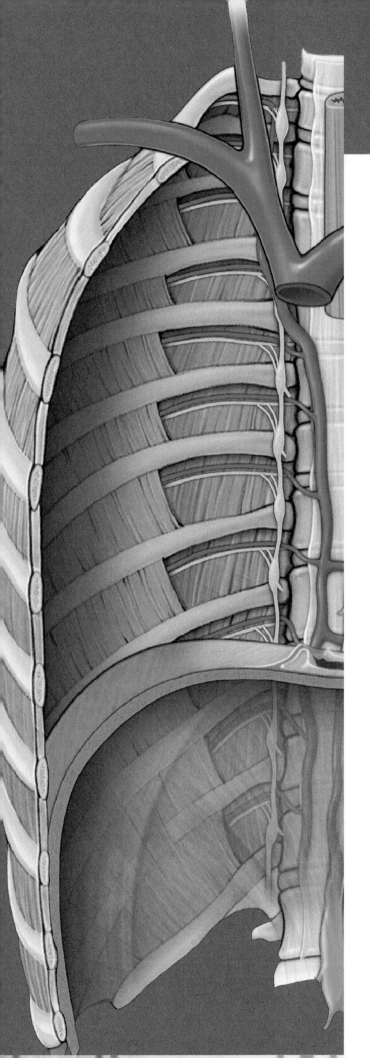

ARM

ARM

BIG PICTURE

The arm (brachium) consists of the humerus, which articulates distally with the forearm (antebrachium) through the elbow complex. The elbow complex consists of three bones: humerus, ulna, and radius. The articulations of these bones result in three separate joints that share a common synovial cavity, enabling the forearm to flex, extend, pronate, and supinate on the humerus.

ACTIONS OF THE ELBOW COMPLEX

The **articulations of the humerus, radius, and ulna** in the elbow result in the following actions:

- **Flexion and extension** (Figure 31-1A)
  - **Humeroulnar joint.** Articulation between the trochlear notch of the ulna and the trochlea of the humerus.
  - **Humeroradial joint.** Articulation between the head of the radius and the capitulum of the humerus.
- **Pronation and supination** (Figure 31-1B and C)
  - **Proximal radioulnar joint.** Articulation between the head of the radius and the radial notch of the ulna.

MUSCLES OF THE ARM

BIG PICTURE

The muscles of the arm are divided by their fascial compartments (anterior and posterior), and may cross one or more joints. Identifying the joints that the muscles cross and the side on which they cross can provide useful insight into the actions of these muscles (Table 31-1).

MUSCLES OF THE ANTERIOR COMPARTMENT OF THE ARM

The muscles in the anterior compartment of the arm are primarily **flexors (of the shoulder or elbow or both)** because of their anterior orientation (Figure 31-1D). The **musculocutaneous nerve (C5–C7)** innervates the muscles in the anterior compartment of the arm. However, each muscle does not necessarily receive each spinal nerve level between C5 and C7. The following muscles are located in the anterior compartment of the arm:

- **Coracobrachialis muscle.** Attaches between the **coracoid process of the scapula** and the midshaft of the **humerus**. The coracobrachialis muscle crosses anteriorly to the glenohumeral joint and, therefore, contributes to **shoulder flexion**. It receives its innervation from the **musculocutaneous nerve (C5–C7)** and its blood supply via branches of the **axillary artery**.

- **Brachialis muscle.** Attaches between the anterior aspect of the **humerus** and the **coronoid process** and the **tuberosity of the ulna**, crossing the anterior elbow joint. The brachialis muscle acts on the ulna (humeroulnar joint), and therefore, it produces **flexion of the elbow**. As with the other muscles in the anterior compartment of the arm, the **musculocutaneous nerve (C5–C6)** provides innervation. However, the **radial nerve (C7)** innervates a small, lateral portion of the muscle. Blood is supplied to the muscle by branches from the brachial artery.

- **Biceps brachii muscle.** Consists of two heads that attach to the **supraglenoid tubercle** (long head) and the **coracoid process** (short head). The biceps brachii muscle converges to insert on the **radial tuberosity**. The biceps brachii crosses anterior to the **glenohumeral joint and the elbow**, primarily producing **flexion** in both joints. Because the distal attachment is to the radius, the biceps brachii will also produce **supination** due to movement of the **radioulnar joints**. The biceps brachii receives its innervation from the **musculocutaneous nerve (C5–C6)** and its blood supply from branches originating from the brachial artery.

MUSCLES OF THE POSTERIOR COMPARTMENT OF THE ARM

The muscles in the posterior compartment of the arm are primarily **extensors of the shoulder and elbow** because of their posterior orientation (Figure 31-1E). The **radial nerve (C6–C8)** innervates the muscle in the posterior compartment of the arm.

- **Triceps brachii muscle.** Consists of three heads. The long head attaches to the **infraglenoid tubercle of the scapula**, and the medial and lateral heads attach to the **posterior humerus**. The three heads converge to attach to the **olecranon process of the ulna**. The long head produces **shoulder extension** and **elbow extension**. The other two heads produce **elbow extension** only. The **radial nerve (C6–C8)** provides innervation to the muscle, and the **profunda brachii** and **superior ulnar collateral arteries** provide most of the blood supply to the triceps brachii muscle.

▼ **Biceps tendinopathy** is a condition that results most frequently in painful sensations in the anterior region of the shoulder. The long head of the biceps is often irritated by overhead motions and excessive or repetitive lifting, resulting in inflammation of the tendon and the peripheral structures. The pain is most frequently described as occurring between the greater and lesser tubercles of the humerus in the intertubercular groove, where the long head travels. The condition is exacerbated by the actions of the biceps (e.g., shoulder flexion, elbow flexion, and supination) and is frequently confused with pathologies of the rotator cuff. ▼

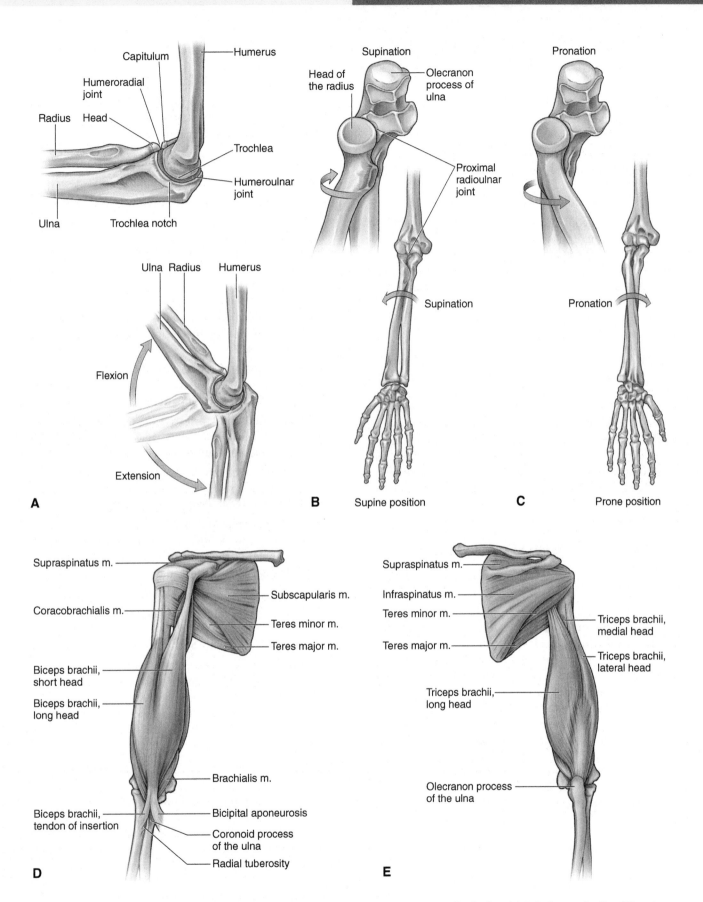

Figure 31-1: A. Lateral view of the elbow demonstrating bony landmarks and articulations. Radioulnar joint during supination (**B**) and pronation (**C**). Anterior (**D**) and posterior (**E**) views of the brachial muscles.

TERMINAL BRANCHES OF THE BRACHIAL PLEXUS IN THE ARM

BIG PICTURE

Innervation to the anterior and posterior compartments of the arm originates from the lateral and posterior cord, giving rise to the **musculocutaneous and radial nerves**, respectively. Both nerves are mixed and provide motor and sensory innervation.

MUSCULOCUTANEOUS NERVE

The musculocutaneous nerve pierces the **coracobrachialis muscle**, innervating it as it passes, and descends through the arm between the **biceps brachii** and **brachialis muscles**, supplying both muscles. The musculocutaneous nerve pierces the deep fascia just distal to the elbow to become the **lateral cutaneous nerve of the forearm** (Figure 31-2A).

RADIAL NERVE

The radial nerve descends posterior to the humerus with the deep artery of the arm **(profundus brachii)**, supplying motor innervation to the **triceps brachii** (Figure 31-2B). It provides cutaneous innervation via the **posterior cutaneous nerve of the arm**, the **inferior lateral cutaneous nerve of the arm**, and the **posterior cutaneous nerve of the forearm**. The radial nerve pierces the intermuscular septum laterally, anteriorly to the lateral epicondyle between the brachialis and the brachioradialis. It then descends to the posterior compartment of the forearm, providing motor innervation, and to the dorsum of the hand, providing cutaneous innervation.

MEDIAL CUTANEOUS NERVE OF THE ARM

The medial cutaneous nerve of the arm branches from the medial cord and, as its name implies, supplies the anteromedial skin of arm.

▽ **Radial neuropathy** (also known as "**Saturday night palsy**") is a condition caused by compression of the radial nerve on the posterior aspect of the humerus, where it spirals around. The compression usually is caused when an intoxicated person falls asleep with the posterior arm being compressed by the edge of a desk, bar, chair, or bench. Although the injury occurs at the posterior humerus, symptoms are identified in the forearm. Symptoms include the inability to extend the wrist and fingers (wrist drop) and loss of sensation to the posterior portion of the hand. In contrast, radial neuropathy occurring in the axillary region due to the improper use of crutches will result in weakness in the triceps muscles because the injury is more proximal. ▽

VASCULARIZATION OF THE ARM

BIG PICTURE

The blood supply to the arm is initiated from the axillary artery. The axillary artery becomes the brachial artery after crossing the inferior border of the teres major muscle.

BRACHIAL ARTERY

The brachial artery courses through the medial side of the anterior compartment of the arm, supplying the muscles of the anterior compartment (Figure 31-2C). This is accomplished through the following arteries:

▪ **Deep artery of arm (profunda brachii artery, deep brachial artery).** Descends, with the radial nerve, posterior to the humerus and through the radial groove to supply the triceps brachii muscle (Figure 31-2D). The deep artery of the arm bifurcates midarm into the **radial** and **middle collateral arteries**. Often, an anastomosis forms with the **posterior circumflex humeral artery**.

  • **Radial collateral artery.** Courses anteriorly to the lateral epicondyle of the humerus, where it forms an anastomosis with the **radial recurrent artery** from the forearm.

  • **Middle collateral artery.** Courses along the posterior humerus, where it forms an anastomosis with the **recurrent interosseous artery** from the forearm.

▪ **Superior ulnar collateral artery.** Courses posteriorly to the medial epicondyle of the humerus, where it forms an anastomosis with the **posterior ulnar recurrent artery** from the forearm.

▪ **Inferior ulnar collateral artery.** Bifurcates around the medial epicondyle of the humerus, where it forms an anastomosis with the **anterior recurrent ulnar artery** and the **middle collateral artery from the forearm**.

The brachial artery terminates anteriorly to the elbow, where it bifurcates into the **radial and ulnar arteries**.

VEINS OF THE ARM

The veins in the arm consist of a superficial and a deep venous system. The **superficial venous system** consists of the basilic vein, which is located medially, and the cephalic vein, which is located laterally. The deep venous system may consist of two or three veins that course with each artery. Most veins in the arm drain into the axillary vein.

▽ A **humeral fracture** is a common fracture of the upper limb and is often caused by a fall. Most humeral fractures heal without surgical intervention; however, depending on the severity of the injury, surgical intervention may be necessary. A patient with a humeral fracture may occasionally develop additional damage to peripheral structures such as the radial nerve. Because of the proximity of the radial nerve to the humerus, the radial nerve may be involved in a displaced humeral fracture, resulting in transient or permanent damage to the radial nerve. ▽

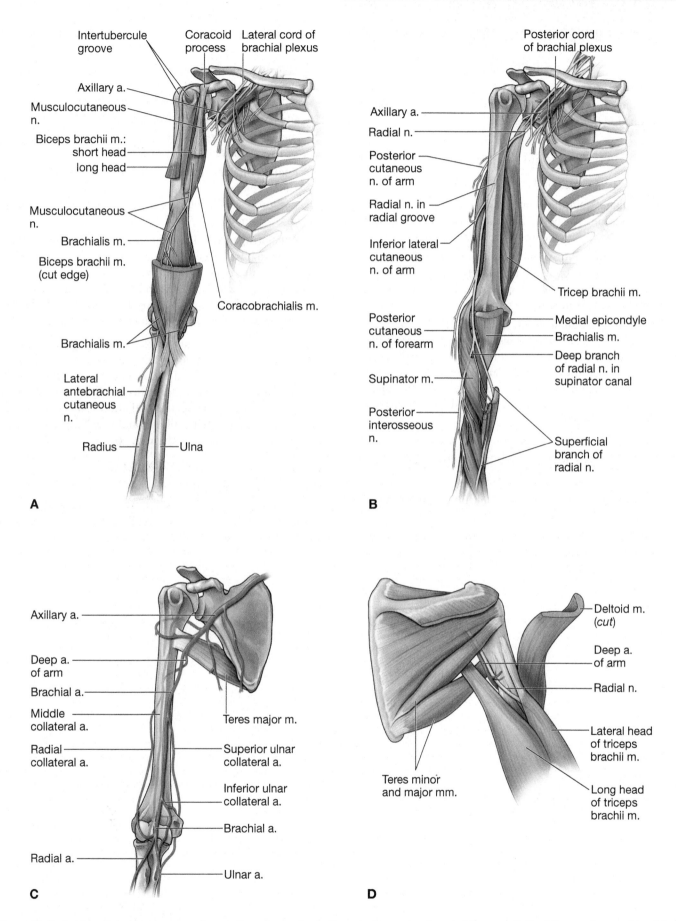

Figure 31-2: A. Musculocutaneous nerve innervation of muscles in the anterior compartment of the arm. **B**. Radial nerve innervation of the muscles in the posterior compartment of the arm and forearm. **C**. Arterial supply to the brachium. **D**. Triangular interval demonstrating the course of the deep artery of the arm and the radial nerve.

JOINTS CONNECTING THE ARM AND FOREARM

BIG PICTURE

The elbow consists of articulations between the humerus, radius, and ulna, and provides flexion, extension, supination, and pronation of the forearm.

ELBOW COMPLEX

The elbow is composed of the following joints (Figure 31-3A):

Humeroulnar joint. Synovial hinge joint producing flexion and extension. The boney articulations include the trochlea of the humerus and the trochlear notch of the ulna.

Humeroradial joint. Synovial gliding joint that works in conjunction with the humeroulnar joint to produce flexion and extension. Boney articulations include the capitulum of the humerus and the head of the radius.

Proximal radioulnar joint. Synovial pivot joint that is mechanically linked with the distal radioulnar joint and produces supination and pronation. The boney articulations include the head of the radius and the radial notch of the ulna. The proximal radioulnar joint is interdependent with the distal radioulnar joint and will be discussed in greater detail in Chapter 32.

LIGAMENTOUS AND CAPSULAR SUPPORT OF THE ELBOW

The following ligaments and the capsule provide support to the elbow joint (Figure 31-3B–D):

Capsule. The humeroulnar, humeroradial, and proximal radioulnar joints are enclosed in a single capsule. The capsule is loose, which accommodates the high degree of motion; however, the capsule is reinforced with ligaments.

Ligaments. Most hinge joints in the body, including the humeroulnar joint, contain medial and lateral collateral ligaments to enhance medial and lateral stability.

- **Ulnar collateral (medial collateral).** Fibers run from the medial epicondyle of the humerus to the proximal ulna, providing medial stability of the elbow complex.
- **Radial collateral (lateral collateral).** Fibers run from the lateral epicondyle to the annular ligament and the olecranon process of the ulna, providing lateral stability of the elbow complex.
- **Annular ligament.** The annular ligament is a circular ligament that is attached to the anterior and posterior surfaces of the radial notch of the ulna and forms a ring that encompasses the radial head. The inner surface of the ligament is covered with cartilage and provides stability for the humeroulnar joint while allowing supination and pronation.

▽ Ligaments as well as muscles reinforce the elbow complex. However, a sufficient longitudinal force on the radius may result in the head of the radius being pulled through the annular ligament, an injury known as **distraction injury** (or more commonly as "**nursemaid's elbow**"). This injury occurs most frequently in small children and usually when a parent unexpectedly lifts a child by the arm. ▼

OLECRANON BURSA

The most important bursa associated with the elbow complex is the olecranon bursa, which is located between the capsule of the elbow complex and the triceps tendon. The olecranon bursa diminishes friction between the two surfaces as they cross over each other.

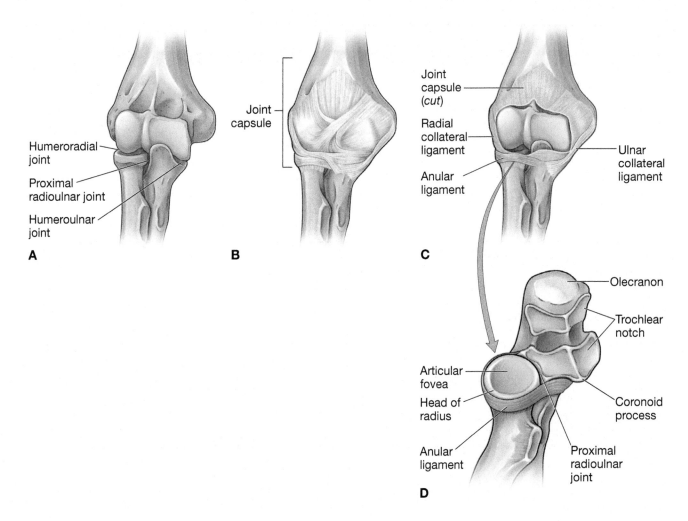

Figure 31-3: A. Joints of the elbow. **B**. Joint capsule of the elbow. **C**. Joint capsule cut and open revealing the articulations. **D**. Superior view of the proximal radioulnar joint demonstrating the annular ligament.

TABLE 31-1. Muscles of the Arm

Muscle	Proximal Attachment	Distal Attachment	Action	Innervation
Anterior compartment of the arm				
Biceps brachii	Long head: supraglenoid tubercle Short head: coracoid process	Radial tuberosity	Flexion of shoulder and flexion and supination of elbow	Musculocutaneous n. (C5–C6)
Brachialis	Distal anterior surface of humerus	Coronoid process and tuberosity of ulna	Flexion of the elbow	Musculocutaneous n. (C5–C6) & radial n. (C7)
Coracobrachialis	Coracoid process of scapula	Medial, midshaft surface of humerus	Flexion of shoulder	Musculocutaneous n. (C5–C7)
Posterior compartment of the arm				
Triceps brachii	Long head: infraglenoid tubercle Lateral head: posterior humerus Medial head: posterior humerus	Olecranon process of ulna	Extension of shoulder and elbow	Radial n. (C6–C8)

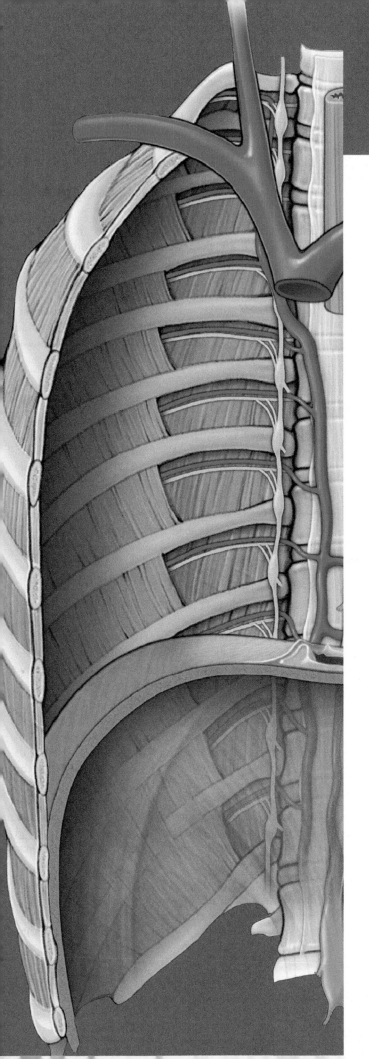

FOREARM

MUSCLES OF THE FOREARM

BIG PICTURE

The forearm (antebrachium) consists of the radius and ulna. Proximally, the forearm articulates with the humerus through the elbow complex (humeroulnar and humeroradial joints). Distally, the forearm articulates with the carpal bones through the wrist complex, enabling a wide array of actions. The muscles of the forearm that act upon the elbow, wrist complex, and the digital joints are organized into two fascial compartments, similar to those of the arm muscles. The anterior compartment contains flexor muscles and the posterior compartment contains extensor muscles.

ACTIONS OF THE WRIST

The configuration of the wrist complex allows for **motion in two planes** (Figure 32-1A):

- **Flexion**
- **Extension**
- **Radial deviation (abduction)**
- **Ulnar deviation (adduction)**

FOREARM MUSCLES OF THE ANTERIOR COMPARTMENT

The actions produced by the muscles in the anterior compartment of the forearm depend upon which joints the muscles cross. Some muscles cross the elbow, wrist, digits, and perhaps a combination of each. The muscles in the anterior compartment of the forearm have the following similar features:

- **Common attachment.** Medial epicondyle of the humerus.
- **Common innervation.** Median nerve with minimal contribution from the ulnar nerve.
- **Common action.** Flexion.

The vascular supply to the anterior forearm muscles is from branches of the ulnar and radial arteries.

The **muscles in the anterior compartment** of the forearm are divided into three groups: superficial, intermediate, and deep.

Superficial group (Figure 32-1B)

- **Pronator teres muscle.** Possesses two heads and crosses the elbow complex. The humeral head of the pronator teres muscle attaches to the medial epicondyle and the supraepicondylar ridge of the humerus, and the ulnar head attaches to the coronoid process. Distally, the pronator teres muscle attaches to the midshaft of the radius. The pronator teres muscle primarily produces pronation at the forearm. The median nerve provides innervation (C6–C7) to the pronator teres muscle.

- **Flexor carpi radialis muscle.** Attaches to the medial epicondyle and the base of metacarpals 2 and 3. The primary action of the flexor carpi radialis muscle is wrist flexion and radial deviation. The median nerve (C6–C7) supplies innervation to this muscle.

- **Palmaris longus muscle.** Attaches to the medial epicondyle of the humerus and courses superficially over the **flexor retinaculum** to the **palmar aponeurosis** in the hand. The primary action of the palmaris longus muscle is to resist shearing forces of the palmar aponeurosis; it is also considered a wrist flexor. Innervation is provided by the median nerve (C7–C8). *It is important to note that the palmaris longus muscle may be absent on one or both sides in some individuals.*

- **Flexor carpi ulnaris muscle.** Possesses two heads. A humeral head attaches to the medial epicondyle of the humerus, and an ulnar head attaches to the olecranon process. Both heads come together and attach to the pisiform, hamate and base of metacarpal 5. The flexor carpi ulnaris muscle crosses both the elbow and the wrist complex, producing weak elbow flexion, wrist flexion, and ulnar deviation. It is innervated by the ulnar nerve (C7–T1).

Intermediate group (Figure 32-1C)

- **Flexor digitorum superficialis muscle.** Possesses two heads. A humeral head attaches proximally to the medial epicondyle and a radial head attaches to the radius. The flexor digitorum superficialis muscle attaches distally to the middle phalanges of digits 2 to 5. The muscle primarily produces flexion at the wrist and at the metacarpophalangeal and proximal interphalangeal joints. Innervation is provided by the median nerve (C8–T1). The four tendons of the flexor digitorum superficialis muscle cross under the **flexor retinaculum** at the wrist and enter the hand through the **carpal tunnel**.

Deep group (Figure 32-1D)

- **Flexor pollicis longus muscle.** Attaches proximally to the radius and the interosseous membrane and to the distal phalanx of the thumb. The flexor pollicis longus muscle produces flexion at the metacarpophalangeal and interphalangeal joints of digit 1 and is innervated by the anterior interosseous nerve from the median nerve (C7–C8).

- **Flexor digitorum profundus muscle.** Attaches proximally to the ulna and interosseous membrane and travels across the wrist complex and attaches distally to the distal phalanges of digits 2 to 5. The flexor digitorum profundus muscle produces flexion at the wrist as well as flexion of the metacarpophalangeal and proximal and distal interphalangeal joints of digits 2 to 5. The lateral half of the flexor digitorum profundus muscle is innervated by the **anterior interosseous nerve** from the median nerve (C8–T1), and the medial half of the muscle is innervated by the **ulnar nerve** (C8–T1).

- **Pronator quadratus muscle.** Courses horizontally from the distal anterior surface of the ulna to the distal anterior surface of the radius. The pronator quadratus muscle produces pronation, and is innervated by the **anterior interosseous nerve** from the median nerve (C7–C8).

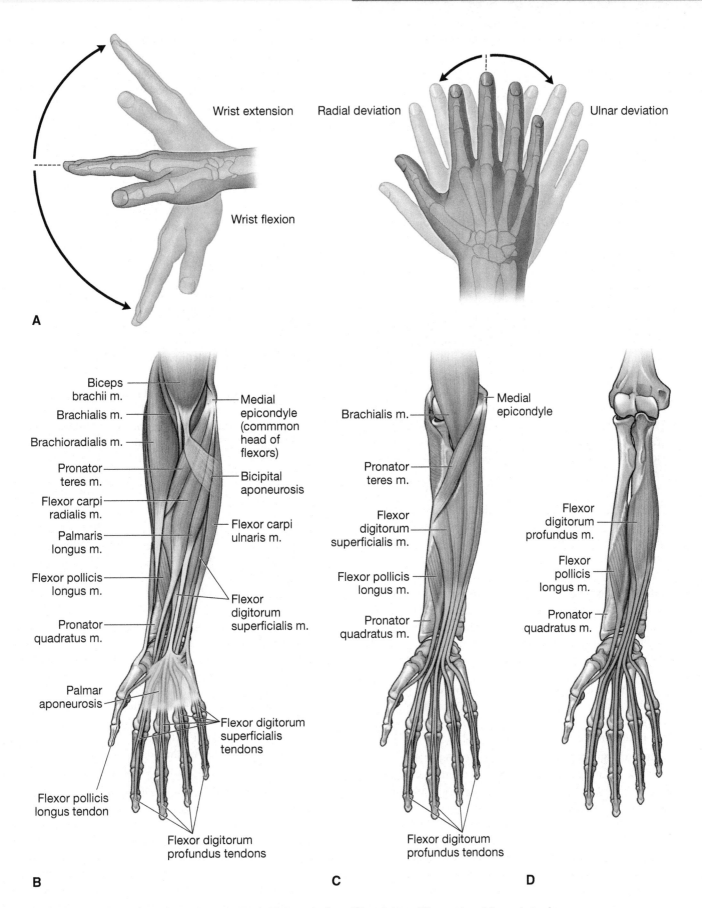

Figure 32-1: A. Actions of the wrist joint. Superficial (**B**) intermediate (**C**) and deep (**D**) muscles of the anterior forearm.

FOREARM MUSCLES OF THE POSTERIOR COMPARTMENT

The actions produced by the muscles in the posterior compartment of the forearm depend upon which joints the muscles cross. Some muscles cross the elbow, wrist, and digits, and perhaps a combination of each. The muscles in the posterior compartment of the forearm have the following similar features (Table 32-1):

Common attachment. Lateral epicondyle of the humerus.

Common innervation. Radial nerve.

Common action. Extension.

The vascular supply to the muscles of the posterior compartment is from branches of the ulnar and radial arteries.

The **muscles in the posterior compartment** are divided into superficial and deep groups.

Superficial group (Figure 32-2A and B)

- **Brachioradialis muscle.** Attaches to the lateral supracondylar ridge of the humerus and the **styloid process of the radius**. The brachioradialis muscle produces elbow flexion (primarily in the midpronated position). It also is important for stabilization of the elbow complex during rapid movements of flexion and extension. The brachioradialis muscle is innervated by the radial nerve (C5–C6).

- **Extensor carpi radialis longus muscle.** Attaches to the lateral supracondylar ridge of the humerus and the dorsal surface of the base of metacarpal 2. The extensor carpi radialis longus muscle produces extension and radial deviation of the wrist and is innervated by the radial nerve (C6–C7).

- **Extensor carpi radialis brevis muscle.** Attaches to the lateral epicondyle of the humerus and the dorsal surface of the base of metacarpals 2 and 3. The muscle produces extension and radial deviation of the wrist and is innervated by the deep branch of the posterior interosseous nerve (C7–C8).

- **Extensor digitorum muscle.** Attaches to the lateral epicondyle of the humerus and the dorsal digital expansions of digits 2 to 5. Intrinsic muscles of the hand, the lumbricals and the dorsal and palmar interossei muscles, also attach to the dorsal digital expansion. Intertendinous connections on the dorsum of the hand may be present, connecting the tendons, but the location and the number of connections are highly variable. The extensor digitorum can extend all of the joints it crosses (wrist and digits 2–5). It is innervated by the posterior interosseous nerve (C7–C8).

- **Extensor digiti minimi muscle.** Attaches to the lateral epicondyle of the humerus and the dorsal digital expansion of digit 5. The primary action of the extensor digiti minimi muscle is extension of digit 1, but it will also assist with wrist extension. The muscle is innervated by the posterior interosseous nerve (C7–C8).

- **Extensor carpi ulnaris muscle.** Attaches proximally to the lateral epicondyle of the humerus and the posterior ulna and distally to the base of metacarpal 5. The extensor carpi ulnaris muscle produces extension and ulnar deviation of the wrist and is innervated by the posterior interosseous nerve (C7–C8).

- **Anconeus muscle.** Attaches to the lateral epicondyle of the humerus and the olecranon. The anconeus muscle contributes to elbow extension as well as controls the ulna during pronation. It is innervated by the radial nerve (C6–C8).

Deep group (Figure 32-2C)

- **Supinator muscle.** Attaches proximally to the lateral epicondyle of the humerus and the supinator crest of the ulna. Distally, the supinator muscle attaches to the lateral surface of the radius and contributes to supination of the forearm. The posterior interosseous nerve from the radial nerve (C6–C7) innervates the supinator muscle. The posterior interosseous nerve travels between the muscle fibers that attach to the humerus and ulna as it travels distally into the musculature of the posterior compartment.

- **Abductor pollicis longus muscle.** Attaches proximally to the ulna, radius, and interosseous membrane. Distally, the abductor pollicis longus muscle attaches to the base of metacarpal 1 and abducts the carpometacarpal joint of digit 1. The muscle is innervated by the posterior interosseous nerve (C7–C8).

- **Extensor pollicis longus muscle.** Attaches proximally to the posterior surface of the ulna and the interosseous membrane. Distally, the extensor pollicis longus muscle attaches to the base of the proximal phalanx of digit 1. The muscle produces extension of the metacarpophalangeal, carpometacarpal, and interphalangeal joints of the thumb. It is innervated by the posterior interosseous nerve (C7–C8).

- **Extensor pollicis brevis muscle.** Attaches to the radius and the interosseous membrane proximally and to the base of the proximal phalanx of digit 1 distally. The extensor pollicis brevis muscle produces extension of the metacarpophalangeal and carpometacarpal joints of the thumb. The muscle is innervated by the posterior interosseous nerve (C7–C8).

- **Extensor indicis muscle.** Attaches proximally to the posterior surface of the ulna and interosseous membrane. Distally, the extensor indicis muscle attaches to the dorsal digital expansion of digit 2. The extensor indicis muscle contributes to extension of the index finger, allowing it to be extended independent of the other fingers. It also will assist with wrist extension. The extensor indicis muscle is innervated by the posterior interosseous nerve (C7–C8).

▼ **Lateral epicondylitis (tennis elbow)** is a condition caused by the overuse of the extensor muscles that attach to the lateral epicondyle. This injury is seen in almost 50% of tennis players (hence, the name "tennis elbow"); however, it can affect anyone who participates in repetitive activity. A person with lateral epicondylitis will typically experience pain over the lateral epicondyle. The etiology of the pain is microtears of the proximal attachment of the extensor muscles. A similar condition called "golfer's elbow" occurs at the medial epicondyle and is most commonly seen in golfers. ▼

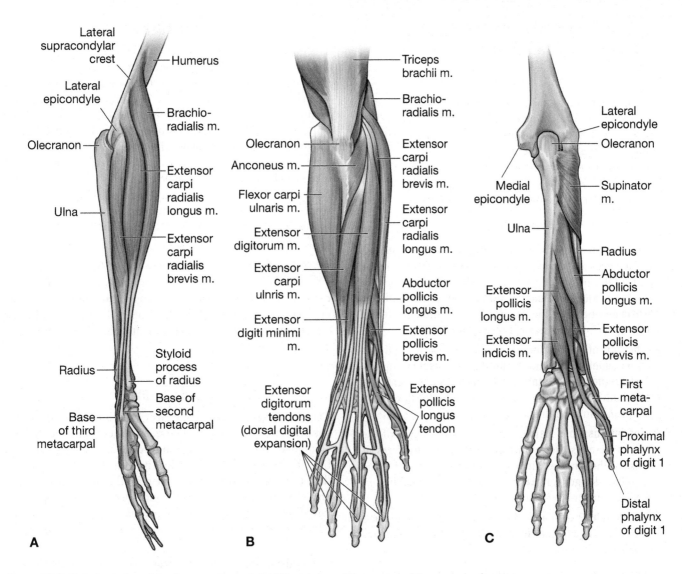

Figure 32-2: A. Lateral view of the forearm. Superficial (**B**) and deep (**C**) muscles of the posterior forearm.

TERMINAL BRANCHES OF THE BRACHIAL PLEXUS IN THE FOREARM

BIG PICTURE

The median, ulnar, and radial nerves provide innervation to the anterior and posterior compartments of the forearm. The median nerve innervates all but one and a half muscles (flexor carpi ulnaris and half of the flexor digitorum profundus muscles) in the anterior compartment of the forearm, which are innervated by the ulnar nerve. The posterior compartment of the forearm is innervated entirely by the radial nerve.

MEDIAN NERVE

The median nerve arises from the medial and lateral cords of the brachial plexus and travels with the brachial artery along the medial side of the arm (Figure 32-3A). In the elbow, the median nerve courses through the **cubital fossa**, deep to the **bicipital aponeurosis** and between the two heads of the **pronator teres**, to enter the anterior compartment of the forearm.

- **Main branch of median nerve.** Courses between the **flexor digitorum superficialis** and the **profundus muscles**, supplying the **superficial and intermediate muscles of the anterior forearm**, with the exception of the ulnar half of the flexor digitorum profundus and the flexor carpi ulnaris.

- **Anterior interosseous nerve.** Once the main branch of the median nerve exits the two heads of the pronator teres, it gives rise to the anterior interosseous nerve, innervating the flexor pollicis longus, pronator quadratus and radial half of the flexor digitorum profundus muscles.

- **Palmar branch.** Proximal to the wrist complex, the median nerve gives rise to a **palmar branch**, which delivers cutaneous innervation to the medial side of the palm.

The median nerve continues distally to travel through the **carpal tunnel** to enter the hand.

▽ **Pronator syndrome** is caused by the entrapment of the median nerve between the two heads of the pronator teres muscle. Depending on the severity of the injury, pronator syndrome can result in varying motor and sensory changes. Regardless of the severity of the injury, the motor and sensory changes occur in the distribution of the median nerve. ▽

▽ **Anterior interosseous syndrome** is the result of entrapment of the anterior interosseous nerve due to tendinous bands, fractures, or compression by the pronator teres muscle (Figure 32-3B). The result is weakness or loss of the muscles innervated by the anterior interosseous nerve. As a result, patients are unable to make the "ok" sign and instead form a triangle between the thumb and index finger. There is no sensory loss involved with this syndrome. ▽

ULNAR NERVE

The ulnar nerve courses posteriorly to the medial epicondyle of the humerus in the osseous groove, into the anterior compartment of the forearm between the two heads of the flexor carpi ulnaris muscle (Figure 32-3B). The ulnar nerve continues through the anterior compartment of the forearm supplying only two muscles, the **flexor carpi ulnaris** and the **ulnar half of the flexor digitorum profundus**. Proximal to the wrist, the ulnar nerve gives rise to two cutaneous branches, a **dorsal branch and a palmar branch**, which provide cutaneous innervation to the dorsal medial side of the hand and the medial side of the palm, respectively. The ulnar nerve continues into the hand superficial to the carpal tunnel and courses through **Guyon's canal** by the pisiform bone to enter the hand.

▽ **Cubital tunnel syndrome** is caused by compression or irritation of the ulnar nerve as it passes under the medial epicondyle. Symptoms are usually tingling and numbness in the cutaneous distribution of the ulnar nerve. In severe cases, muscle weakness may be apparent, with atrophy of the hypothenar eminence. ▽

RADIAL NERVE

The radial nerve enters the forearm, anterior to the lateral epicondyle, and travels distally between the brachialis and the brachioradialis muscles, where it bifurcates into a deep terminal branch and a superficial terminal branch (Figure 32-3C). The deep terminal branch becomes the posterior interosseous nerve, and the superficial terminal branch becomes the superficial radial nerve. The posterior cutaneous nerve of the forearm, which branches in the arm, provides sensory innervation to the posterior forearm.

- **Posterior interosseous nerve.** Pierces between the two heads of the supinator muscle to innervate the muscles in the posterior compartment of the forearm, excluding the brachioradialis and extensor carpi radialis longus, which are innervated by the radial nerve prior to its bifurcation.

- **Superficial radial nerve.** Courses along the brachioradialis muscle and then through the anatomical snuffbox to provide cutaneous innervation to the dorsum of the hand.

MEDIAL CUTANEOUS NERVE OF THE FOREARM

The medial cutaneous nerve of the forearm branches from the medial cord, and as its name implies, it supplies the medial skin of the forearm.

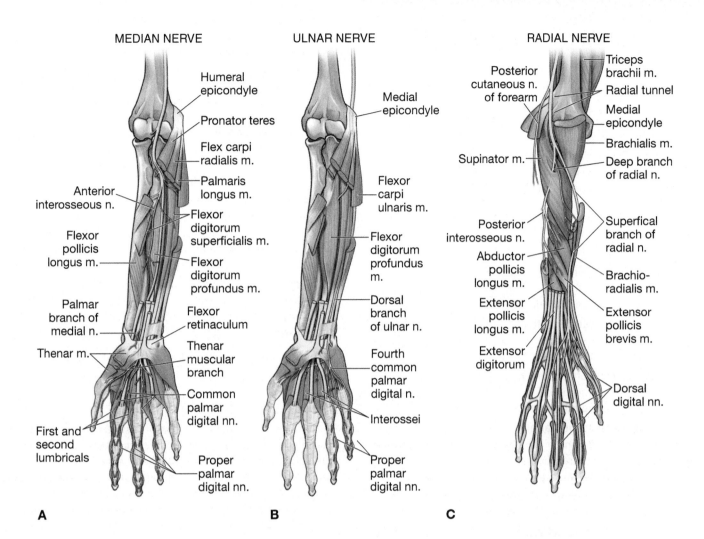

Figure 32-3: A. Median nerve. **B**. Ulnar nerve. **C**. Radial nerve.

VASCULARIZATION OF THE FOREARM

BIG PICTURE

The brachial artery extends from the inferior border of the teres major muscle, giving rise to several branches that supply blood to the anterior and posterior compartments of the arm. The brachial artery bifurcates into the ulnar and radial arteries at the radioulnar joint. The radial and ulnar arteries and their tributaries supply blood to the anterior and posterior compartments of the forearm and extend distally into the hand.

ULNAR ARTERY

The ulnar artery travels through the cubital fossa and continues between the **flexor carpi ulnaris** and the **flexor digitorum profundus muscles**, supplying the medial muscles of the anterior compartment of the forearm. Along the way, the ulnar artery gives rise to the following **branches** (Figure 32-4A and B):

- **Superior ulnar recurrent artery.** Courses in a superior direction anterior to the medial epicondyle and forms an anastomosis with the **inferior ulnar collateral artery**.

- **Inferior ulnar recurrent artery.** Courses in a superior direction posterior to the medial epicondyle and forms an anastomosis with the **superior ulnar collateral artery**.

- **Common interosseous artery.** Courses toward the interosseous membrane and bifurcates into the anterior and posterior interosseous branches.
  - **Anterior interosseous artery.** Travels along the anterior surface of the interosseous membrane, pierces the membrane, and supplies the deep extensor muscles.
  - **Posterior interosseous artery.** Travels along the posterior surface of the interosseous membrane and supplies the superficial extensors. The posterior interosseous artery contributes to the recurrent interosseous artery, which anastomoses with the vascular network on the posterior side of the elbow.
    - **Recurrent interosseous artery.** Travels in a superior direction, posterior to the elbow complex, and forms an anastomosis with the **middle collateral artery**.

The ulnar artery terminates as the deep and superficial ulnar palmar arches of the hand.

RADIAL ARTERY

The radial artery travels through the cubital fossa along the lateral side of the forearm, deep to the brachioradialis, and supplies the lateral forearm muscles. In the proximal forearm, the radial artery gives rise to the radial recurrent artery (Figure 32-4A and B).

- **Radial recurrent artery.** Courses anteriorly to the lateral epicondyle of the humerus to anastomose with the radial collateral artery and supplies the muscles on the lateral side of the forearm.

The radial artery terminates in the hand as the deep and superficial radial palmar arches.

VEINS OF THE FOREARM

The veins in the forearm consist of a superficial and a deep venous system. The superficial system consists of the **basilic vein**, located medially, and the **cephalic vein**, located laterally. Anterior to the elbow complex, the **median cubital vein** forms a connection between the basilic and cephalic veins. The deep venous system may consist of two or three veins that course with each artery.

▽ The **median cubital vein** is part of the superficial venous system. Because of its location, the median cubital vein frequently is used to draw venous blood and for vascular access. ▼

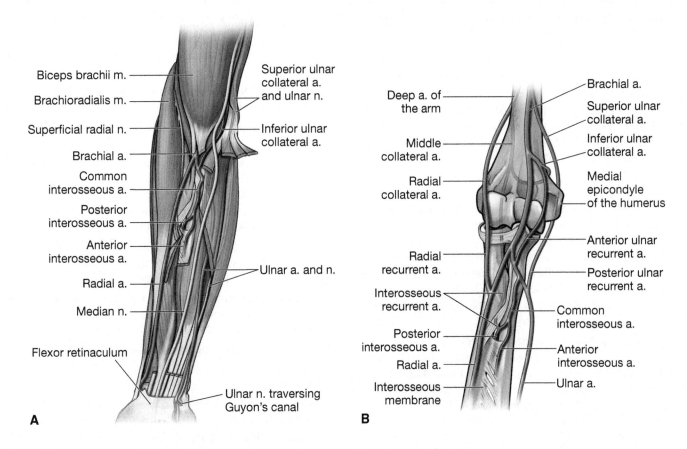

Figure 32-4: A. Arteries and nerves of the anterior forearm. **B**. Arteries of the elbow and forearm.

JOINTS CONNECTING THE FOREARM AND HAND

BIG PICTURE

The proximal and distal radioulnar joints form synovial pivot joints that provide pronation and supination of the forearm. The wrist complex is very flexible because of the synovial joint between the radius and the proximal row of carpal bones (**radiocarpal joint**) and the proximal and the distal row of carpal bones (**midcarpal joint**).

DISTAL RADIOULNAR JOINT

The **proximal** and **distal radioulnar joints** produce supination and pronation. They are mechanically linked; one joint is unable to move without the other (Figure 32-5A and B). The **distal radioulnar joint** consists of a **synovial pivot joint** between the **ulnar notch of the radius**, the **articular disc**, and the **head of the ulna**. The articular disc and its extensive fibrous connections are frequently referred to as the **triangular fibrocartilage complex** (*often referred to as the TFCC*).

- **Dorsal and palmar radioulnar ligaments.** The distal radioulnar joint is supported by two ligaments that originate from the dorsal and palmar aspects of the **ulnar notch of the radius** and extend to the base of the **styloid process of the ulna**. These ligaments form the margins for the triangular fibrocartilage complex.

- **Interosseous membrane.** The interosseous membrane is a wide sheet of connective tissue that connects the radius and ulna and functions to support both the proximal and distal radioulnar joints. The arrangement of the fibers allows for the transmission of forces from the hand and radius to the ulna.

WRIST COMPLEX

The wrist complex consists of the radiocarpal and midcarpal joints that result in wrist flexion and extension and in radial and ulnar deviation (Figure 32-5A and B).

- **Radiocarpal joint.** Articulation between the radius and the radioulnar disc (triangular fibrocartilage complex) with the proximal row of carpal bones (scaphoid, lunate, and triquetrum).

- **Midcarpal joint.** Articulation between the proximal row of carpal bones (scaphoid, lunate, and triquetrum) with the distal row of carpal bones (trapezium, trapezoid, capitate, and hamate).

The radiocarpal and midcarpal joints share similar ligamentous and capsular support because most of the structures that support the radiocarpal joint also cross the midcarpal joint. These joints consist of a fairly loose, but strong **capsule reinforced with the following ligaments**:

- **Palmar radiocarpal ligament.** Reinforces the anterior capsule and attaches proximally to the distal radius and distally to the scaphoid, lunate, triquetrum, and capitate.

- **Palmar ulnocarpal ligament.** Attaches proximally to the ulnar styloid process and the triangular fibrocartilage complex and distally to the lunate and triquetrum.

- **Dorsal radiocarpal ligament.** Reinforces the posterior capsule and attaches proximally to the distal radius and distally to the scaphoid, lunate, and triquetrum.

- **Ulnar collateral ligament.** Attaches from the ulnar styloid process to the triquetrum and pisiform.

- **Radial collateral ligament.** Attaches from the radial styloid process to the scaphoid and trapezium.

- **Intercarpal ligaments.** Interconnects carpal bones within and between rows.

▽ A **Colles' fracture** is a distal radial fracture that is usually caused by falling on an outstretched arm, resulting in a visual deformity proximal to the wrist complex. The fracture most often occurs about 1 to 2 inches proximal to the radiocarpal joint. ▽

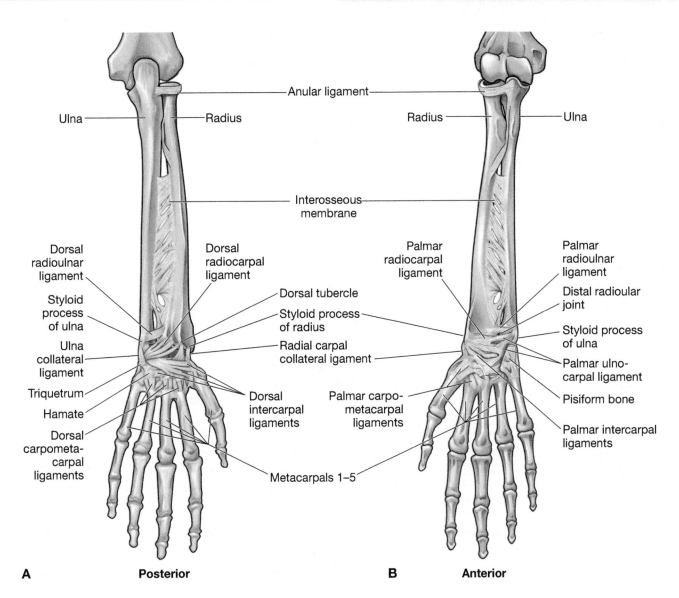

Anular ligament

Ulna

Radius

Radius

Ulna

Interosseous membrane

Dorsal radioulnar ligament

Dorsal radiocarpal ligament

Palmar radiocarpal ligament

Palmar radioulnar ligament

Distal radioular joint

Styloid process of ulna

Ulna collateral ligament

Triquetrum

Hamate

Dorsal carpometacarpal ligaments

Dorsal tubercle

Styloid process of radius

Radial carpal collateral igament

Dorsal intercarpal ligaments

Styloid process of ulna

Palmar ulno-carpal ligament

Pisiform bone

Palmar intercarpal ligaments

Palmar carpo-metacarpal ligaments

Metacarpals 1–5

A   **Posterior**

B   **Anterior**

Figure 32-5: Posterior (**A**) and anterior (**B**) views of the wrist joint.

TABLE 32-1. Muscles of the Forearm

Muscle	Proximal Attachment	Distal Attachment	Action	Innervation
Anterior forearm				
Pronator teres	Humeral head: medial epicondyle and supracondylar ridge of humerus Ulnar head: coronoid process of ulna	Midshaft of radius	Pronation and flexion of elbow	Median n. (C6–C7)
Flexor carpi radialis	Medial epicondyle of humerus	Metacarpals 2 and 3	Flexion and radial deviation of wrist; w elbow flexion	
Palmaris longus		Palmar aponeurosis	Flexion of wrist, weak elbow flexion, and tightens palmar aponeurosis	Median n. (C7–C8)
Flexor carpi ulnaris	Humeral head: medial epicondyle Ulnar head: olecranon and posterior border of ulna	Pisiform, hamate, and metacarpal 5	Weak elbow flexion, wrist flexion, ulnar deviation	Ulnar n. (C7–T1)
Flexor digitorum superficialis	Medial epicondyle, coronoid process of the ulna and anterior border of the radius	Lateral surfaces of the middle phalanx of digits 2–5	Flexion of wrist, and the metacarpophalangeal and proximal interphalangeal joints	Median n. (C8–T1)
Flexor digitorum profundus	Medial surfaces of proximal ulna and interosseous membrane	Distal phalanges of digits 2–5	Flexion of joints from wrist to distal interphalangeal joints	Medial part: ulnar n. (C8–T1) Lateral part: Anterior interosseous n. from median n. (C8–T1)
Flexor pollicis longus	Radius and interosseous membrane	Distal phalanx of digit 1	Flexion of the thumb	Anterior interosseous n. from median n. (C7–C8)
Pronator quadratus	Distal anterior ulna	Distal anterior radius	Pronation of elbow	

TABLE 32-1. Muscles of the Forearm (*Continued*)

Muscle	Proximal Attachment	Distal Attachment	Action	Innervation
Posterior forearm				
Anconeus	Lateral epicondyle of humerus	Olecranon process of the ulna	Extension of elbow	Radial n. (C6–C8)
Brachioradialis	Lateral supracondylar ridge of humerus	Styloid process of the radius	Flexion of elbow	Radial n. (C5–C6)
Extensor carpi radialis longus		Metacarpal 2	Extension and radial deviation of wrist	Radial n. (C6–C7)
Extensor carpi radialis brevis	Lateral epicondyle of humerus	Metacarpals 2 and 3		Posterior interosseous n. (C7–C8), the continuation of deep branch of radial n.
Extensor digitorum		Dorsal digital expansion of digits 2–5	Extension of wrist and digits	
Extensor digiti minimi		Dorsal digital expansion of digit 5	Extension of digit 5	
Extensor carpi ulnaris	Lateral epicondyle of humerus and posterior ulna	Metacarpal 5	Extension and ulnar deviation of wrist	
Supinator	Lateral epicondyle and supinator crest of ulna	Lateral surface of radius	Supination of forearm	Posterior interosseous n. (C6–C7)
Abductor pollicis longus	Ulna, radius, and interosseous membrane	Metacarpal 1	Abduction of thumb	Posterior interosseous n. (C7–C8), the continuation of deep branch of radial n.
Extensor pollicis brevis	Radius and interosseous membrane	Proximal phalanx of digit 1	Extension of thumb at metacarpophalangeal and carpometacarpal joints	
Extensor pollicis zlongus	Ulna and interosseous membrane	Distal phalanx of digit 1	Extension of thumb	
Extensor indicis		Dorsal digital expansion of digit 2	Extension of digit 2	

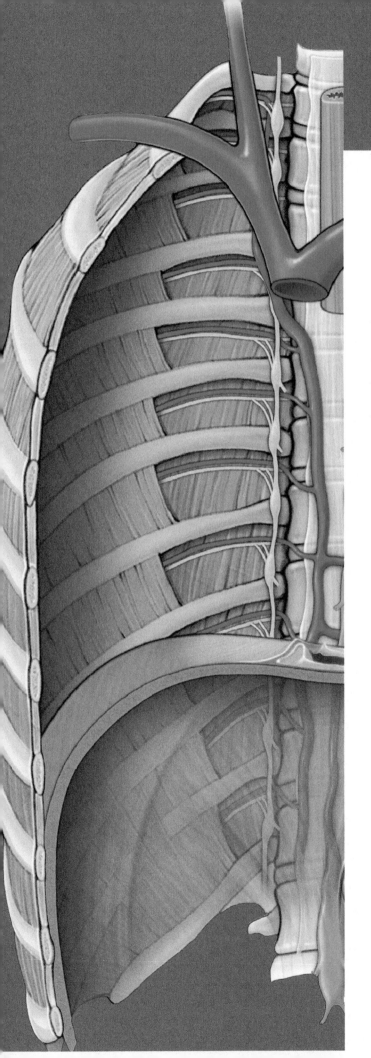

HAND

ORGANIZATION OF THE FASCIA OF THE HAND

BIG PICTURE

The fascia of the hand is continuous with the fascia of the forearm (antebrachium). In the hand, the fascia varies in thickness and divides the hand into five separate compartments that correspond with the five digits and have similar blood supply, innervation, and actions.

FASCIAL LAYERS OF THE PALMAR SIDE OF THE HAND (FIGURE 33-1A)

- **Palmar aponeurosis.** Located over the palm of the hand and covers the flexor tendons and deeper structures of the hand. The palmar aponeurosis extends distally and becomes continuous with the fibrous digital sheaths.
- **Fibrous digital sheaths.** Form a tunnel that encloses the flexor tendons of digits 2 to 5 and the tendon of the flexor pollicis longus muscle and their associated synovial sheaths.
- **Flexor retinaculum (transverse carpal ligament).** Forms a roof over the concavity created by the carpal bones, forming a tunnel (i.e., the carpal tunnel). The median nerve and the tendons of the flexor digitorum superficialis, flexor digitorum profundus, and flexor pollicis longus muscles, and their associated synovial sheaths, pass through this tunnel. The flexor retinaculum anchors medially to the pisiform and the hook of the hamate. Laterally, the flexor retinaculum is anchored to the scaphoid and trapezium.
- **Transverse palmar ligament.** Continuous with the extensor retinaculum from the dorsal side of the wrist and wraps around, anteriorly, to form a fascial band around the flexor tendons. This ligament should not be confused with the flexor retinaculum, which is located deeper to the transverse palmar ligament.

FASCIAL COMPARTMENTS OF THE PALMAR SIDE OF THE HAND

The fascial layers divide the palmar side of the hand into the following five compartments (Figure 33-1A):

- **Thenar compartment.** Contains three muscles that act on digit 1 (thumb).
- **Hypothenar compartment.** Contains three muscles that act on digit 5.
- **Central compartment.** Located between the thenar and hypothenar compartments and contains the flexor tendons and the lumbrical muscles.
- **Adductor compartment.** Contains the adductor pollicis muscle.
- **Interosseous compartment.** Located between the metacarpals and contains the dorsal and palmar interossei muscles.

FASCIAL LAYERS OF THE DORSAL SIDE OF THE HAND

- **Extensor retinaculum.** Continuous with the fascia of the forearm and attached laterally to the radius and medially to the triquetrum and pisiform bones. The extensor retinaculum works to retain the tendons that are near the bone while allowing proximal and distal gliding of the tendons (Figure 33-1B).
- **Dorsal digital expansions.** An aponeurosis covering the dorsum of the digits and attaches distal to the distal phalanx. Proximally and centrally, the extensor digitorum, extensor digiti minimi, extensor indicis, and extensor pollicis brevis muscles attach to the dorsal digital expansion. Laterally, the lumbricals and the dorsal and palmar interossei muscles attach. The small intrinsic muscles that attach laterally are responsible for delicate finger movements that would not be possible with the extensor digitorum, flexor digitorum superficialis, and profundus muscles alone. Because of the attachment of the muscles and the location of the hood, the small intrinsic muscles will produce flexion at the metacarpophalangeal joint while extending the interphalangeal joints.

FASCIAL COMPARTMENTS OF THE DORSAL SIDE OF THE WRIST

The extensor retinaculum of the hand divides the dorsum of the wrist into the following six compartments:

- **Compartment 1.** Contains the abductor pollicis longus and extensor pollicis brevis muscles.
- **Compartment 2.** Contains the extensor carpi radialis longus and brevis muscles.
- **Compartment 3.** Contains the extensor pollicis longus muscles.
- **Compartment 4.** Contains the extensor digitorum and extensor indicis muscles.
- **Compartment 5.** Contains the extensor digiti minimi muscles.
- **Compartment 6.** Contains the extensor carpi ulnaris muscles.

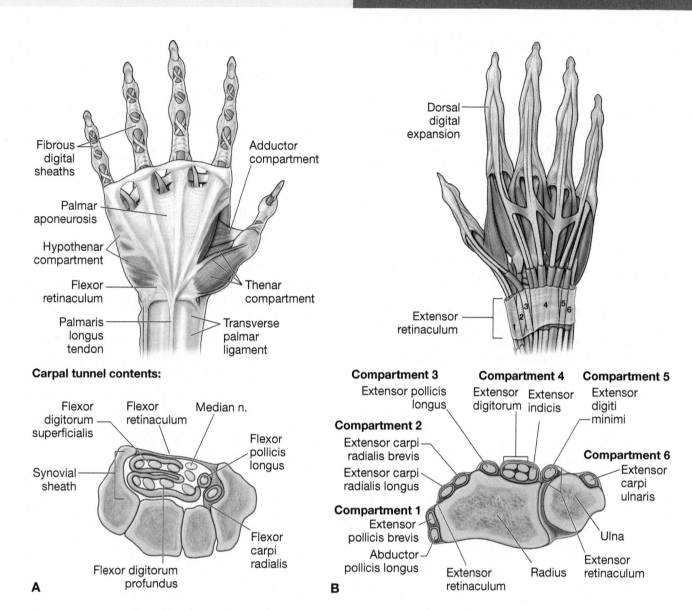

Figure 33-1: A. Fascia of the palm of the hand and carpal tunnel. **B**. Fascia of the posterior hand and extensor compartments.

ACTIONS OF THE FINGERS AND THUMB

BIG PICTURE

The hand consists of five digits (four fingers and a thumb). The thumb is considered digit 1; index finger is digit 2; middle finger is digit 3, ring finger is digit 4, and the little finger is digit 5. There are 19 bones and 19 joints in the hand distal to the carpal bones. Each digit has **carpometacarpal, metacarpophalangeal**, and **interphalangeal joints**.

ACTIONS OF THE FINGERS

The finger joints and associated movements are as follows (Figure 33-1C):

- **Carpometacarpal joints.** Sliding joints that allow for gliding and rotation.
- **Metacarpophalangeal joints.** Condylar joints that allow for flexion and extension as well as abduction and adduction. Movements of abduction and adduction are described in relation to digit 3 (the middle finger). All movements away from digit 3 are considered abduction, and movements toward digit 3 are considered adduction. Rotation is limited because of the collateral ligaments.
- **Interphalangeal joints.** Hinge joints that allow for flexion and extension. There is a proximal interphalangeal joint and a distal interphalangeal joint for digits 2 to 5. They are often referred to as PIP and DIP, respectively.

ACTIONS OF THE THUMB

The thumb joints and associated movements are as follows (Figure 33-1D):

- **Carpometacarpal joint.** Saddle joint that allows for opposition and reposition.
- **Metacarpophalangeal joint.** Hinge joint that allows for flexion and extension.
- **Interphalangeal joint.** Hinge joint that allows for flexion and extension. There is only one interphalangeal joint for digit 1 (the thumb).

The thumb is rotated 90 degrees to digits 2 to 5. Therefore, abduction and adduction occur in the sagittal plane, and flexion and extension occur in the coronal plane.

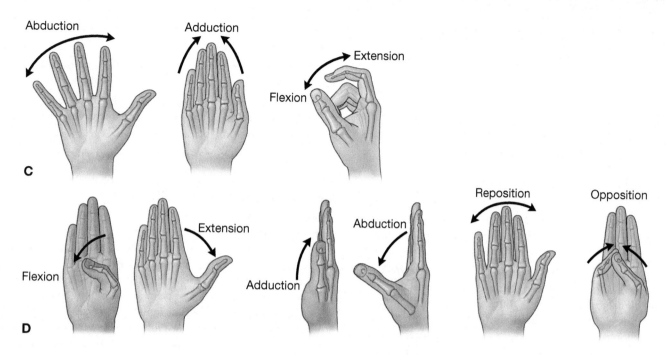

Figure 33-1: (*Continued*) **C**. Actions of digits 2–4. **D**. Actions of digit 1 (thumb).

MUSCLES OF THE HAND

BIG PICTURE

Muscles that act on the joints of the hand can be either extrinsic (originating outside the hand) or intrinsic (originating within the hand), and they may act on a single joint or on multiple joints. The result is movement of multiple joints for activities such as functional grasping or writing (Table 33-1).

MUSCLES IN THE THENAR COMPARTMENT

Thenar muscles have a common innervation from the **recurrent branch (motor branch) of the median nerve (C8–T1)**, except for the deep head of the flexor pollicis brevis muscle. The **thenar muscles** consist of the following (Figure 33-2A):

- **Abductor pollicis brevis muscle.** Attaches to the flexor retinaculum, scaphoid, and trapezium, proximal phalanx and dorsal digital expansion of the thumb. The abductor pollicis brevis muscle produces abduction of the thumb at the metacarpophalangeal joint.

- **Flexor pollicis brevis muscle.** Attaches to the carpals, flexor retinaculum and proximal phalanx of digit 1. Produces flexion of digit 1 and is innervated by the deep branch of the ulnar nerve (C8–T1).

- **Opponens pollicis muscle.** Attaches to the trapezium, flexor retinaculum, and metacarpal 1. The opponens pollicis muscle opposes and flexes the thumb.

MUSCLES IN THE HYPOTHENAR COMPARTMENT

Hypothenar muscles have a common innervation from the **deep branch of the ulnar nerve (C8–T1)** (Figure 33-2A).

- **Abductor digiti minimi muscle.** Attaches to the pisiform, pisohamate ligament, and proximal phalanx of digit 5. Produces abduction of digit 5 at the metacarpophalangeal joint.

- **Flexor digiti minimi brevis muscle.** Attaches to the hamate, flexor retinaculum, and the proximal phalanx of digit 5. Produces flexion of digit 5 at the metacarpophalangeal joint.

- **Opponens digiti minimi muscle.** Attaches to the hamate, flexor retinaculum, and metacarpal 5. The opponens digiti minimi muscle produces lateral rotation of metacarpal 5.

MUSCLES IN THE CENTRAL COMPARTMENT

Consists of the tendons of flexor digitorum superficialis and profundus and lumbrical muscles (Figure 33-2A and B).

- **Lumbrical muscles.** Attach to the tendons of the flexor digitorum profundus and to the lateral sides of the dorsal digital expansions of digits 2 to 5. The medial two lumbricals are innervated by the **deep branch of the ulnar nerve (C8–T1)**, and the lateral two lumbricals are innervated by the **digital branches of the median nerve (C8–T1)**. As a result of the insertion on the dorsal digital expansion, the lumbricals flex the metacarpophalangeal joints and extend the proximal and distal interphalangeal joints of digits 2 to 5.

MUSCLES IN THE ADDUCTOR COMPARTMENT

- **Abductor pollicis muscle.** Consists of a transverse and oblique head, both innervated by the **deep branch of the ulnar nerve (C8–T1)** (Figure 33-2A). As its name implies, this muscle adducts the thumb.

  - **Transverse head.** Attaches to metacarpal 3 and the proximal phalanx and the dorsal digital expansion of the thumb.

  - **Oblique head.** Attaches to the capitate bone, metacarpals 2 and 3 and the proximal phalanx and dorsal digital expansion of the thumb.

MUSCLES IN THE INTEROSSEI COMPARTMENT

All the interossei muscles share a common innervation from the **deep branch of the ulnar nerve (C8–T1)**.

- **Dorsal interossei muscle.** Attaches to the sides of the metacarpals and base of the proximal phalanx and dorsal digital expansions (Figure 33-2C). The dorsal interossei muscle abducts at the metacarpophalangeal joint. However, because the dorsal interossei muscle attaches to the dorsal digital expansion, it can also produce flexion of the metacarpophalangeal joint and extension at the proximal and distal interphalangeal joints. *A helpful acronym to remember the dorsal interossei muscle is "**DAB**," where "D" represents dorsal and "AB" represents abduction.*

- **Palmar interossei muscle.** Attaches to the sides of the metacarpals they act on and to the base of the proximal phalanx and the dorsal digital expansions (Figure 33-2D). The palmar interossei muscle adducts the metacarpophalangeal joint. However, because the muscle attaches to the dorsal digital expansion, it can also produce flexion of the metacarpophalangeal joint and extension at the proximal and distal interphalangeal joints. *A helpful acronym to remember the palmar interossei muscle is "**PAD**," where "P" represents palmar and "AD" represents adduct.*

▽ **Injury to the flexor tendons** is commonly caused by a cut that causes one or both of the flexor tendons that attach to the digits to have impaired flexion. If the flexor digitorum profundus tendon is cut, the distal interphalangeal joint will be unable to flex, whereas the proximal interphalangeal joint will flex due to the intact flexor digitorum superficialis. This injury may be seen in an athlete whose finger is caught in the jersey of an opponent, causing tearing or avulsion of the flexor digitorum profundus from the distal phalanx. The injury is often referred to as "**jersey finger.**" ▽

▽ **Extensor tendon injuries** in the region of the dorsal digital expansion into the distal phalanx may occur due to their superficial location. The injury usually occurs when an extended distal interphalangeal joint is forcefully flexed, causing tearing or even rupture. The result is the inability to extend the distal interphalangeal joint. The injury is often referred to as "**mallet finger.**" ▽

▽ **Injury to the dorsal digital expansion at the central slip** is due to tearing or avulsing of the central slip of the dorsal digital expansion from the middle phalanx. The injury is usually due to forceful flexion of an extended proximal interphalangeal joint, and the result is the inability to extend the proximal interphalangeal joint. If untreated, a **boutonnière deformity** may occur. A boutonnière deformity is flexion of the proximal interphalangeal joint, with hyperextension of the distal interphalangeal and metacarpophalangeal joints. ▽

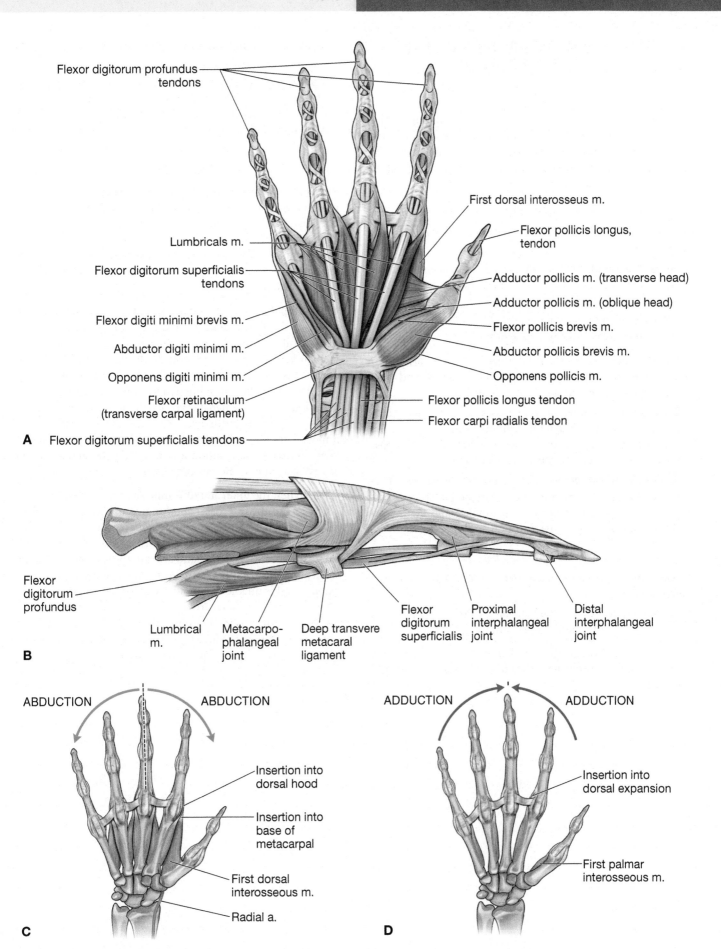

Figure 33-2: A. Muscles of the palm of the hand. **B**. Lumbrical muscles. **C**. Doral interossei muscles. **D**. Palmar interossei muscles.

TERMINAL BRANCHES OF THE BRACHIAL PLEXUS IN THE HAND

BIG PICTURE

The median nerve innervates lumbricals 1 and 2 and the thenar muscles (excluding the deep head of the flexor pollicis brevis). The ulnar nerve provides the remaining motor innervation to the hand. The superficial radial nerve, median nerve, and ulnar nerve provide sensory innervation to the hand.

ULNAR NERVE

In the forearm, the ulnar nerve gives rise to a dorsal branch and a palmar branch (Figure 33-3A).

- **Dorsal branch of ulnar nerve.** Provides cutaneous innervation to the medial side of the dorsum of the hand, digit 5, and the ulnar half of digit 4.
- **Palmar branch of ulnar nerve.** Provides cutaneous innervation to the medial palmar surface of the hand.

The ulnar nerve enters the hand superficial to the carpal tunnel, laterally to the pisiform with the ulnar artery, and then bifurcates into a deep and a superficial branch.

- **Deep branch of ulnar nerve.** Crosses the palm in a **fibro-osseous tunnel (Guyon's tunnel)** and supplies the hypothenar compartment, adductor pollicis, dorsal interossei, palmar interossei, and two medial lumbricals.
- **Superficial branch of ulnar nerve.** Supplies the palmaris brevis and then bifurcates into the common and the proper palmar digital branches to travel along digit 5 and the medial side of digit 4 to supply the surrounding skin.

MEDIAN NERVE

Proximal to the carpal tunnel, the **median nerve** gives rise to a palmar branch (Figure 33-3B).

- **Palmar branch of the median nerve.** Provides cutaneous innervation to the lateral palmar surface of the hand.

After passing through the carpal tunnel, the median nerve branches into the **recurrent branch** and the **palmar digital branches**.

- **Recurrent branch of the median nerve.** Innervates the thenar muscles of the hand.
- **Palmar digital nerves.** Travels along the first three digits and the lateral side of the fourth, supplying the lateral two lumbricals, the palmar skin of the first three digits, and the lateral side of the fourth digit.

RADIAL NERVE

The **superficial branch of the radial nerve** enters the hand by passing superficially to the **anatomical snuffbox**, and supplies the skin on the dorsal side of the first three digits (Figure 33-3C). The radial nerve has no motor innervation to intrinsic muscles of the hand and only innervates the extrinsic muscles that send tendons from muscles that originate in the posterior forearm to the thumb and digits.

▽ **Carpal tunnel syndrome** is a condition caused by swelling of the flexor digitorum superficialis, profundus, and flexor pollicis longus tendons, resulting in pressure on the median nerve. Repetitive motions of the fingers and wrist, hormonal changes, and vibration can be causes of tendon swelling. The result is tingling, numbness, and pain in the cutaneous distribution of the median nerve (lateral side). In more severe cases, atrophy of the thenar eminence may be present. It is important to remember that the palmar cutaneous branch provides cutaneous innervation to the lateral palm and should be spared in a patient who has carpal tunnel syndrome. ▽

▽ When the **ulnar nerve** is injured in the hand, there is a loss of the interossei and lumbricals 3 and 4, and clawing of digits 4 and 5 may become apparent due to an imbalance of the extrinsic and intrinsic muscles. Because the extrinsic extensors of the hand are not opposed by the intrinsic flexors of the hand, the metacarpophalangeal joint hyperextends and is unable to extend the proximal and distal interphalangeal joints. The proximal and distal interphalangeal joints continue to flex because the extrinsic flexors are not opposed by the intrinsic extensors of the distal and proximal interphalangeal joints. The result is extension of the metacarpophalangeal joint and flexion of the proximal and distal interphalangeal joints. ▽

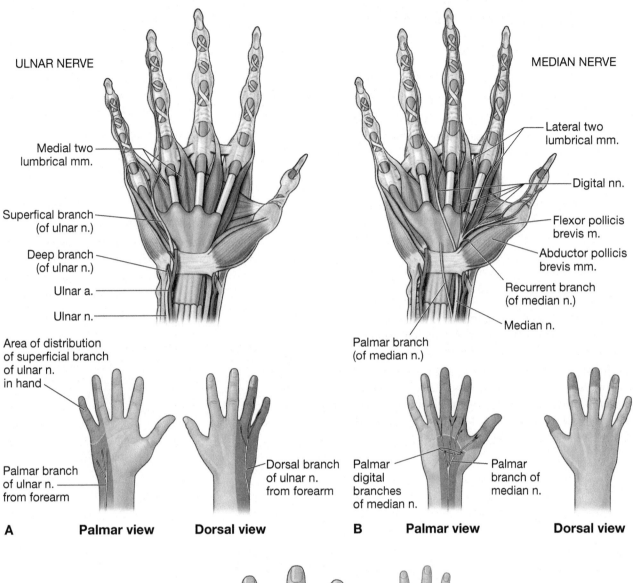

ULNAR NERVE

Medial two lumbrical mm.

Superfical branch (of ulnar n.)

Deep branch (of ulnar n.)

Ulnar a.

Ulnar n.

Area of distribution of superficial branch of ulnar n. in hand

Palmar branch of ulnar n. from forearm

Dorsal branch of ulnar n. from forearm

A   **Palmar view**   **Dorsal view**

MEDIAN NERVE

Lateral two lumbrical mm.

Digital nn.

Flexor pollicis brevis m.

Abductor pollicis brevis mm.

Recurrent branch (of median n.)

Median n.

Palmar branch (of median n.)

Palmar digital branches of median n.

Palmar branch of median n.

B   **Palmar view**   **Dorsal view**

RADIAL NERVE

Anatomical snuff box

Superfical branch (of radial n.)

C

Dorsal view

Palmar view

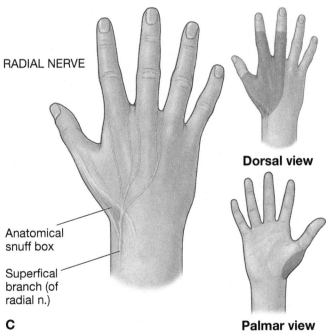

Figure 33-3: A. Ulnar nerve. **B**. Median nerve. **C**. Radial nerve.

VASCULARIZATION OF THE HAND

BIG PICTURE

The blood supply to the hand is provided by the radial and ulnar arteries, which give rise to a superficial and a deep palmar arch and to smaller tributaries as they travel distally to the tips of the fingers. The blood is returned to the axillary and subclavian veins via a deep and superficial venous system. The deep venous system follows the arteries.

ULNAR ARTERY

The ulnar artery, with the ulnar nerve, enters the hand lateral to the pisiform, where it gives rise to the deep palmar branch and becomes the principal contributor to the superficial palmar arch (Figure 33-4A).

- **Deep palmar branch.** Curves medially around the hook of the hamate to the deep layer of the palm, where it anastomoses with the deep palmar arch of the radial artery. It also gives rise to the **palmar metacarpal arteries**, which in turn anastomose with the **common palmar digital arteries** and bifurcate into the **proper palmar digital arteries.**
- **Superficial palmar arch.** Anastomoses with the palmar branch of the radial artery just deep to the palmar aponeurosis, where it gives rise to the common palmar digital arteries. These arteries then bifurcate to become the proper palmar digital arteries that supply the digits.

RADIAL ARTERY

The radial artery courses through the anatomical snuff box, contributes to the dorsal carpal arch, and then travels deep into the hand, becoming the principal contributor to the deep palmar arch (Figure 33-4B).

- **Deep palmar arch.** Travels deep to the adductor pollicis and anastomoses with the deep palmar branch of the ulnar artery, giving rise to the palmar metacarpal arteries.
- **Dorsal carpal arterial arch.** Courses along the dorsal side of the wrist, giving rise to the dorsal metacarpal and dorsal digital arteries.

VEINS OF THE HAND

The hand contains a **deep** and a **superficial venous system**. The deep veins are named according to the arteries they follow (Figure 33-4C). The superficial venous system drains into the **dorsal venous arch**. From the dorsal venous arch, the radial side of the hand drains to the cephalic vein and the ulnar side of the hand drains to the basilic vein.

▽ **Raynaud's syndrome** is a condition caused by a vascular spasm that most commonly involves the fingers but occasionally the toes as well. The spasms can be caused by cold or stress and result in numbness, burning pain, color changes, and tingling of one or more fingers. ▽

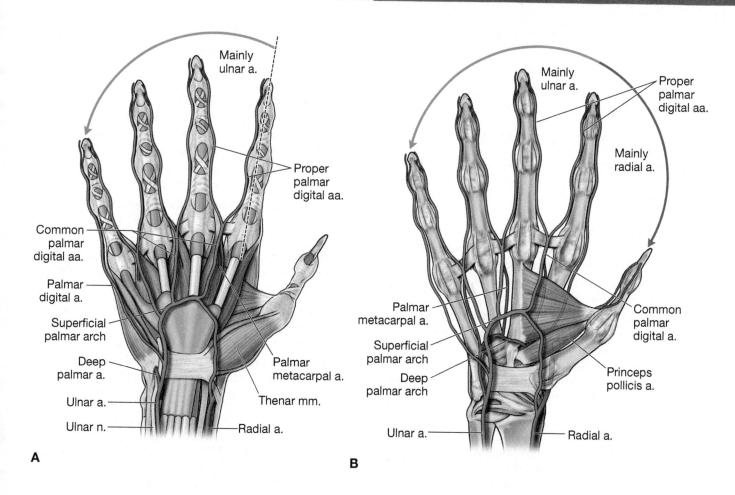

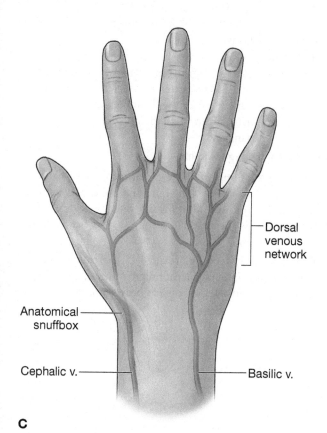

Figure 33-4: A. Superficial palmar arch. **B**. Deep palmar arch. **C**. Veins of the hand.

JOINTS OF THE HAND

BIG PICTURE

Each digit has a carpometacarpal, metacarpophalangeal, and one (digit 1) or two interphalangeal joints (digits 2–5).

JOINTS OF DIGITS 2 TO 4

Carpometacarpal joints. Articulate between the distal row of carpal bones and the metacarpals of digits 2 to 4. The carpometacarpal joints are classified most often as plane synovial joints (gliding) (Figure 33-5C).

Metacarpophalangeal joints. Articulate between the metacarpal head and the base of the proximal phalanx. The metacarpophalangeal joints are classified as a synovial condyloid joint that produces flexion and extension and abduction and adduction.

Interphalangeal joints. Digits 2 to 4 each have two interphalangeal joints (proximal and distal). Articulation of the interphalangeal joints occurs between the head of the proximal phalanx and the base of the phalanx distal to the proximal phalanx. The interphalangeal joints are classified as a synovial hinge that produces flexion and extension.

JOINTS OF DIGIT 1 (THUMB)

Carpometacarpal joints. A saddle joint that produces flexion and extension and abduction and adduction. The carpometacarpal joints also produce some rotation when combined with other motions of the thumb (Figure 33-5C).

Metacarpophalangeal joints. Articulate between the head of the first metacarpal and the base of the first proximal phalanx. The metacarpophalangeal joints are classified as synovial condyloid joints that produce flexion and extension and abduction and adduction.

Interphalangeal joints. Articulates between the head of the proximal phalanx and the base of the distal phalanx. The interphalangeal joints are classified as synovial hinge joints that produce flexion and extension.

LIGAMENTOUS AND CAPSULAR SUPPORT

Carpometacarpal joints. Supported by articular capsules and dorsal, palmar, and interosseous ligaments (Figure 33-5C).

Metacarpophalangeal. Supported by a capsule as well as a palmar and two collateral ligaments, a volar plate, and the deep transverse metacarpal ligament.

- **Palmar ligaments.** Located between the collateral ligaments on the palmar side and support the palmar side of the joint.
- **Collateral ligaments.** Attached proximally to the sides of the metacarpal and run distally in an anterior direction to attach to the phalanges. These ligaments are important for stability of the medial and lateral joint capsules.
- **Volar plate.** Structure that increases joint congruence. It is composed of fibrocartilage and is connected to the proximal phalanx and the joint capsule.
- **Deep transverse metacarpal ligaments.** Connect the metacarpal heads of digits 2 to 4 as well as connect laterally to the volar plate.

Interphalangeal joints. A joint capsule, a volar plate, and two collateral ligaments support both the proximal and distal interphalangeal joints.

- **Volar plate.** Supports and reinforces the joint capsule of the interphalangeal joints.
- **Collateral ligaments.** Located on the medial and lateral sides of the capsule and provide medial and lateral support throughout the proximal and distal interphalangeal movement.

MUSCULAR SUPPORT

Interphalangeal joints (Figure 33-5A and B)

- **Palmar side.** The flexor digitorum superficialis and the flexor digitorum profundus muscles will cross anteriorly at the proximal interphalangeal joint, whereas only the flexor digitorum profundus will cross anteriorly at the distal interphalangeal joint.
- **Dorsal side.** At the proximal and distal interphalangeal joint, the extensor digitorum, extensor digiti minimi, extensor pollicis longus, extensor pollicis brevis, lumbricals, and interossei muscles will support the dorsal aspect of the joint primarily through their attachment to the dorsal digital expansion.

Metacarpophalangeal joints

- **Palmar side.** Support to the metacarpophalangeal joints is provided by the flexor digitorum superficialis, flexor digitorum profundus, lumbricals, interossei, flexor digiti minimi brevis, flexor pollicis longus, and flexor pollicis brevis muscles.
- **Dorsal side.** Support is provided by the extensor digitorum, extensor indicis, extensor digiti minimi, extensor pollicis longus, and extensor pollicis brevis muscles.

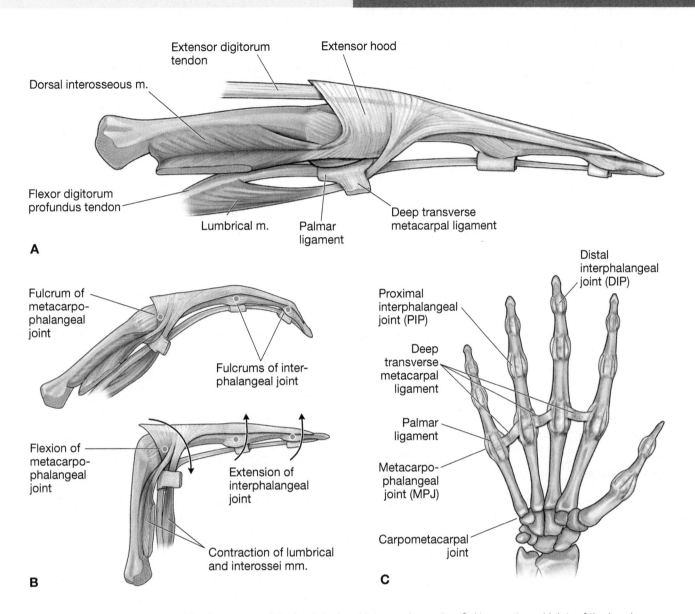

Figure 33-5: A. Extensor expansion. **B**. Movements of the lumbrical and interossei muscles. **C**. Ligaments and joints of the hand.

TABLE 33-1. Muscles of the Hand

Muscle	Proximal Attachment	Distal Attachment	Action	Innervation
Palmaris brevis	Palmar aponeurosis and flexor retinaculum	Dermis on the ulnar side of hand	Tenses the skin over the hypothenar muscles	Ulnar n., superficial branch (C8–T1)
Thenar muscles				
Abductor pollicis brevis	Flexor retinaculum, scaphoid, and trapezium	Proximal phalanx of digit 1	Abduction of thumb	Median n., recurrent branch (C8–T1)
Flexor pollicis brevis	Deep head: trapezium and flexor retinaculum Superficial head: trapezoid and capitate		Flexion of digit 1 (metacarpophalangeal joint)	
Opponens pollicis	Trapezium and flexor retinaculum	Metacarpal 1	Medial rotation of thumb and flexion of metacarpal of digit 1	
Adductor compartment				
Adductor pollicis	Oblique head: metacarpals 2 and 3 and capitate Transverse head: metacarpal 3	Proximal phalanx of digit 1	Adduction of thumb	Ulnar n., deep branch (C8–T1)
Hypothenar muscles				
Abductor digit minimi	Pisiform bone, pisohamate ligament, and tendon of flexor carpi ulnaris	Proximal phalanx of digit 5	Abduction of digit 5	Ulnar n., deep branch (C8–T1)
Flexor digiti minimi brevis	Hook of hamate and flexor retinaculum		Flexion of metacarpophalangeal joint of digit 5	
Opponens digiti minimi		Metacarpal 5	Lateral rotation of metacarpal 5	
Central compartment				
Lumbricals 1 and 2	Lateral two tendons of flexor digitorum profundus	Lateral sides of dorsal digital expansions for digits 2–5	Flexes metacarpophalangeal joints and extends interphalangeal joints	Median n. (C8–T1)
Lumbricals 3 and 4	Medial two tendons of flexor digitorum profundus			Ulnar n., deep branch (C8–T1)
Dorsal interossei 1–4	Adjacent sides of metacarpals	Dorsal digital expansions and base of proximal phalanges of digits 2–4	Abducts digits (DAB) and flexes metacarpophalangeal joints and extends interphalangeal joints	Ulnar n., deep branch (C8–T1)
Palmar interossei 1–3	Metacarpals 2, 4, and 5	Dorsal digital expansions and base of proximal phalanges of digits 2, 4, and 5	Adducts digits (PAD) and flexes metacarpophalangeal joints and extends interphalangeal joints	

STUDY QUESTIONS

Directions: Each of the numbered items or incomplete statements is followed by lettered options. Select the **one** lettered option that is **best** in each case.

1. Which of the following structures is the only boney connection between the axial and appendicular skeleton?

   A. Clavicle

   B. Humerus

   C. Radius

   D. Scapula

   E. Ulna

2. A 38-year-old construction worker sees his healthcare provider because of shoulder pain. Physical examination reveals a dislocated glenohumeral joint. Radiographic imaging reveals a tear in the muscles that stabilize the glenohumeral joint. Identify the muscle most likely injured in this patient.

   A. Biceps brachii muscle

   B. Infraspinatus muscle

   C. Pectoralis minor muscle

   D. Serratus anterior muscle

   E. Triceps brachii muscle

3. A 41-year-old executive sees her physician because of chronic spasm of the scalene muscles due to stress and depression. The physician determines that she has thoracic outlet syndrome. The scalene muscle spasms most likely affect which region of the brachial plexus?

   A. Branches

   B. Cords

   C. Divisions

   D. Roots

   E. Trunks

4. A 46-year-old woman sees her healthcare provider with a complaint of pain over the anterolateral forearm. Clinical examination reveals no muscle weakness in the patient's upper limb, but notes problems with the right lateral cutaneous nerve of the forearm. Which of the following is the most likely activity resulting in this patient's injury?

   A. Avulsion of the medial epicondyle of the humerus

   B. Fracture in the midhumeral region

   C. Hypertrophy of the coracobrachialis muscle

   D. Tendon inflammation on the lateral epicondyle of the humerus

   E. Venipuncture of the right cephalic vein in the antebrachial fossa

5. The radial and ulnar arteries most likely arise from the bifurcation of which artery?

   A. Axillary

   B. Brachial

   C. Cephalic

   D. Subclavian

6. The upper subscapular, lower subscapular, and thoracodorsal nerves branch from which cord of the brachial plexus?

   A. Anterior cord

   B. Lateral cord

   C. Medial cord

   D. Posterior cord

7. Which of the following is the limb muscle in which its motor neuron origin resides in a cranial nerve (CN)?

   A. Levator scapulae

   B. Pectoralis minor

   C. Rhomboid major

   D. Serratus anterior

   E. Trapezius

8. You watch a friend as he is doing pushups and notice the medial border of his right scapula protruding from his thorax more than it protrudes on his left side. Which muscle is weakened on your friend's right side that is causing this protrusion?

   A. Pectoralis major muscle

   B. Serratus anterior muscle

   C. Trapezius muscle

   D. Triceps brachii muscle

9. A paralabral cyst arising from a detached inferior glenoid labrum tear compresses neurovascular structures coursing through the quadrangular space. If this condition were to become chronic, which of the following findings would most likely be revealed on an MRI?

   A. Atrophy in the deltoid muscle

   B. Atrophy in the biceps brachii muscle

   C. Atrophy of the pectoralis major muscle

   D. Impingement of the ulnar nerve

   E. Impingement of the radial nerve

   F. Impingement of the medial nerve

10. The suprascapular and dorsal scapular arteries form a collateral circuit on the posterior side of the scapula with which of the following branches of the axillary artery?

    A. Anterior circumflex humeral artery

    B. Circumflex scapular artery

    C. Posterior circumflex humeral artery

    D. Thoracodorsal artery

11. Which of the following muscles can flex, extend, and abduct the glenohumeral joint?

   A. Biceps brachii muscle

   B. Deltoid muscle

   C. Latissimus dorsi muscle

   D. Pectoralis major muscle

   E. Triceps brachii muscle

12. A 41-year-old construction worker visits his healthcare provider because of an infected cutaneous laceration in his hand. Bacteria entering the lymph via the lesion will next pass through which lymph nodes?

   A. Apical nodes

   B. Central nodes

   C. Humeral nodes

   D. Pectoral nodes

13. The boundaries of the three parts of the axillary artery are determined by its relationship to which muscle?

   A. Pectoralis major muscle

   B. Pectoralis minor muscle

   C. Teres major muscle

   D. Teres minor muscle

14. Which of the following muscles flexes the glenohumeral and elbow joints and supinates the radioulnar joints?

   A. Coracobrachialis muscle

   B. Biceps brachii muscle

   C. Brachialis muscle

   D. Triceps brachii muscle

15. A 17-year-old patient sees his healthcare provider with a complaint of weakness with elbow flexion and numbness on the lateral side of the forearm. Which of the flowing nerves is most likely damaged?

   A. Axillary nerve

   B. Median nerve

   C. Musculocutaneous nerve

   D. Radial nerve

   E. Ulnar nerve

16. Which of the following nerves courses between the brachialis and brachioradialis muscles?

   A. Axillary nerve

   B. Median nerve

   C. Musculocutaneous nerve

   D. Radial nerve

   E. Ulnar nerve

17. The superior ulnar collateral artery forms a collateral circuit with which of the following arteries?

   A. Anterior interosseous artery

   B. Anterior ulnar recurrent artery

   C. Middle collateral artery

   D. Posterior ulnar recurrent artery

   E. Radial collateral artery

18. A patient is diagnosed with a peripheral nerve injury that weakens his ability to extend his elbow, wrist, and fingers. Which area of this patient's upper limb will most likely experience cutaneous deficit as a result of this injury?

   A. Anterior forearm

   B. Lateral forearm

   C. Medial forearm

   D. Posterior forearm

19. A 49-year-old woman is diagnosed with carpal tunnel syndrome. Which tendon of the following muscles would most likely be associated with carpal tunnel syndrome?

   A. Flexor carpi radialis muscle

   B. Flexor carpi ulnaris muscle

   C. Flexor pollicis longus muscle

   D. Palmaris longus muscle

   E. Pronator teres muscle

   F. Pronator quadratus muscle

20. Which of the following muscles flexes the wrist and the metacarpophalangeal and the proximal and distal interphalangeal joints of digits 2 to 5?

   A. Flexor carpi radialis muscle

   B. Flexor carpi ulnaris muscle

   C. Flexor digitorum profundus muscle

   D. Flexor digitorum superficialis muscle

21. Which of the following muscles flexes the metacarpophalangeal joints, but extends the interphalangeal joints of digits 2 to 5?

   A. Flexor digitorum profundus muscle

   B. Lumbrical muscle

   C. Flexor digitorum superificialis muscle

   D. Palmaris brevis muscle

22. Which of the following arteries supplies blood to the deep extensor muscles of the forearm?

    A. Anterior interosseous artery

    B. Posterior interosseous artery

    C. Radial collateral artery

    D. Radial recurrent artery

23. The radiocarpal joint includes the distal end of the radius, the triangular fibrocartilage complex, the scaphoid bone, the triquetrum bone, and which of the following carpal bones?

    A. Capitate

    B. Hamate

    C. Lunate

    D. Trapezium

24. Which of the following fascial layers forms the roof of the carpal tunnel?

    A. Fibrous digital sheaths

    B. Flexor retinaculum

    C. Palmar aponeurosis

    D. Transverse palmar ligament

25. Inflammation in Guyon's canal will most likely result in weakness in which of the following movements?

    A. Adduction of digits 2 to 5

    B. Adduction of the thumb

    C. Flexion of the wrist

    D. Extension of the wrist

26. Compression of the median nerve in the carpal tunnel leads to weakness in the thenar muscles and the first and second lumbricals as well as cutaneous deficits in which of the following regions?

    A. Lateral dorsal surface of the hand

    B. Medial dorsal surface of the hand

    C. Palmar surface of digit 5

    D. Palmer surface of digits 2 and 3

27. Which of the following arteries courses through the anatomical snuffbox?

    A. Deep palmar arch

    B. Radial artery

    C. Superficial palmar arch

    D. Ulnar artery

ANSWERS

1—A: The clavicle connects the manubrium of the sternum to the acromion of the scapula. All other support is through muscles and ligaments.

2—B: The rotator cuff muscle group stabilizes the glenohumeral joint. The tendons of these muscles reinforce the ligaments of the glenohumeral joint capsule. The tendons of the long head of the biceps and triceps brachii muscles attach to the supraglenoid and infraglenoid tubercles, but do not significantly contribute to stability of the glenohumeral joint.

3—D: The roots of the brachial plexus pass between the anterior and middle scalene muscles. Spasm of these muscles can cause entrapment of the roots of the plexus.

4—E: The musculocutaneous nerve innervates the anterior compartment of the arm. The cephalic vein courses in the antebrachial fossa adjacent to the lateral cutaneous nerve of the forearm. Therefore, a venipuncture of the cephalic vein may injure the adjacent cutaneous branch of the musculocutaneous nerve. Avulsion of the medial epicondyle would affect forearm flexors, and a midhumeral fracture would affect the radial nerve. Hypertrophy of the coracobrachialis muscle would affect the entire musculocutaneous nerve and result in the cutaneous presentation, but would also negatively affect motor activity. Lateral epicondyle inflammation would affect forearm extensors.

5—B: The brachial artery bifurcates just distal to the elbow to form the radial and ulnar arteries.

6—D: The upper subscapular, lower subscapular, and thoracodorsal nerves branch off of the posterior cord in the axilla, just anterior to the subscapularis muscle.

7—E: The trapezius muscle is the only upper limb muscle innervated by the spinal accessory nerve (CN XI).

8—B: The serratus anterior muscle stabilizes the medial border of the scapula against the thorax. When in a pushup position, the medial border of the scapula is pushed away from the thorax, making the weakness more apparent.

9—A: The axillary nerve courses through the quadrangular space with the posterior humeral circumflex artery. Therefore, compression of the axillary nerve would weaken the deltoid muscle and thus weaken shoulder abduction.

10—B: The circumflex scapular artery courses through the triangular space to form a collateral circuit with the suprascapular and dorsal scapular arteries.

11—B: The deltoid muscle inserts in the deltoid tuberosity of the humerus. However, its origins include the lateral third of the clavicle (flexion), acromion (abduction), and scapular spine (extension).

12—C: Lymph traveling from the arm will first course through the humeral nodes before continuing toward the central and apical nodes of the axilla.

13—B: The boundaries of the three parts of the axillary artery are determined by its relationship to the pectoralis minor muscle.

14—B: The biceps brachii muscle flexes the glenohumeral and elbow joints because of its anterior position. Its attachment to the radial tuberosity also allows it to supinate the radioulnar joints.

15—C: When the musculocutaneous nerve is damaged, the biceps brachii and brachialis muscles are weakened or paralyzed. The skin innervated by the lateral cutaneous nerve of the forearm would feel tingly or numb.

16—D: The radial nerve courses between the brachialis and brachioradialis muscles on the lateral side of the brachium after piercing the intermuscular septum.

17—D: The superior ulnar collateral artery anastomoses with the posterior ulnar recurrent artery from the ulnar artery that is posterior to the medial epicondyle.

18—D: Damage to the radial nerve would cause the weakness in the triceps brachii muscle and extension of the elbow. This damage would cause deficits in the cutaneous field of the radial nerve in the posterior forearm.

19—C: The tendon of the flexor pollicis longus muscle courses through the carpal tunnel with the tendons of the flexor digitorum superficialis and the flexor digitorum profundus muscles and the median nerve.

20—C: The flexor digitorum profundus muscle flexes the wrist and the metacarpophalangeal and the proximal and distal interphalangeal joints of digits 2 to 5.

21—B: The lumbrical muscles cross anterior to the metacarpophalangeal joints, then insert on the extensor expansion. It is this orientation that allows the muscles to flex the metacarpophalangeal joints and extend the interphalangeal joints.

22—B: The posterior interosseous artery branches from the common interosseous artery, courses along the anterior surface of the interosseous membrane, and pierces the membrane to supply the deep extensor muscles.

23—C: The radiocarpal joint is formed by the distal end of the radius, the triangular fibrocartilage complex, and the proximal row of the carpal bones. The lunate bone is included in the proximal carpals.

24—B: The flexor retinaculum anchors to the hamate, pisiform, trapezium, and scaphoid bones to enclose the tendons of the flexor digitorum superficialis, the flexor digitorum profundus, and the flexor pollicis longus muscles and the median nerve.

25—A: The ulnar nerve courses through Guyon's canal. Compression of the nerve will cause weakness in the muscles it innervates, including the palmer interosseous muscles, which are responsible for adduction of digits 2 to 5.

26—D: The digital branches of the median nerve send cutaneous branches to the skin of digits 2 and 3 and half of 4 primarily on the palmar side of the hand after the median nerve passes through the carpal tunnel. The palmar branch of the median nerve that innervates the lateral skin of the palm branches proximal to the carpal tunnel and would not be involved.

27—B: The radial artery courses through the anatomical snuffbox. The radial pulse can be felt at this site.

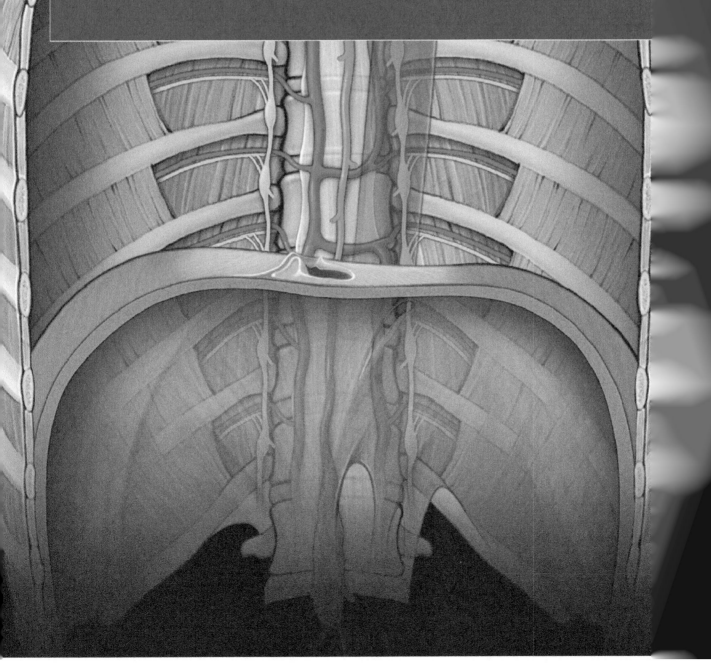

SECTION 7

LOWER LIMB

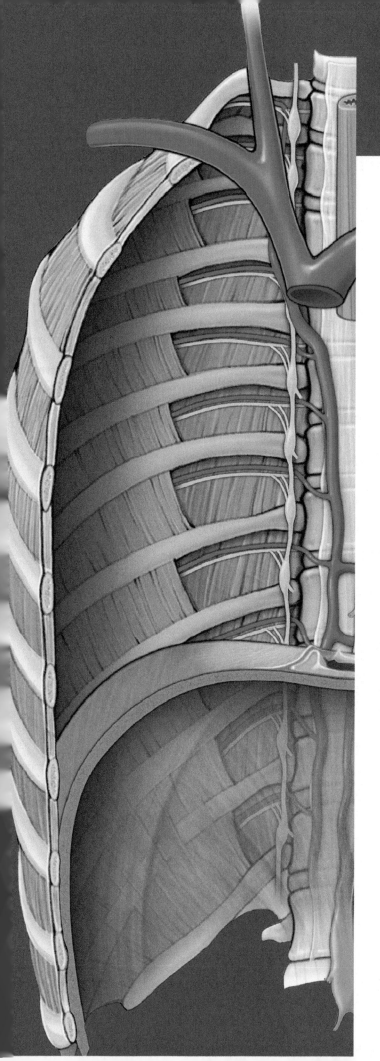

CHAPTER 34

OVERVIEW OF THE LOWER LIMB

BONES OF THE PELVIC REGION AND THIGH

BIG PICTURE

The bones of the skeleton provide a framework that serves as an attachment for soft tissues (e.g., muscles). The bony structure of the gluteal region and thigh, from proximal to distal, consists of the pelvis, femur, patella, tibia, and fibula (Figure 34-1A). Synovial joints and fibrous ligaments serve to connect bones together.

PELVIS

The pelvis is an irregularly shaped bone consisting of right and left pelvic bones. The pelvic bones articulate posteriorly with the sacrum, via the **sacroiliac joints**, and anteriorly with each other at the **pubic symphysis** (Figure 34-1A and B). Each pelvic bone has three components: **ilium, ischium**, and **pubis**. The **acetabulum** is a large cup-shaped structure at the junction where the ilium, ischium, and pubis fuse. The acetabulum protrudes laterally for articulation with the head of the femur bone. The three bony components of the pelvis form an opening, called the **obturator foramen**.

ILIUM The ilium is the most superior and the largest bone of the three components of the pelvis.

- **Iliac crest.** The entire superior margin of the ilium is thick and forms a prominent crest, which is palpable. The iliac crest is the site for muscle attachment and fascia of the abdomen, back, and lower limb. The iliac crest terminates anteriorly at the **anterior superior iliac spine** and posteriorly at the **posterior superior iliac spine**, both providing a site for muscle attachments.

- **Anterior inferior iliac spine.** A rounded protuberance located just inferior to the anterior superior iliac spine on the anterior surface of the ilium. The anterior inferior iliac spine serves as a site for the attachment of muscles and ligaments.

- **Posterior inferior iliac spine.** A less prominent spine along the posterior border of the sacral surface of the ilium.

- **Iliac fossa.** Has an anteromedial surface of the wing, which is concave and forms a large fossa. The iliac fossa serves as a site for muscle attachment.

ISCHIUM The ischium is the posterior and inferior component of the pelvic bone.

- **Ischial tuberosity.** The most prominent feature of the ischium, a large tuberosity on the posteroinferior aspect of the bone. The ischial tuberosity is an important site for the attachment of muscles of the lower limb, primarily the hamstrings, and for supporting the body in a seated position.

- **Ischial ramus.** Projects anteriorly to join with the inferior ramus of the pubis.

- **Ischial spine.** A prominent spine that separates the lesser sciatic notch from the greater sciatic notch.

PUBIS The pubis is the anterior and inferior part of the pelvic bone. The pubis has a body and two arms called rami.

- **Pubic tubercle.** A rounded crest on the superior surface of the pubis.

- **Superior pubic ramus.** Projects posterolaterally from the body and joins with the ilium and ischium at its base, positioned toward the acetabulum.

- **Inferior pubic ramus.** Projects laterally and inferiorly to join with the ramus of the ischium.

FEMUR

The femur is located in the thigh and is the longest bone of the body. The following landmarks are located on the femur (Figure 34-1A and C):

- **Head.** A spherically shaped knob on the proximal end of the femur that articulates with the acetabulum of the pelvic bone. The head is characterized by a nonarticular **fovea** on its medial surface, which serves as an attachment for the **foveolar ligament** (ligament of the head of the femur).

- **Neck.** A cylindrical part of the bone that connects the head to the shaft of the femur. The neck has a unique superomedial projection from the shaft at an angle of about 125 degrees and a slight forward projection.

- **Greater trochanter.** Extends superiorly from the shaft of the femur, just lateral to the site where the neck joins the shaft. The greater trochanter is a major attachment site for muscles.

- **Lesser trochanter.** A smaller but prominent conically shaped protuberance. The lesser trochanter projects posteromedially from the shaft, just inferior to the junction with the neck. The lesser trochanter serves as an attachment for muscles.

- **Linea aspera.** A distinct vertical ridge on the posterior aspect of the femoral shaft that serves as an attachment for muscles.

- **Pectineal line.** Curves medially under the lesser trochanter and around the shaft of the femur to merge with the linea aspera.

- **Medial and lateral condyles.** Both condyles lie at the distal aspect of the femur and articulate with the tibia to form the knee joint. The medial supracondylar lines terminate at a prominent adductor tubercle, which lies just superior to the medial condyle.

- **Medial and lateral epicondyles.** Rounded eminences on the medial and lateral surfaces of either condyle, which serve as the attachment for the collateral ligaments of the knee joint.

- **Popliteal fossa.** The posterior surface of the distal shaft of the femur forms the floor of the popliteal fossa and its margins.

PATELLA

The patella (knee cap) is the largest sesamoid bone in the body. It is formed in the tendon of the quadriceps femoris muscle as it crosses the anterior surface of the knee joint (Figure 34-1A). The patella has a unique triangular shape.

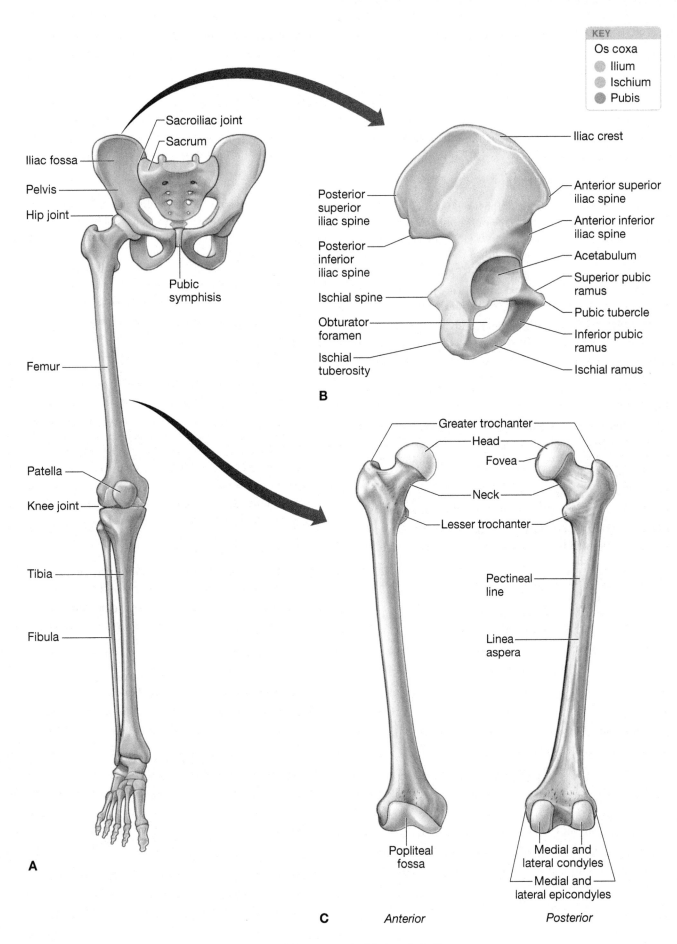

Figure 34-1: A. Skeleton of the lower limb. **B**. Osteology of the os coxa (pelvic bone). **C**. Femur.

BONES OF THE LEG AND FOOT

BIG PICTURE

The bony structure of the leg and foot, from proximal to distal, consists of the tibia, fibula, 7 tarsals, 5 metatarsals, and 14 phalanges (Figure 34-2A). The tibia and fibula are bound together by a tough, fibrous sheath known as the **interosseous membrane**.

TIBIA

The tibia is the medial and larger of the two bones of the leg (the fibula is the other bone). It is also the only bone that articulates with the femur at the knee joint (Figure 34-2A and B). The following landmarks are found on the tibia:

- **Medial and lateral condyles.** Lie on the proximal aspect of the tibia. The medial and lateral condyles are flattened in the horizontal plane and overhang the shaft of the tibia.
- **Intercondylar eminence.** The intercondylar region of the tibial plateau lies between the articular surfaces of the medial and lateral condyles and narrows centrally at a raised site to form the eminence. The intercondylar eminence contains sites for attachments of the **anterior and posterior cruciate ligaments**.
- **Tibial tuberosity.** A palpable inverted triangular area that lies on the proximal and anterior part of the shaft. The tibial tuberosity is a large tuberosity and is the site of attachment for the **patellar ligament**.
- **Soleal line.** A roughened and oblique line that lies posteriorly on the tibia, serving as an attachment for the soleus muscle.
- **Anterior border.** A sharp and palpable ridge on the anterior surface of the tibia that descends from the tibial tuberosity, down the tibial shaft.
- **Interosseous border.** A vertical ridge that descends along the lateral surface of the tibia. The interosseous border is the attachment site for the interosseous membrane, which is located between the tibia and the fibula.
- **Medial malleolus.** The most distal aspect of the tibia. The medial malleolus is shaped like a rectangular box with a bony protuberance on the medial side. The medial malleolus articulates with the talus, a tarsal bone, to form a large part of the ankle joint.
- **Fibular notch.** The lateral surface of the distal end of the tibia at the site where the fibula articulates.

FIBULA

The fibula is lateral to the tibia and has limited involvement in weight-bearing activity. The fibular shaft is much narrower than the shaft of the tibia and is mainly enclosed by muscles (Figure 34-2A and B). The following landmarks are found on the fibula:

- **Head.** The most proximal aspect of the fibula. The head is a globe-shaped expansion that articulates with the tibia.
- **Neck.** The neck of the fibula separates the head from the fibular shaft.

- **Interosseous border.** A subtle vertical ridge that descends along the medial surface of the fibula. The interosseous border is the attachment site for the interosseous membrane, which is located between the fibula and the tibia.
- **Lateral malleolus.** The most distal end of the fibula, which expands to form a spade-shaped lateral malleolus. The lateral malleolus articulates with the lateral surface of the talus, forming the lateral part of the ankle joint.

FOOT

The foot is the region of the lower limb, distal to the ankle joint, and is subdivided into three sections: ankle, metatarsus, and digits. There are five digits in the foot, beginning medially with the great toe and four lateral digits (Figure 34-2A and C). The foot has a superior surface, the dorsum, and an inferior surface, the plantar surface. The three groups of bones that comprise the foot are the tarsals, metatarsals, and phalanges.

- **Tarsal bones.** There are seven tarsal bones, arranged in a proximal and distal group with an intermediate bone on the medial side of the foot. The proximal group consists of the calcaneus bone and the talus bone; the intermediate bone is the navicular bone; and the distal group consists of the cuboid and the three cuneiform bones.
  - **Talus bone.** The most superior bone of the foot; sits on top of the calcaneus bone. The talus bone articulates with the tibia and fibula to form the ankle joint, as well as articulating with the intermediate navicular bone on the medial side of the foot.
  - **Calcaneus bone.** The largest of the tarsal bones. The calcaneus bone forms the heel posteriorly; anteriorly, it protrudes forward to articulate with the cuboid bone on the lateral side of the foot.
  - **Navicular bone.** Articulates with the talus bone posteriorly and with the cuneiform bones anteriorly.
  - **Medial, intermediate, and lateral cuneiform bones.** Are part of the distal group of tarsal bones. The cuneiform bones articulate with the navicular bone and the three medial metatarsal bones.
  - **Cuboid bone.** Articulates with the calcaneus bone and with the two most lateral metatarsal bones.
- **Metatarsal bones.** There are five metatarsal bones in the foot, numbered one through five, from medial to lateral. The first metatarsal is associated with the great toe and is the shortest and the thickest bone; the second metatarsal is the longest bone. Each metatarsal has a head at the distal end, a base at the proximal end, and a shaft between the head and the base.
- **Phalanges.** The phalanges are the very short bones of the digits. Each digit, except the great toe, has three phalanges (the great toe has only two phalanges). The phalanges are named by location as proximal, middle, and distal for the lateral four digits, and only proximal and distal for the great toe. Each phalange has a base (proximal), a shaft, and a head (distal).

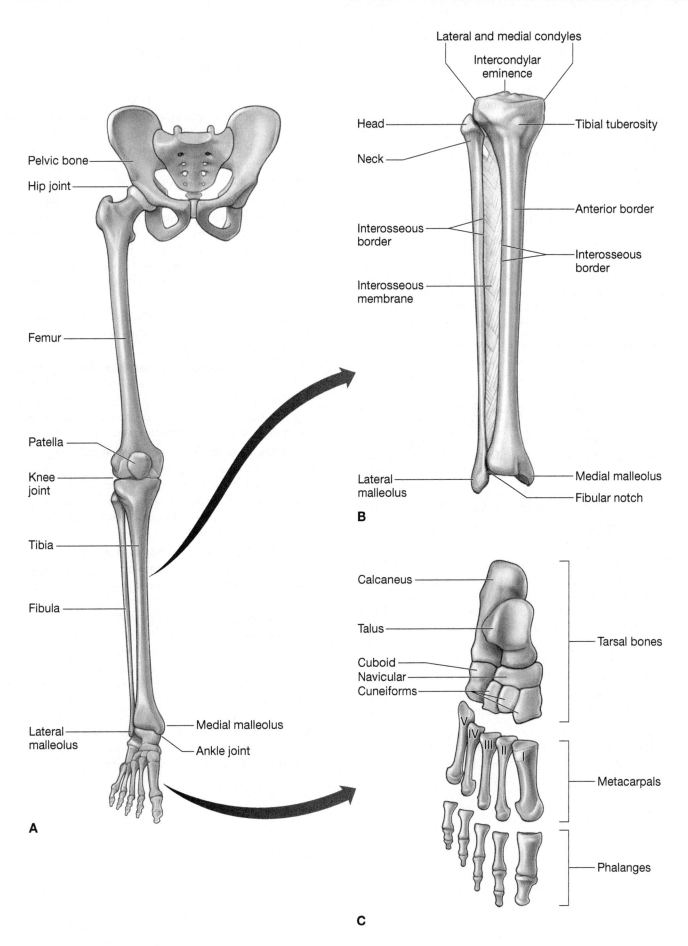

Figure 34-2: A. Skeleton of the lower limb. **B**. Tibia and fibula. **C**. Osteology of the foot.

FASCIAL PLANES AND MUSCLES

BIG PICTURE

Two facial layers, known as the superficial fascia and the deep fascia, are located between the skin and bone of the lower limb. The deep fascia divides the lower limb into anterior, medial, and posterior compartments of the thigh (fascia lata) and anterior, lateral, and posterior compartments of the leg (crural fascia). Muscles are organized into these compartments and possess common attachments, innervation, and actions.

FASCIA OF THE LOWER LIMB

The lower limb consists of superficial and deep fascia (Figure 34-3A and B).

- **Superficial fascia.** Referred to as the subcutaneous or hypodermis layer and located deep to the skin. The superficial fascia primarily contains fat and superficial veins (i.e., greater and small saphenous veins), lymphatics, and cutaneous nerves.

- **Deep fascia.** Lies deep to the superficial fascia and primarily contains muscles, nerves, vessels, and lymphatics. The deep fascia of the lower limb connects or is continuous with the inguinal ligament, pubic bone, and Scarpa's fascia of the inferior abdominal wall. The deep fascia develops intermuscular septae that extend to the bones, dividing the thigh and leg into three compartments. Each compartment contains muscles that perform similar movements and have a common innervation.

MUSCLES OF THE LOWER LIMB

The muscles of the lower limb are organized into the following groups:

- **Gluteal muscles.** The muscles of the gluteal region primarily act on the hip joint, producing extension, medial rotation, lateral rotation, and abduction. In addition to producing motion, the muscles of the gluteal region are important for stability of the trunk and hip joint and for locomotion. These muscles consist of the gluteus maximus, gluteus medius, gluteus minimus, piriformis, superior gemellus, inferior gemellus, obturator internus, quadratus femoris, and tensor fascia lata muscles.

- **Thigh muscles.** The deep fascia divides the thigh into anterior, medial, and posterior compartments, with common actions and innervation (Figure 34-3A).

  - **Muscles of the anterior compartment of the thigh.** Consist of the psoas major, psoas minor, iliacus, sartorius, and quadriceps muscles. The quadriceps muscle group consists of the rectus femoris, vastus lateralis, vastus medialis, and vastus intermedius muscles. Most of the muscles share common actions (extension of the knee and flexion of the hip) and innervation (femoral nerve).

  - **Muscles of the medial compartment of the thigh.** Consist of the pectineus, adductor longus, adductor magnus, adductor brevis, gracilis, and obturator externus muscles. Most of the muscles share common actions (adduction of the hip) and innervation (obturator nerve).

  - **Muscles of the posterior compartment of the thigh.** Consist of the hamstring group of muscles, which consists of the semitendinosus, semimembranosus, and biceps femoris muscles. Most of the muscles share common actions (hip extension and knee flexion), a common proximal attachment (ischial tuberosity), and innervation (tibial nerve).

- **Leg muscles.** The deep fascia divides the leg into anterior, lateral, and posterior compartments with common attachments, actions, and innervation (Figure 34-3B).

  - **Muscles of the anterior compartment of the leg.** Consist of the tibialis anterior, extensor hallucis longus, and fibularis tertius muscles. Many of these muscles share common actions (dorsiflexion and inversion at the ankle and extension of the digits) and innervation (deep fibular nerve).

  - **Muscles of the posterior compartment of the leg.** Consist of the gastrocnemius, plantaris, soleus, popliteus, flexor hallucis longus, flexor digitorum longus, and tibialis posterior muscles. Many of these muscles share common actions (plantarflexion at the ankle and flexion of the digits) and innervation (tibial nerve).

  - **Muscles of the lateral compartment of the leg.** Consist of the fibularis (peroneus) longus and fibularis (peroneus) brevis muscles. Many of these muscles share common actions (plantarflexion and eversion at the ankle) and innervation (superficial fibular nerve).

- **Foot muscles.** There are four layers of intrinsic muscles of the foot. Layer 1 consists of the abductor digiti minimi, flexor digitorum brevis, and abductor hallucis muscles. Layer 2 consists of the lumbricals and the quadratus plantae muscles. Layer 3 consists of the flexor digiti minimi, flexor hallucis brevis, and adductor hallucis muscles. Layer 4 consists of the plantar and dorsal interossei muscles. The extensor digitorum brevis and extensor hallucis brevis muscles are the only two intrinsic muscles on the dorsum of the foot.

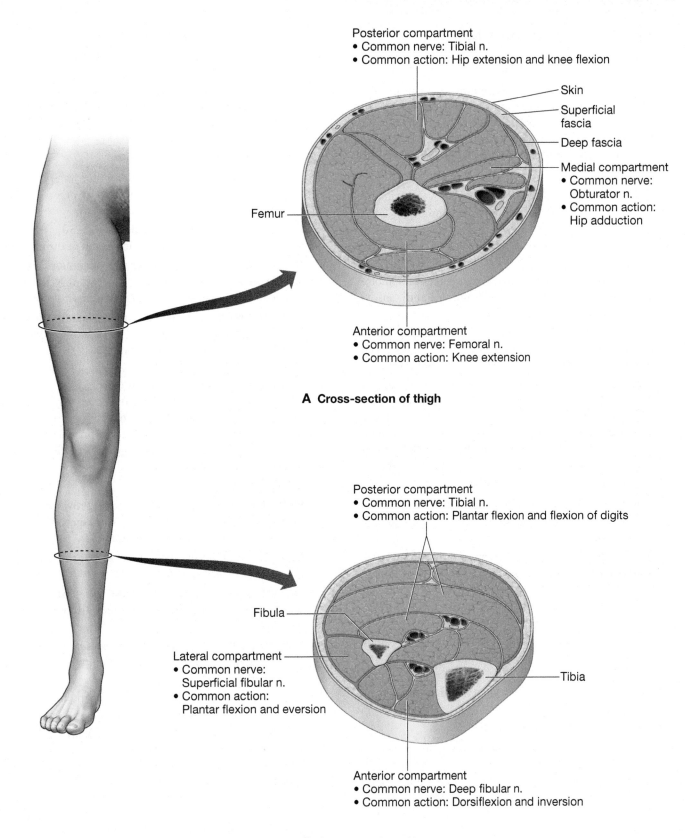

Posterior compartment
• Common nerve: Tibial n.
• Common action: Hip extension and knee flexion

Skin

Superficial fascia

Deep fascia

Medial compartment
• Common nerve: Obturator n.
• Common action: Hip adduction

Femur

Anterior compartment
• Common nerve: Femoral n.
• Common action: Knee extension

A Cross-section of thigh

Posterior compartment
• Common nerve: Tibial n.
• Common action: Plantar flexion and flexion of digits

Fibula

Lateral compartment
• Common nerve: Superficial fibular n.
• Common action: Plantar flexion and eversion

Tibia

Anterior compartment
• Common nerve: Deep fibular n.
• Common action: Dorsiflexion and inversion

B Cross-section of leg

Figure 34-3: A. Cross-section of the thigh (**A**) and leg (**B**).

INNERVATION OF THE LOWER LIMB

BIG PICTURE

The lower limb receives sensory and motor innervation from anterior rami that originate from spinal nerve levels L1–S4. The anterior rami form two networks of nerves, referred to as the **lumbar plexus** (L1–L4) and the **sacral plexus** (L4–S4), which are connected via the lumbosacral trunk (L4–L5) (Figure 34-4).

LUMBAR PLEXUS

The lumbar plexus originates from ventral rami of L1–L4. It provides motor and sensory contributions to the anterior and medial compartment of the leg as well as to the abdominal wall and pelvic areas.

Iliohypogastric nerve (L1). Provides a portion of the motor innervation to the abdominal wall muscles and sensory innervation to the pubic region.

Ilioinguinal nerve (L1). Provides a portion of the motor innervation to the abdominal wall muscles and sensory innervation to the superior medial thigh and pubic area.

Genitofemoral nerve (L1–L2). Possesses two branches, the genital branch and the femoral branch. The **genital branch** provides motor innervation to the cremasteric muscle (male only) and sensory innervation to the skin of the anterior scrotum or labia majora. The **femoral branch** provides sensory innervation to the skin over the anterior region of the thigh.

Posterior division of the lumbar plexus

• **Lateral cutaneous nerve of the thigh (L2–L3).** Provides sensory innervation to the skin over the lateral region of the thigh.

• **Femoral nerve (L2–L4).** Provides motor innervation to the anterior compartment of the thigh, including the quadriceps femoris muscle group. The femoral nerve provides sensory innervation via sensory branches, which include the following nerves:

  • **Anterior and medial cutaneous nerves of the thigh.** Provide innervation to the skin of the anterior and medial thigh.

  • **Saphenous nerve.** Provides innervation to the skin on the medial leg.

Anterior division of the lumbar plexus

• **Obturator nerve (L2–L4).** Provides motor innervation to the medial compartment of the thigh (excluding the pectineus and hamstring portion of the adductor magnus muscle). In addition, the obturator nerve provides sensory innervation via the cutaneous branch of the obturator nerve to the skin over the medial region of the thigh.

SACRAL PLEXUS

Most of the sacral plexus (L4–S4) is divided into anterior and posterior divisions. The anterior division provides the primary motor innervation to the posterior compartment of the thigh, leg, and foot. The posterior division provides the primary motor innervation to the anterior and lateral compartments of the leg.

Anterior division of the sacral plexus

• **Pudendal nerve (S2–S4).** Provides motor innervation to the muscles of the pelvic floor and sensory innervation to the skin of the perineum, penis, and clitoris.

• **Posterior femoral cutaneous nerve (S1–S3).** Provides sensory innervation to the posterior region of the thigh.

• **Nerve to the superior gemellus and obturator internus muscles (L5–S2).** Provides motor innervation to the superior gemellus muscle and obturator internus muscles.

• **Nerve to the inferior gemellus and quadratus femoris muscles (L4–S1).** Provides innervation to the inferior gemellus and quadratus femoris muscles.

• **Tibial nerve (L4–S3).** Provides motor innervation to the hamstring muscles (excluding the short head of the biceps femoris muscle), hamstring head of the adductor magnus muscle, posterior compartment of the leg, and plantar muscles of the foot. The tibial nerve provides sensory innervation to the posterolateral region of the leg and the plantar surface of the foot.

Posterior division

• **Perforating cutaneous nerve (S2–S3).** Provides sensory innervation to the skin over the inferior aspect of the gluteus maximus muscle (inferior gluteal fold).

• **Nerve to the piriformis muscle (S1–S2).** Provides motor innervation to the piriformis muscle.

• **Posterior femoral cutaneous nerve (S1–S3).** Provides sensory innervation to the skin over the posterior region of the thigh.

• **Superior gluteal nerve (L4–S1).** Provides motor innervation to the gluteus medius, gluteus minimus, and tensor fascia latae muscles.

• **Inferior gluteal nerve (L5–S2).** Provides motor innervation to the gluteus maximus muscle.

• **Common fibular (peroneal) nerve (L4–S2).** Provides motor innervation to the short head of the biceps femoris muscle, muscles in the anterior and lateral compartments of the leg, and the intrinsic muscles on the dorsum of the foot. The common fibular nerve provides sensory innervation to the lateral compartment of the leg, the dorsum of the foot, and a small area between the first and second digits.

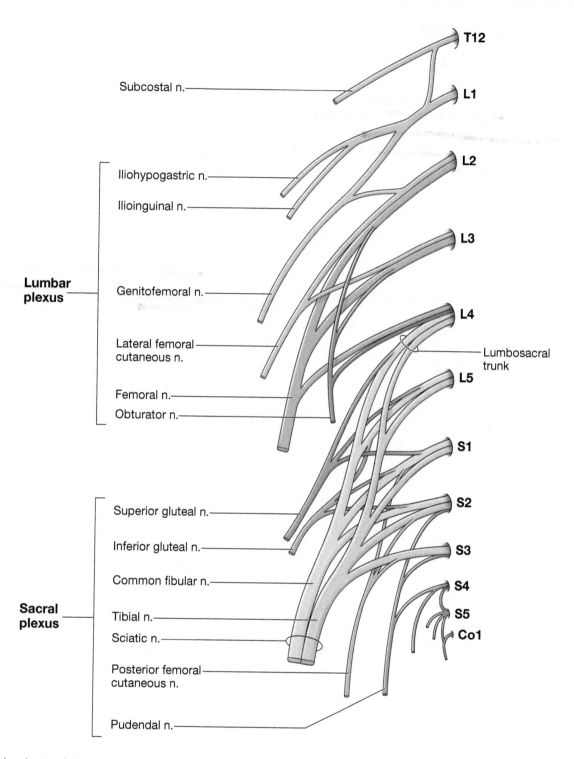

Figure 34-4: Lumbosacral plexus.

SENSORY INNERVATION

BIG PICTURE

Branches of the lumbar and sacral plexuses supply the sensory innervation to the lower limb, providing **cutaneous** and **dermatomal distributions**. The cutaneous distribution consists of multiple nerve root levels carried by a single nerve, whereas the dermatomal distribution consists of innervation from a single nerve root carried by multiple nerves.

DERMATOMAL DISTRIBUTION

The dermatomal innervation comes from multiple peripheral nerves that carry the same spinal levels. For example, the spinal root of L2 and L3 provides cutaneous innervation to the anterior, medial, and lateral parts of the thigh. In other words, all sensory neurons originating in this region will terminate at either the L2 or the L3 spinal nerve level.

Some principal dermatomes to remember are as follows (Figure 34-5A):

L1. Skin overlying the inguinal region.

L4. Skin overlying the anteromedial leg.

L5. Skin overlying the anterolateral leg.

S1. Skin overlying fifth digit and the calcaneus.

S2. Skin overlying the posterior thigh.

CUTANEOUS NERVES

Peripheral nerves supply cutaneous innervation to an area of skin on the surface of the body. For example, the lateral region of the thigh receives its cutaneous innervation from the L2 and L3 nerve roots. However, the sensory neurons from L2 and L3 are distributed to the lateral part of the thigh via the lateral cutaneous nerve of the thigh. The principal cutaneous nerves supplying the skin of the lower limb are illustrated in Figure 34-5B.

▼ The two distributions, the dermatomes and the cutaneous nerves, provide different sensory patterns although they are supplied by the same peripheral nerve root. This is especially evident when examining patients with nerve root injuries. A patient with a **lesion of a cutaneous nerve** presents differently than does a patient with a **lesion of a spinal root** (dermatomal distribution). If the lateral cutaneous nerve of the thigh is cut, there will be a loss of sensation on the lateral side of the thigh, and the anterior and medial sides of the thigh will remain intact. Therefore, the result is partial loss of the L2 and L3 dermatome. In contrast, if the L2 and L3 nerve roots are cut, the result is loss of sensation of the skin served by the L2 and L3 dermatome, including the anterior and medial thigh. ▼

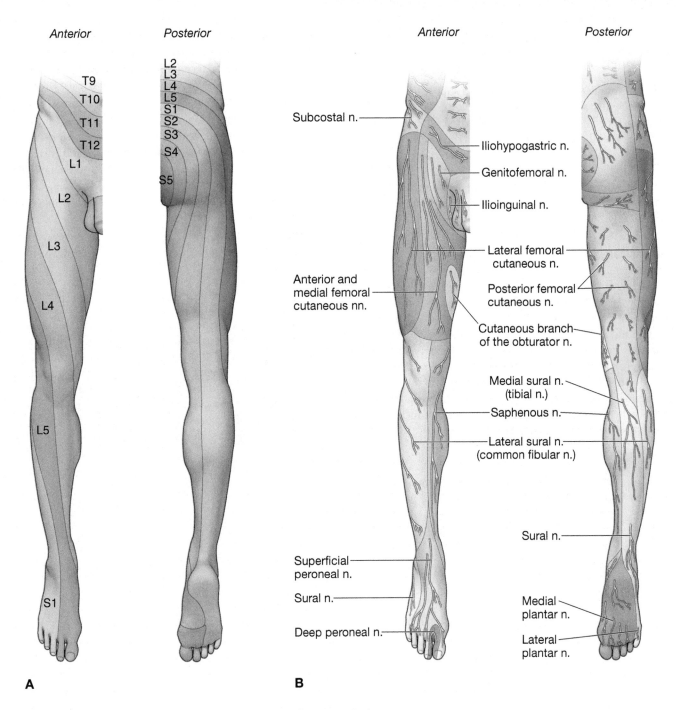

Figure 34-5: Dermatomal (**A**) and cutaneous (**B**) innervation of the lower limb.

VASCULARIZATION

BIG PICTURE

The common iliac arteries provide blood supply to both lower limbs. Each iliac artery bifurcates into an external and an internal iliac artery. The internal iliac artery primarily supplies blood to the pelvic and gluteal regions, whereas the external iliac supplies blood to the remainder of the lower limbs. Blood is returned to the heart via a superficial and a deep venous system. The deep venous system follows the arteries and shares the same name as the adjacent artery (i.e., popliteal artery and vein).

ARTERIES

The **common iliac artery** bifurcates into the internal and external iliac arteries (Figure 34-6A). The internal iliac artery gives rise to the obturator artery and to the superior and inferior gluteal arteries. The external iliac artery becomes the femoral artery as it passes the inguinal ligament and enters the thigh. The femoral artery gives rise to the deep artery of the thigh and continues distally to become the popliteal artery behind the knee joint. The popliteal artery bifurcates into the anterior and posterior tibial arteries, which travel distally into the leg. These arteries continue into the dorsal and plantar surfaces of the foot. Smaller vessels throughout the lower limb branch from the larger vessels to supply muscle, bone, and joints.

VEINS

Generally, the superficial and deep venous system of the lower limb drains into the internal and external iliac veins before reaching the inferior vena cava. The **deep veins** follow the arteries and usually consist of two or more veins that wrap around the accompanying artery (**vena comitantes**). The **superficial veins** originate in the foot and primarily consist of the **great** and **small saphenous veins**.

- **Great saphenous vein.** Originates along the great toe (digit 1) from the **dorsal venous arch**. The great saphenous vein courses anterior to the **medial malleolus** and travels along the medial side of the lower limb, medial to the **medial epicondyle** of the femur. It traverses an opening, called the **saphenous opening**, in the fascia lata and drains into the **femoral vein** (Figure 34-6B).

- **Small saphenous vein.** Travels along the lateral foot and ascends the posterior region of the leg to drain into the popliteal vein.

▽ A **coronary arterial bypass graft** is a type of surgery performed when blood must be rerouted or bypassed around a clogged coronary artery. The surgeon removes a segment of a healthy vessel from another part of the body to serve as the bypass. The **great saphenous vein** of the thigh is a source of a graft in which one end of the vein is grafted above the blocked area (often to the aortic arch) and the other end is grafted below the blocked area. Thus, the great saphenous vein "detours" blood, "bypassing" the blocked part of the coronary artery, and supplies the myocardium distal to the blocked artery. ▽

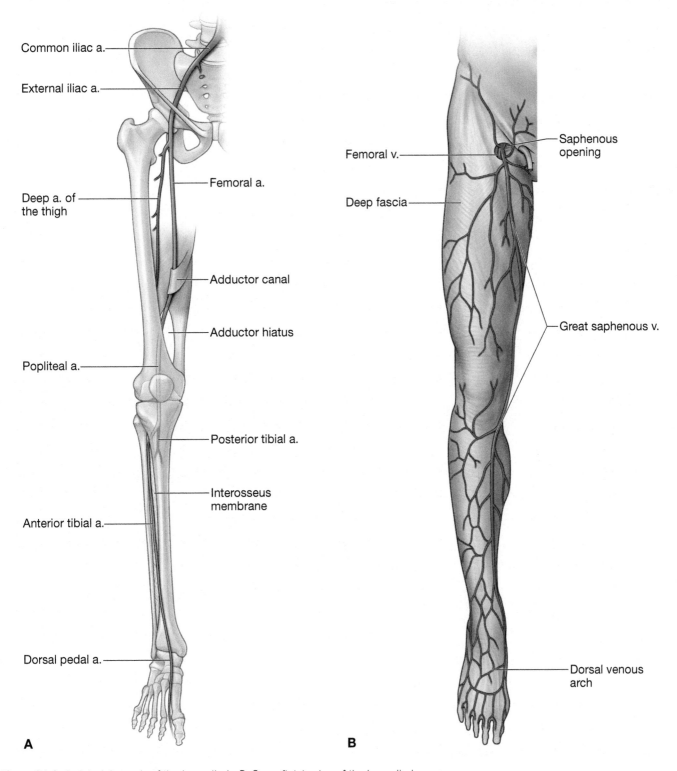

Figure 34-6: A. Arterial supply of the lower limb. **B**. Superficial veins of the lower limb.

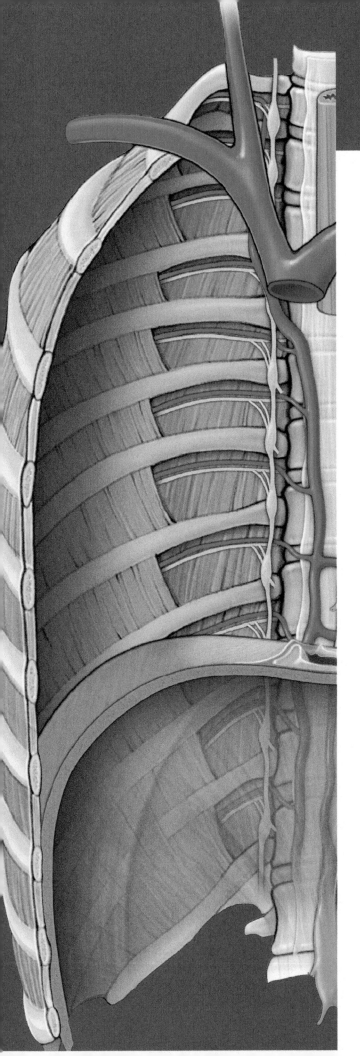

CHAPTER 35

GLUTEAL REGION AND HIP

GLUTEAL REGION

BIG PICTURE

The bony component of the gluteal (buttocks) region consists of two pelvic bones (os coxae) joined anteriorly by the symphysis pubis and posteriorly by the sacrum. Each os coxa is composed of three fused bones: ilium, ischium, and pubis. The bones of the gluteal region contain foramina (notches), which serve as conduits for nerves and blood vessels that travel between the pelvis, gluteal region, perineum, and lower limb. Muscles of the gluteal region primarily act on the hip joint.

ACTIONS OF THE HIP JOINT

The hip joint is a synovial, ball-and-socket joint. The "ball" is the head of the femur, and the "socket" is the acetabulum of the pelvic bone. The motions of the hip joint are as follows (Figure 35-1A):

- **Flexion.** Movement anterior in the sagittal plane.
- **Extension.** Movement posterior in the sagittal plane.
- **Abduction.** Movement away from the midline in the frontal plane.
- **Adduction.** Movement toward the midline in the frontal plane.
- **Medial rotation.** Movement toward the midline in the transverse or axial plane.
- **Lateral rotation.** Movement away from the midline in the transverse or axial plane.
- **Circumduction.** A combination of hip joint motions that produces a circular motion.

MUSCLES OF THE GLUTEAL REGION

BIG PICTURE

The muscles of the gluteal region primarily act on the hip joint, producing extension, medial rotation, lateral rotation, and abduction. In addition to producing motion, the muscles of the gluteal region are important for stability of the hip joint as well as for locomotion.

GLUTEAL MUSCLES (FIGURE 35-1B)

- **Gluteus maximus muscle.** Attaches proximally on the ilium behind the posterior gluteal line, the sacrum, the coccyx, and the sacrotuberous ligament; distally, the muscle attaches at the iliotibial tract and the gluteal tuberosity of the femur. The gluteus maximus muscle is a powerful extensor of a flexed femur at the hip joint and a lateral stabilizer of the hip joint. The inferior gluteal nerve (L5, S1, S2) innervates this muscle.
- **Gluteus medius muscle.** Attaches proximally on the ilium between the anterior and posterior gluteal lines; distally, the muscle attaches on the greater trochanter of the femur. The gluteus medius muscle abducts and medially rotates the femur at the hip joint. In addition, the gluteus medius holds the pelvis secure over the stance leg, preventing pelvic drop on the opposite swing side during gait. The superior gluteal nerve (L4, L5, S1) innervates this muscle.

- **Gluteus minimus muscle.** Attaches proximally on the ilium between the anterior and posterior gluteal lines; distally, the muscle attaches on the greater trochanter of the femur. The action of the gluteus minimus muscle is the same as that of the gluteus medius—it abducts the femur at the hip joint, holding the pelvis secure over the stance leg and preventing pelvic drop on the opposite swing side during gait and hip medial rotation. The superior gluteal nerve (L5, S1, S2) innervates this muscle.
- **Tensor fascia lata muscle.** Attaches more anteriorly than the other muscles of the gluteal region. The tensor fascia lata muscle attaches proximally at the lateral aspect of the iliac crest between the anterior superior iliac spine and the tubercle of the crest. Distally, the muscle attaches to the **iliotibial tract of the fascia lata**, which extends to the tibia. Because of its distal insertion, the tensor fascia lata muscle is unique when compared to the other gluteal muscles. The main action of this muscle is to stabilize the knee in extension and in hip flexion. The superior gluteal nerve (L4, L5, S1) innervates this muscle.

DEEP HIP ROTATOR MUSCLES

The deep hip rotator muscles all have several common characteristics—they are deep to the gluteal muscles, they arise from the pelvis, they share common attachments around the greater trochanter of the femur, and they laterally rotate the hip. The deep hip rotator muscles are as follows (Figure 35-1C):

- **Piriformis muscle.** Attaches proximally on the anterior surface of the sacrum; distally, the muscle attaches at the greater trochanter of the femur. The piriformis muscle laterally rotates the femur at the hip joint. The nerve to the piriformis muscle (S1, S2) innervates this muscle.
- **Superior gemellus muscle.** Attaches proximally at the ischial spine; distally, the muscle attaches on the greater trochanter of the femur. The superior gemellus muscle laterally rotates the femur at the hip joint. The nerve to the obturator internus and superior gemellus muscles (L5, S1, S2) innervates this muscle.
- **Obturator internus muscle.** Attaches proximally on the deep surface of the obturator membrane and surrounding bone; distally, the muscle attaches at the greater trochanter of the femur. The obturator internus muscle laterally rotates the femur at the hip joint. The nerve to the obturator internus and superior gemellus muscles (L5, S1, S2) innervates this muscle.
- **Inferior gemellus muscle.** Attaches proximally on the ischial tuberosity; distally, the muscle attaches at the greater trochanter of the femur. The inferior gemellus muscle laterally rotates the femur at the hip joint. The nerve to the inferior gemellus and quadratus femoris muscles (L4, L5, S1) innervates this muscle.
- **Quadratus femoris muscle.** Attaches proximally at the lateral aspect of the ischium just anterior to the ischial tuberosity; distally, the muscle attaches on the intertrochanteric crest. The quadratus femoris muscle laterally rotates the femur at the hip joint. The nerve to the inferior gemellus and quadratus femoris muscles (L4, L5, S1) innervates this muscle.

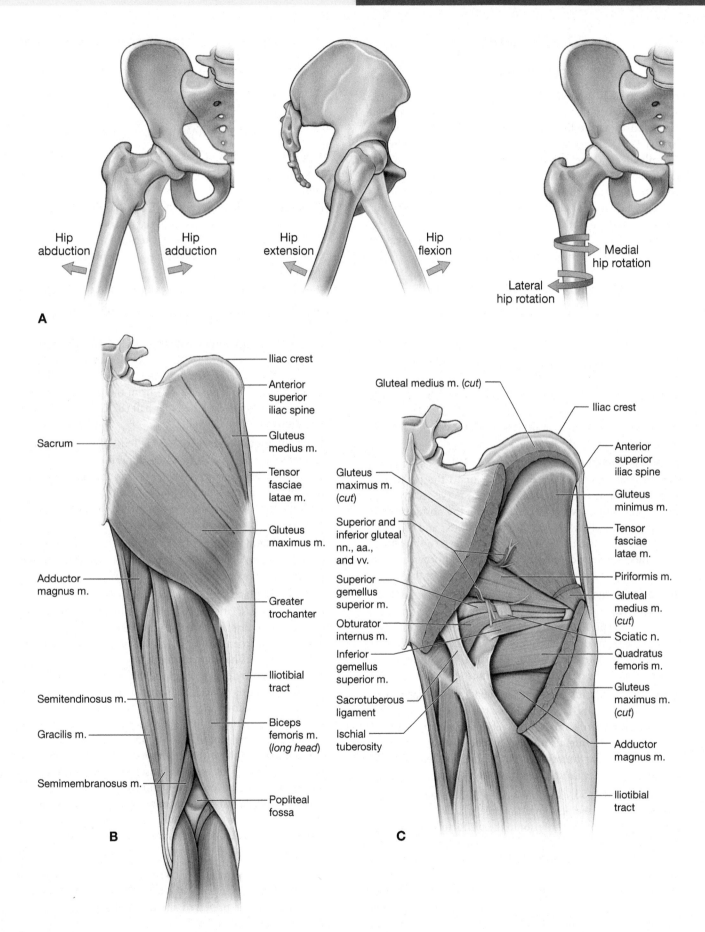

Figure 35-1: A. Actions of the hip joint. The right gluteal region illustrating the posterior view of the superficial gluteal muscles (**B**) and the deep gluteal muscles (**C**).

SACRAL PLEXUS

BIG PICTURE

The lower limb is innervated by the ventral rami from nerve roots L1–S4, which form two separate networks of nerves and are referred to as the lumbar plexus (L1–L4) and the sacral plexus (L4–S4). The lumbar plexus communicates with the sacral plexus via the lumbosacral trunk (L4, L5), which descends into the pelvic cavity to contribute to the sacral plexus. Most of the sacral plexus is divided into anterior and posterior divisions. The anterior division provides the primary motor innervation to the posterior compartment of the thigh and leg. The posterior division provides the primary motor innervation to the anterior and lateral compartments of the leg.

ANTERIOR DIVISION OF THE SACRAL PLEXUS

The anterior division of the sacral plexus consists of ventral rami from L4 to S4, which form a network and give rise to the following five nerves (Figure 35-2A):

- **Pudendal nerve (S2–S4).** Exits the pelvis via the greater sciatic foramen, enters the gluteal region, and courses to the perineum through the lesser sciatic foramen. The pudendal nerve provides motor innervation to the muscles of the pelvic floor and sensory innervation to the skin of the perineum, penis, and clitoris.
- **Posterior femoral cutaneous nerve (S1–S3).** Exits the pelvis via the greater sciatic foramen, inferior to the piriformis muscle. The posterior femoral cutaneous nerve receives half of its innervation levels (S1 and S2) from the posterior division of the sacral plexus and the other half (S2 and S3) from the anterior division. The nerve remains deep to the gluteal maximus muscle and emerges at the inferior border, providing sensory innervation to the posterior region of the thigh.
- **Nerve to the superior gemellus and obturator internus muscles (L5–S2).** Exits the pelvis via the greater sciatic foramen, inferior to the piriformis muscle, and provides motor innervation to the superior gemellus muscle. The nerve then reenters the pelvis, via the lesser sciatic foramen, providing motor innervation to the obturator internus muscle.
- **Nerve to the inferior gemellus and quadratus femoris muscles (L4–S1).** Exits the pelvis via the greater sciatic foramen, inferior to the piriformis, and travels along the deep surface of the superior gemellus muscle and the obturator internus tendon, providing innervation to the inferior gemellus and quadratus femoris muscles on their deep surface.
- **Tibial nerve (L4–S3).** The tibial nerve (a division of the sciatic nerve) exits the pelvis via the greater sciatic foramen to enter the gluteal region inferior to the piriformis muscle. The nerve descends along the posterior aspect of the thigh, providing motor innervation to the hamstring muscles (excluding the short head of the biceps femoris muscle) and a hamstring head of the adductor magnus muscle in the medial compartment of the thigh. The tibial nerve descends through the popliteal fossa and enters the posterior compartment of the leg, deep to the gastrocnemius and soleus muscles. It provides motor innervation to the posterior compartment of the leg as well as to the plantar muscles of the foot. Sensory branches provide cutaneous innervation to the posterolateral region of the leg and the lateral region of the foot.

POSTERIOR DIVISION OF THE SACRAL PLEXUS

The posterior division of the sacral plexus consists of ventral rami from L4–S3, which form a network and give rise to the following six nerves:

- **Perforating cutaneous nerve (S2–S3).** Pierces the sacrotuberous ligament and travels to the inferior edge of the gluteus maximus muscle, providing sensory innervation to the skin over the inferior aspect of the gluteus maximus (inferior gluteal fold).
- **Nerve to piriformis muscle (S1–S2).** Travels directly from the plexus to the piriformis muscle, providing motor innervation without leaving the pelvic cavity.
- **Posterior femoral cutaneous nerve (S1–S3).** Exits the pelvis via the greater sciatic foramen, inferior to the piriformis muscle. The posterior femoral cutaneous nerve receives half of its innervation levels (S1–S2) from the posterior division of the sacral plexus and the other half (S2–S3) from the anterior division.
- **Superior gluteal nerve (L4–S1).** Exits the pelvis via the greater sciatic foramen and travels superior to the piriformis muscle and between the gluteus medius and minimus muscles, providing innervation to both muscles. The superior gluteal nerve continues anteriorly, providing motor innervation to the tensor fascia latae muscle.
- **Inferior gluteal nerve (L5–S2).** Exits the pelvis via the greater sciatic foramen and travels inferior to the piriformis, providing motor innervation to the gluteus maximus muscle.
- **Common fibular (peroneal) nerve (L4–S2).** Is the smallest division of the sciatic nerve (half the size of the tibial nerve). The common fibular nerve exits the pelvis via the greater sciatic foramen to enter the gluteal region inferior to the piriformis muscle. The nerve descends along the posterior aspect of the thigh, providing motor innervation to the short head of the biceps femoris muscle. The common fibular nerve descends to the popliteal fossa and curves laterally around the neck of the fibula to bifurcate into the superficial and deep fibular nerves, providing motor innervation to the lateral and anterior compartments of the leg, respectively. Furthermore, the superficial fibular nerve provides sensory innervation to the anterolateral region of the leg and the dorsum of the foot. The deep fibular nerve provides sensory innervation to a small area between digits 1 and 2. In addition, a branch from the common fibular nerve (lateral sural nerve) provides cutaneous innervation to the superior lateral leg.

▼ The gluteal region is a common site for **intermuscular injections**. Specifically, the superior lateral portion of the gluteal region is the preferred site to avoid injuring structures such as the sciatic nerve. ▼

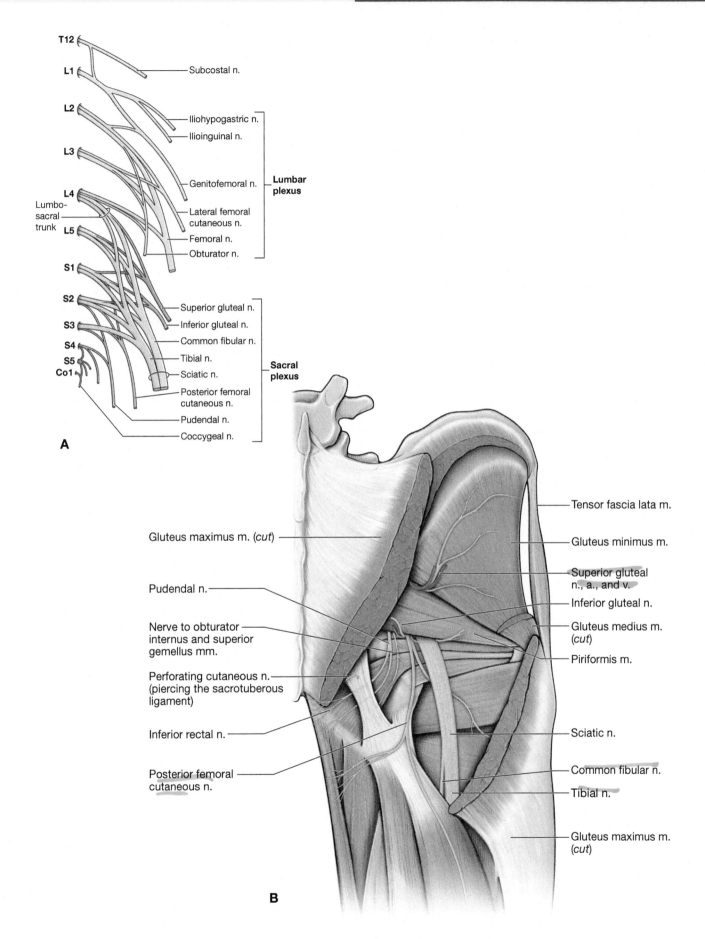

Figure 35-2: A. Schematic of the lumbosacral plexus. **B**. Neurovascular structures of the gluteal region.

VASCULARIZATION OF THE GLUTEAL REGION

BIG PICTURE

The descending aorta bifurcates into the common iliac artery, which provides the blood supply to the lower extremities. The common iliac artery bifurcates into the internal and external iliac arteries. The external iliac artery becomes the femoral artery. The internal artery is the major blood supply to the pelvis and gluteal region.

ARTERIES OF THE GLUTEAL REGION

Internal iliac artery. The superior and inferior gluteal arteries branch from the internal iliac artery and travel with the superior and inferior gluteal nerves.

- **Superior gluteal artery.** Travels between the lumbosacral trunk and the first sacral ramus or between the first and second sacral rami to exit the pelvis through the greater sciatic foramen. In the pelvis, the superior gluteal artery supplies blood to the obturator internus and piriformis muscles. In the gluteal region, the superior gluteal artery supplies blood to the muscles and skin in the gluteal region, including the tensor fascia latae muscle.

- **Inferior gluteal artery.** The terminal branch of the internal iliac artery; travels between the first and second or the second and third sacral rami. The inferior gluteal artery supplies blood to the muscles of the gluteal region and forms anastomoses with blood vessels surrounding the hip joint.

VEINS OF THE GLUTEAL REGION

The veins of the gluteal region generally follow the arteries; therefore, arteries and veins have the same name, except one is an artery and one is a vein. The veins of the lower extremities generally drain into the internal and external iliac veins. Specifically, the superior and inferior gluteal veins drain into the internal iliac vein, which in turn drains into the common iliac vein and terminates in the inferior vena cava before reaching the heart. Because of its association with the femoral artery, additional vascular supply will be discussed in Chapter 36.

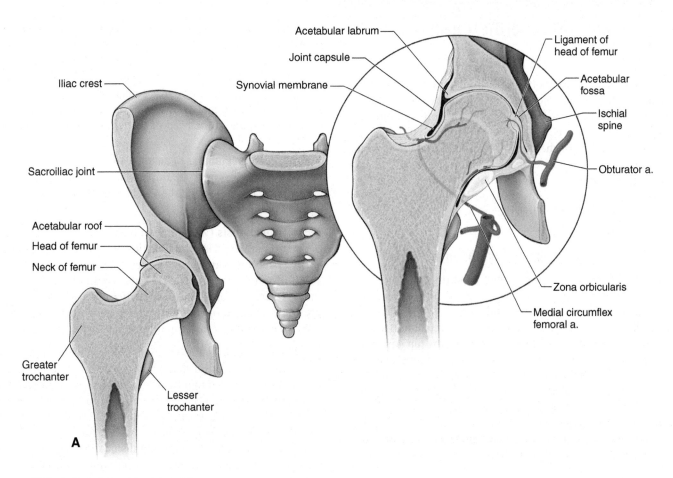

Figure 35-3: A. Structure of the hip joint.

JOINTS OF THE GLUTEAL REGION

BIG PICTURE

The hip joint is a synovial, ball-and-socket joint that allows for a great deal of freedom, including flexion and extension, abduction and adduction, medial and lateral rotation, and circumduction of the femur. The role of the hip joint is to provide support for the weight of the head, arms, and trunk during static postures (standing) and dynamic movements (walking and running). In addition to the hip joint, the gluteal region also contains the sacroiliac joint and the pubic symphysis, which connect the pelvic bones together as well as connecting the pelvic bones to the spine (i.e., the sacrum).

STRUCTURE OF THE HIP JOINT

The articulating surface of the **pelvic bone (os coxa)** is a concave socket that is composed of three fused bones, the ilium, ischium, and pubis, called the **acetabulum** (Figure 35-3A). The acetabulum is horseshoe-shaped fossa. The acetabulum articulates with the **head of the femur**. In addition, the hip joint has a wedge-shaped fibrocartilaginous ring around the periphery of the acetabulum (**acetabular labrum**), which increases stability by deepening the socket and increasing the concavity of the articulating surface. The wedge shape of the acetabular labrum also assists in maintaining contact of the acetabulum with the femoral head. The **medial circumflex femoral artery** provides the principal blood supply to the hip.

LIGAMENTOUS SUPPORT OF THE HIP JOINT

The following structures provide ligamentous support to the hip joint (Figure 35-3B):

- **Joint capsule.** Is strong and extends like a sleeve from the acetabulum to the base of the neck of the femur. The joint capsule possesses circular fibers, which form a ring around the neck of the femur, called the **zona orbicularis**. The capsule contains three capsular ligaments: two anterior ligaments and one posterior ligament. The ligaments of the hip primarily become taught with extension of the hip and permit little, if any, distraction between the articulating surfaces.
  - **Anterior ligaments**
    - **Iliofemoral ligament (Y ligament, or ligament of Bigelow).** A fan-shaped ligament that resembles an inverted "Y." The iliofemoral ligament extends from the anterior iliac spine and bifurcates to attach to the intertrochanteric line of the femur.
    - **Pubofemoral ligament.** Attaches from the anterior aspect of the pubic ramus and extends posteriorly to attach to the anterior surface of the intertrochanteric fossa.
  - **Posterior ligament**
    - **Ischiofemoral ligament.** Attaches from the superior acetabular rim and labrum to the inner surface of the greater trochanter.

- **Ligament of the head of the femur (ligamentum teres).** Attaches to the head of the femur (fovea). The ligament courses deep to the transverse acetabular ligament to attach to the acetabular notch. The ligament of the head of the femur does not appear to play a major role in stability of the hip joint, but rather serves as a conduit for the secondary arterial supply to the head of the femur from the **obturator artery**.

- **Transverse acetabular ligament.** Completes the circle of the acetabular labrum by spanning the acetabular notch and forming a foramen for the passage of the ligament of the head of the femur.

▽ **Osteoporosis** (weakening of bones) is a common condition in the geriatric population and predisposes this segment of the population to a higher risk of hip fractures as a result of trauma (e.g., falls). The most common sites that sustain fractures are at the neck of the femur, at the level of the greater trochanter, or inferior to the greater trochanter. Treatment may result in a total hip replacement. ▼

▽ The hip joint is a strong joint; however, with considerable force, the head of the femur can be dislocated from the acetabulum. Most commonly, the head of the femur is forced in a posterior dislocation, resulting in a flexed and internally rotated femur. When this occurs, the patient is unable to move. Occasionally, additional damage such as pelvic fractures and nerve injury occurs around the joint. **Hip dislocations** usually are the result of an automobile accident or a fall from a high surface. ▼

BURSAE

Bursae are synovial sacs filled with synovial fluid. They are found at areas in the tissue at which friction would otherwise develop. Bursae serve as small cushions because they decrease the friction between two moving structures, such as tendon and bone. The two most important bursae in the gluteal region are as follows:

- **Subtendinous iliac bursa.** Separates the iliacus and psoas major muscles from the anterior joint capsule.

- **Trochanteric bursa.** Separates the gluteus maximus muscle from the greater trochanter.

▽ **Bursitis of the hip** occurs when the trochanteric bursa becomes inflamed due to overuse or trauma, resulting most frequently in pain over the area of the greater trochanter. Hip bursitis is one of the most common causes of hip pain. ▼

SACROILIAC JOINT

The sacroiliac joint is a plane synovial joint that connects the sacrum with the bilateral pelvic bones. This joint transmits forces from the vertebral column to the pelvic bones and lower limbs. The combination of strong ligaments and the irregular shape of the articulating surfaces increases the stability of the sacroiliac joint. In some individuals this joint may be fused, and in others it will allow minimal movement.

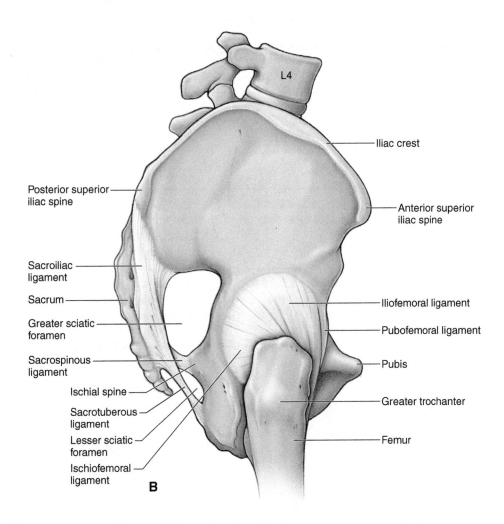

Figure 35-3: B. The right hip illustrating the lateral view of the ligaments of the hip joint.

TABLE 35-1. Muscles of the Gluteal Region

Muscle	Proximal Attachment	Distal Attachment	Action	Innervation
Gluteal region				
Tensor fascia lata	Lateral aspect of crest of ilium between anterior superior iliac spine and tubercle of crest	Iliotibial tract of fascia lata	Stabilizes knee in extension	Superior gluteal n. (L4, L5, S1)
Gluteus maximus	Ilium behind posterior gluteal line, sacrum, coccyx, and sacrotuberous ligament	Iliotibial tract and gluteal tuberosity of femur	Powerful extensor of flexed femur at hip joint; lateral stabilizer of hip and knee joints	Inferior gluteal n. (L5, S1, S2)
Gluteus medius	Ilium between anterior and posterior gluteal lines	Greater trochanter	Abducts femur at hip joint; holds pelvis secure over stance leg and prevents pelvic drop on opposite swing side during walking; hip internal rotation	Superior gluteal n. (L4, L5, S1)
Gluteus minimus	Ilium between anterior and inferior gluteal lines			
Piriformis	Anterior surface of sacrum	Greater trochanter	Laterally rotates the hip joint	Nerve to piriformis m. (S1, S2)
Superior gemellus	Ischial spine			Nerve to obturator internus and superior gemellus mm. (L5, S1, S2)
Obturator internus	Deep surface of obturator membrane and surrounding bone			
Inferior gemellus	Ischial tuberosity			
Quadratus femoris	Lateral aspect of ischium just anterior to ischial tuberosity	Intertrochanteric crest		Nerve to inferior gemellus and quadratus femoris mm. (L4, L5, S1)

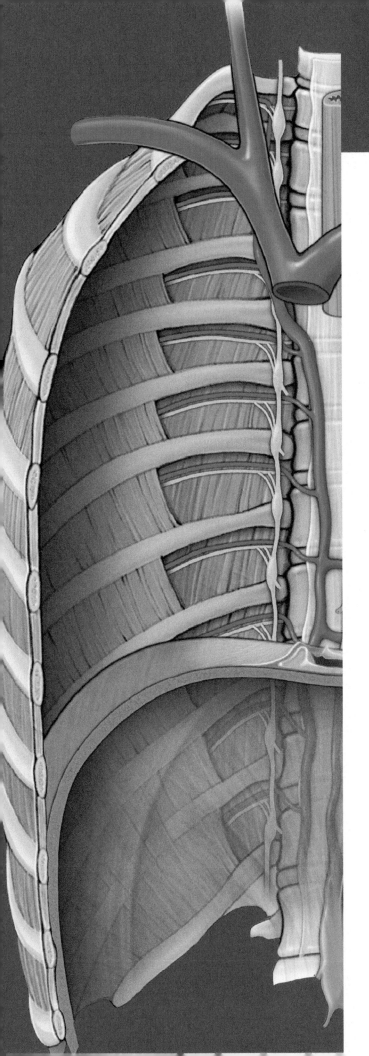

CHAPTER 36

THIGH

THIGH

BIG PICTURE

The bone between the hip and the knee is the femur. It is the longest and strongest bone in the body. The femur articulates proximally with the acetabulum and distally with the tibia and patella. The knee joint is formed by articulations of the femur, tibia, and patella. The knee joint enables flexion, extension, and minimal rotation of the femur and tibia. Also, it plays an important role in supporting the weight of the body during static positions and dynamic movement during gait.

ACTIONS OF THE KNEE COMPLEX

The articulations between the femur, tibia, and patella form the knee joint and enable the following actions (Figure 36-1A):

- **Flexion.** Movement in the sagittal plane, decreasing the knee joint angle.
- **Extension.** Movement in the sagittal plane, increasing the knee joint angle.
- **Medial rotation.** Movement toward the midline in the transverse or axial plane.
- **Lateral rotation.** Movement away from the midline in the transverse or axial plane.

MUSCLES OF THE THIGH

BIG PICTURE

The muscles of the thigh are divided by their fascial compartments (anterior, medial, and posterior) and may cross the hip or knee joint (Figure 36-1B). Identifying which joints the muscles cross and the side on which they cross can provide useful insight into the actions of these muscles (Table 36-1).

MUSCLES OF THE ANTERIOR COMPARTMENT OF THE THIGH

The muscles in the anterior compartment of the thigh are primarily **flexors of the hip** or **extensors of the knee** because of their anterior orientation (Figure 36-1C). The **femoral nerve (L2–L4)** innervates these muscles; however, each muscle does not necessarily receive each spinal nerve level between L2 and L4.

- **Iliopsoas musculature.** Originates from two muscles, the **psoas major and iliacus muscles, which join to form a common tendon.** The psoas major muscle attaches along vertebrae T12–L5, discs, and the iliacus within the iliac fossa. Both the psoas and iliacus muscles join together as they course deep to the inguinal ligament and insert onto the lesser trochanter of the femur. The main action of these muscles is to flex and laterally rotate the thigh at the hip joint. Innervation to the psoas major muscle is via the anterior rami of L1, L2, and L3, whereas innervation to the iliacus is through the femoral nerve (anterior rami of L2 and L3).

- **Sartorius muscle.** Attaches proximally to the anterior superior iliac spine. The distal insertion of the sartorius muscle is medial to the tibial tuberosity, contributing to the **pes anserinus.** Pes anserinus ("goose's foot") is a term used to describe the conjoined tendons of the sartorius, gracilis, and semitendinosus muscles; their common insertion is medial to the tibial tuberosity. The action of the sartorius muscle is to flex, abduct, and laterally rotate the thigh at the hip joint and flex the leg at the knee joint. The femoral nerve (L2 and L3) innervates this muscle.

- **Quadriceps femoris muscle group.** A four-headed muscle in the anterior compartment of the thigh and is a strong extensor muscle of the knee. There are four separate muscles in this group, each with distinct origins. However, all four parts of the quadriceps femoris muscle attach to the patella, via the quadriceps tendon, and then insert onto the tibial tuberosity. The femoral nerve (L2–L4) innervates the quadriceps femoris muscle group. The four separate muscles are as follows:

  - **Rectus femoris muscle.** Attaches on the anterior inferior iliac spine and to the quadriceps femoris tendon. The rectus femoris muscle flexes the thigh at the hip joint and extends the leg at the knee joint.

  - **Vastus lateralis muscle.** Attaches proximally at the intertrochanteric line and the lateral lip of the linea aspera; distally, the muscle attaches to the quadriceps femoris tendon. The vastus lateralis muscle extends the leg at the knee joint.

  - **Vastus medialis muscle.** Attaches proximally at the intertrochanteric line and the lateral lip of the linea aspera; distally, the muscle attaches to the quadriceps femoris tendon. The vastus medialis muscle extends the leg at the knee joint.

  - **Vastus intermedius muscle.** Attaches proximally along the anterior and lateral surfaces of the upper two-thirds of the femoral shaft; distally, the muscle attaches to the quadriceps femoris tendon. The vastus intermedius muscle extends the leg at the knee joint.

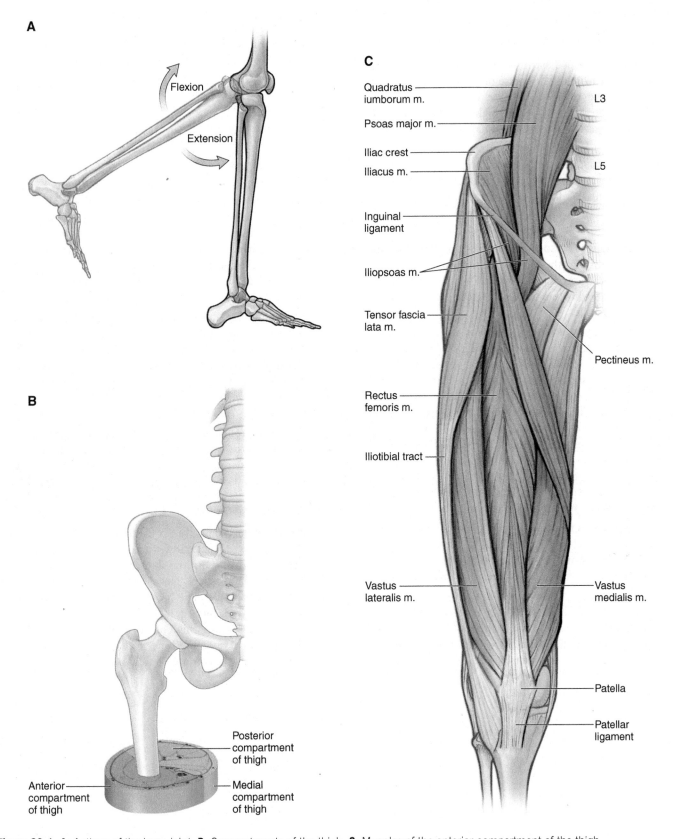

A

Flexion

Extension

B

Posterior compartment of thigh

Anterior compartment of thigh

Medial compartment of thigh

C

Quadratus iumborum m.

Psoas major m.

Iliac crest

Iliacus m.

Inguinal ligament

Iliopsoas m.

Tensor fascia lata m.

Rectus femoris m.

Iliotibial tract

Vastus lateralis m.

L3

L5

Pectineus m.

Vastus medialis m.

Patella

Patellar ligament

Figure 36-1: A. Actions of the knee joint. **B**. Compartments of the thigh. **C**. Muscles of the anterior compartment of the thigh.

MUSCLES OF THE MEDIAL COMPARTMENT OF THE THIGH

The muscles in the medial compartment of the thigh are primarily **adductors of the hip** because of their medial orientation. The **obturator nerve (L2–L4)** innervates most of the muscles in the medial compartment of the thigh. However, each muscle does not necessarily receive each spinal nerve level between L2 and L4 (Figure 36-2A and B).

Pectineus muscle. Attaches to the pectineal line of the pubis and the posterior surface of the proximal femur. The pectineus muscle adducts and flexes the thigh at the hip joint. The femoral nerve (L2 and L3) innervates this muscle, with occasional branches from the obturator nerve.

Adductor longus muscle. Attaches proximally to the body of the pubis; distally, the muscle attaches on the linea aspera. The adductor longus muscle adducts and medially rotates the thigh at the hip joint. The obturator nerve (L2–L4) innervates this muscle.

Adductor magnus muscle. Consists of an **adductor part** and a **hamstring part**. Proximally, the adductor part attaches to the ischiopubic ramus, and the hamstring part attaches to the ischial tuberosity. Distally, the adductor part of the muscle attaches on the linea aspera, and the hamstring part attaches on the adductor tubercle. The adductor magnus muscle is the largest and deepest muscle of the muscles of the medial compartment of the thigh. It adducts and medially rotates the thigh at the hip joint. The **obturator nerve (L2–L4)** innervates the adductor part of the muscle, and the **tibial division of the sciatic nerve (L4)** and the **obturator nerve (L2 and L3)** innervate the hamstring part of the muscle.

Adductor brevis muscle. Attaches proximally to the inferior pubic ramus and the linea aspera. The adductor brevis muscle adducts and medially rotates the thigh at the hip joint. The obturator nerve (L2–L4) innervates this muscle.

Gracilis muscle. Attaches to the inferior pubic ramus and the medial surface of the proximal shaft of the tibia (pes anserinus). The gracilis muscle adducts the thigh at the hip joint and flexes the leg at the knee joint. The obturator nerve (L2 and L3) innervates this muscle.

Obturator externus muscle. Attaches to the external surface of the obturator membrane, adjacent bone, and trochanteric fossa. The obturator externus muscle laterally rotates the femur at the hip joint. The obturator nerve (L3 and L4) innervates this muscle.

MUSCLES OF THE POSTERIOR COMPARTMENT OF THE THIGH

The muscles in the posterior compartment of the thigh are primarily **extensors of the hip** or **flexors of the knee** because of their posterior orientation. The **tibial nerve (L4–S3)** innervates the muscles in the posterior compartment of the thigh, with the exception of the short head of the biceps femoris muscles (common fibular nerve). Muscles in this compartment do not receive all the innervation levels from the tibial nerve; rather, they receive innervation from the spinal nerve level between L5 and S2 (Figure 36-2C).

Semitendinosus muscle. Attaches proximally to the ischial tuberosity and the medial surface of the proximal tibia (pes anserinus). The semitendinosus muscle extends and medially rotates the thigh at the hip joint. In addition, the muscle flexes and medially rotates the leg at the knee joint. The tibial division of the sciatic nerve (L5–S2) innervates this muscle.

Semimembranosus muscle. Attaches proximally to the ischial tuberosity and the medial tibial condyle. The semimembranosus muscle flexes and medially rotates the leg at the knee joint and extends and medially rotates the thigh at the hip joint. The tibial division of the sciatic nerve (L5–S2) innervates this muscle.

Biceps femoris muscle. Consists of two heads (**long and short heads**). Proximally, the long head attaches on the ischial tuberosity, and the short head attaches to the lateral lip of the linea aspera. Distally, the muscle attaches to the head of the fibula. The biceps femoris muscle flexes and medially rotates the leg at the knee joint and extends and medially rotates the thigh at the hip joint. The tibial nerve (L5–S2) innervates the long head, and the short head is innervated by the common fibular nerve (L5–S2).

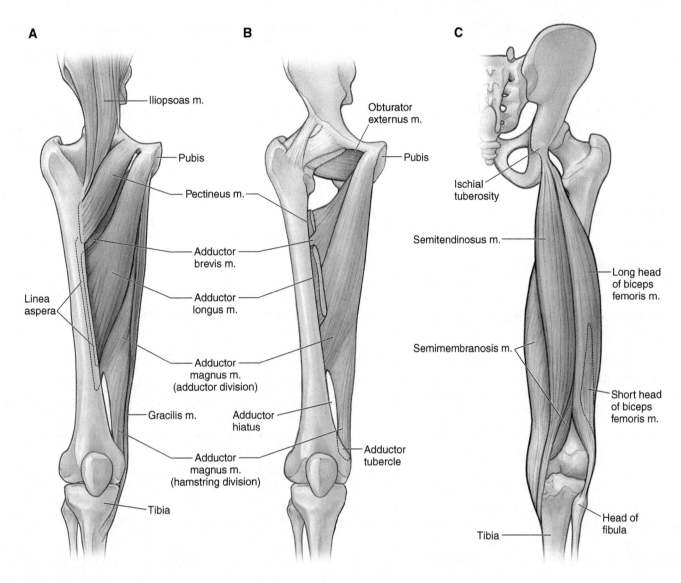

Figure 36-2: A. Superficial view of muscles of the medial compartment of the thigh. **B**. Deep view of muscles of the medial compartment of the thigh. **C**. Muscles of the posterior compartment of the thigh (hamstrings).

FEMORAL TRIANGLE

BIG PICTURE

The femoral triangle is an area in the inguinal region that is shaped like an upside-down triangle. The femoral triangle contains the femoral nerve, artery, and vein, and the lymphatics. The femoral triangle is an area in the inguinal region that is shaped like an upside-down triangle and is bordered by the sartorius muscle, adductor longus muscle and inguinal ligament. The femoral triangle contains the following structures from lateral to medial (Figure 36-3A):

- **Femoral nerve.** Originates as a branch of the lumbar plexus. The femoral nerve is not contained within the femoral sheath.
- **Femoral artery.** Continuation of the external iliac artery. The femoral artery is located midway between the anterior superior iliac spine and the pubic symphysis.
- **Femoral vein.** Continues as the external iliac vein.
- **Lymphatics.**

Often, the acronym **NAVL** is used to represent the orientation of the structures of the femoral triangle. The inferior portion of the femoral triangle communicates with a facial canal (adductor canal) that runs deep to the sartorius muscle.

LUMBAR PLEXUS

BIG PICTURE

The lower limb is innervated by the ventral rami from nerve roots L1–S4, which form two separate networks of nerves that are referred to as the lumbar plexus (L1–L4) and the sacral plexus (L4–S4). (The sacral plexus is discussed in Chapter 35.) The lumbar plexus consists of ventral rami from the L2 to L4 levels of the spinal cord, which exit the intervertebral foramina and course along the posterior abdominal wall, en route to the anterolateral abdominal wall and lower limb. The ventral rami are divided into anterior and posterior divisions, corresponding to the anterior (ventral) muscles (flexors) and the posterior (dorsal) muscles (extensors), similar to those in the brachial plexus (Figure 36-3B).

ABDOMINOPELVIC CUTANEOUS BRANCHES OF THE LUMBAR PLEXUS

Branches from the lumbar plexus provide motor and sensory contributions not only to the anterior and medial compartments of the leg but also to the abdominal wall and pelvic areas. Branches are summarized as follows:

- **Iliohypogastric and ilioinguinal nerves (L1).** Emerge from the lateral border of the psoas major muscle and pierce the transverse abdominus muscle. The iliohypogastric and ilioinguinal nerves course anteriorly between the transverse abdominus and the internal oblique muscles, contributing to the motor innervation of the muscles of the abdominal wall (internal oblique and transversus abdominus muscles). In addition, the **iliohypogastric nerve** provides sensory innervation to the pubic region and the posterolateral gluteal skin. The **ilioinguinal nerve (L1)** provides sensory innervation to the superior medial thigh, root of the penis, anterior scrotum, mons pubis, and the labium majus.

- **Genitofemoral nerve (L1–L2).** Pierces the psoas major muscle and divides into two branches, the genital branch and the femoral branch. The genital branch enters the deep inguinal ring, providing motor innervation to the cremasteric muscle (male only) and sensory innervation to the skin of the anterior scrotum (or mons pubis) and the labium majus. The femoral branch passes behind the inguinal ligament to enter the femoral triangle, providing sensory innervation to the skin over the femoral triangle (superior and anterior region of the thigh).

LOWER LIMB BRANCHES OF THE LUMBAR PLEXUS

The lumbar plexus is divided into posterior and anterior divisions, according to posterior and anterior muscle groups (Figure 36-3C).

- **Posterior division**
  - **Lateral cutaneous nerve of the thigh (L2–L3).** Emerges from the lateral border of the psoas major muscle, crossing the iliacus muscle, and enters the thigh just medial to the anterior superior iliac spine, providing sensory innervation to the lateral thigh.
  - **Femoral nerve (L2–L4).** Emerges as the largest branch of the lumbar plexus, deep to the lateral border of the psoas major muscle, and passes deep to the inguinal ligament. The femoral nerve then enters the anterior compartment of the thigh, where it provides motor innervation to the quadriceps femoris muscle group, the sartorius muscle, and part of the pectineus muscles. In addition, the femoral nerve provides sensory innervation via the following **sensory branches**:
    - **Anterior and medial cutaneous nerve of the thigh.** Branches off the femoral nerve and supplies the skin of the anterior and medial thigh.
    - **Saphenous nerve.** Branches off the femoral nerve and courses with the femoral artery and vein into the **adductor canal**. The femoral artery and femoral vein traverse the **adductor hiatus** and enter the **popliteal fossa**. In contrast, the saphenous nerve exits the adductor canal and supplies the skin on the medial region of the leg.

- **Anterior division**
  - **Obturator nerve (L2–L4).** Emerges from the medial border of the psoas major muscle, passing posterior to the common iliac artery, and enters the medial compartment of the thigh through the obturator foramen. The obturator nerve provides motor innervation to the medial compartment of the thigh (excluding the pectineus and hamstring portion of the adductor magnus muscle). In addition, the obturator nerve provides sensory to the medial region of the thigh.

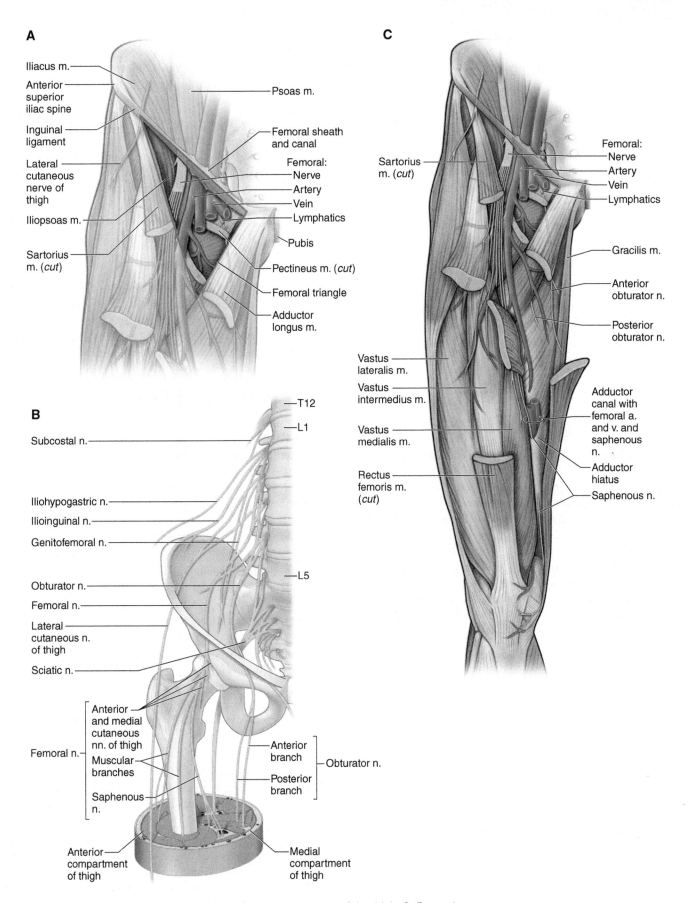

Figure 36-3: A. Femoral triangle. **B**. Innervation of the compartments of the thigh. **C**. Femoral nerve.

VASCULARIZATION OF THE THIGH

BIG PICTURE

The blood supply to the lower extremity initiates from the descending aorta, which divides into the common iliac arteries. The common iliac arteries divide into the external and internal iliac arteries. The external iliac artery passes deep to the inguinal ligament to become the **femoral artery**, serving as the primary blood supply to the lower limb. The internal iliac artery gives rise to the **obturator artery**, which also contributes to blood supply of the lower limb.

OBTURATOR ARTERY

Branches from the internal iliac artery exit the pelvis through the obturator foramen and bifurcate into anterior and posterior divisions. The two divisions circle the obturator foramen and supply blood, proximally, to the muscles inserting in the area. In addition, a branch enters the hip joint at the acetabular notch and travels through a conduit in the ligament of the head of the femur to supply blood to the femoral head.

FEMORAL ARTERY

Several arteries originate from the femoral artery, including the following (Figure 36-4A and B):

■ **Deep artery of the thigh (profunda femoris artery).** The largest branch of the femoral artery. The deep artery of the thigh branches posteriorly in the femoral triangle and travels between the adductor longus and brevis muscles and the adductor longus and magnus muscles. **Perforating branches** of the deep artery of the thigh pierce through the adductor magnus muscle and are the major supplier of blood to the three compartments of the thigh. Distally, the deep artery of the thigh anastomoses with branches of the popliteal artery. Other **branches that originate from the deep artery of the thigh** include the following:

  • **Lateral circumflex femoral artery.** Branches from the lateral side of the deep artery of the thigh (sometimes directly from the femoral artery) and travels laterally to branch into ascending, transverse, and descending branches.

    • **Ascending branch.** Ascends deep to the tensor fasciae latae muscle and connects with the medial femoral circumflex artery.

    • **Transverse branch.** Pierces the vastus lateralis muscle and wraps around the proximal shaft of the femur to anastomose with other vessels (i.e., medial femoral circumflex artery) to supply blood to the proximal femur and hip.

    • **Descending branch.** Descends laterally and pierces the vastus lateralis muscle. The descending branch distally connects with the popliteal artery.

  • **Medial circumflex femoral artery.** Branches medially from the deep artery of the thigh (may branch from femoral artery) and passes around the shaft of the femur. The medial circumflex femoral artery and branches of the lateral circumflex femoral artery supply the hip joint.

The main trunk of the femoral artery follows the adductor canal distally. It travels through an opening in the distal attachment of the adductor magnus (adductor hiatus) muscle and enters the popliteal fossa (posterior to the knee) to become the **popliteal artery**.

▽ **Coronary angioplasty** is a procedure that is frequently performed on patients who have obstructed blood flow to the heart musculature or who are experiencing a myocardial infarction. The femoral triangle allows easy access to a major blood vessel, the femoral artery. Because of the easy access, small balloons can be threaded through the femoral artery, at the femoral triangle, to coronary vessels in the heart. The balloon is then expanded to open the blood vessel. ▽

POPLITEAL ARTERY

The popliteal artery is the continuation of the femoral artery after it traverses the adductor hiatus (Figure 36-4A and B). The branches form an anastomotic vascular supply to the knee and are named according to their relationship to each other (superior lateral and medial genicular arteries and inferior lateral and medial genicular arteries).

LYMPHATICS OF THE INGUINAL AREA

The lymphatics of the thigh are organized into the following inguinal lymph nodes:

■ **Superficial inguinal nodes.** Nine to eleven nodes that run parallel and distal to the inguinal ligament. The superficial inguinal nodes receive lymph from the external genitalia, anal canal, gluteal region, and the inferior abdominal wall; they receive most of the lymph from the superficial vessels of the lower limb. The superficial inguinal nodes drain to the external iliac nodes.

■ **Deep inguinal nodes.** One to three nodes located medial to the femoral vein. The deep inguinal nodes receive lymph from deep vessels associated with the femoral vessels and glans penis or clitoris. Also, the deep inguinal nodes drain into the external iliac nodes.

■ **Popliteal nodes.** Six small nodes located in the popliteal fat. The popliteal nodes receive lymph from the knee and the deep vessels associated with the tibial vessels of the leg. Vessels from the popliteal nodes ascend the thigh to drain into the deep inguinal nodes, eventually reaching the external iliac nodes.

VEINS OF THE THIGH

The veins of the thigh consist of a superficial and a deep venous system. The superficial system consists of the great saphenous vein.

■ **Great saphenous vein.** Originates from the dorsal venous arch in the foot on the medial side. The great saphenous vein ascends the leg and thigh on the medial side and pierces the fascia latae of the thigh, forming the saphenous opening to drain in the femoral vein.

■ The deep venous system consists of as many as three veins that course with each artery. Most of the veins in the thigh drain into the femoral vein.

A

B

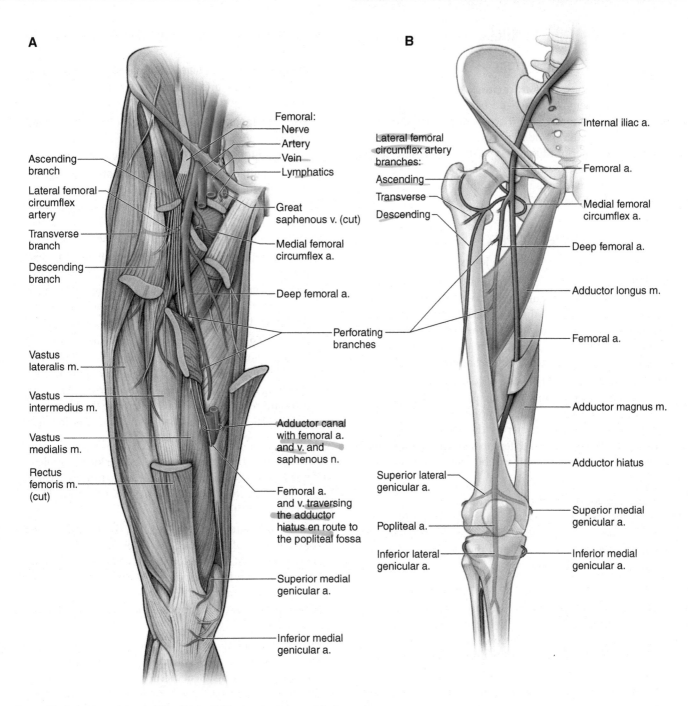

Femoral:
- Nerve
- Artery
- Vein
- Lymphatics

Ascending branch

Lateral femoral circumflex artery

Transverse branch

Descending branch

Great saphenous v. (cut)

Medial femoral circumflex a.

Deep femoral a.

Perforating branches

Vastus lateralis m.

Vastus intermedius m.

Vastus medialis m.

Rectus femoris m. (cut)

Adductor canal with femoral a. and v. and saphenous n.

Femoral a. and v. traversing the adductor hiatus en route to the popliteal fossa

Superior medial genicular a.

Inferior medial genicular a.

Lateral femoral circumflex artery branches:
- Ascending
- Transverse
- Descending

Internal iliac a.

Femoral a.

Medial femoral circumflex a.

Deep femoral a.

Adductor longus m.

Femoral a.

Adductor magnus m.

Adductor hiatus

Superior lateral genicular a.

Popliteal a.

Inferior lateral genicular a.

Superior medial genicular a.

Inferior medial genicular a.

Figure 36-4: A. Vasculature of the thigh. **B**. Femoral artery and its branches.

KNEE COMPLEX

BIG PICTURE

The knee complex consists of articulations between the femur and the tibia (tibiofemoral joint) and between the femur and the patella (patellofemoral joint). These articulations allow for static positions (standing) and dynamic movements (walking or running). The superior tibiofibular joint is not considered as part of the knee complex because it does not have the same capsule and is biomechanically linked with the ankle; therefore, it will be discussed in Chapter 37.

KNEE COMPLEX

The knee is composed of the following joints:

- **Tibiofemoral joint.** A synovial bicondylar joint with two degrees of motion. Articulations occur between the two condyles of the femur and the two tibial plateaus, producing flexion and extension. In addition, the tibiofemoral joint allows for minimal axial rotation with the pivot point, located medially on the medial tibial plateau.
- **Patellofemoral joint.** Articulation is between the intercondylar notch of the femur and the patella and shares the same joint capsule as the tibiofemoral joint. The patellofemoral joint directly serves the tibiofemoral joint; however, because of the vast differences in clinical problems and pathologies, the two joints will be discussed independently.

▽ **Patellofemoral disorder** is one of the most common knee disorders seen in patients who visit orthopedic clinics. The disorder is usually caused by excessive pressure or malalignment between the patella and the femur, resulting in pain at the patellofemoral joint. ▽

LIGAMENTOUS AND CAPSULAR SUPPORT OF THE KNEE Ligament and capsule support of the knee are critical because of the incongruence of the joint, weight bearing of the joint, and the large range of motion with flexion and extension (Figure 36-5A–C). The ligaments and capsule provide support to the knee joint, as follows:

- **Capsule.** Surrounds the knee joint and includes the patellofemoral joint. The capsule extends from the distal femur to the proximal tibia and contains areas of laxity and recesses to allow for range of motion. The ligaments that support the capsule on all four sides of the joint are as follows:
  - **Oblique popliteal ligament.** Supports the posterior capsule and attaches from the posteromedial tibial condyle to the center of the posterior capsule.
  - **Arcuate ligament.** Supports the posterior capsule and attaches from the posterior fibular head to the intercondylar area of the tibia and the lateral epicondyle of the femur.
  - **Medial collateral ligament (tibial collateral ligament).** Resists valgus forces on the knee (tibia abducting on femur). The medial collateral ligament attaches from the medial epicondyle of the femur to the tibial condyle and proximal shaft. In addition, it has an attachment to the medial meniscus.
  - **Lateral collateral ligament (fibular collateral ligament).** Resists varus forces on the knee (tibia adducting on the femur). The lateral collateral ligament appears as a strong cord and attaches from the lateral femoral epicondyle to the fibular head.
  - **Patellar ligament.** Attaches from the patella to the tibial tuberosity on the proximal anterior surface of the tibia. In addition, superficial fibers of the patellar ligament fan out, attaching to the sides of the tibial tuberosity. These structures are referred to as the medial and the lateral patella retinacula. The patellar ligament transmits forces produced by the quadriceps muscles to the tibia.

In addition to the capsular ligaments, the knee complex also contains ligaments inside the capsule, called the cruciate ligaments. The cruciate ligaments are named by their location of attachment on the tibia.

- **Anterior cruciate ligament.** Attaches on the anterior intercondylar area of the tibia. The anterior cruciate ligament (known as the ACL) twists on itself as it ascends in a posterolateral direction to attach on the posteromedial side of the lateral femoral condyle. The ACL resists anterior translation of the tibia on the femur or posterior translation of the femur on the tibia.
- **Posterior cruciate ligament.** Stronger than the ACL and attaches to the posterior intercondylar area of the tibia. The posterior cruciate ligament ascends in an anteromedial direction to attach on the lateral surface of the medial femoral condyle. The posterior cruciate ligament resists posterior translation of the tibia on the femur or anterior translation of the femur on the tibia.

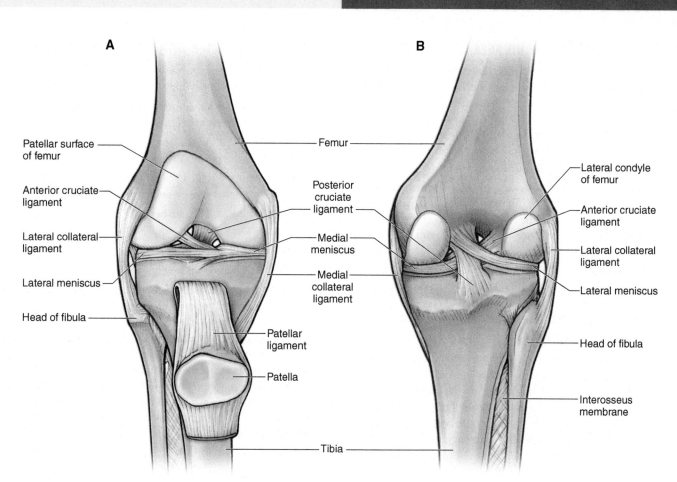

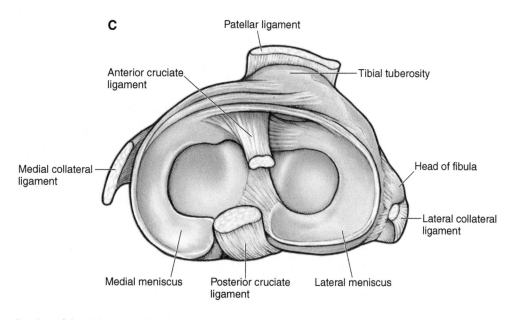

Figure 36-5: A. Anterior view of the right knee joint with the joint capsule open showing the patella reflected inferiorly. Posterior (**B**) and superior (**C**) views of the right knee joint.

MENISCI OF THE KNEE JOINT

The knee contains two fibrocartilaginous structures, one over the medial tibial plateau, the medial meniscus, and one over the lateral tibial plateau, the lateral meniscus (Figure 36-5A–C). Both are crescent shaped and do not complete a full circle. The menisci are also wedge shaped and, medially, are thin. However, laterally, the menisci are thicker, which increases the concavity of the articulating surface of the tibia.

- **Medial meniscus.** The anterior portion of the medial meniscus attaches to the anterior intercondylar area of the tibia and the posterior portion to the posterior intercondylar area. Its peripheral border is attached to the capsule, and a portion is attached to the medial collateral ligament.

- **Lateral meniscus.** Forms four-fifths of a complete circle. The anterior portion of the lateral meniscus attaches to the anterior intercondylar area of the tibia, and the posterior portion attaches to the posterior intercondylar area.

▽ A **tear of the ACL** is usually seen in patients who participate in sports that require cutting movements with deceleration (e.g., soccer, football). The mechanism of injury is usually deceleration of the body on an outstretched leg with lateral rotation of the femur on a fixed tibia. Treatment is often surgery using an autograft (i.e., tissue is taken from the patient to replace the patient's ligament). Frequently, a midpatellar ligament graft is used. ▼

BURSAE ASSOCIATED WITH THE KNEE JOINT COMPLEX

The knee joint has many bursae (sacs of synovial fluid) to decrease frictional forces. The most important bursae are as follows:

- **Suprapatellar bursa.** Located between the quadriceps tendon and the anterior femur.

- **Subpopliteal bursa.** Located between the popliteus muscle and the lateral femoral condyle.

- **Gastrocnemius bursa.** Located between the medial head of the gastrocnemius muscle and the medial femoral condyle.

- **Prepatellar bursa.** Located between the skin and the anterior patella.

- **Subcutaneous infrapatellar bursa.** Located between the patellar ligament and the tibial tubercle.

▽ **Prepatellar bursitis (commonly known as "housemaid's knee")** is caused by inflammation or bursitis of the superficial infrapatellar bursa between the skin and the patellar ligament. The mechanism of injury can be from direct impact or from an irritation to the knee that occurs over time. The condition is often seen in individuals whose occupations require them to place pressure on the knees, such as carpet layers or people who wash the floor on their hands and knees (thus the term housemaid's knee), resulting in pain over the patellar ligament. ▼

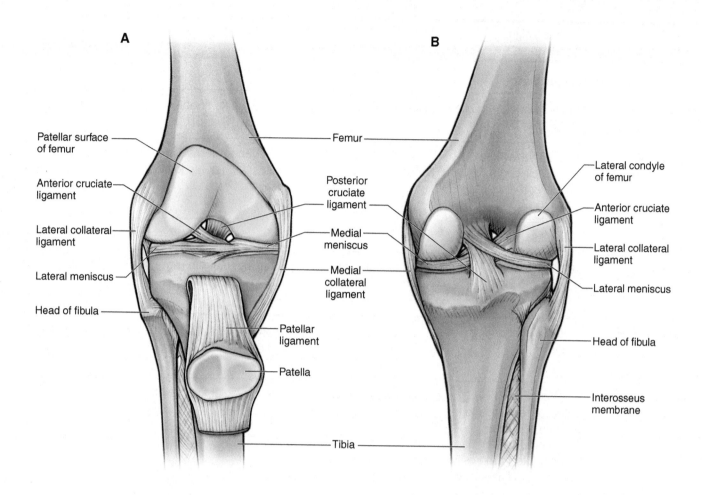

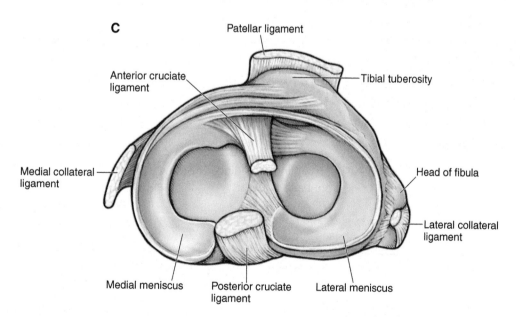

Figure 36-5: A. Anterior view of the right knee joint with the joint capsule open showing the patella reflected inferiorly. Posterior (**B**) and superior (**C**) views of the right knee joint.

TABLE 36-1. Muscles of the Thigh

Muscle	Proximal Attachment	Distal Attachment	Action	Innervation
Anterior compartment of the thigh				
Psoas minor	T12–L1 vertebral bodies and discs	Pectin pubis	Lumbar spine flexion, posterior pelvic tilt	Anterior rami (L1)
Psoas major	T12–L5 transverse processes, vertebral bodies and discs	Lesser trochanter of femur	Flexes and externally rotates thigh at hip joint; flexes trunk (psoas major)	Anterior rami (L1–L3)
Iliacus	Iliac fossa			Femoral n. (L2, L3)
Sartorius	Anterior superior iliac spine	Inferomedial to tibial tuberosity (pes anserinus)	Flexes thigh at hip joint and flexes leg at knee joint	Femoral n. (L2, L3)
Rectus femoris	Anterior inferior iliac spine		Flexes thigh at hip joint and extends leg at knee joint	Femoral n. (L2–L4)
Vastus lateralis	Lateral part of intertrochanteric line, margin of greater trochanter, lateral margin of gluteal tuberosity, lateral lip of linea aspera	Quadriceps femoris tendon	Extends leg at knee joint	
Vastus medialis	Medial part of intertrochanteric line, pectineal line, medial lip of linea aspera, medial supracondylar ridge			
Vastus intermedius	Femur: upper two-thirds of anterior and lateral surfaces			
Medial compartment of the thigh				
Pectineus	Pectineal line	Oblique line extending from base of lesser trochanter to linea aspera on posterior surface of proximal femur	Adducts and flexes thigh at hip joint	Femoral n. (L2, L3)
Adductor longus	Body of pubis	Linea aspera	Adducts and medially rotates thigh at hip joint	Obturator n. (anterior division) (L2–L4)
Adductor brevis	Body of pubis and inferior pubic ramus			Obturator n. (anterior division) (L2, L3)

TABLE 36-1. Muscles of the Thigh (*Continued*)

Muscle	Proximal Attachment	Distal Attachment	Action	Innervation
Adductor magnus	Adductor part: ischiopubic ramus Hamstring part: ischial tuberosity	Adductor part: linea aspera Hamstring part: Adductor tubercle	Adducts and medially rotates thigh at hip joint	Adductor part: obturator n. (L2–L4) Hamstring part: tibial division of sciatic n. (L4) and obturator n. (L2, L3)
Gracilis	Body and inferior ramus of pubic bone	Medial surface of proximal shaft of tibia (pes anserinus)	Adducts thigh at hip joint and flexes leg at knee joint	Obturator n. (L2, L3)
Obturator externus	External surface of obturator membrane and adjacent bone	Trochanteric fossa	Laterally rotates hip	Obturator n. (posterior division) (L3, L4)
Posterior compartment of the thigh				
Semitendinosus	Ischial tuberosity	Medial surface of proximal tibia (pes anserinus)	Flexes leg at knee joint and extends thigh at hip joint; medially rotates thigh at hip joint and leg at knee joint	Tibial division of sciatic n. (L5–S2)
Semimembranosus		Medial and posterior surface of medial tibial condyle		
Biceps femoris	Long head: ischial tuberosity Short head: lateral lip of linea aspera	Head of fibula	Knee flexion Hip extension Lateral rotation of hip and knee	Long head: tibial division of sciatic n. (L5–S2) Short head: common fibular division of sciatic n. (L5–S2)

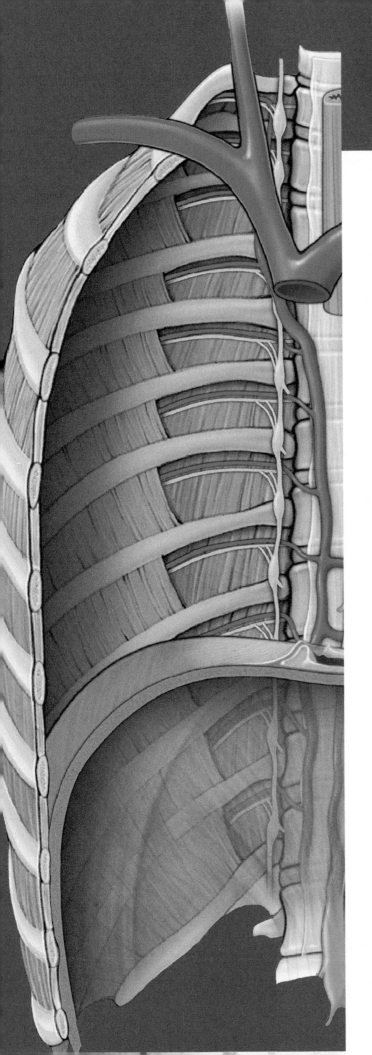

LEG

MUSCLES OF THE LEG

BIG PICTURE

The leg consists of the tibia and fibula. Proximally, the tibia of the leg articulates with the femur of the thigh through the knee joint. Distally, the tibia and fibula of the leg articulate with the talus bone of the foot through the ankle joint. The muscles of the leg that act on the knee and ankle as well as on the joints of the foot are organized into three fascial compartments, similar to those of the thigh muscles (Figure 37-1A). The anterior compartment primarily contains muscles that produce extension (dorsiflexion) and inversion; the posterior compartment primarily contains muscles that produce flexion (plantarflexion) and inversion; and the lateral compartment primarily contains muscles that produce flexion (plantarflexion) and eversion.

ACTIONS OF THE ANKLE

The ankle (talocrural) joint consists of articulations between the tibia and talus (tibiotalar joint) and the fibula and talus (talofibular joint) and allows for motion primarily in the saittal plane, as (Figure 37-1B) follows:

- **Plantar flexion (flexion).** Movement in which the angle between the leg and foot increases.
- **Dorsiflexion (extension).** Movement in which the angle between the leg and foot decreases.

The subtalar joint is formed by articulations between the talus and the calcaneus and allows for motion primarily in the coronal plane, as follows:

- **Inversion (pronation).** Movement in which the plantar surface of the foot faces medially.
- **Eversion (supination).** Movement in which the plantar surface of the foot faces laterally.

MUSCLES OF THE ANTERIOR COMPARTMENT OF THE LEG

The muscles of the anterior compartment of the leg produce numerous actions because some muscles cross the ankle, foot, and digits, and perhaps a combination of each of these joints (Table 37-1). The muscles in the anterior compartment of the leg have the following similar features:

- **Common innervation.** Deep fibular nerve.
- **Common action.** Dorsiflexion.
- **Common vascular supply.** Anterior tibial artery.

The following muscles are located in the anterior compartment of the leg (Figure 37-1C):

- **Tibialis anterior muscle.** Attaches proximally to the tibia and interosseous membrane; distally, it attaches to the medial cuneiform and the base of metatarsal 1. The tibialis anterior muscle dorsiflexes the foot at the ankle joint and inverts the foot. The deep fibular nerve (L4 and L5) innervates this muscle.
- **Extensor digitorum longus muscle.** Attaches proximally on the fibula and lateral tibial condyle; distally, it attaches to the dorsal digital expansions into digits 2 to 5. The extensor digitorum longus muscle extends lateral digits 2 to 4 and dorsiflexes the foot at the ankle joint. The deep fibular nerve (L5 and S1) innervates this muscle.
- **Extensor hallucis longus muscle.** Attaches proximally on the fibula and interosseous membrane; distally, it attaches to the distal phalanx of the great toe. The extensor hallucis longus muscle extends the great toe and dorsiflexes the foot. The deep fibular nerve (L5 and S1) innervates this muscle.
- **Fibularis (peroneus) tertius muscle.** Attaches proximally to the distal part of the fibula; distally, it attaches to the base of metatarsal 5. The fibularis tertius muscle dorsiflexes and everts the foot. The deep fibular nerve (L5 and S1) innervates this muscle.

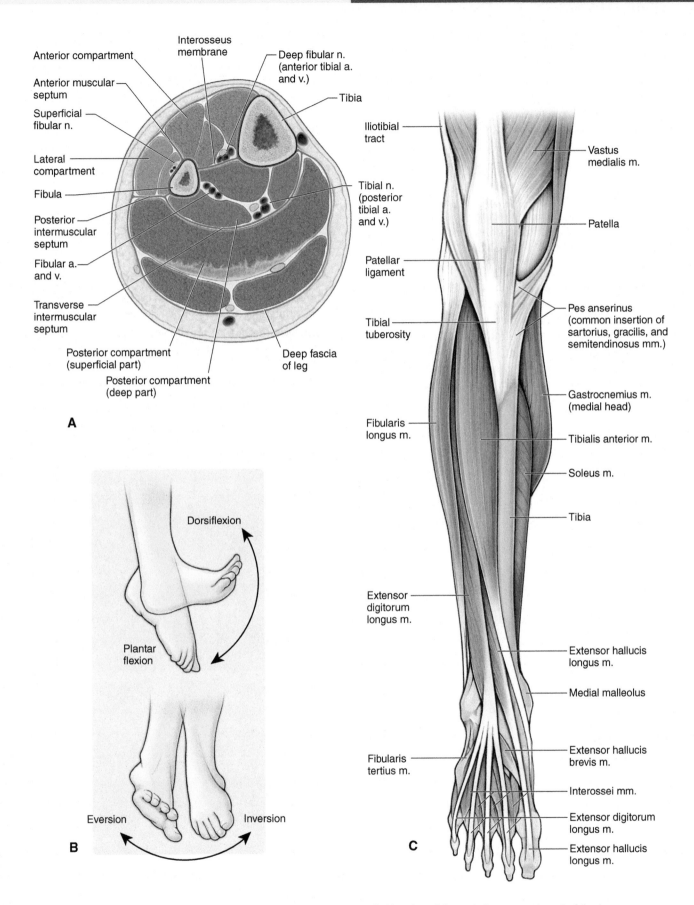

Figure 37-1: A. Cross-section of the right leg. **B.** Movements of the ankle. **C.** Muscles of the anterior compartment of the leg.

DORSUM OF THE FOOT

Muscles and their associated tendons cross the anterior surface of the ankle and insert in the foot. In addition, the following intrinsic muscles are located on the dorsal surface of the foot:

- **Extensor digitorum brevis muscle.** Attaches proximally to the lateral calcaneus; distally, it attaches to the dorsal surface of digits 2 to 4. The extensor digitorum brevis muscle extends digits 2 to 4. The deep fibular nerve (S1 and S2) innervates this muscle.

- **Extensor hallucis brevis muscle.** Attaches proximally to the lateral calcaneus; distally, it attaches to the dorsal surface of the great toe. The extensor hallucis brevis muscle extends the great toe. The deep fibular nerve (S1 and S2) innervates this muscle.

▽ The term "**shin splints**" is often an all-inclusive term used to describe pain in the anterior compartment of the leg. Most commonly, shin splints are caused by physical activity in which the foot is lowered to the ground following heel strike (such as occurs when running and especially when running downhill). The pain is due to inflammation of the periosteum of the tibia. In more severe cases, shin splints can result in stress fractures. ▽

MUSCLES OF THE LATERAL COMPARTMENT OF THE LEG

The muscles of the lateral compartment of the leg produce numerous actions because some muscles cross the ankle, foot, and digits and perhaps a combination of each of these joints. The muscles in the lateral compartment of the leg have the following similar features:

- **Common innervation.** Superficial fibular nerve.

- **Common action.** Plantarflexion and eversion.

- **Common vascular supply.** Anterior tibial and fibular arteries.

The following muscles are located in the lateral compartment (Figure 37-2):

- **Fibularis (peroneus) longus muscle.** Attaches proximally to the upper surface of the fibula; distally, it attaches to the medial cuneiform bone and the base of metatarsal 1. The fibularis longus muscle plantarflexes and everts the foot. The superficial fibular nerve (L5, S1, S2) innervates this muscle.

- **Fibularis (peroneus) brevis muscle.** Attaches proximally to the upper surface of the fibula; distally, it attaches to the base of metatarsal 5. The fibularis brevis muscle plantarflexes and everts the foot. The superficial fibular (peroneal) nerve (L5, S1, S2) innervates this muscle.

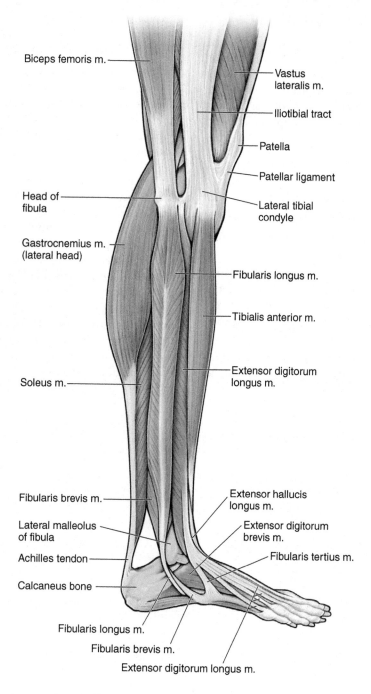

Figure 37-2: Muscles of the lateral compartment of the leg.

MUSCLES OF THE POSTERIOR COMPARTMENT OF THE LEG

The muscles of the posterior compartment of the leg produce numerous actions because some muscles cross the ankle, foot, and digits, and perhaps a combination of each of these joints. The muscles in the posterior compartment of the leg are divided into a superficial group and a deep group and have the following similar features:

- **Common innervation.** Tibial nerve.
- **Common action.** Plantarflexion.
- **Common vascular supply.** Posterior tibial, fibular, and popliteal arteries.

The **muscles in the superficial group** of the posterior compartment of the leg are as follows (Figure 37-3A and B):

- **Gastrocnemius muscle.** Attaches proximally to the femoral condyles; distally, it attaches to the calcaneus bone via the calcaneal tendon. The gastrocnemius muscle plantarflexes the foot and flexes the knee. The tibial nerve (S1 and S2) innervates this muscle.
- **Plantaris muscle.** Attaches proximally to the upper surface of the fibula; distally, it attaches to the posterior surface of the calcaneus via the calcaneal tendon. The plantaris muscle plantarflexes the foot and flexes the knee. The tibial nerve (S1 and S2) innervates this muscle.
- **Soleus muscle.** Attaches proximally to the posterior aspect of the tibia (soleal line) and the posterior aspect of the fibular head and shaft; distally, it attaches to the posterior surface of the calcaneus via the calcaneal tendon. The soleus muscle plantarflexes the foot. The tibial nerve (S1 and S2) innervates this muscle.

▽ The calcaneal (Achilles) tendon is a large ropelike band of fibrous tissue in the posterior ankle that connects the calf muscles (gastrocnemius and soleus muscles) to the calcaneus bone. When the calf muscles contract, the calcaneal tendon tightens and pulls the heel, resulting in standing on tiptoe; therefore, it is important in activities such as walking and jumping. **Rupture of the calcaneal tendon** usually is caused by a forceful push-off during an activity such as sprinting when running or jumping in a game of basketball. The result is tearing of the tendon that connects the gastrocnemius and soleus muscles to the calcaneus bone. Bruising usually is apparent, and a visible bulge forms in the posterior region of the leg because of muscle shortening. Surgical intervention is the most common treatment. ▼

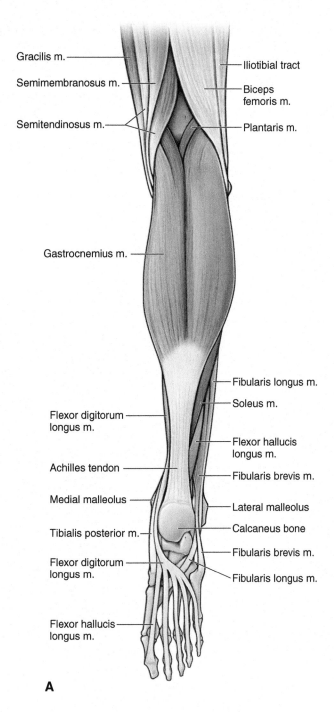

Gracilis m.

Semimembranosus m.

Semitendinosus m.

Iliotibial tract

Biceps femoris m.

Plantaris m.

Gastrocnemius m.

Fibularis longus m.

Soleus m.

Flexor digitorum longus m.

Flexor hallucis longus m.

Achilles tendon

Fibularis brevis m.

Medial malleolus

Lateral malleolus

Tibialis posterior m.

Calcaneus bone

Fibularis brevis m.

Flexor digitorum longus m.

Fibularis longus m.

Flexor hallucis longus m.

A

Figure 37-3: Muscles of the posterior compartment of the leg: (**A**) superficial dissection,

The **muscles of the deep group** of the posterior compartment of the leg are as follows (Figure 37-3B and C):

Popliteus muscle. Attaches proximally to the posterior surface of the proximal tibia; distally, it attaches to the lateral femoral condyle. The popliteus muscle unlocks the knee joint (it laterally rotates the femur on a fixed tibia). The tibial nerve (L4, L5, S1) innervates this muscle.

Flexor hallucis longus muscle. Attaches proximally to the posterior surface of the fibula and the interosseous membrane; distally, it attaches to the distal phalanx of the great toe. The flexor hallucis longus muscle flexes the great toe. The tibial nerve (S2 and S3) innervates this muscle.

Flexor digitorum longus muscle. Attaches proximally to the tibia; distally, it attaches to the distal phalanges of digits 2 to 5. The flexor digitorum longus muscle flexes digits 2 to 5. The tibial nerve (S2 and S3) innervates this muscle.

Tibialis posterior muscle. Attaches proximally to the interosseous membrane and the tibia and fibula; distally, it attaches to the navicular bone, all cuneiform bones, and metatarsals 2 to 4. The tibialis posterior muscle inverts and plantarflexes the foot, providing support to the medial arch of the foot during walking. The tibial nerve (L4 and L5) innervates this muscle.

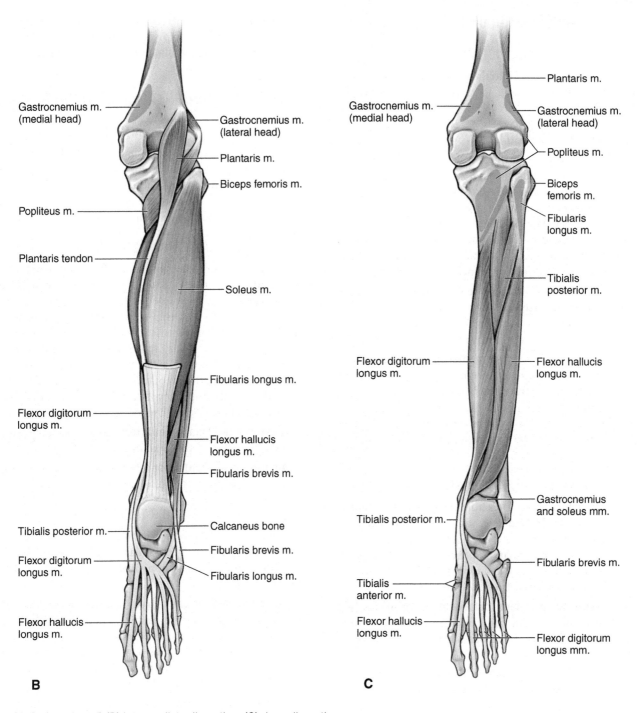

Figure 37-3: (*continued*) (**B**) intermediate dissection, (**C**) deep dissection.

INNERVATION OF THE LEG

BIG PICTURE

The sciatic nerve bifurcates into the tibial and common fibular nerves, near the popliteal fossa, to innervate muscles in the leg and foot.

TIBIAL NERVE

The tibial nerve arises from the anterior division of the sacral plexus (L4–S3), descends through the popliteal fossa, and courses deep to the soleus muscle to innervate the superficial and deep group of muscles in the posterior compartment of the leg (Figure 37-4A). The tibial nerve descends the posterior region of the leg and enters the foot inferior to the medial malleolus to innervate the plantar surface of the foot. The tibial nerve has muscular and sensory branches.

Muscular branches. The tibial nerve innervates the muscles in the posterior compartment of the leg (gastrocnemius, plantaris, soleus, popliteus, flexor hallucis longus, flexor digitorum longus, and tibialis posterior muscles).

Sensory branch. Gives rise to the **medial sural nerve**. Branches from the tibial nerve in the popliteal fossa and descends superficial to the gastrocnemius muscle to join the sural communicating branch from the lateral sural nerve. The medial sural nerve then becomes the sural nerve. The sural nerve provides sensory innervation to the posterolateral region of the leg and foot.

COMMON FIBULAR (PERONEAL) NERVE

The common fibular nerve arises from the posterior division of the sacral plexus (L4–S2) and descends in an inferolateral direction, across the popliteal fossa to the fibular head (Figure 37-4A and B). Just distal to the fibular head, the common fibular nerve bifurcates into the deep fibular and superficial fibular nerves. The following branches originate from the common fibular nerve:

Lateral sural nerve (lateral cutaneous nerve of the leg). Originates from the common fibular nerve and courses superficially to provide cutaneous innervation to the proximal lateral region of the leg.

Deep fibular nerve. Originates at the bifurcation of the common fibular nerve and courses into the anterior compartment of the leg. The nerve descends deep to the extensor digitorum longus and courses along the anterior interosseous membrane with the anterior tibial artery. The deep fibular nerve provides motor innervation to the muscles in the **anterior compartment of the leg** (tibialis anterior, extensor digitorum longus, extensor hallucis longus, and fibularis tertius muscles). The deep fibular nerve continues distally across the dorsum of the foot, and provides motor innervation to the extensor digitorum brevis and extensor hallucis brevis muscles and provides cutaneous innervation to the **skin between digits 1 and 2.**

Superficial fibular nerve. Originates at the bifurcation of the common fibular nerve. The superficial fibular nerve descends through the lateral compartment of the leg, providing motor innervation to the **fibularis (peroneus) longus and brevis muscles**. The nerve pierces the deep fascia to enter the anterior compartment of the leg, where it provides cutaneous innervation to the distal **anterolateral leg and dorsum of the foot.**

VASCULARIZATION OF THE LEG

BIG PICTURE

The popliteal artery is a continuation of the femoral artery. It courses through the popliteal fossa on the posterior side of the knee and bifurcates into the anterior and posterior tibial arteries at the inferior border of the popliteus muscle. The anterior and posterior tibial arteries supply blood to the leg and foot.

ANTERIOR TIBIAL ARTERY

The anterior tibial artery originates from the popliteal artery and courses anteriorly through a proximal opening in the interosseous membrane to enter the anterior compartment of the leg (Figure 37-4B). Distally, the anterior tibial artery courses in the anterior compartment with the deep fibular nerve, crosses the anterior ankle, and continues as the dorsalis pedis artery. The anterior tibial artery supplies blood to structures in the anterior compartment of the leg as well as partial blood supply to the lateral compartment. At the level of the distal tibia, the anterior tibial artery gives rise to the following arteries:

Anterior lateral malleolar artery. Courses laterally across the ankle joint to join the fibular artery.

Anterior medial malleolar artery. Courses laterally across the ankle to join the posterior tibial artery.

POSTERIOR TIBIAL ARTERY

The posterior tibial artery originates from the popliteal artery and quickly gives rise to the fibular artery (Figure 37-4A and B). The posterior tibial and fibular arteries descend deep to the soleus muscle.

Posterior tibial artery. Curves medially, as it courses inferiorly, giving rise to the posterior lateral malleolar artery, which joins the arterial network around the ankle. The posterior tibial artery supplies blood to the posterior compartment of the leg and continues distally, under the medial malleolus, to supply blood to the foot.

Fibular artery. Descends along the posterior region of the leg, traversing laterally to continue its descent along the medial side of the fibula. The fibular artery provides blood supply to the posterior and lateral compartments of the leg. Distally, the artery gives off a perforating branch, which courses through the inferior aspect of the interosseous membrane to join the arterial network around the ankle. In addition, the fibular artery gives rise to the posterior lateral malleolar artery, which also joins the atrial network around the ankle joint.

▽ **Anterior compartment syndrome** can be caused by a tibial fracture or a high-velocity blow to the anterior compartment of the leg, resulting in increased pressure in the anterior compartment of the leg. Because the fascia covering the

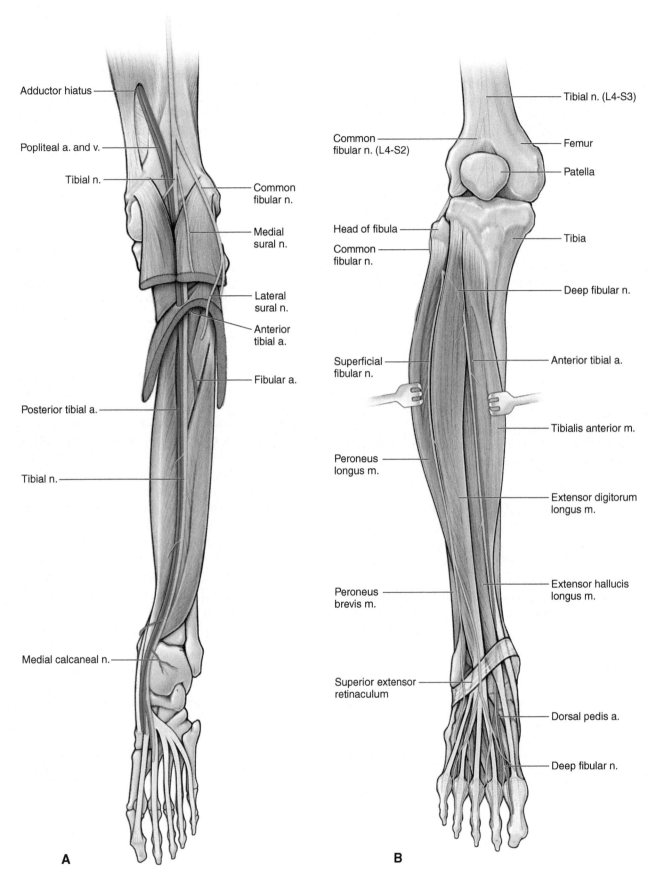

Figure 37-4: A. Posterior view of the leg showing the tibial nerve and the posterior tibial artery. **B**. Anterior view of the leg showing the common fibular nerve and the anterior tibial artery.

anterior compartment is unable to expand, pressure continues to build, causing restricted blood flow and eventual necrosis of tissues. If untreated, anterior compartment syndrome can result in amputation of the limb. Treatment varies; in more severe cases, a fasciotomy is performed and the fascia covering the anterior compartment is cut to relieve the pressure. ▼

VEINS OF THE LEG

The veins in the leg consist of a superficial and a deep venous system. The superficial system consists of the **great saphenous vein**, located medially, and the **small saphenous vein**, located posterolaterally. The great saphenous vein originates from the medial side of the dorsal venous arch in the foot and drains in the femoral vein. The small saphenous vein originates from the lateral side of the dorsal venous arch in the foot and drains in the popliteal vein. The deep venous system consists of as many as three veins, which course with each artery.

JOINTS OF THE LEG

BIG PICTURE

The boney components of the leg include the tibia and the fibula, which articulate via the proximal and distal tibiofibular joints. Distally, the tibia and fibula articulate with the talus, forming the ankle (talocrural) joint. The ankle joint is a combination of articulations between the tibia and the talus (tibiotalar joint) as well as the fibula and the talus (talofibular joint) (Figure 37-5A and B).

PROXIMAL TIBIOFIBULAR JOINT

The proximal tibiofibular joint consists of articulations between the proximal tibia and the fibula. It is a synovial joint with a joint capsule, which is separate from the knee joint and is reinforced by the following structures:

- **Anterior tibiofibular ligaments.** Connect between the anterior tibia and the fibula.
- **Posterior tibiofibular ligaments.** Connect between the posterior tibia and the fibula.
- **Interosseous membrane.** Supports both the proximal and the distal tibiofibular joints. The interosseous membrane is strong and consists of multiple small fibers that join the tibia and fibula, proximal to distal.

DISTAL TIBIOFIBULAR JOINT

The distal tibiofibular joint consists of the articulation between the distal tibia and the fibula. It is a syndesmosis joint that plays an important role in maintaining a stable mortise. The following structures reinforce the distal tibiofibular joint:

- **Anterior tibiofibular ligaments**
- **Posterior tibiofibular ligaments**
- **Interosseous membrane**

ANKLE JOINT

The ankle (talocrural) joint is a synovial hinge joint that allows plantarflexion and dorsiflexion. The ankle joint consists of articulations between the tibia and the talus (tibiotalar joint) as well as between the fibula and the talus (talofibular joint). The articulation between the tibia and the fibula (distal tibiofibular joint) forms a mortise into which the talus fits. The ligaments of the distal tibiofibular joint reinforce the mortise.

LIGAMENTOUS SUPPORT The ankle joint has a fairly weak capsule that is primarily supported by the medial and lateral collateral ligaments.

- **Medial collateral ligament (deltoid ligament).** A fan-shaped ligament that attaches to the medial malleolus of the tibia and the navicular, talus, and calcaneus bones. The medial collateral ligament is strong and will often avulse the medial malleolus before tearing. This ligament prevents medial distraction (eversion) and excessive range of motion.
- **Lateral collateral ligament.** Consists of three separate bands. The combination of ligamentous bands prevents lateral distraction (inversion) and excessive range of motion. The bands consist of the following ligaments:
  - **Anterior talofibular ligament.** Connect between the lateral malleolus of the fibula and the posterior talus.
  - **Posterior talofibular ligament.** Connect between the lateral malleolus of the fibula and the posterior talus.
  - **Calcaneofibular ligament.** Connect between the lateral malleolus of the fibula and the calcaneus.

▽ Injury to the anterior talofibular ligament due to excessive inversion with plantarflexion is the most common **ankle injury**. The second most common ankle injury occurs at the calcaneofibular ligament and results in anterolateral rotary instability of the ankle joint. The severity of **ankle sprains** is variable. ▼

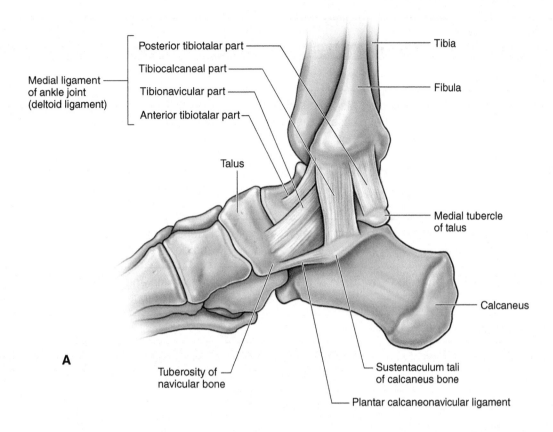

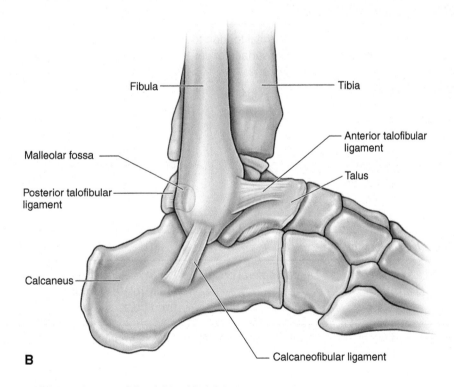

Figure 37-5: (**A**) Medial and (**B**) lateral views of the right ankle joint.

TABLE 37-1. Muscles of the Leg

Muscle	Proximal Attachment	Distal Attachment	Action	Innervation
Anterior compartment of the leg				
Tibialis anterior	Tibia and interosseous membrane	Medial cuneiform and base of metatarsal 1	Dorsiflexion of foot at ankle joint; inversion of foot	Deep fibular n. (L4, L5)
Extensor digitorum longus	Fibula and lateral tibial condyle	Via dorsal digital expansions into digits 2–5	Extension of lateral digits 2–5 and dorsiflexion of foot	Deep fibular n. (L5, S1)
Extensor hallucis longus	Fibula and interosseous membrane	Distal phalanx of great toe	Extension of great tow and dorsiflexion of foot	
Fibularis (peroneus) tertius	Distal part of fibula	Base of metatarsal 5	Dorsiflexion and eversion of foot	
Lateral compartment of the leg				
Fibularis (peroneus) longus	Upper surface of fibula	Medial cuneiform and base of metatarsal 1	Eversion and plantarflexion of foot	Superficial fibular n. (L5, S1, S2)
Fibularis (peroneus) brevis	Lower surface of fibula	Base of metatarsal 5		
Posterior compartment of the leg (superficial group)				
Gastrocnemius	Medial head: superior to medial femoral condyle Lateral head: superior to lateral femoral condyle	Via calcaneal tendon to posterior surface of calcaneus bone	Plantarflexes foot and flexes knee	Tibial n. (S1, S2)
Plantaris	Superior to lateral femoral condyle			
Soleus	Posterior aspect of tibia (soleal line) and posterior aspect of fibular head and shaft		Plantarflexes the foot	
Posterior compartment of the leg (deep group)				
Popliteus	Posterior surface of proximal tibia	Lateral femoral condyle	Unlocks knee joint; laterally rotates femur on fixed tibia	Tibial n. (L4, L5, S1)
Flexor hallucis longus	Posterior surface of fibula and interosseous membrane	Distal phalanx of great toe	Flexes great toe	Tibial n. (S2, S3)
Flexor digitorum longus	Tibia	Distal phalanges of digits 2–5	Flexes digits 2–5	
Tibialis posterior	Interosseous membrane, tibia, and fibula	Navicular, all cuneiform bones, and metatarsals 2–4	Inversion and plantarflexion of foot; support of medial arch of foot during walking	Tibial n. (L4, L5)

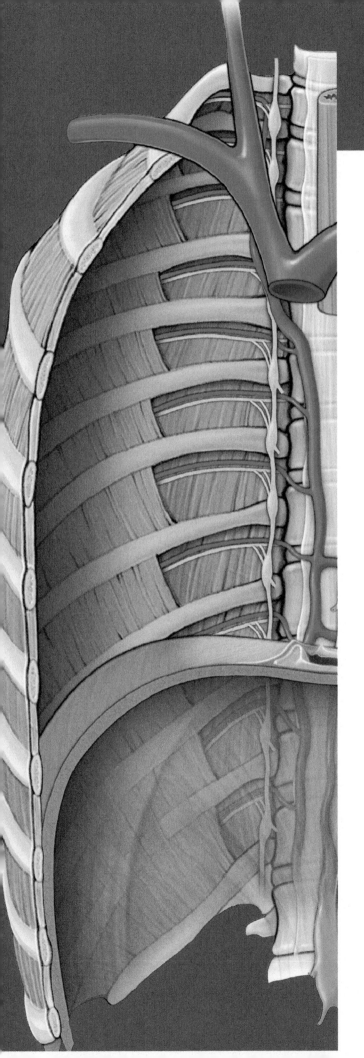

CHAPTER 38

FOOT

JOINTS OF THE DIGITS AND FASCIA OF THE FOOT

BIG PICTURE

The foot is connected to the leg by the ankle (talocrural) joint, which is an articulation between the tibia, fibula, and talus. The foot consists of 7 tarsal bones, 5 metatarsal bones, and 14 phalanges. Motion at the digits for abduction and adduction is defined by an imaginary line along the long axis of the second digit, unlike the hand in which the long axis runs along the third digit. Each digit, with the exception of the great toe, consists of three phalanges (proximal, middle, and distal); the great toe has two phalanges (proximal and distal). The articulations between the bones of the foot create multiple joints. The muscles that move these joints are divided into two groups, intrinsic and extrinsic foot muscles. The intrinsic muscles originate and attach in the foot, whereas the extrinsic muscles originate in the leg and insert in the foot, creating motion at multiple joints.

JOINTS OF THE DIGITS

The several bony articulations within the foot assist in accommodating uneven surfaces during weight-bearing activities. These motions of the foot are accomplished via the following joints (Figure 38-1):

- **Metatarsophalangeal joints.** Consist of articulations between the metatarsals and the proximal phalanges. The metatarsophalangeal joints allow **flexion and extension** and **abduction and adduction**.

- **Interphalangeal joints.** Consist of articulations between the phalanges, resulting in five proximal and four distal interphalangeal joints, which allow for **flexion and extension**.

FASCIAL STRUCTURES OF THE FOOT (FIGURE 38-1)

- **Plantar aponeurosis.** Radiates from the **calcaneus bone** toward the digits. The plantar aponeurosis is a very thick fascia that invests the muscles of the plantar surface of the foot.

- **Superior extensor retinaculum.** Attaches from the anterior border of the fibula to the tibia, proximal to the ankle joint. The superior extensor retinaculum holds the tendons of the tibialis anterior, extensor hallucis longus, extensor digitorum longus, and fibularis (peroneus) tertius muscles next to the structures of the anterior ankle during contraction.

- **Inferior extensor retinaculum.** A "Y-shaped" structure that attaches laterally to the superior surface of the calcaneus bone and courses medially to attach to the medial malleolus and the medial side of the plantar aponeurosis. The inferior extensor retinaculum serves to tether the tendons of the fibularis (peroneus) tertius, extensor digitorum longus, extensor hallucis longus, and anterior tibialis muscles.

- **Flexor retinaculum.** Attaches between the medial malleolus and calcaneus bones, forming the roof of the **tarsal tunnel**. The tendons of the tibialis posterior, flexor digitorum longus, and flexor hallucis longus muscles as well as tibial nerve and posterior tibial artery pass through the tarsal tunnel to the enter into the plantar surface of the foot.

- **Fibular retinacula.** Tethers the tendons of the fibularis (peroneus) longus and brevis muscles on the lateral side of the ankle as they course inferior to the lateral malleolus bone.

- **Dorsal digital expansions.** An aponeurosis covering the dorsum of the digits that attaches proximally to the middle phalanx (digits 2–5) or proximal phalanx (digit 1), via the central band, and distally to the distal phalanx, via the lateral bands. The extensor digitorum longus and brevis muscles and the extensor hallucis longus and brevis muscles attach proximally and centrally to the dorsal digital expansion. The lumbricals and the dorsal and plantar interossei attach on the free edges. Because of the attachment of the muscles and the location of the dorsal digital expansion, the small intrinsic muscles produce flexion at the metatarsophalangeal joint while extending the interphalangeal joints.

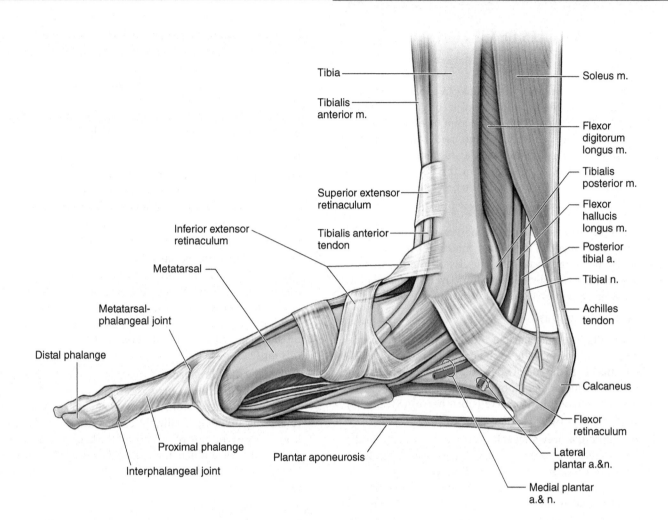

Figure 38-1: Medial view of the fascia of the right foot.

MUSCLES OF THE FOOT

BIG PICTURE

Muscles that act on the joints of the foot can either be extrinsic (originating outside the foot) or intrinsic (originating within the foot), and they may act on a single joint or multiple joints. The result is movement of multiple joints used to accommodate uneven surfaces or for activities such as running or jumping. The intrinsic muscles are discussed in this section, and the extrinsic muscles are discussed in Chapter 37.

FOOT MUSCLES

The medial plantar or lateral plantar nerves originate from the tibial nerve and innervate the plantar muscles of the foot. The deep fibular nerve innervates the muscles on the dorsal side. These foot muscles are divided into four layers on the plantar surface and one group on the dorsal surface (Table 38-1).

LAYER 1 (FIGURE 38-2A)

- **Abductor digiti minimi muscle.** Attaches proximally to the medial and lateral tubercle of the calcaneal tuberosity; distally, it attaches to the lateral side of the base of proximal phalanx 5. The abductor digiti minimi muscle abducts and flexes digit 5. The lateral plantar nerve (tibial nerve) (S1, S2, S3) innervates this muscle.

- **Flexor digitorum brevis muscle.** Attaches proximally to the medial tubercle of the calcaneal tuberosity; distally, it attaches to both sides of the middle phalanges of digits 2 to 5. The flexor digitorum brevis muscle flexes the lateral four digits. The medial plantar nerve (tibial nerve) (S1 and S2) innerves this muscle.

- **Abductor hallucis muscle.** Attaches proximally to the medial tubercle of the calcaneal tuberosity and distally to the medial side of the base of the proximal phalanx 1. The abductor hallucis muscle abducts and flexes the great toe. The medial plantar nerve (tibial nerve) (S1 and S2) innervates this muscle.

LAYER 2 (FIGURE 38-2B)

- **Lumbricals.** Attach proximally to the tendons of the flexor digitorum longus muscle and distally to the medial aspect of the dorsal digital expansions of digits 2 to 5. The lumbricals flex the proximal phalanges and extend the middle and distal phalanges. The medial plantar nerve (tibial nerve) innervates the first lumbrical, and the deep branch of the lateral plantar nerve (tibial nerve) innervates lumbricals 2 to 4 (S2 and S3).

- **Quadratus plantae muscle.** Attaches proximally to the medial and lateral margin of the plantar surface of the calcaneus bone and distally to the posterolateral margin of the tendon of the flexor digitorum longus muscle. The quadratus plantae muscle assists the flexor digitorum longus muscle in flexing the lateral four digits. The lateral plantar nerve (tibial nerve) (S1, S2, S3) innervates this muscle.

LAYER 3 (FIGURE 38-2C)

- **Flexor digiti minimi brevis muscle.** Attaches proximally to the base of metatarsal 5 and distally to the base of proximal phalanx 5. The flexor digiti minimi brevis muscle flexes the proximal phalanx of digit 5. The superficial branch of the lateral plantar nerve (tibial nerve) (S2 and S3) innervates this muscle.

- **Flexor hallucis brevis muscle.** Attaches proximally to the plantar surface of the cuboid and lateral cuneiform bones and distally to both sides of the base of the proximal phalanx of the **hallux** (great toe, or digit 1). The flexor hallucis brevis muscle flexes the proximal phalanx of the hallux. The medial plantar nerve (tibial nerve) (S1 and S2) innervates this muscle.

- **Adductor hallucis muscle.** Proximally, the oblique head of this muscle attaches to the bases of metatarsals 2 to 4 and the transverse head attaches to metatarsophalangeal joints (plantar ligaments); distally, the entire muscle attaches to the lateral side of the base of phalanx of the hallux. The adductor hallucis muscle adducts the hallux. The deep branch of the lateral plantar nerve (tibial nerve) (S2 and S3) innervates this muscle.

LAYER 4 (FIGURE 38-2D)

- **Plantar interossei muscle.** Attaches proximally to the bases and medial sides of metatarsals 3 to 5 and distally to the medial sides of the dorsal digital expansions of the proximal phalanges 3 to 5. The plantar interossei muscle adducts digits 3 to 5 and flexes the metatarsophalangeal joints of digits 3 to 5. The deep branch of the lateral plantar nerve (tibial nerve) (S2 and S3) innervates this muscle.

- **Dorsal interossei muscle.** Attaches proximally to the adjacent sides of metatarsals 1 to 5 and distally to the medial side of the dorsal digital expansions of proximal phalanx 2 and to the lateral sides of proximal phalanges 2 to 4. The dorsal interossei muscle abducts digits 2 to 4 and flexes the metatarsophalangeal joints. The deep branch of the lateral plantar nerve (tibial nerve) (S2 and S3) innervates this muscle.

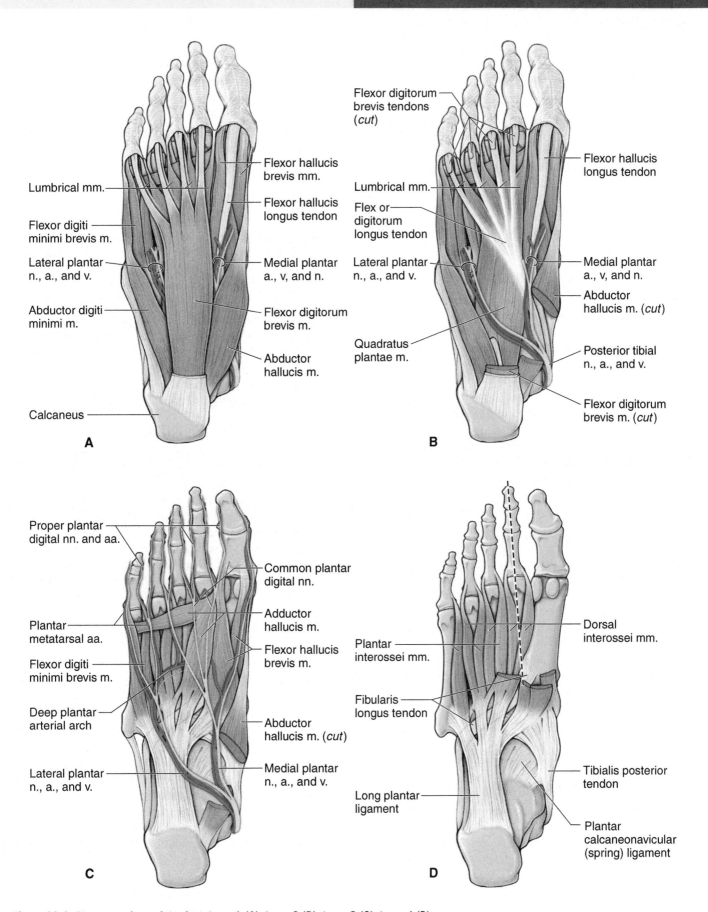

Figure 38-2: Plantar surface of the foot: layer 1 (**A**); layer 2 (**B**); layer 3 (**C**); layer 4 (**D**).

INNERVATION OF THE FOOT

BIG PICTURE

The **tibial nerve** enters the foot inferior to the medial malleolus through the **tarsal tunnel**, giving rise to the medial calcaneal branch (sensory). The nerve then bifurcates into the **medial and lateral plantar nerves** to supply motor and sensory innervation to the plantar surface of the foot. The **deep fibular nerve** supplies motor innervation to the dorsum of the foot as well as sensory innervation to a small area between the first and second digits. Sensory innervation to the medial and lateral sides of the foot is provided by the **saphenous nerve** (from the femoral nerve) and the **sural nerve** (from the tibial nerve) (Figure 38-3A and B).

MEDIAL CALCANEAL NERVE

The medial calcaneal nerve originates from the tibial nerve as it courses through the **tarsal tunnel**, providing sensory innervation to the posterior plantar surface of the foot over the heel. The medial calcaneal branch does not provide any motor innervation.

MEDIAL PLANTAR NERVE

The medial plantar nerve is one of two terminal branches of the tibial nerve and travels deep to the abductor hallucis muscle (the other terminal branch is the lateral plantar nerve). The medial plantar nerve courses adjacent to the flexor digitorum brevis muscle, providing motor innervation to the first lumbrical and the abductor hallucis, flexor hallucis brevis, and flexor digitorum brevis muscles. The medial plantar nerve also provides sensory innervation to the medial plantar surface of the foot.

LATERAL PLANTAR NERVE

The lateral plantar nerve is one of the terminal branches of the tibial nerve that courses from the medial side of the foot laterally toward the head of metatarsal 5, deep to the flexor digitorum brevis muscle. The lateral plantar nerve continues forward and bifurcates into a deep branch and a superficial branch. This nerve provides motor innervation to lumbricals 2 to 4 and the plantar and dorsal interossei, adductor hallucis, quadratus plantae, flexor digiti minimi brevis, and abductor digiti minimi muscles. The lateral plantar nerve provides sensory innervation to the lateral aspect of the plantar surface of the foot.

VASCULARIZATION OF THE FOOT

BIG PICTURE

The **posterior tibial artery** courses through the tarsal tunnel, inferior to the medial malleolus bone, and bifurcates into the medial and lateral plantar arteries to supply blood to the plantar surface of the foot. On the dorsal side, the **anterior tibial artery** crosses the anterior ankle joint as the dorsalis pedis artery, which

supplies blood to the dorsal side of the foot. The blood of the foot is returned to the **femoral veins** via the deep and superficial venous systems. The deep venous system follows the arteries.

DORSALIS PEDIS ARTERY

The dorsalis pedis artery courses distally on the dorsum of the foot and provides the following branches:

- **Medial and lateral tarsal arteries.** Branch medially and laterally from the dorsalis pedis artery, supplying blood to the tarsals and adjacent structures, and contributes to the arterial network surrounding the ankle.
- **Arcuate artery.** A lateral branch from the dorsalis pedis artery that branches into the dorsal metatarsal arteries, providing blood to adjacent structures of metatarsals and digits 2 to 5. In addition, the arcuate artery communicates with the plantar surface of the foot through perforating branches.
- **First dorsal metatarsal.** The dorsalis pedis artery courses distally to the base of metatarsals 1 and 2 and bifurcates into the first dorsal metatarsal artery and the deep plantar artery, which perforates through to the plantar surface of the foot. The first dorsal metatarsal artery supplies blood to adjacent structures around the first metatarsal and the great toe.

LATERAL PLANTAR ARTERY

The lateral plantar artery originates from the posterior tibial artery inferior to the medial malleolus bone and courses laterally between the quadratus plantae and flexor digitorum brevis muscles. The lateral plantar artery courses distally along the medial edge of the abductor digiti minimi muscle and curves medially, forming the deep plantar arch. The deep plantar arch provides blood to the adjacent structures of metatarsals and digits 2 to 5, the lateral side of metatarsal 1, and the great toes, via plantar metatarsal arteries and digital branches. The terminal end of the deep plantar arch joins the deep plantar artery of the dorsalis pedis artery.

MEDIAL PLANTAR ARTERY

The medial plantar artery originates from the posterior tibial artery, inferior to the medial malleolus bone. It courses distally along the medial edge of the abductor hallucis muscle, supplying blood to the adjacent structures on the medial side of the first metatarsal and the great toe, via the first plantar metatarsal artery and digital branches.

VEINS OF THE FOOT

The foot contains **deep and superficial venous systems**. The deep veins are named according to the arteries they follow. The superficial venous system drains into the **dorsal venous arch**. The medial side of the foot drains into the great saphenous vein from the dorsal venous arch and the lateral side of the foot into the small saphenous vein.

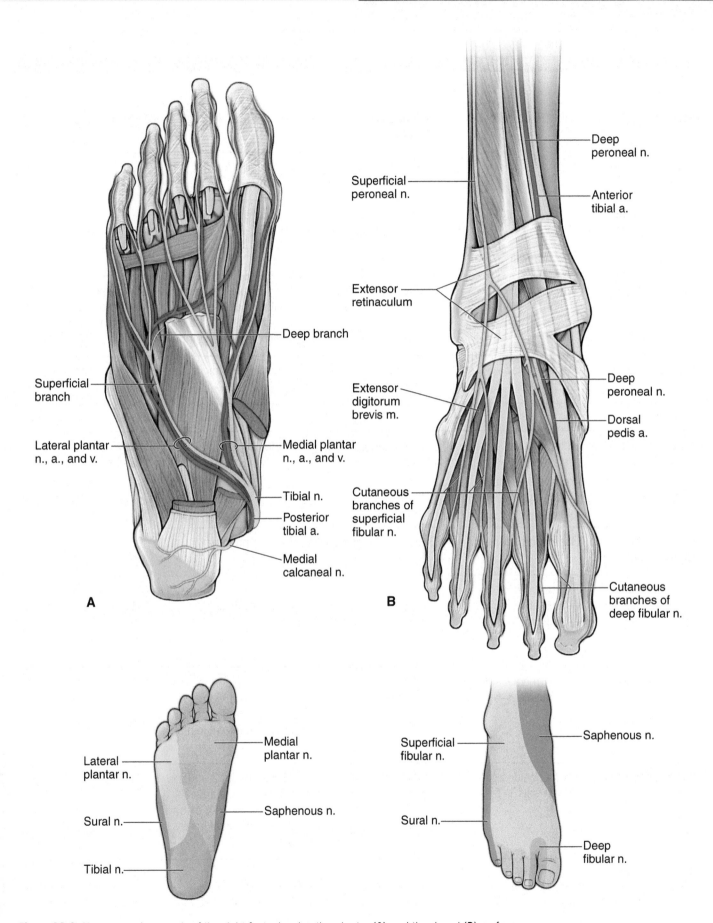

Figure 38-3: Neurovascular supply of the right foot, showing the plantar (**A**) and the dorsal (**B**) surfaces.

TABLE 38-1. Muscles of the Foot

Muscle	Proximal Attachment	Distal Attachment	Action	Innervation
Layer 1				
Abductor digiti minimi	Calcaneal tuberosity	Lateral base of proximal phalanx 5	Abduct and flex digit 5	Lateral plantar n. (S1–S3)
Flexor digitorum brevis		Both sides of middle phalanges digits 2–5	Flex digits 2–5	Medial plantar n. (S1–S2)
Abductor hallucis		Medial side of base of proximal phalanx digit 1	Abduct and flex great toe	
Layer 2				
Lumbricals	Tendons of flexor digitorum longus	Medial dorsal digital expansion digits 2–5	Flex metatarsophalangeal joint and extend proximal and distal interphalangeal joints	Lumbrical 1: medial plantar n. (S2–S3); lumbricals 2–4: lateral plantar n. (S2–S3)
Quadratus plantae	Plantar surface of calcaneus	Tendon of flexor digitorum longus	Assists flexor digitorum to flex digits 2–5	Lateral plantar nerve (S1–S3)
Layer 3				
Adductor hallucis	Oblique head: base of metatarsals 2–4 Transverse head: metatarsophalangeal joints	Lateral side, base of phalanx 1	Adducts digit 1	Lateral plantar nerve (S2–S3)
Flexor digiti minimi brevis	Base of metatarsal 5	Proximal phalanx digit 5	Base of proximal phalanx 5	
Flexor hallucis brevis	Plantar surface of cuboid and lateral cuneiform	Proximal phalanx digit 1	Flex metatarsophalangeal joint digit 1	Medial plantar n. (S1–S2)
Layer 4				
Plantar interossei	Bases, medial sides of metatarsals 3–5	Medial sides of dorsal digital expansions digits 3–5	Adduct digits 3–5; flex metatarsophalangeal joints digits 3–5; extend interphalangeal joints digits 3–5	Lateral plantar nerve (S2–S3)
Dorsal interossei	Adjacent sides of metatarsals 1–5	Medial side of dorsal digital expansions digits 2–4	Abduct digits 2–4; flex metatarsophalangeal joints digits 2–4; extend interphalangeal joints digits 2–4	
Dorsum of foot				
Extensor digitorum brevis	Lateral calcaneus	Dorsal digital expansion digits 2–4	Extend digits 2–4	Deep fibular nerve (S1–S2)
Extensor hallucis brevis		Dorsal digital expansion digits 1	Extend digit 1	

STUDY QUESTIONS

Directions: Each of the numbered items or incomplete statement is followed by lettered options. Select the **one** lettered option that is **best** in each case.

1. A 42-year-old man is admitted to the emergency department in shock and requires a saphenous cut-down to receive an infusion. To isolate the great saphenous vein in the ankle region, you would most likely determine its location in which of the following areas?

   A. Anterior to the lateral malleolus

   B. Anterior to the medial malleolus

   C. Posterior to the lateral malleolus

   D. Posterior to the medial malleolus

2. A lesion to the lateral cutaneous nerve of the thigh would most likely represent a lesion in which area?

   A. Dermatome

   B. Cutaneous field

3. The hip is a synovial joint composed of articulations between which of the following structures?

   A. Femoral head and acetabulum

   B. Femur and tibia

   C. Ilium and sacrum

   D. Obturator foramen and pelvic outlet

   E. Pubis and ischium

4. A 17-year-old boy is admitted to the emergency department after being involved in a motorcycle accident. He has a compound fracture in his right leg, and a thin bone is protruding out of the lateral aspect of his leg. Which of the following bones is most likely seen protruding through the skin of this boy's leg?

   A. Calcaneus

   B. Femur

   C. Fibula

   D. Tibia

5. A 29-year-old man is diagnosed with paralysis of the left piriformis muscle. Which of the following actions is the most likely difference between the left and right foot during gait?

   A. Left foot points more laterally

   B. Left foot points more medially

   C. Right foot points more laterally

   D. Right foot points more medially

6. A 72-year-old woman is brought to the emergency department after falling in her home. Radiographic studies show that she has fractured her hip. A serious complication of fractures of the femoral neck in the elderly is avascular necrosis of the femoral head. Avascular necrosis usually results from rupture of arteries to the head of the femur. Which of the following is the most likely artery damaged in this patient?

   A. Acetabular branch of the obturator artery

   B. Deep circumflex iliac artery

   C. Inferior gluteal artery

   D. Medial circumflex femoral artery

   E. Lateral circumflex femoral artery

   F. Superior circumflex iliac artery

7. A 34-year-old man is diagnosed with a left internal iliac artery aneurism. As a result, he presents with a left superior gluteal nerve lesion and an accompanying gait disorder. While walking, this patient would most likely compensate by flexing his trunk to the

   A. left, to lift his left lower limb so that his left foot can be lifted off the ground

   B. left, to lift his right lower limb so that his right foot can be lifted off the ground

   C. right, to lift his left lower limb so that his left foot can be lifted off the ground

   D. right, to lift his right lower limb so that his right foot can be lifted off the ground

8. When administering an intramuscular gluteal injection in the superior–lateral quadrant, the healthcare provider would most likely avoid injury to which of the following nerves?

   A. Femoral nerve

   B. Genitofemoral nerve

   C. Inguinal nerve

   D. Obturator nerve

   E. Sciatic nerve

9. A 17-year-old football player complains of severe knee pain after being tackled from the side. When the knee is flexed, the tibia can be moved anteriorly. Rupture or tearing of which of the following ligaments would most likely account for this observation?

   A. Anterior cruciate ligament

   B. Fibular collateral ligament

   C. Lateral meniscus

   D. Medical meniscus

   E. Posterior cruciate ligament

   F. Tibial collateral ligament

10. The hospital vascular team physician is instructed to place a central venous line in a patient's femoral vein. The femoral artery is palpated to determine the location of the femoral triangle contents. The contents of the femoral triangle, from lateral to medial, are the

A. femoral artery, femoral vein, femoral nerve, lymphatics

B. femoral nerve, femoral artery, femoral vein, lymphatics

C. femoral vein, femoral artery, femoral nerve, lymphatics

D. lymphatics, femoral nerve, femoral artery, femoral vein

E. lymphatics, femoral vein, femoral artery, femoral nerve

11. Most of the muscles of the medial thigh compartment are innervated by the obturator nerve. The exception is the vertical division of the adductor magnus muscle, which is innervated by which of the following nerves?

A. Common fibular (peroneal) nerve

B. Deep fibular (peroneal) nerve

C. Femoral nerve

D. Superficial fibular (peroneal) nerve

E. Tibial nerve

12. The posterior compartment of the thigh primarily receives its blood supply from branches of which of the following arteries?

A. Deep femoral artery

B. Femoral artery

C. Inferior gluteal artery

D. Medial circumflex femoral artery

E. Popliteal artery

13. The biceps femoris muscle receives its name because it has two origins. One attachment is to the linea aspera of the femur. The other attachment is to the

A. anterior inferior iliac spine

B. greater trochanter

C. ischial spine

D. ischial tuberosity

E. lesser trochanter

F. tibial tuberosity

14. During a physical examination, the muscles of the lower limb are tested. For the purpose of this question, only the right lower limb will be considered. You place the patient's leg so that the right knee is bent with the foot resting on the floor. Which muscle group are you testing when you instruct your patient to straighten his leg against resistance?

A. Anterior leg (shin) muscles

B. Anterior thigh (quadriceps) muscles

C. Medial thigh muscles

D. Lateral leg muscles

E. Posterior leg (calf) muscles

F. Posterior thigh (hamstrings) muscles

15. A 28-year-old man sees his healthcare provider because he is having difficulty with dorsiflexion and has a diminished dorsalis pedis pulse. These symptoms are most likely attributable to swelling in which compartment of the leg?

A. Anterior compartment of the leg

B. Dorsal surface of the foot

C. Lateral compartment of the leg

D. Plantar surface of the foot

E. Posterior compartment of the leg

16. A 17-year-old boy is admitted to the emergency department with a leg fracture. He fell off his motorcycle and tore the interosseous membrane and fractured the proximal fibula. On examination, the patient is found to have decreased cutaneous sensation over the distal lateral aspect of his right leg and over the dorsal aspect of his right foot, with sparing of the space between his first and second digits. The primary motor abnormality you are most likely to observe would be decreased

A. dorsal flexion

B. eversion of the foot

C. inversion of the foot

D. knee flexion

E. knee extension

F. plantar flexion

17. During a physical examination, a 24-year-old woman is instructed to lie supine on the examination table. During the procedure, she is instructed to resist allowing the healthcare provider to pull her feet downward into plantarflexion. The patient presents with right-sided weakness in this task. Which of the following nerves is most likely responsible for this muscle weakness in this patient?

A. Deep fibular (peroneal) nerve

B. Femoral nerve

C. Lateral plantar nerve

D. Medial plantar nerve

E. Superficial fibular (peroneal) nerve

F. Tibial nerve

18. Which of the following actions would you most likely expect to be the weakest if your patient has a lesion of the tibial nerve in the popliteal fossa?

A. Dorsiflexion of the ankle

B. Extension of the hip

C. Extension of the digits

D. Flexion of the knee

E. Flexion of the digits

19. A 20-year-old woman stepped on a nail and it penetrated the plantar surface of her bare foot, injuring the lateral plantar nerve. Which of the following muscles would most likely be rendered nonfunctional?

A. Abductor hallucis muscle

B. Dorsal interossei muscles

C. First lumbrical muscle

D. Flexor digitorum brevis muscle

E. Flexor hallucis brevis muscle

20. A 38-year-old man is admitted to the emergency department after being involved in an automobile accident. He is unable to abduct or adduct his toes. If this patient has a deficit from a spinal cord lesion, which of the following spinal cord levels is most likely affected by this injury?

A. L1–L2

B. L3–L4

C. L5–S2

D. S1–S2

E. S2–S3

21. A 45-year-old woman is admitted to the emergency department after being involved in an automobile accident. She is experiencing pain, but is conscious. She can feel sensation in the groin, anteromedial leg, and great toe, but not in the calcaneal region. The physicians in the emergency department are concerned that this patient may have a spinal cord lesion at which level?

A. L1

B. L2

C. L3

D. L4

E. L5

F. S1

G. S2

H. S3

ANSWERS

1—B: The great saphenous vein is formed from the dorsal venous arch on the dorsum of the foot. The great saphenous vein then courses anterior to the medial malleolus of the tibia, up the medial aspect of the leg.

2—B: A dermatome is an area of skin supplied by a single spinal cord level. However, a cutaneous field is an area of skin supplied by more than the spinal cord level. The lateral cutaneous nerve of the thigh receives its innervation from spinal cord levels L2 and L3, which represents a cutaneous field.

3—A: The hip is a synovial ball-and-socket joint. It is composed of articulations between the head of the femur and the acetabulum of the os coxa.

4—C: The fibula is the lateral bone in the leg articulating with the tibia. Therefore, the most likely bone seen protruding through the skin would be the fibula. The tibia is not a thin bone and is located more medially. The femur is not located in the leg but in the thigh. The calcaneus is the heal bone.

5—B: The action of the piriformis muscle is lateral rotation of the hip. Therefore, during gait, the left piriformis muscle would laterally rotate the left hip so that the toe points anteriorly. However, in a patient with a paralyzed left piriformis muscle, the toe points more medially because there is a weaker counter contraction from the piriformis muscle.

6—D: The medial circumflex artery is the principal blood supply to the neck and head of the femur. Therefore, in patients with hip fractures, it is the most likely artery that is affected by the injury. The acetabular branch of the obturator artery may also be affected, but its vascular supply diminishes with age.

7—B: The left superior gluteal nerve innervates the gluteus medius and minimus muscles. While walking, the left gluteal muscles will stabilize the pelvis so that the right limb does not droop when swinging. However, if there is a lesion on the superior gluteal nerve, the left gluteal muscles are not functioning, and therefore, the right hip drops. To compensate, the patient will laterally flex the spine to the left so that the right foot will be higher off the ground when walking.

8—E: The sciatic nerve exits the pelvis inferior to the piriformis muscle. Intramuscular gluteal injections may damage the sciatic nerve, and therefore, the ideal position is in the superior–lateral quadrant, which is farthest from the nerve.

9—A: The anterior cruciate ligament prevents anterior translation of the tibia on the femur. Therefore, if the anterior cruciate ligament is injured, the tibia, is able to move anteriorly on the femur; this is referred to as an "anterior drawer sign." If the posterior cruciate ligament was damaged, the tibia would be able to move posteriorly on the femur.

10—B: The contents of the femoral triangle are, from lateral to medial, the femoral nerve, femoral artery, femoral vein, and femoral lymph nodes. The acronym *NAVL* may be helpful in remembering the order where the first letter of each word represents the femoral nerve, artery, vein, and lymphatics.

11—E: The vertical division of the adductor magnus muscle is also known as the hamstring division because it receives the same innervation of the hamstring muscles via the tibial nerve.

12—A: The deep femoral artery gives rise to perforating arteries, which pierce through openings in the adductor magnus muscle insertion and provide the primary vascular supply to the posterior thigh compartment. The inferior gluteal artery provides some vascular supply but not as much as the deep femoral artery.

13—D: The biceps femoris muscle is a part of the hamstring group. Each hamstring muscle originates on the ischial tuberosity, which includes the long head of the biceps femoris muscle.

14—B: When the patient is instructed to straighten his leg against resistance, the anterior thigh (quadriceps) muscles are primarily responsible for this action. Therefore, you are not only testing the anterior thigh muscles but also the femoral nerve.

15—A: The anterior compartment of the leg contains the tibialis anterior and other muscles that dorsiflex the ankle. In addition, the anterior tibial artery courses with the deep fibular nerve in the anterior compartment of the leg. Therefore, if compartment syndrome has affected the anterior compartment of the leg, then the dorsiflexors of the ankle would be negatively affected and the dorsalis pedis pulse would diminish.

16—B: The decreased cutaneous sensation of this patient is in the field of the superficial fibular (peroneal) nerve with sparing of the deep fibular (peroneal) nerve (space between digits 1 and 2). Therefore, with injuries involving the superficial fibular (peroneal) nerve, the muscles of the lateral compartment of the leg would be affected, and therefore, eversion of the foot would be weakened.

17—A: When the feet are pulled downward, the dorsiflexion muscles in the anterior compartment of the leg are being tested. If the patient exhibits weakness in this task, the most likely explanation then is lack of innervation from the deep fibular (peroneal) nerve.

18—E: The tibial nerve innervates the posterior compartment of the leg and intrinsic muscles of the feet. Therefore, if there is a nerve lesion within the popliteal fossa, the nerves innervating the posterior muscles of the leg would then be affected. These muscles include the flexor digitorum muscles.

19—B: The dorsal interossei are muscles innervated by the lateral plantar nerve. The other muscles listed as choices (i.e., first lumbrical, abductor hallucis, flexor digitorum brevis, and flexor hallucis brevis muscles) are all innervated by the medial plantar nerve.

20—E: The intrinsic muscles of the feet are innervated by the lateral plantar nerves. The lateral plantar nerve primarily carries motor innervation from the S2–S3 spinal cord levels. L1–L2 would result in weak hip flexion. L3–L4 would result in weak knee extension. L5–S2 would result in weak hip extension and knee flexion, and S1–S2 would result in weak dorsiflexion and plantar flexion.

21—F: The dermatome associated with the calcaneal region is S1. The L1 dermatome is in the groin, and L4 is in the antero-medial leg and great toe. Dermatomes S2 and S3 are posterior to the thigh and gluteal regions.

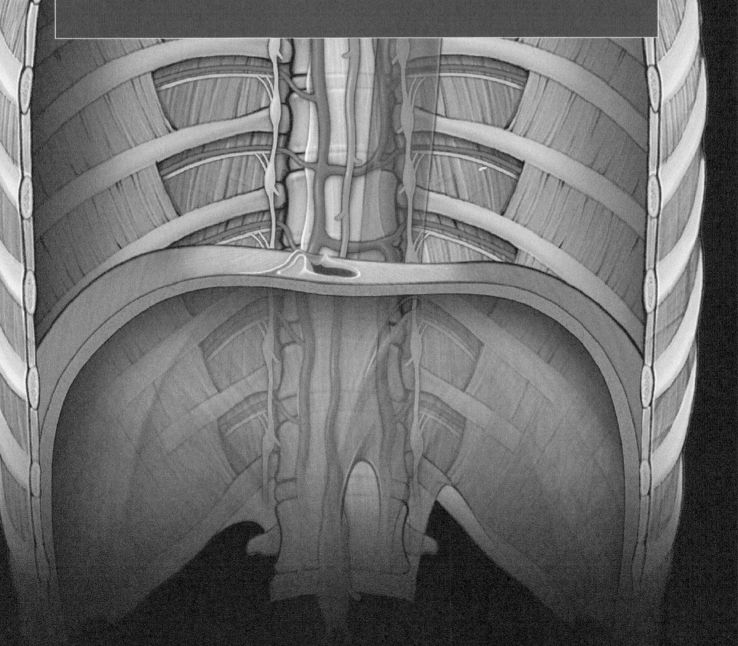

SECTION 8

FINAL EXAMINATION

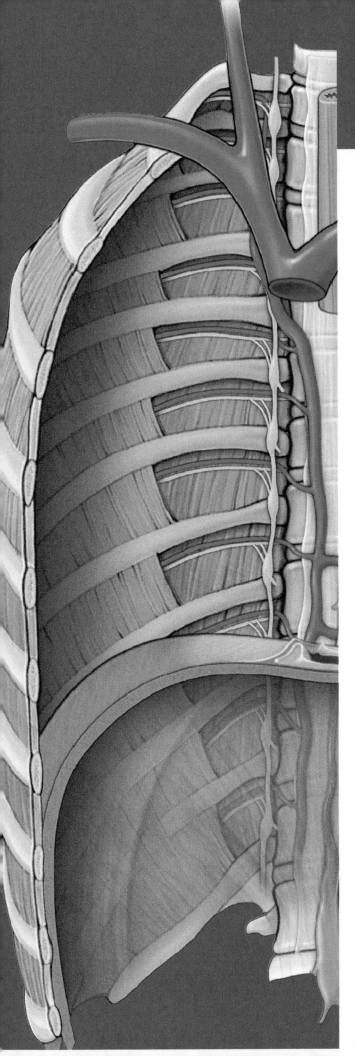

CHAPTER 39

STUDY QUESTIONS AND ANSWERS

DIRECTIONS

Each of the numbered items or incomplete statements is followed by lettered options. Select the **one** lettered option that is **best** in each case.

1. Which of the following structures is the only osteologic connection between the axial skeleton and the upper limb skeleton?

   A. Clavicle

   B. Humerus

   C. Scapula

   D. Ulna

2. Damage to which of the following muscles would most likely decrease the stability of the glenohumeral joint?

   A. Biceps brachii muscle

   B. Infraspinatus muscle

   C. Pectoralis major muscle

   D. Serratus anterior muscle

   E. Triceps brachii muscle

3. Spasm of the scalene muscles may entrap which region of the brachial plexus?

   A. Cords

   B. Divisions

   C. Roots

   D. Terminal branches

   E. Trunks

4. A 43-year-old man is experiencing spasms of the coraco-brachialis muscle, which impinges the nerve that courses through it. Which movement is most likely affected by this muscle spasm?

   A. Elbow extension

   B. Elbow flexion

   C. Shoulder abduction

   D. Shoulder adduction

   E. Shoulder extension

   F. Shoulder flexion

5. The radial and ulnar arteries arise from the bifurcation of which artery?

   A. Axillary

   B. Brachial

   C. Cephalic

   D. Deep brachial

   E. Subclavian

6. Damage to the posterior cord of the brachial plexus would most likely result in weakness of which of the following muscles?

   A. Coracobrachialis

   B. Flexor carpi radialis

   C. Latissimus dorsi

   D. Pectoralis major

   E. Supraspinatus

7. Which muscle of the upper limb is innervated by the spinal accessory nerve [cranial nerve (CN) XI]?

   A. Levator scapulae

   B. Rhomboid major

   C. Serratus anterior

   D. Splenius capitis

   E. Trapezius

8. When a 45-year-old woman performs pushups, the medial border of the right scapula protrudes from her thorax more than it protrudes on the left side. Which nerve is most likely injured, resulting in this observation?

   A. Lateral pectoral nerve

   B. Long thoracic nerve

   C. Medial pectoral nerve

   D. Suprascapular nerve

   E. Thoracodorsal nerve

9. A 34-year-old woman is diagnosed with quadrangular space syndrome, a rare abnormality localized within the posterior shoulder region. Which of the following structures would most likely be compressed in this patient?

   A. Axillary nerve and anterior humeral circumflex artery

   B. Axillary nerve and deep brachial artery

   C. Axillary nerve and posterior humeral circumflex artery

   D. Radial nerve and anterior humeral circumflex artery

   E. Radial nerve and deep brachial artery

   F. Radial nerve and posterior humeral circumflex artery

10. The suprascapular and dorsal scapular arteries form a collateral circuit on the posterior side of the scapula with which of the following branches of the axillary artery?

    A. Anterior humeral circumflex artery

    B. Circumflex scapular artery

    C. Posterior humeral circumflex artery

    D. Thoracoacromial artery

    E. Thoracodorsal artery

11. A 77-year-old patient is diagnosed with nerve entrapment, consistent with a herniated disc on the C5 spinal nerve. Which of the following muscles is most likely affected by this herniation?

    A. Deltoid

    B. Flexor carpi ulnaris

    C. Latissimus dorsi

    D. Pectoralis minor

    E. Triceps brachii

12. Which of the following anatomic regions will most likely NOT contribute lymph to the thoracic duct?

    A. Left large toe

    B. Left thigh

    C. Left thumb

    D. Right large toe

    E. Right thigh

    F. Right thumb

13. The boundaries of the three parts of the axillary artery are determined by its relationship to which of the following muscles?

    A. Pectoralis major

    B. Pectoralis minor

    C. Teres major

    D. Teres minor

14. A 24-year-old woman comes to the physician because of weakness in elbow flexion and numbness on the lateral side of the forearm. A lesion in which of the following nerves would most likely result in these symptoms?

A. Axillary

B. Median

C. Musculocutaneous

D. Radial

E. Ulnar

15. The superior ulnar collateral artery forms a collateral circuit with which of the following arteries?

A. Anterior ulnar recurrent

B. Anterior interosseous

C. Middle collateral

D. Posterior ulnar recurrent

E. Radial collateral

16. A 39-year-old man is diagnosed with a peripheral nerve injury that weakens his ability to extend his elbow, wrist, and fingers. Which area of this patient's upper limb will cause cutaneous deficit because of this injury?

A. Anterior compartment of the forearm

B. Lateral compartment of the forearm

C. Medial compartment of the forearm

D. Posterior compartment of the forearm

17. A 28-year-old woman is diagnosed with carpal tunnel syndrome. Which of the following tendons course through the carpal tunnel?

A. Flexor carpi radialis

B. Flexor carpi ulnaris

C. Flexor pollicis longus

D. Extensor carpi radialis longus

E. Extensor carpi ulnaris

F. Extensor pollicis longus

18. A 50-year-old woman has difficulty moving her thumb toward the palmar surface of the digiti minimi (fifth digit). She also experiences pain over the palmar surface of the thumb, index, and the middle digits. Pressure and tapping over the lateral portion of the flexor retinaculum causes tingling of the thumb and the 2nd and 3rd digits, indicating nerve damage. The damaged nerve that results in motor and sensory deficits most likely travels via which of the following routes?

A. Between the flexor digitorum superficialis and pro-fundus muscles

B. Between the two heads of the flexor carpi ulnaris muscle

C. Superficial to the flexor retinaculum

D. Through the coracobrachialis muscle

E. Through the supinator muscle

19. The radiocarpal joint includes the distal end of the radius, the triangular fibrocartilage complex, the scaphoid bone, the triquetrum bone, and the

A. capitate

B. hamate

C. lunate

D. trapezium

20. Which of the following fascial layers forms the roof of the carpal tunnel?

A. Extensor retinaculum

B. Fibrous digital sheaths

C. Flexor retinaculum

D. Palmar aponeurosis

E. Transverse palmar ligament

21. A 26-year-old woman is diagnosed with inflammation within Guyon's canal. She will most likely experience weakness when performing which of the following actions?

A. Abduction of the thumb

B. Adduction of digits 2 to 5

C. Flexion of the wrist

D. Radial deviation of the wrist

22. Compression of the median nerve in the carpal tunnel results in weakness in the thenar muscles and the first and second lumbricals. In which of the following areas would the patient most likely experience cutaneous deficits?

A. Lateral dorsal surface of the hand

B. Lateral palmar surface of the hand

C. Medial dorsal side of the hand

D. Medial palmar side of the hand

23. Which of the following arteries course through the anatomical snuffbox?

A. Deep palmar arch artery

B. Radial artery

C. Superficial palmar arch artery

D. Ulnar artery

24. Which of the following muscles is responsible for flexion of the metacarpophalangeal joints and extension of the interphalangeal joints of digits 2 to 5?

A. Doral interossei muscles

B. Lumbrical muscles

C. Palmar interossei muscles

D. Palmaris brevis muscles

E. Palmaris longus muscles

25. An avulsion fracture results when a bone fragment is pulled from its parent bone by forceful distraction of a tendon or ligament. An avulsion fracture of the ischial tuberosity most likely results from forceful contraction of which of the following muscles?

A. Adductors

B. Gluteals

C. Hamstrings

D. Iliopsoas

E. Quadriceps femoris

26. A 51-year-old man experiences a loss of skin sensation along the medial compartment of the thigh. No other areas of skin are affected. Which of the following best describes the area of deficit?

A. L2 dermatome

B. L3dermatome

C. L4 dermatome

D. Cutaneous field of the femoral nerve

E. Cutaneous field of the obturator nerve

F. Cutaneous field of the saphenous nerve

27. A 55-year-old man has difficulty extending his hip while walking up a flight of stairs. He experiences no cutaneous deficits. Which damaged nerve is most likely responsible for causing this man's symptoms?

A. Inferior gluteal nerve

B. S1 nerve root

C. S2 nerve root

D. Superior gluteal nerve

28. A 33-year-old man's pelvis drops on the right side when he steps with his right foot. He has no cutaneous deficits. Which nerve lesion is most likely causing this problem?

A. Common peroneal nerve

B. Femoral nerve

C. Inferior gluteal nerve

D. Obturator nerve

E. Superior gluteal nerve

F. Tibial nerve

29. Which of the following structures serves as a common attachment for the external rotator muscles of the hip?

A. Greater trochanter

B. Inferior pubic ramus

C. Ischial spine

D. Ischial tuberosity

E. Lesser trochanter

F. Superior pubic ramus

30. Which of the following is the most stable position of the hip as a result of the tension generated in the capsular ligaments?

A. Abduction

B. Adduction

C. Flexion

D. Extension

31. Which of the following quadriceps femoris muscles flexes the femur at the hip joint?

A. Rectus femoris

B. Vastus intermedius

C. Vastus medialis

D. Vastus lateralis

32. A 31-year-old woman has a lesion on the femoral nerve. Which of the following actions will the patient most likely have difficulty performing?

A. Knee extension

B. Knee flexion

C. Hip abduction

D. Hip adduction

E. Hip extension

F. Hip flexion

33. The hamstring musculature receives its primary vascular supply from which branch of the deep femoral artery?

A. Medial circumflex femoral branch

B. Lateral circumflex femoral branch

C. Perforating branches

D. Transverse branch

34. Which of the following ligaments primarily resist posterior translation of the tibia on the femur?

A. Anterior cruciate

B. Lateral collateral

C. Medial collateral

D. Posterior cruciate

35. Which of the following nerves can be found deep to the soleus muscle?

A. Common fibular

B. Deep fibular

C. Femoral

D. Obturator

E. Superficial fibular

F. Tibial

36. A 22-year-old man experiences weakness in ankle dorsi-flexion and numbness of the skin between digits 1 and 2 of his foot. Which of the following nerves is most likely damaged, resulting in these observed deficits?

A. Common fibular nerve

B. Deep fibular nerve

C. Femoral nerve

D. Obturator nerve

E. Superficial fibular nerve

F. Tibial nerve

37. Which of the following arteries courses through the proximal part of the interosseous membrane?

A. Anterior tibial

B. Fibular

C. Popliteal

D. Posterior tibial

38. "Rolling the ankle" is a common injury that causes excessive inversion and plantar flexion of the foot. Which of the following ligaments would experience the most damage during such an event?

A. Anterior talofibular

B. Anterior tibiofibular

C. Deltoid

D. Posterior tibiofibular

39. Which of the following fascial structures forms the roof of the tarsal tunnel?

A. Inferior extensor retinaculum

B. Fibular retinaculum

C. Flexor retinaculum

D. Superior extensor retinaculum

E. Talofibular ligament

40. A 51-year-old woman has difficulty abducting and adducting digits 2 to 5 of her foot. In addition, the skin on the plantar surface of digits 4 and 5 is numb. Which of the following nerves is most likely damaged, resulting in these deficits?

A. Deep peroneal

B. Lateral plantar

C. Medial plantar

D. Superficial peroneal

E. Tibial

41. Which of the following arteries most likely gives rise to the deep plantar arch?

A. Arcuate

B. Dorsal pedal

C. First dorsal metatarsal

D. Lateral plantar

E. Medial plantar

42. Which of the following joints is most likely responsible for inversion and eversion of the foot?

A. Calcaneonavicular joint

B. Metatarsophalangeal joint

C. Subtalar joint

D. Talofibular joint

E. Tibiotalar joint

43. A 62-year-old man with portal hypertension caused by alcoholic cirrhosis is taken to the emergency department. He has been vomiting blood as a result of hemorrhage of the gastroesophageal plexus of veins. Which other veins would most likely be enlarged in this patient?

A. Gonadal

B. Iliac

C. Pudendal

D. Rectal

E. Suprarenal

44. In a healthy person, blood from the pulmonary veins flows next into which of the following structures?

A. Aortic arch

B. Left atrium

C. Left ventricle

D. Lungs

E. Pulmonary arteries

F. Pulmonary veins

G. Right atrium

H. Right ventricle

45. A Doppler echocardiogram evaluates blood flow, speed, and the direction of blood within the heart and also screens for any leakage of the four valves. If heart function during systole was being studied, which valves would the Doppler echocardiogram detect to be open?

A. Mitral and aortic

B. Mitral and pulmonary

C. Mitral and tricuspid

D. Pulmonary and aortic

E. Pulmonary and mitral

F. Pulmonary and tricuspid

46. A 52-year-old man visits the office of his family physician. On auscultation, a systolic murmur is heard in the right second intercostal space adjacent to the sternum. What is the most likely cause of this murmur?

A. Prolapsed aortic valve

B. Prolapsed mitral valve

C. Prolapsed tricuspid valve

D. Stenotic aortic valve

E. Stenotic mitral valve

F. Stenotic tricuspid valve

47. An 8-year-old boy is diagnosed with aortic coarctation beyond the left subclavian artery. Aortic coarctation is a congenital abnormality most commonly diagnosed at birth, but it occasionally remains undetected until later in life. Collateral circulation through which of the following vessels is most likely responsible for this coarctation to have remained undetected for so long?

    A. Azygos vein

    B. Axillary artery

    C. Intercostal arteries

    D. Internal thoracic vein

48. A 52-year-old woman is diagnosed with gastric cancer. During surgery to remove the cancerous tissue, regional lymph nodes were removed to assist in staging the cancer. Lymph nodes associated with which of the following vessels were most likely sampled from this patient?

    A. Celiac artery

    B. External iliac arteries

    C. Inferior mesenteric artery

    D. Portal vein

    E. Right renal vein

    F. Superior mesenteric vein

49. A 25-year-old woman involved in a motor vehicle accident is brought to the emergency department complaining of abdominal pain. Radiographic imaging of her abdomen reveals a hematoma in the retroperitoneal space. Trauma to which abdominal structure is most likely responsible for this finding?

    A. Jejunum

    B. Liver

    C. Pancreas

    D. Esophagus

    E. Transverse colon

50. A 55-year-old man undergoes a colonoscopy, which reveals multiple polyps in the descending and sigmoid colons. Because polyps may develop into cancer, the polyps or the regions of the bowel with multiple polyps are often surgically removed. The surgeon will most likely ligate which of the following arteries when removing the affected portion of bowel?

    A. Celiac trunk

    B. External iliac artery

    C. Inferior mesenteric artery

    D. Internal iliac artery

    E. Superior mesenteric artery

51. The primary vascular supply to the uterus is most likely from branches of which of the following arteries?

    A. External iliac

    B. Femoral

    C. Gonadal

    D. Internal iliac

    E. Pudendal

52. During sexual intercourse, male ejaculation is associated with innervation provided by which of the following nerves?

    A. Genitofemoral

    B. Ilioinguinal

    C. Lesser splanchnic

    D. Pelvic splanchnic

    E. Sacral splanchnic

53. A 49-year-old woman visits her physician with a complaint of loss of the ability to sense temperature and touch on the right side of the anterior tongue. She says that she has all sensations of taste. Which additional finding might you also observe in this patient?

    A. Adducted eye

    B. Loss of corneal reflex

    C. Reduced gag reflex

    D. Tongue deviation during protrusion

    E. Weakness in the masseter muscle

54. A radiographic image of the brain of an 84-year-old woman reveals a berry aneurysm in the anterior communicating cerebral artery. The aneurysm is most likely adjacent to which of the following arteries?

    A. Anterior cerebral

    B. Basilar

    C. Middle cerebral

    D. Posterior communicating

    E. Vertebral

55. A 26-year-old woman goes to a clinic because she has noticed a loss of cutaneous sensation on one side of her face. Which cranial nerve is most likely affected that results in this patient's condition?

    A. Abducens

    B. Trigeminal

    C. Facial

    D. Glossopharyngeal

    E. Vagus

56. A 51-year-old woman is experiencing ptosis and mydriasis of the left eye. Which additional finding would most likely be present in this patient (assume the left side of the head for each of the following)?

A. Inability to look laterally

B. Inability to accommodate the lens

C. Loss of salivary glands

D. Loss of sweat glands to the face

E. Reduced gag reflex

F. Reduced production of tears

57. When the physician is testing cranial nerves, the patient is often asked to stick the tongue straight out of the mouth. Which of the following muscles is most likely responsible for this action?

A. Anterior digastricus

B. Genioglossus

C. Mylohyoid

D. Palatoglossus

E. Posterior digastricus

58. The maxillary artery gains entrance to the pterygopalatine fossa and eventually the nasal cavity and infraorbital canal via which of the following structures?

A. Foramen rotundum

B. Foramen spinosum

C. Mandibular foramen

D. Pterygomaxillary fissure

E. Superior orbital fissure

59. The pterygopalatine ganglion most likely houses postganglionic neuronal cell bodies for visceral motor parasympathetic components of which of the following cranial nerves?

A. CN III

B. CN V

C. CN VII

D. CN IX

E. CN X

60. During general surgical procedures, anesthetics and muscle relaxants are used routinely. However, these drugs may decrease nerve stimulation to skeletal muscles, including the intrinsic muscles of the larynx, which results in closure of the vocal folds. In such cases, laryngeal intubation is necessary. Because of the effect of the anesthetics, which of the following intrinsic muscles of the larynx will most likely NOT maintain an open glottis?

A. Cricothyroid

B. Lateral cricoarytenoid

C. Posterior cricoarytenoid

D. Thyroarytenoid

E. Transverse arytenoid

61. During an inferior alveolar nerve block, the dentist must avoid damaging the inferior alveolar artery, which enters the mandibular foramen posterior to its associated nerve. The inferior alveolar artery originates in which of the following arteries?

A. Facial

B. Infraorbital

C. Lingual

D. Maxillary

E. Supraorbital

62. A 37-year-old woman complains of hoarseness of several weeks' duration. Upon further examination, the physician determines that the patient has partial paralysis of her vocal cords. Radiographic studies confirm an aortic arch aneurysm. Which of the following most accurately describes the relationship between the patient's symptoms and hoarseness and this further finding?

A. Direct contact of the aneurysm with the trachea in the superior mediastinum

B. Injury to that part of the sympathetic chain that provides sensory innervation to the larynx

C. Irritation of the left phrenic nerve as it crosses the arch of the aorta on its way to the diaphragm

D. Pressure of the aneurysm on the esophagus in the posterior mediastinum

E. Pressure on the left recurrent laryngeal nerve, which wraps around the aortic arch

63. A 55-year-old man visits his physician because he is experiencing paralysis of all of the extraocular eye muscles and a loss of sensation of the root of the nose, upper eyelid, and forehead. Examination shows an abolition of the corneal reflex, but the patient's vision is not impaired. The most likely cause of this condition would be a fracture of which of the following structures?

A. Foramen rotundum

B. Internal acoustic meatus

C. Superior orbital fissure

D. Pterygopalatine fossa

E. Maxillary sinus

64. A 26-year-old woman involved in an automobile accident was thrown into the windshield and sustained a deep gash to her face, just lateral to her upper lip. The facial artery was severed, resulting in substantial arterial bleeding. At which location, apart from the wound itself, would pressure most likely be placed to inhibit the bleeding in this patient?

A. Internal carotid artery just inferior to the mandible

B. Medial canthus of the eye

C. Midpoint of the neck just posterior to the sternocleidomastoid muscle

D. Skin overlying the mandible just anterior to the masseter muscle attachment

E. Temporal region anterior to the ear

65. When looking through an otoscope, the physician is able to view the tympanic membrane. Which structure is most likely attached to the center of the tympanic membrane on its internal surface?

A. Cochlea

B. Incus

C. Malleus

D. Stapes

E. Tensor tympani muscle

66. An 84-year-old woman is brought to the emergency department because her son thinks she has had a stroke because of the paralysis on the right side of the woman's body. Neurologic studies show that an intracerebral hemorrhage has interrupted the blood supply to the posterior part of the frontal lobe, the parietal lobe, and medial portion of the temporal lobe of the left cerebral hemisphere. Which vessel most likely caused the stroke in this patient?

A. Anterior cerebral artery

B. Middle cerebral artery

C. Posterior cerebral artery

D. Middle meningeal artery

E. Vertebral artery

67. In the cervical region, the phrenic nerve courses along the anterior surface of which of the following muscles?

A. Anterior scalene

B. Middle scalene

C. Posterior scalene

D. Sternocleidomastoid

E. Trapezius

68. A 30-year-old woman has become anemic because she has been having severe anterior epistaxis on the nasal septum. An ear, nose, and throat specialist has been called to consult about the woman's bleeding. It is necessary to surgically ligate the nasal arteries in this patient. The specialist must consider arterial branches from the maxillary and ophthalmic arteries as well as which other artery?

A. Ascending pharyngeal

B. External carotid

C. Facial

D. Internal carotid

E. Lingual

69. A 2-year-old boy is diagnosed with torticollis involving the right sternocleidomastoid muscle. Which of the following anatomic changes is most likely to occur in this patient?

A. Head extended backward in the midline

B. Head flexed forward in the midline

C. Head rotated to the left

D. Head rotated to the right

70. A 46-year-old woman is diagnosed with a tumor of the parotid gland. Which of the following functions is most likely to be disrupted by this lesion (assume the left side for each choice)?

A. Corneal sensation

B. Elevation of the shoulder

C. Facial sensation

D. Protrusion of the tongue

E. Taste to the anterior tongue

F. Wrinkling of the forehead

71. To clinically test the superior oblique muscle of the eye, the physician would most likely have the patient look

A. laterally

B. laterally and then downward

C. laterally and then upward

D. medially

E. medially and then upward

F. medially and then downward

72. During a physical examination, the patient is instructed to look laterally and then upward. Which extraocular muscle is being tested in this patient?

A. Inferior oblique

B. Inferior rectus

C. Lateral rectus

D. Medial rectus

E. Superior oblique

F. Superior rectus

73. A 63-year-old woman visits her physician for a routine physical examination. During the examination, the physician touches the patient's scalp with a pin near the hairline to test for cutaneous sensation. Which of the following nerves is the physician most likely testing?

A. CN IV

B. CN V

C. CN VI

D. CN VII

E. CN VIII

F. CN IX

G. CN X

74. A 23-year-old man is brought to the emergency department after being involved in an automobile accident. Examination shows that the patient has an intracranial hemorrhage resulting from lateral trauma to the skull in the region of the pterion. Which of the following is the most likely location for the hemorrhage?

A. Immediately superficial to the dura mater

B. Immediately deep to the dura mater

C. Within the subarachnoid space

D. Within the brain parenchyma

75. A 32-year-old man with carcinoma of the testis undergoes exploratory surgery to biopsy lymph nodes. Which of the following lymph nodes is being sampled to determine if the cancer has metastasized via the lymphatic system?

A. External iliac

B. Femoral

C. Internal iliac

D. Paraaortic

E. Superficial inguinal

76. A 20-year-old woman is brought to the emergency department after being involved in an automobile accident. Physical examination reveals hypotension and tenderness along the left midaxillary line. Radiographic imaging reveals a large swelling below the left costal margin, and ribs 9 and 10 are fractured near their angles. Which of the following abdominal organs was most likely injured as a result of this accident?

A. Descending colon

B. Left kidney

C. Pancreas

D. Spleen

E. Stomach

77. During sexual arousal, an erection is caused by a dilation of arteries filling the erectile tissue of the penis. These arteries are innervated by which of the following nerves?

A. Genitofemoral

B. Iliohypogastric

C. Parasympathetic

D. Pudendal

E. Sympathetic

78. The functional significance of the marginal artery of Drummond is anastomosis among which of the following vessels?

A. Arteries supplying the colon

B. Arteries supplying the liver

C. Lymphatics draining the kidneys

D. Lymphatics draining the pancreas

E. Veins draining the bladder

F. Veins draining the posterior abdominal wall

79. A 40-year-old man undergoes a vasectomy. After the procedure, when the patient has an orgasm during sexual intercourse, he will most likely

A. no longer have an ejaculate

B. still have an ejaculate, and the ejaculate will contain sperm

C. still have an ejaculate, but the ejaculate will not contain sperm

80. A 70-year-old-man has a 90% blockage at the origin of the inferior mesenteric artery. This blockage rarely results in intestinal angina because of collateral arterial supply. Which of the following arteries is the most likely additional source of blood to the descending colon?

A. Left gastroepiploic

B. Middle colic

C. Sigmoid

D. Splenic

E. Superior rectal

81. The mesoappendix is a fold of mesentery that contains an artery that is most likely a direct branch of which of the following?

A. Celiac trunk

B. Ileocolic artery

C. Middle colic artery

D. Right colic artery

E. Superior mesenteric artery

82. A 50-year-old woman is diagnosed with severe obstructive jaundice. Blockage of which of the following structures would most likely result in her condition?

A. Common hepatic duct

B. Pancreatic duct

C. Parotid duct

D. Submandibular duct

E. Thoracic duct

83. A 22-year-old man is admitted to the emergency department after being stabbed with a knife. The laceration is 8-cm long and involves the right cheek, from the right ear to near the corner of the mouth. Which of the following structures is most likely injured?

A. Lingual artery

B. Mandibular branch of facial nerve

C. Parotid duct

D. Submandibular duct

E. Superficial temporal artery

84. A 61-year-old man is diagnosed with an acute stroke. His primary deficit is a partial loss of the visual field as a result of a lesion in the occipital lobe. Which of the following arteries is most likely to be involved?

A. Anterior cerebral

B. Internal carotid

C. External carotid

D. Middle cerebral

E. Posterior cerebral

85. A 4-year-old boy is taken to the pediatrician because of recurrent ear infections. Tubes were placed in the tympanic membranes in the boy's ears 3 days ago, and he is now complaining of difficulty in tasting sweet foods. Which nerve was most likely disrupted during the insertion of the tubes that resulted in these findings?

A. Chorda tympani

B. Greater petrosal

C. Lesser petrosal

D. Vagus

E. Vestibulocochlear

86. A 4-year-old girl is brought to the pediatrician because she has pain in the left ear. Examination reveals acute otitis media. Which nerve is responsible for conducting the painful sensation from the internal surface of the tympanic membrane to the brain?

A. CN VII

B. CN VIII

C. CN IX

D. CN X

E. CN XI

87. An emergency cricothyroidotomy is warranted when an airway collapses or when severe laryngoedema occurs. Which of the following is the most accurate description of the location of the cricothyroid membrane?

A. Immediately inferior the cricoid cartilage

B. Immediately inferior to the hyoid bone

C. Immediately inferior to the thyroid cartilage

D. Immediately superior to the hyoid bone

E. Immediately superior to the thyroid cartilage

88. A 49-year-old woman visits her physician because of severe nose bleeds. Which major blood supply to the nasal cavity would need to be occluded to correct this patient's condition?

A. Ethmoidal artery

B. Facial artery

C. Greater palatine artery

D. Sphenopalatine artery

E. Superior labial artery

89. A 52-year-old man is brought to the emergency department because he is experiencing severe chest pain in the mediastinum. He says that 3 weeks ago he was treated for an abscess in the left mandibular molar. Studies determine that the chest pain is the result of an infection in the mediastinum. Which of the following is the most likely space that infection spread through to course from the mandibular region to the mediastinum?

A. Carotid

B. Masticator

C. Pretracheal

D. Retropharyngeal

E. Suprasternal

90. A 14-year-old girl arrives at the dentist's office to have a cavity in her lower right incisor filled. Which nerve will the dentist most likely block before beginning the procedure?

A. CN V-1

B. CN V-2

C. CN V-3

91. A radiologist is conducting a contrast study of the pulmonary circulation on a 41-year-old man. What is the most likely number of veins observed entering the left atrium?

A. Two

B. Three

C. Four

D. Five

E. Six

92. A 55-year-old woman undergoes surgery of the lateral abdominal wall. The surgeon entering the cavity will be careful to avoid injury to vessels and nerves within the abdominal wall. The vessels and nerves will most likely be located deep to which of the following structures?

A. External oblique muscle

B. Internal oblique muscle

C. Superficial fascia

D. Transverse abdominis muscle

E. Transversalis fascia

93. During surgery of a 60-year-old man, the anterior rectus muscle sheath between the xiphoid process and the umbilicus is incised. In this region, the rectus sheath is derived from which of the following muscles?

A. External oblique muscle

B. External and internal oblique muscles

C. Internal oblique muscle

D. Internal oblique and transverse abdominis muscles

E. Transverse abdominis muscle

94. In a healthy person, blood from the left ventricle would most likely flow next into which of the following structures?

A. Aortic arch

B. Left atrium

C. Left ventricle

D. Right atrium

E. Right ventricle

F. Pulmonary arteries

G. Pulmonary veins

95. One aspect of the physical examination is measuring the jugular venous pressure (JVP). The JVP appears as a pulse in the neck by the external jugular vein. Therefore, the JVP is produced by the venous system, not the arterial system, because of the right atrial contraction. There are no valves in the superior vena cava. Therefore, during diastole, some blood is pushed, in a pulsating fashion, back out of the right atrium and up the superior vena cava, all the way to the external jugular vein. The JVP is only pathologic if the pulse is observed too high up the neck, indicating an overload or backup of blood entering the heart. An abnormally high JVP can be caused by several conditions. Which of the following conditions is most likely to cause an abnormally high JVP?

A. Left-sided heart failure

B. Mitral valve prolapse (regurgitation or backflow of blood)

C. Right atrial fibrillation (uncoordinated contraction)

D. Tricuspid valve stenosis (narrowing)

96. A 22-year-old man visits his physician and is diagnosed with a herniated disc impinging the spinal nerve that exits inferior to the C6 vertebra. Pain from the impinged nerve would most likely radiate to which cutaneous region?

A. Lateral shoulder

B. Lateral surface of digit 5

C. Medial surface of the elbow

D. Medial surface of the manubrium

E. Palmar surface of digit 3

F. Palmar surface of the thumb

97. A 51-year-old man visits his physician with a complaint of back pain that the man says resulted from bending over and picking up a heavy box without bending his knees. Which of the following muscles was most likely injured in this patient?

A. Iliocostalis

B. Latissimus dorsi

C. Rhomboid major

D. Serratus posterior inferior

E. Trapezius

98. A 61-year-old woman visits her physician with a complaint of shortness of breath. Physical examination reveals cyanosis and an enlarged right ventricle. Which of the following structures is most likely obstructed in this patient?

A. Bronchial arteries

B. Bronchioles

C. Coronary arteries

D. Coronary sinus

E. Pulmonary arteries

F. Pulmonary veins

99. Heart murmurs are abnormal heart sounds caused by turbulent blood flow. They are often associated with pathologic heart valves. The murmurs are generally organized into the following categories:

- *Systolic murmurs* occur during ventricular contraction.

- *Diastolic murmurs* occur during atrial contraction (ventricular relaxation and filling).

The two common causes of murmurs are valve stenosis and valve regurgitation (prolapse):

- *Valve stenosis* occurs when the valve becomes narrower. During contraction, the blood is forced through a smaller opening and the flow becomes turbulent, causing the extra heart sound.

- *Valve regurgitation* occurs when the valve is unable to close completely and thus becomes incompetent, allowing blood to flow in reverse, back through the valve. This murmur occurs when the affected valve is supposed to be closed.

Which of the following would most likely present as a diastolic murmur?

A. Aortic valve stenosis

B. Mitral valve regurgitation

C. Pulmonary valve stenosis

D. Tricuspid valve stenosis

100. *Neisseria meningitidis* and *Streptococcus pneumoniae* are the leading causes of bacterial meningitis. To confirm diagnosis of bacterial meningitis, cerebrospinal fluid (CSF) is most likely obtained from which of the following regions?

A. Epidural space

B. Intervertebral foramen

C. Subarachnoid space

D. Subdural space

E. Subpial space

ANSWERS

1–A: The clavicle connects the manubrium of the sternum to the acromion of the scapula.

2–B: The infraspinatus is a rotator cuff muscle that stabilizes the glenohumeral joint. The tendons of the rotator cuff reinforce the ligaments of glenohumeral joint capsule.

3–C: The roots of the brachial plexus pass between the anterior and middle scalene muscles. Spasm of these muscles may entrap the brachial plexus roots.

4–B: The musculocutaneous nerve courses through the coracobrachialis muscle and innervates the anterior compartment of the arm. Muscles of the anterior arm include the biceps brachii and brachialis; both of these muscles flex the elbow. The triceps muscle extends the elbow and is innervated by the radial nerve. Shoulder abduction is produced by the deltoid muscle, which is innervated by the axillary nerve. Shoulder adduction, extension, and flexion are produced by the latissimus dorsi muscle and the pectoralis major muscle with innervation from the thoracodorsal nerve and pectoral nerves, respectively.

5–B: The brachial artery bifurcates just distal to the elbow to form the radial and ulnar arteries.

6–C: The upper subscapular, lower subscapular, and thoracodorsal nerves branch off the posterior cord in the axilla, just anterior to the subscapularis muscle. The thoracodorsal nerve innervates the latissimus dorsi muscle. All other muscles are innervated by anterior divisions of the brachial plexus.

7–E: The trapezius muscle is innervated by the spinal accessory nerve (CN XI). The levator scapulae and rhomboid major muscles are innervated by the dorsal scapular nerve (C5), and the splenius capitis is innervated by cervical dorsal rami.

8–B: The serratus anterior muscle stabilizes the medial border of the scapula against the thorax. A pushup position pushes the medial border of the scapula away from the thorax, causing the weakness to become more apparent. The long thoracic nerve (C5–C7) innervates the serratus anterior muscle.

9–C: Quadrangular space syndrome results when the muscles surrounding the quadrangular space, mainly the teres major and minor and the long head of the triceps brachii, compress the posterior humeral circumflex artery and axillary nerve.

10–B: The circumflex scapular artery courses through the triangular space to form a collateral circuit with the suprascapular and dorsal scapular arteries.

11–A: The C5 spinal nerve level is a principal contributor to the axillary nerve, which innervates the deltoid muscle.

12–F: The right side of the head, neck, and thorax and the right upper limb drain lymph into the right lymphatic duct. All other parts of the body drain lymph into the thoracic duct. Therefore, the right thumb is the only structure in the choices listed that does *not* drain into the thoracic duct.

13–B: The boundaries of the three parts of the axillary artery are determined by its relationship to the pectoralis minor muscle.

14–C: When the musculocutaneous nerve is damaged, the biceps brachii and brachialis muscles are weakened or paralyzed. In addition, the skin on the lateral side of the forearm receives its cutaneous innervation via the lateral cutaneous nerve of the forearm, a branch of the musculocutaneous nerve.

15–D: The superior ulnar collateral artery anastomoses with the posterior ulnar recurrent artery from the ulnar artery, posterior to the medial epicondyle.

16–D: Damage to the radial nerve would cause the weakness in the triceps brachii muscle and extension of the elbow. This damage would cause deficits in the cutaneous field of the radial nerve in the posterior compartment of the forearm.

17–C: The flexor pollicis longus tendon courses through the carpal tunnel along with the flexor digitorum superficialis and flexor digitorum profundus tendons and the median nerve.

18–A: The median nerve courses between the flexor digitorum superficialis and profundus muscles en route to the carpal tunnel. The ulnar nerve courses between the two heads of the flexor carpi ulnaris muscle; the palmar branch of the median nerve courses superficial to the flexor retinaculum; the musculocutaneous nerve courses through the coracobrachialis muscle; and the radial nerve courses through the supinator muscle.

19–C: The radiocarpal joint is formed by the distal end of the radius, the triangular fibrocartilage complex, and the proximal row of the carpal bones. The lunate bone is included in the proximal carpals.

20–C: The flexor retinaculum anchors to the hamate, pisiform, trapezium, and scaphoid bones to enclose the tendons of the flexor digitorum superficialis, flexor digitorum profundus, and flexor pollicis longus muscles and the median nerve.

21–B: The ulnar nerve courses through Guyon's canal. Compression of the nerve will cause weakness in the muscles it innervates, including the palmer interosseus muscles, which are responsible for adduction of digits 2 to 5.

22–D: The palmar digital branches of the median nerve send cutaneous branches to the skin of the medial side of the palm after the median nerve passes through the carpal tunnel. The palmar branch of the median nerve that innervates the lateral skin of the palm branches proximal to the carpal tunnel.

23–B: The radial artery courses through the anatomical snuffbox, where a radial pulse can be felt.

24–B: The lumbrical muscles cross anterior to the metacarpophalangeal joints, then insert on the extensor expansion. It is this orientation that allows the muscles to flex the metacarpophalangeal joints and extend the interphalangeal joints.

25–C: The ischial tuberosity is the attachment site for the hamstring muscles on the ischial tuberosity. The adductor muscles attach to the pubis; the gluteal muscles attach to the ilium; the quadriceps femoris muscle attaches to the femur; and the iliopsoas muscle attaches to the lumber vertebrae and ilium.

26–E: The cutaneous field of the obturator nerve only covers the skin of the medial compartment of the thigh. The L3–L4 dermatome also covers part of the medial compartment of the thigh, but extends over the distal anterior compartment of the thigh as well.

27–A: The patient is having difficulty extending his hip from a flexed position. This action is largely performed by the gluteus maximus muscle. Damage to the inferior gluteal nerve would weaken the gluteus maximus without causing cutaneous deficit. Damage to either the S1 or S2 nerve roots may weaken the gluteus maximus, but their dermatomes would also be affected.

28–E: The gluteus medius and minimus muscles abduct the hip and hold the pelvis over the stance limb (limb that is on the ground during gate), preventing drop on the opposite swing side when walking. Damage to the superior gluteal nerve would weaken both muscles without a cutaneous deficit.

29–A: Five of the six external hip rotator muscles attach to some aspect of the greater trochanter of the femur. The quadratus femoris attaches near the greater trochanter on the intertrochanteric crest.

30–D: The capsular ligaments of the hip are pulled taut during extension of the hip, decreasing distraction between articular surfaces and stabilizing the joint.

31–A: The rectus femoris muscle is the only quadricep muscles that crosses the hip joint, originating on the anterior inferior iliac spine. A muscle must cross a joint to produce an action at that joint.

32–A: The femoral nerve innervates the anterior compartment of the thigh (quadricep muscles group). The muscles of the anterior compartment of the thigh are the primary knee extensors.

33–C: The perforating arteries branch off the deep femoral artery and pierce through the adductor magnus muscle as it inserts on the linea aspera. These arteries are the primary arterial supply to hamstring musculature.

34–D: The posterior cruciate ligament ascends from the posterior element of the superior tibia to the femur in the joint capsule of the knee. This orientation makes it very strong so that it is able to resist posterior translation of the tibia on the femur. In contrast, the anterior cruciate ligament resists anterior translation of the tibia on the femur.

35–F: The tibial nerve descends through the posterior part of the leg between the soleus and the deep posterior muscles of the leg.

36–B: The muscles of the anterior compartment of the leg produce dorsiflexion of the ankle. The deep fibular nerve innervates the anterior compartment and supplies the skin between digits 1 and 2 of the foot.

37–A: The anterior tibial artery branches from the popliteal artery, then courses through the proximal part of the interosseous membrane to enter the anterior compartment of the leg.

38–A: The lateral side of the ankle experiences the most damaging strain when the ankle is rolled or over-inverted and plantar flexed. The anterior talofibular ligament is located on the lateral side of the ankle and will experience the most damage.

39–C: The flexor retinaculum forms the roof of the tarsal tunnel between the calcaneus and the medial malleolus.

40–B: The dorsal and plantar interossei are responsible for adduction and abduction of the small toes. The lateral plantar nerve supplies the dorsal and plantar interossei as well as the plantar skin of digits 4 and 5.

41–D: The lateral plantar artery gives rise to the deep plantar arch. The terminal end of the deep plantar arch joins the deep plantar branch of the dorsalis pedis artery.

42–C: The subtalar joint allows for movement primarily in the coronal plane.

43–D: Cirrhosis of the liver may result in portal hypertension because of the backup of venous blood from the gut. Therefore, congested blood results in engorged veins in the portocaval anastomoses, such as the periumbilical veins, rectal veins, and the gastroesophageal veins.

44–B: The right atrium of the heart collects systemic and coronary deoxygenated blood. The right ventricle pumps blood through the pulmonary arteries to the lungs to become oxygenated. Oxygenated blood then returns from the lungs to the left atrium of the heart via the pulmonary veins. Therefore, blood from the pulmonary veins will most likely flow next into the left atrium.

45–D: During systole, both ventricles contract. As pressure increases, the atrioventricular valves are forced shut and the semilunar valves (pulmonary and aortic) open to enable blood to flow out of the pulmonary arteries and aorta.

46–D: The aortic valve is auscultated in the right second intercostal space adjacent to the sternum. This patient has a systolic murmur, and therefore, a stenotic or narrowed valve would be heard during systole. In contrast, a murmur of a prolapsed aortic valve would most likely be heard during diastole.

47–C: The coarctation (narrowing) is beyond the left subclavian artery, and therefore, blood flowing from the aortic arch to the thoracic aorta is restricted. Blood is shunted through the subclavian arteries to the internal thoracic arteries, where blood next flows into the anterior intercostal arteries, through the posterior intercostal arteries, and retrograde enters the thoracic aorta.

48–A: Lymph flows along the course of arteries within the abdominopelvic cavity. The primary blood supply to the stomach is via branches from the celiac artery; therefore, lymph nodes associated with the celiac artery must be biopsied in this patient.

49–C: Imaging reveals the hematoma to be in the retroperitoneal space. The only structure from the list of choices (i.e., jejunum, liver, pancreas, esophagus, and transverse colon) that is located in the retroperitoneal space is the pancreas.

50–C: The polyps are located in the hindgut (descending and sigmoid colon). The primary arterial supply to the hindgut is via the inferior mesenteric artery.

51–D: The uterine and vaginal arteries provide the primary arterial supply to the uterus. Both arteries are branches from the internal iliac artery.

52–E: Ejaculation is under sympathetic innervation. Sympathetic neurons responsible for ejaculation begin at the L2 level of the spinal cord, course down the sympathetic trunk, exit via the sacral splanchnic nerves, and course to the ductus deferens and smooth muscle of the urethra. Pelvic splanchnics transport the parasympathetic neurons that are responsible for erection. Remember, *Point* (*Parasympathetic*) and *Shoot* (*Sympathetic*).

53–E: Temperature and touch to the anterior tongue is provided by CN V-3, and taste is sensed via the chorda tympani nerve (CN VII, the facial nerve). If the lesion results from loss of touch but not taste, the lesion is proximal to the chorda tympani union to the lingual branch of CN V-3. Therefore, muscles of mastication, such as the masseter, would be affected.

54–A: The anterior communicating artery is located between the paired anterior cerebral arteries. A berry aneurysm is a saclike outpouching in the anterior communicating cerebral artery.

55–B: The trigeminal nerve (CN V) is responsible for general sensory innervation of the face.

56–B: The oculomotor nerve (CN III) is affected in this patient, resulting in ptosis (droopy eyelid due to no tone in the levator palpebrae superioris) and mydriasis (dilatation of the pupil due to loss of the papillary constrictor muscle). The oculomotor nerve is also responsible for innervating the ciliary muscles, causing an inability to accommodate the lens. The abducens nerve (CN VI) innervates the lateral rectus (look laterally); the facial nerve (CN VII) innervates the lacrimal and salivary glands; and the glossopharyngeal nerve (CN IX) innervates the parotid salivary gland and is part of the gag reflex.

57–B: The genioglossus muscle attaches to the internal surface of the mental symphysis of the mandible and into the tongue. Therefore, contraction results in protrusion of the tongue. The palatoglossus is the only other tongue muscle listed as a choice, and it elevates the root of the tongue.

58–D: The maxillary artery branches off the external carotid artery in the infratemporal fossa. The maxillary artery courses through the pterygomaxillary fissure, pterygopalatine fossa, and sphenopalatine foramen into the nasal cavity.

59–C: The facial nerve (CN VII) provides parasympathetic innervation to both the pterygopalatine and submandibular ganglia. CN III (oculomotor nerve) provides parasympathetic innervation for the ciliary ganglion. CN IX (glossopharyngeal nerve) provides parasympathetic innervation for the otic ganglion and CN X (vagus nerve) for intramural ganglia. CN V (trigeminal nerve) does not have parasympathetic neurons originating in its nuclei; however, it does provide a pathway for parasympathetics on which to "hitch-hike."

60–C: The posterior cricoarytenoid muscles abduct the vocal ligaments, whereas the other muscles listed as choices (i.e., cricothyroid, lateral cricoarytenoid, thyroarytenoid, and transverse arytenoid) adduct or tense the vocal ligaments.

61–D: The inferior mandibular artery originates from the maxillary artery in the infratemporal fossa.

62–E: The left recurrent laryngeal nerve courses deep to the aortic arch and can affect its functioning.

63–C: The superior orbital fissure transmits CNN III, IV, V-1, VI, and the superior ophthalmic vein. Therefore, damage to CNN III, IV, and VI accounts for the paralysis of extraocular muscles, and damage to CN V-1 accounts for loss of sensation to the nose, upper eyelid, and forehead.

64–D: The facial artery originates with the external carotid artery, and after emerging from the submandibular triangle, the artery courses along the lateral corner of the mouth and medial canthus of the eye.

65–C: The malleus attaches into the medial surface of the tympanic membrane.

66–B: The cerebral region affected by the stroke (the parietal lobe and the medial portion of the temporal lobe of the left cerebral hemisphere) is supplied by the middle cerebral artery.

67–A: The phrenic nerve is formed by branches of the C3, C4, and C5 ventral rami and immediately courses vertically along the anterior scalene muscle en route to the thoracic cavity.

68–C: The facial artery gives rise to the superior labial artery, which provides arterial branches to the nasal cavity, including the nasal septum.

69–C: Torticollis causes shortening of the sternocleidomastoid muscle, which causes the head to rotate to the contralateral side. In this case, because the right sternocleidomastoid muscle is affected, the patient will look to his left.

70–F: The facial nerve (CN VII) innervates the frontalis muscle, which is responsible for wrinkling of the forehead. The facial nerve is responsible for taste in the anterior part of the tongue; however, the chorda tympani branches from the main trunk before exiting the stylomastoid foramen. Sensation to both the cornea and the face is provided by the trigeminal nerve (CN V). The genioglossus muscle protrudes the tongue from the mouth and is innervated by the hypoglossal nerve (CN XII).

71–F: When the patient is asked to look medially, the axis of vision is parallel to the contraction axis of the superior oblique muscle. When the superior oblique muscle contracts, the eye looks downward. Therefore, to clinically test the superior oblique muscle, the patient is first instructed to look medially and then to look downward. There are two muscles that cause the eye to look downward: the superior oblique and the inferior rectus. When the patient is instructed to look medially, the biomechanical advantage to looking downward is isolated to the superior oblique muscle, not the inferior rectus muscle.

72–F: When the patient looks laterally, via contraction of the lateral rectus muscle, the axis of vision becomes parallel with contraction axis of the superior rectus muscle. Therefore, when the superior rectus muscle contracts, the eye looks upward. There are two muscles that cause the eye to look upward: the superior rectus and the inferior oblique. When the patient is instructed to look laterally, the biomechanical advantage to looking up is isolated to the superior rectus muscle, not the inferior oblique muscle.

73–B: The physician is testing the trigeminal nerve (CN V). This nerve is responsible for providing general sensory innervation to the anterior scalp and face.

74–A: The patient has an epidural hematoma as a result of rupture of the middle meningeal artery. The middle meningeal artery courses on the internal surface of the skull in the region of the pterion. The lateral trauma most likely caused a skull fracture, which in turn damaged the middle meningeal artery. The middle meningeal artery courses superficial to the dura mater in this location and, as such, bleeds into the epidural space.

75–D: In the abdomen, pelvis, and perineum, lymph flows along the arterial supply of its organ. Therefore, the blood supply for the testis is the testicular artery, which is a branch of the aorta. Paraaortic lymph nodes would be biopsied in a patient who has carcinoma of the testis.

76–D: The spleen is located in the left upper quadrant of the abdomen in midaxillary line. Fractured ribs 9 and 10 would most likely damage the spleen, resulting in significant blood loss and tenderness.

77–C: Pelvic splanchnics transport parasympathetic neurons to the erectile tissue, causing blood vessels to dilate and fill erectile tissue, which causes an erection. Remember, *P*oint (*P*arasympathetic) and *S*hoot (*S*ympathetic).

78–A: The marginal artery of Drummond courses in the mesentery adjacent to the large bowel. This artery serves as the vascular arcade connecting the superior mesenteric artery branches (right and middle colic) with the inferior mesenteric artery branches (left colic and sigmoid).

79–C: A vasectomy ligates the ductus deferens in the spermatic cord. Therefore, during ejaculation, sperm cannot reach the urethra. However, secretions from the seminal vesicles, prostate, and bulbourethral glands will continue to produce and secrete their products into the urethra during ejaculations, and therefore, this man will still have an ejaculate.

80–B: The middle colic artery anastomoses with the inferior mesenteric arterial branches, such as the left colic artery, via the marginal artery of Drummond.

81–B: The mesoappendix is a fold of mesentery that transports the appendicular artery to the appendix. The appendicular artery is a branch of the ileocolic artery.

82–A: The liver produces bile and transports it to the gallbladder for storage via the common hepatic duct. Blockage of the common hepatic duct would most likely result in jaundice (interruption of the drainage of bile from the biliary system). The pancreatic duct joins with the common bile duct, but this would not result in jaundice if blocked. The parotid and submandibular ducts transport saliva in the oral cavity. The thoracic duct transports lymph.

83–C: The buccal branch of the facial nerve (CN VII) and the parotid duct travel in the area of the cheek and can be located by a line drawn from the external acoustic meatus to the corner of the mouth.

84–E: The posterior cerebral artery is the artery that primarily provides vascular supply to the occipital lobe.

85–A: The chorda tympani nerve is a branch from the facial nerve (CN VII) and transports special sensory neurons for taste from the anterior portion of the tongue to the brain. The chorda tympani nerve courses along the internal surface of the tympanic membrane and, therefore, is the nerve most likely injured in the procedure of placing tubes in the tympanic membranes.

86–C: The glossopharyngeal nerve (CN IX) originates in the medulla and exits the skull via the jugular foramen. General sensory and visceral motor fibers enter the petrous part of the temporal bone and enter the middle ear as a tympani plexus of nerves. The tympanic plexus conducts general sensory information from the auditory tube and internal surface of the tympanic membrane to the brain.

87–C: The cricothyroid membrane is just inferior to the thyroid cartilage and superior to the cricoid cartilage.

88–D: The major vascular supply to the anterior septum is the sphenopalatine artery; a branch of this artery supplies the nasal septum. The sphenopalatine artery arises from the maxillary artery, which is a terminal branch of the external carotid artery.

89–D: The major pathway between infections of the neck and the chest is through the retropharyngeal space, a potential space between the prevertebral layer of fascia and the buccopharyngeal fascia surrounding the pharynx.

90–C: The inferior alveolar nerve provides general sensory innervation of the mandibular teeth and branches from CN V-3.

91–C: There are two pairs of pulmonary veins (four veins) that enter the left atrium of the heart.

92–B: The neurovascular plane in the abdominal wall is deep to the internal oblique and superficial to the transverse abdominis muscles.

93–B: The rectus sheath superior to the arcuate line is composed of the aponeuroses from both the external and internal oblique muscles.

94–A: Blood in the left ventricle is oxygenated and is ready to be pumped throughout the systemic circulation via the aorta to provide the body with oxygen.

95–D: The tricuspid valve is the first valve the blood encounters from the venous return to the heart. When this valve is stenotic, blood is pushed back into the venous system, causing an elevated JVP. Left-sided heart failure would more acutely present with pulmonary edema. Mitral valve prolapse will result in blood flowing from the left ventricle into the left atrium, but will not result in an elevated JVP. Right atrial contraction is mostly responsible for the pulsating appearance of the JVP. However, if the atrium has an uncoordinated and random contraction, the JVP would be lower due to weaker atrial contractions and more indistinct due to the uncoordinated rhythm.

96–E: The C7 spinal nerve exits inferior to the C6 vertebra. The C7 dermatome is associated with digit 3, and therefore, pain would radiate to the lateral surface of digit 3.

97–A: The iliocostalis muscle is a part of the erector spinae group, a group of muscles responsible for maintaining an erect vertebral column. When a person bends over, these muscles stretch to accommodate the flexibility. However, this movement weakens the muscle, and thus, when a person attempts to lift a heavy object, the muscle fibers will possibly be injured.

98–E: This patient most likely has pulmonary edema resulting in back flow of blood in the pulmonary arteries. The back flow causes the right ventricle to become enlarged to accommodate for the increased volume of blood. When the patient arrived at the physician's office, symptoms were shortness of breath and cyanosis (discoloration of skin as a result of lack of oxygenated blood).

99–D: If the tricuspid valve is stenotic, then the turbulent flow will occur during atrial contraction, which occurs toward the end of diastole. Aortic valve stenosis will cause a murmur when the left ventricle is contracting, causing it to be systolic. Mitral valve regurgitation will also cause a murmur when the left ventricle is contracting because blood will be forced back through the valve and will be pushed back into the left atrium, causing a systolic murmur. Pulmonary valve stenosis will be similar to aortic valve stenosis in that it will cause a murmur during right ventricular contraction and, therefore, a systolic murmur.

100–C: CSF resides in the subarachnoid space between the arachnoid and pia mater. CSF is obtained through a spinal tap (lumbar puncture) between the L4 and L5 vertebrae.

INDEX

Note: Page numbers followed by "f" and "t" indicate figures and tables, respectively.

CPSIA information can be obtained
at www.ICGtesting.com
Printed in the USA
LVOW02s0350120216

474771LV00007BA/10/P